The Plant Root

and

Its Environment

The Plant Root and Its Environment,

*Proceedings of an institute
sponsored by the Southern Regional Education Board,
held at Virginia Polytechnic Institute and State University,
July 5–16, 1971*

edited by E. W. Carson

Virginia Polytechnic Institute
and State University

University Press of Virginia

Charlottesville

THE UNIVERSITY PRESS OF VIRGINIA
Copyright © 1974 by the Rector and Visitors
of the University of Virginia

First published 1974

ISBN: 0-8139-0411-0
Library of Congress Catalog Card Number: 72-92877
Printed in the United States of America

Contents

PART I

Biological Aspects of the Root and Its Environment

PART II

Soil Physical and Chemical Aspects of the Root Environment

Foreword

THE Southern Regional Education Board has been concerned with strengthening and expanding the opportunities in high-quality education in agriculture and its related sciences. This effort has been carried on through the Council on Higher Education in the Agricultural Sciences founded in 1956. The council has served effectively in formulating policy and providing general guidance for program development.

Currently, the council is guiding a five-year Southern Regional Education Board project supported by the W. K. Kellogg Foundation designed to advance land-grant institutions, agriculture, and agricultural sciences. To guide the planning activities involved in this project effectively, the council membership was organized into four sub-committees, with each assigned a particular area of responsibility. The council subcommittee no. 3, headed by Dr. Ben T. Lanham, Jr., Vice-President for Research, Auburn University, studied the needs and opportunities for advancing scientific knowledge in the land-grant institutions in the region.

The Institute on the Plant Root and Its Environment was a recommendation of this subcommittee, approved by the council for implementation. The concept of the institute was fully supported by the deans of agriculture and directors of the experiment stations of the region as an effective means of further developing faculty competence in an essential area of soil and plant agriculture. Scientists from twenty land-grant colleges of agriculture and the United States Department of Agriculture served as a planning committee to revise the proposed program prepared by Dr. Lanham's committee and plan this institute. This publication of the proceedings of the institute is being made available so that participants and other scientists in the region can use it to expand their programs of research and instruction.

> T. J. HORNE, *Project Director*
> *Agricultural Sciences*
> *Southern Regional Education Board*

Preface

THE two-week institute from which this book resulted was intended to present the existing knowledge of the plant root, its functions and processes, and the many environmental factors that influence these functions. In addition, the institute was designed to assist individuals and institutions in evaluating their own research and educational efforts in these areas.

The large regional planning committee realized that even in a concentrated two-week program not all aspects of the root and its environment could be adequately covered. Therefore, the committee planned a program to cover those pertinent topics, e.g., *Rhizobia*, not already adequately covered by an institute, symposium, or publication. Nevertheless some duplication of material covered elsewhere was necessary to give the essential background and the cohesiveness to provide a self-contained program.

The presentations were before senior scientists and graduate students with very broad and diverse backgrounds of experiences and education. Thus, it is our hope that this publication will be useful in upper division and graduate courses in soils and plant nutrition where students of diverse training are found.

Not all that is desirable can be accomplished and often no more than is feasible. Nevertheless, we do regret the impossibility of including a complete transcript of all discussions and seminars. For example, informal seminars were conducted on Root–Root Interactions, Forest Pathology, Microbial Antagonisms in the Soil, Taxonomy and Morphology of the Genus *Endogone*, Root Exudations, Research Techniques, and Nutrient Supply. In addition, twenty hours of discussion were recorded, transcribed, and studied for use in these proceedings. One cannot convey the excitement of such informal sessions or do justice to the ideas presented and formulated in these discussions.

As with any undertaking of this magnitude, many individuals and agencies make significant contributions, and recognition of all involved is impossible. However, I would like to express appreciation to T. J. Horne of SREB, to the twenty-three members of the SREB Regional Planning Committee, and to the following from Virginia Polytechnic Institute and State University: M. E. Austin, Associate Professor of Horticulture, and W. H. Wills, Professor of Plant Pathology, for their service on the local planning committee; T. B. Hutcheson, Jr., Head, Agronomy Department, H. B. Couch, Head, Plant Pathology and Physiology Department, P. H. Massey, Associate Dean of Agronomic and Plant Science Division, and J. E. Martin, Dean of the College of Agriculture and Life Sciences, for their administrative support. I also wish to express my sincere thanks to my wife for her understanding and many hours of assistance throughout the institute and preparation of this publication.

E. W. CARSON
Director of the Plant Root Institute

Blacksburg, Virginia
April 1973

Contributors

Fred Adams, Department of Agronomy and Soils, Auburn University, Auburn, Alabama 36830.

Louis H. Aung, Department of Horticulture, Virginia Polytechnic Institute and State University, Blacksburg, Virginia 24061.

Stanley A. Barber, Department of Agronomy, Purdue University, Lafayette, Indiana 47907.

A. C. Bennett, Department of Agronomy and Soils, Auburn University, Auburn, Alabama 36830.

J. M. Byrne, Department of Biology, Virginia Polytechnic Institute and State University, Blacksburg, Virginia 24061.

W. A. Campbell, United States Department of Agriculture, Forest Service, Ret., School of Forest Resources, University of Georgia, Athens, Georgia 30601.

C. B. Davey, Department of Forestry, School of Forest Resources, North Carolina State University at Raleigh, Raleigh, North Carolina 27607.

Charles D. Foy, Soil Scientist, United States Department of Agriculture, Agricultural Research Service, Northeastern Region, Agricultural Research Center, Beltsville, Maryland 20705.

J. W. Gerdemann, Department of Plant Pathology, University of Illinois, Urbana, Illinois 61801.

F. F. Hendrix, Jr., Department of Plant Pathology and Plant Genetics, University of Georgia, Athens, Georgia 30601.

A. J. Hiatt, Department of Agronomy, University of Kentucky, Lexington, Kentucky 40506.

James E. Leggett, Department of Agronomy, University of Kentucky, Lexington, Kentucky 40506.

W. L. Lindsay, Department of Agronomy, Colorado State University, Fort Collins, Colorado 80521.

Konrad Mengel, Institut für Pflanzenerährung der Justus Liebig Universität, Giessen, Germany.

M. H. Miller, Department of Land Resource Science, University of Guelph, Guelph, Ontario, Canada.

David P. Moore, Department of Soils, Oregon State University, Corvallis, Oregon 97331.

E. I. Newman, Department of Botany, The University, Bristol, BS8 1UG, England.

Kenneth F. Nielsen, Director of Marketing and Agronomy, Western Co-operative Fertilizers Limited, Box 2500, Calgary 2, Alberta, Canada.

Robert W. Pearson, United States Department of Agriculture, Agriculture Research Service, Soil and Water Conservation Research Division, Alabama Agricultural Experiment Station, Auburn, Alabama 36830.

A. D. Rovira, C.S.I.R.O., Division of Soils, Private Bag No. 1, Glen Osmond, South Australia 5064.

Lewis H. Stolzy, Department of Soils and Plant Nutrition, University of California, Riverside, California 92502.

Howard M. Taylor, United States Department of Agriculture, Agriculture Research Service, Soil and Water Conservation Research Division, Alabama Agricultural Experiment Station, Auburn, Alabama 36830.

Grant W. Thomas, Department of Agronomy, University of Kentucky, Lexington, Kentucky 40506.

Abbreviations and Symbols

The most commonly used abbreviations and symbols are given here. Some may have other uses, and the less frequently used are described when they first appear in a chapter.

a—activity coefficient
ads—adsorbed
Å—Angstrom
atm—atmosphere
ATP—adenosine triphosphate
avg.—average
C—Celsius, e.g., 10C = 10 degrees C
CEC—cation exchange capacity
coef—coefficient
cpm—counts per minute
df—degrees of freedom
deg—degree
diam—diameter
dry wt—dry weight
EDTA—ethylenediaminetetraacetic acid
eq—chemical equivalent, may be combined, e.g., meq
ft-c—foot-candle
g—gram
ha—hectare
hr—hour
HSD—honesty significant difference (Tukey's test)
kg—kilogram
l—liter
LSD—least significant difference
m—meter; mili-, 10^{-3}, may be combined
mm—millimeter
M—molar, moles/liter
meq—milliequivalent
ml—milliliter

mV—millivolt(s)
n—nano, 10^{-9}, may be combined
N—Newton (when not nitrogen)
N—normality
NAD—nicotinamide adinine dinucleotide
NADP—nicotinamide adinine dinucleotide phosphate
NAR—net assimilation rate ($mg \cdot dm^2$ leaves $\cdot$ unit time^{-1})
ns—not significant
ODR—oxygen diffusion rate
OAA—oxaloactate
P—partial pressure; pressure; pressure potential
PEP—phosphoenol pyrurate
ppb—parts per billion
ppm—parts per million
r—radius
R—resistance
RGR—relative growth rate ($g \cdot 100g^{-1} \cdot$ unit time^{-1})
S:R—shoot to root ratio
var.—variety
γ—surface tension
η—viscosity
Δ—difference
μ—micro-, 10^{-6}, micron, may be combined, e.g., μeq
ρ—density
σ—reflection coefficient
τ—matric potential
ψ—water potential
ω—osmotic potential

Part I

Biological Aspects of the

Root and Its Environment

1.   Root Morphology

John M. Byrne

THE purpose of this presentation is not to review in detail the field of root
anatomy. It is my intention to place our current knowledge or lack of it
into a historical perspective, to point out controversies concerning the
organization and function of the root meristem, and to emphasize from a
plant anatomist's point of view problems concerning root growth and
development that await solution.

I. ANGIOSPERM ROOT APEX, 1868–1900

Hanstein (1868, 1870) and his protégé Reinke (1871) were the first to
investigate the apical organization of angiosperm roots. Both men
realized that the meristem of the angiosperm root was multicellular
rather than a single apical cell as Nageli (1858) had described for vascular
cryptograms. After intense investigations of both shoot and root
meristems, in which he used extremely crude histological methods,
Hanstein (1869, 1870) proposed the histogen theory of apical organization. Early workers in the field of plant anatomy utilized this theory
exclusively, although it was based on the misconception that an
obligate histogenic relationship exists between the initial layers
(histogens) and the mature tissue in shoot and root apices. As
recently as the 1950's several workers in both the United States
and Europe were using the histogen theory for the interpretation
of root organization, even though the concept had been discredited
for the shoot apex (Schmidt, 1924).

Hanstein's (1868, 1870) and Reinke's (1871) publications stimulated a
great deal of interest in the structural aspects of the root apex. From 1874
to 1900 a massive research effort was made on the angiosperm root apex
(Fleisher, 1874; Hegelmaier, 1874; Janczewski, 1874; Eriksson, 1876;

Holle, 1876; Traub, 1876; Flahault, 1878; Schwendener, 1882). Many of these studies were broad surveys of plant taxa; their major goals were to describe the apical organization in many taxa and to classify each species into structural types. Five types of apical organization were described for angiosperm roots:

1. Four groups of initials (histogens), one each for the central cylinder, cortex, rootcap, and epidermis; e.g., *Hydrocharis*, *Pistia*
2. Three groups of initials (histogens), one each for the central cylinder and the rootcap and a common group for the cortex and the epidermis; e.g., most monocotyledons
3. Three groups of initials (histogens), one each for the central cylinder and the cortex and a common group for the rootcap and the epidermis; e.g., most dicotyledons
4. Two groups of initials (histogens), one for the central cylinder and a common group of initials for the cortex, epidermis, and rootcap; e.g., *Gossypium* and *Hibiscus*
5. One group of ill-defined initials (histogens), the transversal type with all tissues sharing a common origin; e.g., Fabaceae.

These early studies, reviewed by Kroll (1912) and Schüepp (1926), account for the bulk of information available on the structure of root apices. It should be noted that none of the authors cited above included growth conditions in their papers.

II. ANGIOSPERM ROOT APEX, 1900–1971

The structural aspects of the root apex have received relatively little attention in the past 70 years. Instead of broad surveys, most of the more recent anatomical investigations have been intense developmental studies of one or only a few species. Also, many of the more recent investigations have focused on tissue differentiation rather than apical organization of the root (Young, 1933; Hayward, 1938; Esau, 1940, 1943a; Guttenberg, 1940; Cheadle, 1944; Goodwin and Stepka, 1945; Williams, 1947; Heimsch, 1951; Popham, 1955; Peterson, 1967). In the United States, because of the influence of Esau—who, in turn, was probably influenced by Soueges (1934–39)—the histogen terminology and rationale became less popular. Esau prefers a topographic approach in describing the apical organization of roots; i.e., she proposes that the *promeristem* (1953), or *protomeristem* (1967), be defined as the initials and their most recent derivatives. The difference between these two terms is not conceptual but lexicographical (Jackson, 1953). Also, she names each initial group of tiers or initials according to the mature tissue derived from the group. It will be pointed out later that recent data have placed limitations on this approach to the structure and function of the root apex.

A. Ontogenetic Changes in the Root Apex

Recently, Guttenberg (1947, 1960, 1964) and his students (Guttenberg *et al.*, 1955) have revitalized interest in the root apex. Following a detailed investigation of the embryogeny and subsequent growth of the root in several species, they reported that ontogenetic changes occur in the apical organization of some roots such as *Anoda* and *Helianthus*. As described by Guttenberg, the dormant embryonic roots of these plants have a layered or "closed" pattern of apical organization. During growth, however, certain cells, which are histogenically related to the cortex, participate in rootcap formation. The result of this activity is that the apex becomes "open" during growth and resembles a transversal meristem.

In addition to reporting ontogenetic changes in some roots, Guttenberg has proposed that the promeristem (*urmeristem*) of a root consists of only one cell (*Zentralzelle*) or, at the most, a few cells (Guttenberg, 1947, 1960, 1964). According to his interpretation the *Zentralzelle* behaves like a formative center which may from time to time renew the initials (histogens) of mature tissues. Guttenberg's proposal of the existence of central cells was diametrically opposed to the view held by most plant anatomists and resulted in widespread criticism (Clowes, 1953, 1954, 1958, 1961).

Clowes (1950), working with *Fagus*, also reported histogenic involvement of the cortex with the rootcap in growing beech roots. However, Clowes's interpretation of the promeristem differed from that of Guttenberg. Making use of Schüepp's (1917) *Körper-Kappe* concept to analyze planes of divisions within the root apex, Clowes proposed that the promeristem of *Fagus* is quite large. Up to this time there had been little use made of the *Körper-Kappe* analysis. Schüepp's central idea was that the orientation of the T (or Y) divisions in the root apex could be used to determine the ontogenic relationship of root tissues. Basically, all tissues with the T divisions oriented toward the apex belong to the *Körper* (body), and tissues with T divisions oriented away from the apex are related to the *Kappe* (cap).

In two subsequent investigations Clowes (1953, 1954), utilizing both the *Körper-Kappe* analysis and surgical techniques, strengthened his contention that the promeristem in roots is quite large. Also, more importantly, Clowes (1954) suggested that cell divisions rarely occur within what he originally termed the *minimal construction center* of the root apex. He quickly investigated further the idea that a nonmeristematic group of cells exists within the meristem of the root.

B. Quiescent Center

Using both radioactive labels and more sophisticated staining techniques, Clowes (1956a, 1956b) presented direct evidence for the existence of a hemispherically shaped quiescent center in the roots of *Zea, Vicia,* and *Allium.* Other investigators (Rabideau and Mericle, 1953) had earlier published an autoradiograph of a corn root, which showed a zone of cells that did not assimilate $^{14}CO_2$; these investigators did not comment upon or point out the lack of label in their material. Clowes, unaware of the Rabideau and Mericle publication, was the first to recognize the significance of the unlabeled area and thus was the first to discover the quiescent center.

Initially many people were reluctant to accept the presence of a group of mitotically inactive cells in the root apex. This skepticism concerning the validity of the quiescent center was probably due to the fact that a similar region in the shoot apex, the *méristéme d'attent* (Buvat, 1952), had been discredited (Partanen and Gifford, 1958; Clowes, 1959b).

Clowes and his associates (Clowes, 1956a, 1956b, 1958, 1959a, 1959c, 1963, 1967; Clowes and Juniper, 1964; Hall, Lajtha, and Clowes, 1962; Thompson and Clowes, 1968) have demonstrated conclusively the presence of a quiescent center and, in some cases, have experimented with the region in *Sinapis, Pistia, Eichornia, Zea, Vicia,* and *Allium.* Other investigators have either discovered or confirmed the existence of a quiescent center in *Allium* and *Vicia* (Jensen, 1958); *Euphorbia* (Raju *et al.* 1964); *Glycine* (Miksche and Greenwood, 1966); cultured *Lycopersicon* (Thomas, 1967); cultured *Convolvulus* (Phillips and Torrey, 1970); *Abutilon* (Byrne, 1969); and *Malva* (Byrne and Heimsch, 1970b).

In view of the presence of the quiescent center in every angiosperm root thus far investigated, several basic questions, some of which remain unanswered, have arisen concerning quiescence.

First, when does the quiescent center appear in seedling roots? Depending upon the species, the quiescent center appears a few days after germination (Clowes, 1958) or is present in the dormant seed (Clowes, 1961; Byrne and Heimsch, 1970b). The latter is more difficult to prove because many nuclei in the dormant embryonic root apex may be at the 4C level of DNA synthesis (Avanzi *et al.,* 1963; Davidson, 1966).

Second, how big is the quiescent center? All the evidence suggests that several hundred cells may be quiescent and that the number is proportional to the size of the root apex. Clowes (1961) suggested that several thousand cells may be quiescent in very large apices; and Byrne and

Heimsch (1970b) have reported 180 to 700 quiescent cells in *Malva sylvestris*, the number fluctuating with the size of the apex and cellular adjustments within the apex.

Third, in addition to being mitotically inactive, is the quiescent center metabolically inactive? All the evidence suggests that it is. This evidence is based both on the staining properties (Clowes, 1956a; Jensen, 1958) and the ultrastructure (Clowes and Juniper, 1964; Street *et al.*, 1967) of the quiescent center.

Fourth and most important, what is the function of the quiescent center? This question remains unanswered. Clowes (1959a) has suggested that the quiescent center—

1. may serve as a pool of diploid cells,
2. is necessary for the root to maintain its geometric integrity, or
3. functions as a source or sink for metabolites and growth regulators.

Guttenberg (1964) never accepted the existence of a quiescent center, and recent studies (Camp, 1966; Seago and Heimsch, 1969; Byrne and Heimsch, 1970a) have confirmed that some roots undergo organizational changes similar to those reported by Guttenberg (1947) and his students (Guttenberg *et al.*, 1955). Until recently no investigation had simultaneously considered ontogenetic changes in apical organization and the development of the quiescent center. But since it seemed likely that ontogenetic changes might occur within the quiescent center (Byrne and Heimsch, 1968), such an investigation has recently been completed (Byrne, 1969; Byrne and Heimsch, 1970a, 1970b).

C. Malvaceous Root Apex

The root apices of *Malva sylvestris*, *M. crispa*, and *M. Alcea* and *Abutilon Theophrasti* were investigated. Dormant embryonic and growing roots up to 33 cm long were used for anatomical and autoradiographic analysis.

In those *Malva* species with large embryonic roots, *M. sylvestris* and *M. crispa*, ontogenetic changes similar to those reported previously by Guttenberg were noted. In these two species embryonic roots had a layered or closed pattern of apical organization (Fig. 1.1). As growth proceeded, cells, which might be considered cortical initials, produced cells of the central portion of the rootcap or secondary columella (Fig. 1.2). Also, the outer cortex, like that of *Fagus* (Clowes, 1950) and *Helianthus* (Guttenberg *et al.*, 1955), participated in the formation of the lateral portions of the rootcap (Fig. 1.3). This lack of histogenic integrity in the outer cortex contributed to the changing of the root apex from a closed to an open pattern of organization.

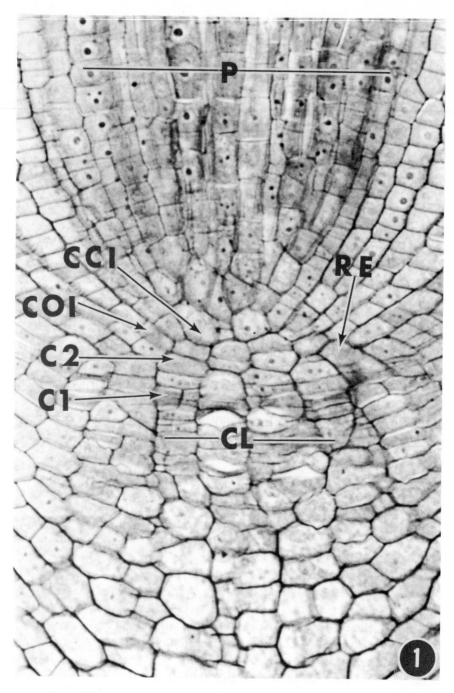

Fig. 1.1.   Malva crispa: median longitudinal section of a dormant embryonic root. Note the distinct boundary between the initials of the central cylinder and the cortex and secdary columella. *CCI:* central cylinder initials; *COI:* cortical initials; *C2:* secondary columella initials; *Cl:* columella initials; *CL:* columella; *RE:* rootcap epidermis initials; *P:* pericycle. × 1310

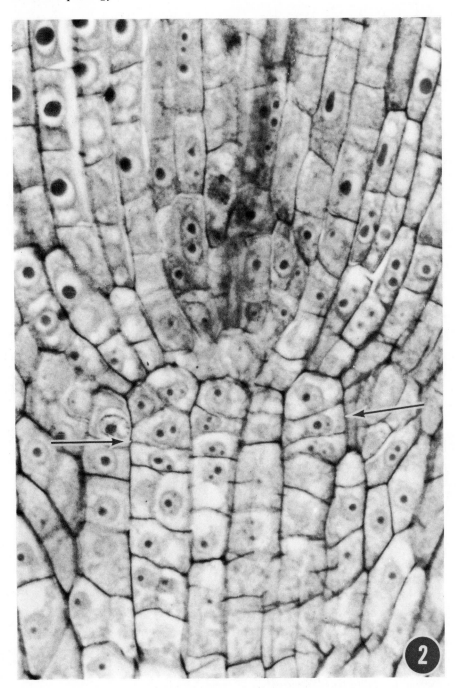

Fig. 1.2. *Malva crispa:* median longitudinal section of a root tip 4 cm long. Note the horseshoe-shaped boundary formed by the columella (*arrows*). × 1310

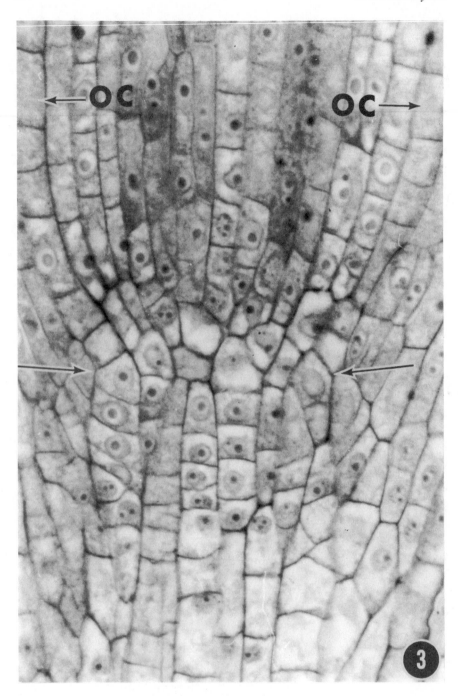

Fig. 1.3. Malva sylvestris: median longitudinal section of a root tip 6 cm long. Note the continuity of the outer cortex (*OC*) with the rootcap (*arrows*). × 1310

In *M. Alcea* and *A. Theophrasti* ontogenetic changes of smaller magnitude were observed. In both species the secondary columella of the rootcap was formed by cells that have been considered cortical initials. Since embryonic roots of *A. Theophrasti* had three layers of cortical initials (Fig. 1.4) and during growth only the outermost layer produced a secondary columella (Fig. 1.5), the apical organization of the long roots remained closed (Fig. 1.6). Embryonic roots of *M. Alcea* had only a single layer of cortical initials, which during growth functioned as secondary columella initials. In this case, the production of the secondary columella resulted in an open pattern of apical organization in the longer roots (Fig. 1.7).

In all four species investigated a quiescent center was present before germination (Fig. 1.8). Furthermore, autoradiography proved that the changes in apical organization occurred outside the quiescent center. This was due to the fact that while the outer cortex was involved in the production of the rootcap, it was not quiescent (Fig. 1.9). Later, in older and longer roots, the quiescent center expanded and incorporated the outer cortex (Fig. 1.10). At no time were the cells involved in the formation of columella quiescent. Data concerning the development of the quiescent center in *M. crispa* are presented in Table 1.1.

To date it would seem that (1) angiosperm roots have a quiescent center and (2) some angiosperm roots change in apical organization during growth. Then—what is the nature of the meristem in angiosperm roots?

Clowes's current concept of the promeristem seems functionally more realistic than that of Guttenberg (1947, 1960, 1964) or Esau (1953, 1967). This interpretation (Clowes, 1956a, 1956b, 1961), that the promeristem comprises meristematic cells on the surface of the hemispherical quiescent center, seems to me more relevant to the way a root grows than relegating the promeristem to a single or a few cells (Guttenberg, 1947, 1960, 1964). Furthermore, as Clowes proposed, the large promeristem makes it easier to understand ontogenetic changes that occur in the apices of some roots.

From a structural point of view, Esau's (1953, 1967) concept of the promeristem, or protomeristem, cannot be discounted. She suggested that the protomeristem comprises various tiers of initials and their most recent derivatives. Since there is a substantial amount of evidence that most of the cells within the protomeristem are quiescent, a functional application of this concept could be misleading. For example, in root apices with a closed pattern of apical organization, such as *A. Theophrasti* or *Sinapis* (Clowes, 1958), the tiers of cells between the central cylinder and rootcap are structurally cortical initials. In reality, however, the cortical initials are quiescent. Even though the cortical and central cylinder initials and their immediate derivatives are quiescent, it seems

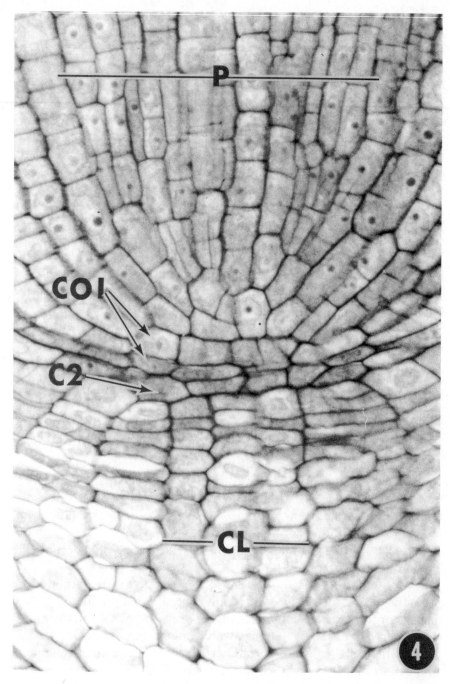

Fig. 1.4. Abutilon theophrasti: median longitudinal section of a dormant embryonic root. Note the three tiers of "cortical" initials (*COI*). The outermost layer (*C2*) functions as secondary columella initials during growth. ×1310

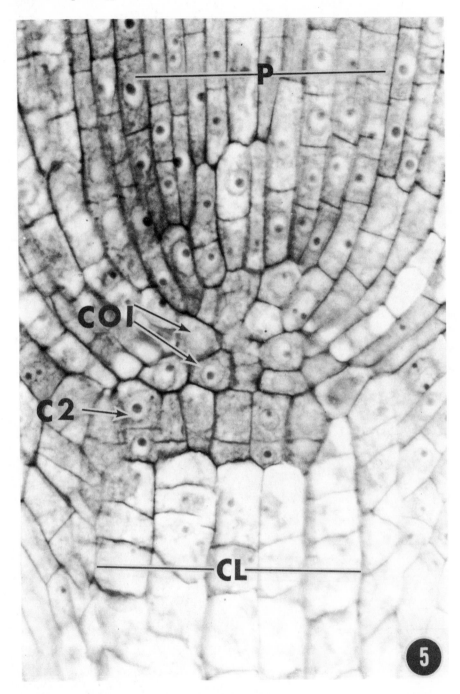

Fig. 1.5. Abutilon theophrasti: median longitudinal section of a root tip 6 cm long. Note the recent transverse divisions in the outermost layer of "cortical" or secondary columella initials (*C2*). × 1310

John M. Byrne

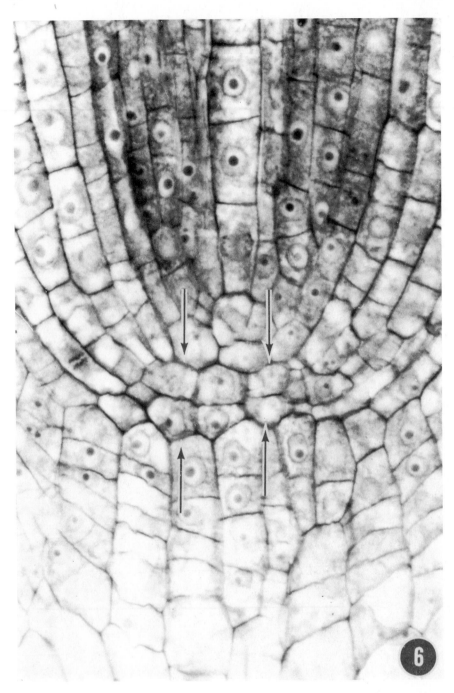

Fig. 1.6. Abutilon theophrasti: median longitudinal section of a root tip 33 cm long. Note the retention of two layers of cortical initials or closed pattern of apical organization (*between arrows*). × 1310

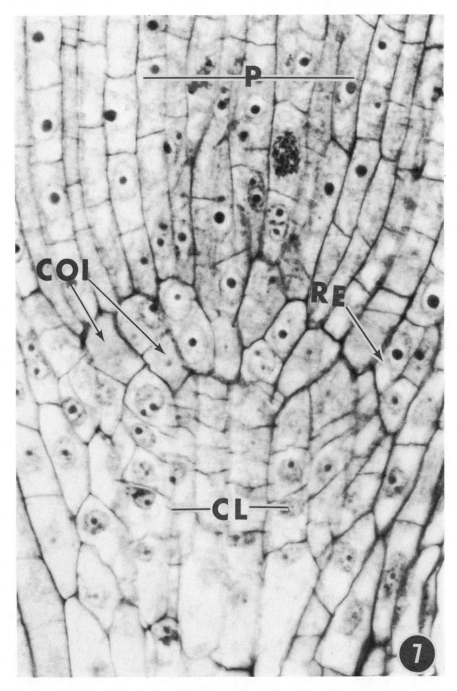

Fig. 1.7. *Malva alcea:* median longitudinal section of a root tip 16 cm long. Note the distinct cortical initials (*COI*) and the origin of the columella (*CL*). × 1310

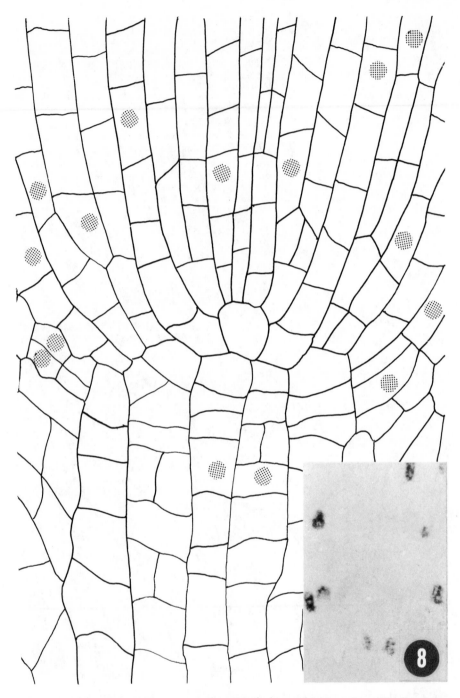

Fig. 1.8.   Malva crispa: schematic representation of a root tip 3 cm long. Shaded circles represent labeled nuclei. × 1310. Insert: autoradiograph of the median section. × 360. The schematic representation is a montage of the median and sections on either side of the median section, as are those in Figs. 1.9 and 1.10

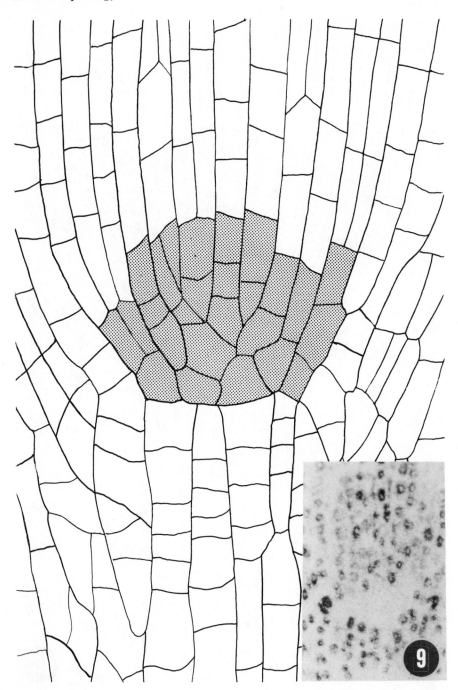

Fig. 1.9. Malva sylvestris: schematic representation of a root tip 6 cm long. Note the outer cortex is not quiescent (*shaded area*). × 1310. Insert: autoradiograph of the median section. × 360

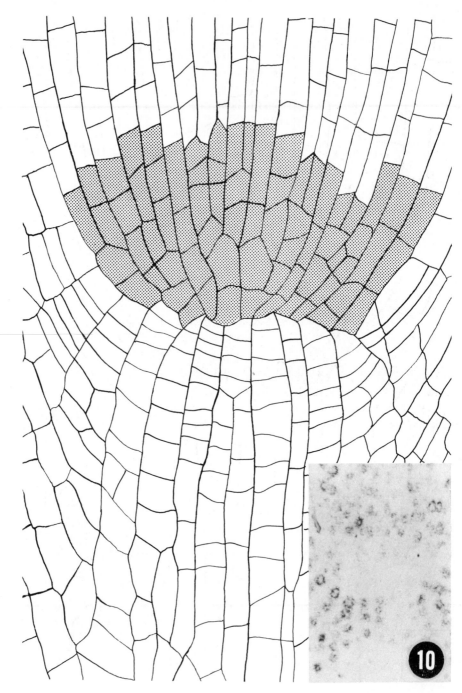

Fig. 1.10. Malva crispa: schematic representation of a root tip 33 cm long. Note the outer cortex is quiescent (*shaded area*). × 1310. Insert: autoradiograph of the median section. × 360

Table 1.1. The quiescent center of *Malva crispa*

Root length (cm)	Avg. diam† (mm)	Age‡ (days)	No. of cells in quiescent center	Comments
Embryo	534	24-hr soak	—	Scanty cytoplasmic labeling
0.2–0.4	490	1.5	—	Scanty cytoplasmic labeling
1.0	290	2.0	—	Cytoplasmic labeling evident
2.0	248	3.0	—	Cytoplasmic labeling evident
3.0	228	3.0	—	Cytoplasmic labeling evident
4.5	221	3.5	—	Scattered labeled nuclei
6.0	230	5.0	251	
9.0	247	7.0	190	
16.0	284	10.0	350	
23.0	264	13.0	582	
33.0	354	23.0	800	

† Measurement made 490 mm from apex of central cylinder; a minimum of 12 root apices per sample.
‡ Age based on time from initial soaking of seeds.

that Esau's protomeristem concept will continue to be useful for descriptive purposes. Furthermore, the presence of the quiescent center in growing roots does not negate the fact that the topographic initials were functional during embryogeny or that under some circumstances they may resume their meristematic role (Clowes, 1959c; Clowes and Stewart, 1967).

III. PRIMARY ROOT DIFFERENTIATION
AND
DEVELOPMENT OF MATURE TISSUE

Unlike the shoot, the root exhibits a simple arrangement of mature primary tissues. Recently, there have been many investigations of the development of root tissues, particularly vascular tissue.

A. Rootcap

The rootcap is a unique plant tissue. Derived from meristematic cells, either independently or commonly with the epidermis, mature rootcap cells exhibit many properties of typical parenchyma. Shielding the meristem is the function usually assigned to the rootcap, but recently Juniper *et al.* (1966) have demonstrated that decapped roots do not respond to gravity. This lack of geotrophic response indicates that the rootcap is more than a shield and may function in growth regulation.

B. Epidermis

With few exceptions the epidermis has a common origin with either the rootcap or the cortex. Structurally, the mature root epidermis is an uniseriate layer of elongate, closely packed cells, many of which form extensions, or root hairs. There is abundant evidence that root hairs function in absorption of water and minerals.

In a classical study of rye (*Secale cereale* L.), Dittmer (1937) showed that a single rye plant exposed 4,300 square feet of surface to the soil. The ability of the epidermis, both hairless and root hair cells, to absorb water has been amply demonstrated (Rosene, 1954, 1955). Although there is some difference of opinion about the presence of a cuticle on the root epidermis (Esau, 1967), one has been reported in *Allium* roots (Scott *et al.*, 1958).

The development of some hairs is acropetal from the base to the apex, and in some plants, particularly grasses, hair-forming cells (trichoblasts) show early differentiation. Recent investigations by Avers (1957, 1958), Kawata and Ishihara (1961), and Cormack (1962) have shown that trichoblasts differ from nontrichoblasts in their enzyme systems and RNA levels before differentiation into root hairs.

C. Cortex

When viewed in transverse section, the cortex generally appears as orderly radial rows of parenchyma cells. This orderliness is a manifestation of the centripetal development of the cortex (Williams, 1947; Guttenberg, 1940, 1943; Heimsch, 1960). Schizogenous (formed by splitting of adjacent cell walls) intercellular spaces of various sizes are present; their appearance is apparently related to the substrate (Peterson, 1967). Functionally, the cortex is considered to be the major site of starch storage in roots lacking secondary growth and to be the intermediary for translocation of water and minerals from the epidermis to the vascular tissue. According to current dogma, the innermost layer of cortex, the endodermis, regulates the movement of materials passing from the cortex to the vascular cylinder. This view of the endodermis is based upon anatomical consideration, i.e., the presence of Casparian strips on the endodermal cells, and upon experimental evidence (Arnold, 1952; Stewart and Sutcliffe, 1959).

D. Vascular Cylinder

The vascular cylinder, delimited by the pericycle, consists of alternately arranged primary xylem and phloem with associated parenchyma. The pericycle, which is potentially meristematic, may function in the initiation of lateral roots and may also participate in the formation of vascular and cork cambia. Primary vascular tissue differentiates centripetally with the protoxylem and protophloem located next to the pericycle. The number of protoxylem poles varies in angiosperms, ranging from two to many; and the number of protoxylem poles may vary within a given root (Guttenberg, 1940; Cheadle, 1944; Reinhard, 1956). The acropetal course of vascular tissue maturation in roots is well documented; the first protophloem elements mature closer to the apex than the first protoxylem elements (Esau, 1943b; Torrey, 1953; Heimsch, 1951; Popham, 1955). Also, there is a direct relationship between the appearance of the mature vascular tissue and the rate of root growth. In fast-growing roots mature vascular tissues are more distal from the apex than in slow-growing roots (Heimsch, 1951; Peterson, 1967).

It is evident that anatomically the angiosperm root is a relatively simple structure and that there is considerable descriptive literature on the angiosperm root. The mass of information available tends to obscure the fact that the causal factors of root growth and development remain a biological riddle. We do not know, for example, what controls the pattern of vascular differentiation. There is some evidence that the control of archy (diarch, triarch, tetrarch, polyarch) in roots is exercised by the apical meristem (Reinhard, 1956; Torrey, 1953, 1957), but the mechanism is unknown. Observations of extremely high DNA levels in differentiating metaxylem elements (Swift, 1950; Clowes, 1959a) and the role of auxins and cytokinins in tracheary element differentiation (Torrey and Fosket, 1970) illustrate the complexity of root growth and development.

IV. CONCLUSION

In my introductory remarks, I promised a look at angiosperm roots. In light of the theme of this volume, I must confess that my introductory declaration was not a very subtle defense mechanism. Although I do not feel obliged to apologize for myself or my past or present colleagues' approach to the root, I am somewhat chagrined by the paucity of environmentally oriented anatomical investigations. Under the circumstances, it is ironical that plant anatomists generally go to great lengths to avoid growing root material in soil. Of the numerous researchers cited in this presentation, only two used soil. The remainder either used a soil substitute or did not bother to report growth conditions. Most contemporary plant anatomists—I suppose it is the influence of the plant physiologists, who I understand also abhor soil—feel obliged to control environmental factors such as light, temperature, and nutrients. Since the soil is highly variable, and since soil-grown roots are difficult to handle histologically, they deem soil substitutes desirable. Paper, *Sphagnum*, vermiculite, water, moist air, and tissue culture are commonly used in lieu of soil. Whether or not these growth techniques produce roots comparable to those grown in soil has not been documented.

References

Arnold, A. 1952. Uber den Funktionsmechanismus der Endodermizellen der Wurzeln. *Protoplasma* 41: 189–211.

Avanzi, S., A. Brunori, F. D'Amato, V. N. Ronchi, and G. T. S. Mugnozza. 1963. Occurrence of 2C (G_1) and 4C (G_2) nuclei in the radicle meristems of dry seeds of *Triticum durum* Desf. Its implications in studies of chromosome breakage and on developmental processes. *Caryologia* 16: 533–88.

Avers, C.J. 1957. An analysis of difference of growth rate of trichoblasts and and hairless cells in the root epidermis of *Phleum pratense. Amer. J. Bot.* 44: 686–90.

——. 1958. Histochemical localization of enzyme activity in the root epidermis of *Phleum pratense. Amer. J. Bot.* 45: 609–13.

Buvat, R. 1952. Structure, évolution, et fonctionnement du méristème apical de quelques dicotylédones. *Ann. Sci. Natur. Bot.*, ser. 13, 2: 198–300.

Byrne, J. M. 1969. The root apex of *Malva.* Ph.D. Thesis. Miami University, Oxford, Ohio.

Byrne, J. M., and C. Heimsch. 1968. The root apex of *Linum. Amer. J. Bot.* 55: 1011–19.

——. 1970a. The root apex of *Malva sylvestris.* I. Structural development. *Amer. J. Bot.* 57: 1179–84.

——. 1970b. The root apex of *Malva sylvestris.* II. The quiescent center. *Amer. J. Bot.* 57: 1179–84.

Camp, R. 1966. The root apex of the Malvaceae. M.A. Thesis. Miami University, Oxford, Ohio.

Cheadle, V. T. 1944. Specialization of vessels within the xylem of each organ in the monocotyledons. *Bot. Gaz.* 98: 535–55.

Clowes, F. A. L. 1950. Root special meristem of *Fagus sylvatica. New Phytol.* 49: 248–68.

——. 1953. The cytogenerative center in roots with broad columellans. *New Phytol.* 52: 48–57.

——. 1954. The promeristem and minimal construction center in grass root apices. *New Phytol.* 53: 108–15.

——. 1956a. Nucleic acids in root apical meristems of *Zea. New Phytol.* 55: 29–34.

——. 1956b. Localization of nucleic acid synthesis in root meristems. *J. Exp. Bot.* 7: 307–12.

——. 1958. Development of quiescent centres in root meristems. *New Phytol.* 57: 84–88.

——. 1959a. Apical meristems of roots. *Biol. Rev.* 34: 501–29.

24   *John M. Byrne*

——. 1959b. Adenine incorporation and cell division in shoot apices. *New Phytol.* 58: 16–19.

——. 1959c. Reorganization of root apices after irradiation. *Ann. Bot.*, n.s. 23: 205–10.

——. 1961. *Apical Meristems.* Oxford: Blackwell Scientific Publications.

——. 1963. X-irradiation of root meristems. *Ann. Bot.*, n.s. 27: 343–52.

——. 1967. The quiescent centre. *Phytomotphology* (Delhi) 17: 132–40.

Clowes, F. A. L., and B. E. Juniper. 1964. The fine structure of the quiescent centre and neighboring tissues of root meristems. *J. Exp. Bot.* 15: 622–30.

Clowes, F. A. L., and H. E. Stewart. 1967. Recovery from dormancy in roots. *New Phytol.* 66: 115–23.

Cormack, R. G. H. 1962. The development of root hairs in angiosperms. II. *Bot. Rev.* 28: 446–64.

Davidson, D. 1966. The onset of mitosis and DNA synthesis in roots of germinating beans. *Amer. J. Bot.* 53: 491–95.

Dittmer, H. J. 1937. A quantitative study of the roots and root hairs of winter rye (*Secale cereale*). *Amer. J. Bot.* 24: 417–20.

Eriksson, J. 1876. Vegetationspunkt der Dikotylenwurzeln. *Bot. Zeit.* 34: 641–44.

Esau, K. 1940. Developmental anatomy of the fleshy storage organ of *Daucus carota. Hilgardia* 13: 175–226.

——. 1943a. Origin and development of primary vascular tissues in seed plants. *Bot. Rev.* 9: 125–206.

——. 1943b. Vascular differentiation in the pear root. *Hilgardia* 15: 299–324.

——. 1953. *Plant Anatomy.* New York: John Wiley and Sons.

——. 1967. *Plant Anatomy.* 2nd ed. New York: John Wiley and Sons.

Flahault, C. 1878. Recherches sur l'acroissement terminal de la racine chez Phanerogames. *Ann. Sci. Natur. Bot.* ser. 6, 6: 1–168.

Fleisher, E. 1874. Beiträge zur Embryologie der Monokotylen and Dikotylen. *Flora*, nos. 24–28.

Goodwin, R. H., and W. Stepka. 1945. Growth and differentiation in the root tip of *Phleum pratense. Amer. J. Bot.* 32: 36–46.

Guttenberg, H. von. 1940. Der primäre Bau der Angiosperm-wurzel. In *Handbuch der Pflanzenanatomie*, ed. K. Linsbauer, vol. 8, no. 39. Berlin: Gebrüder Borteraeger.

——. 1943. Die Physiologishen Scheiden. In *Handbuch der Pflanzenanatomie*, ed. K. Linsbauer, vol. 5, no. 42. Berlin: Gebrüder Borteraeger.

——. 1947. Studien über die Entwicklung des Wurzelvegetationspunktes der Dikotyledonen. *Planta Arch. Wiss. Bot.* 46: 179–222.

——. 1960. Gründzuge der Histogenese höhener Pflanzen. In *Handbuch der Pflanzenanatomie*, ed. K. Linsbauer, vol. 8, pt. 3. Berlin: Gebrüder Borteraeger.

——. 1964. Die Entwicklung der Wurzel. *Phytomorphology* (Delhi) 14: 265–87.

——. J. Burmeister, and H. Brosell. 1955. Studien über die Wurzelvegetation-spunktes der Dicotyledons. II. *Planta Arch. Wiss. Bot.* 46: 179–222.

Hall, E. J., L. G. Lajtha, and F. A. L. Clowes. 1962. The role of the quiescent centre in the recovery of *Vicia faba* roots from irradiation. *Radiat. Bot.* 2: 189–94.

Hanstein, J. 1868. Die Scheitelzell im Vegetationspunkt der Phanerogamen. *Abhandl. Geb. Nat. Math. U. Med.*, pp. 109–34.

——. 1870. Entwicklung des Keimes der Monokotylen und Dikotylen. *Bot. Abhandl. Geb. Morphol.*, ed. Hanstein, vol. 1, no. 1.

Hayward, H. E. 1938. *The Structure of Economic Plants.* New York: Macmillan.

Hegelmaier, F. 1874. Zur Entwicklungsgeschichte monokotyledoner Keime nebst Bemerkungen uber die Bildung der Samerdeckel. *Bot. Zeit.*, no. 49: 44.

Heimsch, C. 1951. Development of vascular tissue in barley roots. *Amer. J. Bot.* 38: 523–37.

——. 1960. A new aspect of cortical development in roots. *Amer. J. Bot.* 47: 195–201.

Holle, H. G. 1876. Über den Vegetationspunkt der Angiospermenwurzeln. *Bot. Zeit.*, no. 16: 17.

Jackson, B. D. 1953. *A Glossary of Botanical Terms.* 4th ed. New York: Hafner.

Janczewski, E. de. 1874. Recherches sur développement des radicelles dans les Phanérogames. *Ann. Sci. Nat. Bot.*, ser. 5, 20: 208–33.

Jensen, W. A. 1958. The nucleic acid and protein content of root tip cells of *Vicia faba* and *Allium cepa. Exp. Cell Res.* 14: 575–83.

Juniper, B. E., S. Groves, B. Landauschachar, and L. J. Audus. 1966. Rootcap and the perception of gravity. *Nature* (London) 209: 93–94.

Kawata, S., and K. Ishihara. 1961. Studies on the distribution of ribonucleic acid in the epidermis of the crown roots of the rice plant. *Crop Sci. Soc. Japan Proc.* 27: 387–91.

Kroll, G. 1912. Kritische Studies über Wurzelhaubentypen für die Entwicklung-sgeschichte. *Bot. Zentbl.*, supplement 28: 134–58.

Miksche, J. P., and M. Greenwood. 1966. Quiescent centre of the primary roots of *Glycine max. New Phytol.* 65: 1–4.

Nägeli, C. 1858. Entstehung und wachtum der Wurzeln. *Wiss. Bot.* 4: 73–160.

Partanen, C. R., and E. Gifford, Jr., 1958. Application of autoradiographic techniques to studies of shoot apices. *Nature* (London) 182: 1747–48.

Peterson, R. L. 1967. Differentiation and maturation of primary tissue in white mustard roots. *Can. J. Bot.* 45: 314–31.

Phillips, H. L., and J. G. Torrey. 1970. The quiescent center in cultured roots of *Convolvulus arvensis. Amer. J. Bot.* 57: 735 (abstract).

Popham, R. A. 1955. Levels of differentiation on primary roots of *Pisum sativum. Amer. J. Bot.* 42: 529–40.

Rabideau, G. S., and L. W. Mericle. 1953. The distribution of C^{14} in the root and shoot of young corn plants. *Plant Physiol.* 28: 329–30.

Raju, M. V. S., T. A. Steeves, and J. M. Naylor. 1964. Developmental studies on *Euphorbia esula* L.: Apices of long and short roots. *Can. J. Bot.* 42: 1615–28.

Reinhard, E. 1956. Ein Vergleich zwischen diarchen and triarchen Wurzeln of *Sinapis alba. Z. Bot.* 44: 505–14.

Reinke, J. 1871. Untersuchungen über Waschstumsgeschichte und Morphologie der Phanerogamenwurzeln. *Bot. Abhandle. Geb. Morphol.*, ed. Hanstein, vol. 1, no. 3.

Rosene, H. F. 1954. A comparative study of the rates of water influx into hairless epidermal surface and root hairs of onion roots. *Physiol. Plant.* 7: 676–86.

——. 1955. The water absorptive capacity of winter rye root hairs. *New Phytol.* 54: 95–97.

Schmidt, A. 1924. Histologische Studien an phanerogamen Vegetationspunkten. *Bot. Arch.* 8: 345–404.

Schüepp, O. 1917. Untersuchungen über Waschustum Formwechesel von Vegetationspunkten. *Jahrb. Wiss. Bot.* 57: 17–79.

——. 1926. Meristeme. In *Handbuch der Pflanzenanatomie*, ed. K. Linsbauer, vol. 4, no. 16. Berlin: Gebrüder Burteraeger.

Schwendener, S. 1882. Über das Scheitwaschstum der Phanerogamen Wurzeln. *S.B. preuss Akad. Wiss.*, pp. 183–99.

Scott, F. M., K. C. Hammer, E. Baker, and E. Bowler. 1958. Electron microscope studies of the epidermis in *Allium cepa. Amer. J. Bot.* 45: 449–61.

Seago, J., and C. Heimsch. 1969. Apical organization in the roots of the convolvulaceae. *Amer. J. Bot.* 56: 131–38.

Souèges, R. 1934–39. *Exposés d'embryologie et de morphologie vegétales.* 10 vols. Paris: Herman.

Stewart, F. C., and J. F. Sutcliffe. 1959. Relation to inorganic salts. In *Plant Physiology*, vol. 2, ed. F. C. Stewart. New York: Academic Press.

Street, H. E., A. Öpik, and F. E. James. 1967. Fine structure of the main axis meristem of cultured tomato roots. *Phytomorphology* 17: 391–401.

Swift, H. 1950. The constancy of deoxyribose nucleic acid in plant nuclei. *Proc. Nat. Acad. Sci.* (U.S.) 36: 643–64.

Thomas, D. R. 1967. The quiescent center in cultured tomato roots. *Nature* (London) 214: 739.

Thompson, J., and F. A. L. Clowes. 1968. The quiescent centre and rates of mitosis in the root meristem of *Allium sativum. Ann. Bot.*, n.s. 32: 1–13.

Torrey, J. G. 1953. The effect of certain metabolic inhibitors on vascular tissue differentiation in isolated pea roots. *Amer. J. Bot.* 40: 525–33.

——. 1957. Auxin control of vascular pattern formation in regenerating pea root meristems grown in vitro. *Amer. J. Bot.* 44: 859–70.

Torrey, J. G., and D. E. Fosket. 1970. Cell division in relation to cytodifferentiation in cultured root segments. *Amer. J. Bot.* 57: 1072–80.

Traub, M. 1876. *Le meristéme primitiv de la racine dans les monocotylédons.* Leiden.

Williams, B. C. 1947. The structure of the meristematic root tip and origin of primary tissues in roots of vascular plants. *Amer. J. Bot.* 34: 455–62.

Young, P. T. 1933. Histogenesis and morphogenesis in the primary root of *Zea mays.* Ph.D. Thesis. Columbia University.

2. Root-Shoot Relationships

Louis H. Aung

HIGHER land plants evolved during the Devonian period some four hundred million years ago (Seward, 1931; Chaloner, 1970). Unicellular organisms most probably acquired the power of cell multiplication before their migration to land, since cell division is a prerequisite for the differentiation and development of organs essential for survival in new environments. Such a capacity resides most prominently in meristematic cells (Clowes, 1961). Concurrent with the differentiation of new tissues and the assumption of separate functions by various organs, the plant must have developed a vascular system, vital for the transport of water and organic and inorganic substances between organs. The evolution of this vascular network to coordinate the various metabolic activities of root and shoot was decisive, as it enabled plants to adapt to a terrestrial existence.

I. TERMINOLOGY AND SCOPE

The terms *root* and *shoot* are employed here in a botanical sense (Esau, 1953a) and refer, respectively, to the entire subterranean and aerial portions of higher seed plants. In this usage, for example, we would exclude underground stems (tubers) of potatoes (*Solanum tuberosum* L.) and Jerusalem artichokes (*Helianthus tuberosus* L.) because they are not aerial organs, although botanically as stem tissues they could be included as part of the shoot system. Such a restriction, while arbitrary, affords a clearer line of demarcation and avoids unnecessary confusion in thinking about root-shoot systems. The term *relationship* shall be used to indicate any influence(s) that may exist between root and shoot.

The topic of root-shoot relationships is a vast one, and we shall therefore limit the consideration of the subject to (1) root and shoot systems, (2) their interrelationships, and (3) the shoot:root ratio and its alteration. The second category will further be subdivided and considered under five headings: (*a*) energy sources, (*b*) gas exchange, (*c*) location of major metabolic activities, (*d*) sites of ion uptake, and (*e*) growth regulators. Perhaps such an approach will impart some degree of order to the selected literature and provide the reader an integrated view of what we call "a plant."

II. ORIGIN AND NATURE OF ROOT AND SHOOT

Ontogenetically, the root and shoot are first discernible as apical meristems in the embryonic structure (Esau, 1953a; Clowes, 1961). Through the morphogenetic activities of these meristems arise the root and shoot organs. Characteristically, the shoot apical meristem produces stem tissues, leaf and floral primordia, but the root meristem is not directly involved with lateral organ formation since lateral roots arise endogenously some distance back of the meristematic apex. In the differentiation of the vascular system, the presence of leaves greatly alters the nature of vascular development of the shoot, but the root vascular system as an axial structure is independent of lateral root development. Another distinctive feature of the root is the presence of an endodermis tissue (Stocking, 1956) which is absent in the shoot (Esau, 1953b).

An appreciation of the architectural differences between the root and the shoot may be gained by reference to the investigations of Clowes (1961) and Popham (1964). In general, the shoot apices of vascular plants have a surface meristem of one or more cells, a central meristem, and a peripheral meristem, in contrast to a single localized subterminal meristem of root apices (Fig. 2.1). Consequently, the root and shoot apices differ strikingly in several ways: (1) only the root meristem is surmounted by a rootcap; (2) only the root apices have a quiescent center; and (3) only the shoot apices have lateral appendages. Clowes (1961) pointed out that the root's lack of lateral appendages makes it a suitable experimental material for biochemical studies. Nonetheless, impressive and valuable information about shoot apices has been obtained, especially in connection with floral induction and the attendant biochemical changes (Bonnett, 1966; Evans, 1969).

Several types of root and shoot systems can be distinguished by their morphological and biological functions (Weaver, 1926; Weaver and Bruner, 1927; Street, 1962; Kramer, 1956). For example, root systems may be feeder roots, support roots, or storage roots, while shoot systems may be leafy shoots, storage shoots, or reproductive (flower- and fruit-bearing) shoots. In a given species of plant, one or more of these types may occur. All these varied structures apparently make possible the

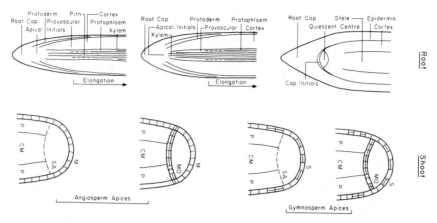

Fig. 2.1. Diagrammatic representation of some types of root and shoot apices of higher plants. Root: *upper left,* broad bean (*Vicia faba*) apex; *middle,* onion (*Allium cepa*) apex; and *upper right,* corn (*Zea mays*) apex. Shoot: *S,* surface meristem; *M,* mantle; *MO,* central mother cells; *C,* cambiumlike zone; *SA,* subapical initials; *CM,* central meristem; and *P,* peripheral meristem. (Redrawn by permission from F. A. L. Clowes, *Apical Meristems,* © 1961, Blackwell Scientific Publication Ltd., and from Popham, 1964)

exploitation of land, air, and sun that assures the perpetuation of higher plant species.

The aerial portions of plant species have received greater attention and study, probably because of their conspicuousness, while the subterranean portions were neglected. Part of this neglect sprang from the difficulty of access to the roots below the solid earth mantle. However, Weaver and associates (1926, 1927), using excavation techniques, have revealed fascinating information about the developmental life patterns of numerous herbaceous plant species under natural growing conditions. Similar information on woody plants is limited, but advances are being made with radioactive tracers and underground root observation laboratories (Whittington, 1969).

III. ROOT-SHOOT INTERRELATIONSHIPS

A. Energy Sources

The ultimate source of energy for the continuance of life on this planet is the sun, and chloroplast-bearing leaves capture much of this energy. The development of the chloroplasts and the maintenance of their structural and functional integrity in higher plants, however, depend upon chemical substances absorbed and transported via the roots. Street (1959) noted that while the whole plant itself is autotrophic in nature, the individual organs composing it are heterotrophic, depending upon their neighbors

for certain assimilated or elaborated substances. Aseptic culture techniques have shown that the growth of excised tissues or organs can be
maintained almost indefinitely if a suitable external source of energy
substances is available (White, 1934; De Ropp, 1947; Street, 1959,
Steward *et al.*, 1969). Nevertheless, one should not lose sight of the fact
that all the available organic substances employed in aseptic culture are
derived from assimilating leaves.

Hicks (1928a, 1928b) studied the distribution pattern of carbon and
nitrogen in the organs of wheat (*Triticum aestivum* L.) at different periods
of its development. She found that the products of assimilation, represented by carbon, rapidly moved from the leaves to the stems, grains, and
roots, while nitrogen derived from absorption by the roots was translocated upward and accumulated in the leaves and seed (Table 2.1).
Furthermore, it seems that most of the grain N is stored as protein-N
(Gericke, 1922).

Table 2.1. Distribution of carbon and nitrogen and C:N ratio in the
organs of wheat during development

Plant		% (dry wt basis)		
Stage†	Organ	C	N	C:N
I. Seedling	Plumule	29.0	7.4	3.9
	Radicle	39.0	3.1	12.5
II. Ten-leaf	Leaves	31.0	4.3	7.2
	Stem	38.5	3.1	12.4
	Roots	37.3	2.2	17.0
III. Ear	Flower	39.0	2.3	17.0
	Stamen	36.5	7.0	5.2
	Ovary	39.1	2.5	15.6
	Seed, embryo	49.7	8.7	5.7
	endosperm	30.7	2.2	14.9
	Leaves	35.6	2.8	12.7
	Stem	31.6	1.5	21.0
	Roots	31.6	1.5	21.0

Source: Hicks, 1928b.
† Stage I = 10 days old; stage II = 5 months; and stage III = near maturation.

Both indirect and direct evidence indicates that the growth of roots
depends upon sources of assimilates such as carbohydrates from the
shoots. One kind of evidence came from excision, decapitation, and girdling of the shoot and observation of their subsequent effects on the
roots. Keeble *et al.* (1930) showed that while the growth of the main root

of corn (*Zea mays* L.) seedlings increased slightly for a few days after shoot removal, the growth of adventitious roots decreased greatly. Maggs (1964, 1965) found that leaf removal in young apple trees led to an increased rate of dry matter production by the remaining leaves and that this material was distributed mainly to the nearest region of utilization. With continuous defoliation, the new stem formed the major "sink" (site of photosynthate storage or utilization), while the roots were most heavily penalized, especially in autumn when root growth was normally greatest. Pruning half the root system in summer reduced growth of all parts except the root, and leaf growth was disproportionately reduced. If pruning was carried out in autumn, stem growth was increased. He further observed that early root growth was at the expense of old stem and root reserves and took place when shoot growth was proceeding at a steady rate in June. Aung and Kelly (1966) observed in tomatoes (*Lycopersicon esculentum* Mill.) that the removal of mature leaves increased the net assimilation rate (NAR) of the remaining leaves and reduced the relative growth rate (RGR) of the stems and roots (Table 2.2). Cooper (1955), on the other hand, found that the increase in the number of tomato roots at first paralleled the growth of the shoots and then sharply declined at the time when fruits were rapidly increasing in number and size. These results indicated that tomato leaves supplied photosynthates

Table 2.2. Influence of partial defoliation on net assimilation rate (NAR) and relative growth rate (RGR) of tomatoes (cv. 'Fireball')

Treatment	NAR ($mg\ dm^{-2}\ day^{-1}$)	RGR (%)	
		Stem	Root
Control	36.8	14.0	11.9
Young leaves removed	29.5	14.1	11.6
Older leaves removed	43.4	12.2	9.8
HSD† at 5% level	9.1	1.7	1.5

Source: Aung and Kelly, 1966.
† HSD refers to Tukey's test at the 5% level of probability.

for root growth and that the supply could be cut off by leaf removal or diminished greatly by fruit competition. Brouwer (1962) also reported similar examples of assimilate supplies and demands between shoot and roots in other plants.

Mason and Maskell (1928a, 1928b) demonstrated that the removal of a ring of bark in cotton (*Gossypium* spp.) caused a considerable reduction in the amount of sugar present below the ring within a period of 7 hours, while sugar accumulated above the ring and in the leaves located 0.6 m above the ring. Although the downward movement of sugar was interrupted by the ring, sugar present below the ring continued to be moved to the roots but at a decreasing rate. In a different manner, Crowther (1934, 1941) showed that increased N application and water supply stimulated leaf and boll growth of cotton but suppressed root growth. He attributed the poor root growth to carbohydrate shortage induced by shoot competition. Eaton (1931) and Eaton and Johann (1944) obtained experimental evidence supporting Crowther's view. They found that defloration and defruiting of cotton plants increased the size of the root system and the content of carbohydrates of the roots (Table 2.3). Nightingale *et al.* (1928) have also shown that N application, while promoting shoot growth, resulted in a diminished amount of carbohydrate in tomatoes.

Table 2.3. Influence of defruiting on growth and carbohydrate content of cotton

Observations	Organs	Control	Defruited
Fresh wt (g)	Shoot	319.00	441.00
	Root, fibrous	164.00	330.00
	primary	19.00	28.00
Total sugar (mg)	Leaves	2.90	2.92
	Root, fibrous	0.10	0.30
	primary	5.13	14.11
Starch (mg)	Leaves	13.2	12.1
	Root, fibrous	1.95	2.17
	primary	13.5	30.2

Source: Eaton and Johann, 1944.

A second type of evidence is derived from the stimulation of photosynthesis by a root system serving as a metabolic sink. Humphries (1962, 1967) found that the root growth rate of beans (*Phaseolus vulgaris* cv. 'Canadian Winter') controlled the rate of movement of photosynthates from the lamina. When roots grew slowly, sugars and starches increased in the lamina; faster root growth was accompanied by a decrease of carbohydrates in the lamina and a higher NAR. Similarly, Thorne and Evans (1964) have shown with two cultivars of beets (*Beta vulgaris*) that the size of the sink determined the NAR (Fig. 2.2).

A third type of evidence came from the pattern of regrowth in perennial species. Lenkel (1927) noted in alfalfa (*Medicago sativa*) that the initiation of new top and root growth in early spring caused partial expenditure of the organic reserves stored in the roots during the previous season. The amount of these stored reserves depended upon the degree and maturity of top growth. He indicated that the quantity of carbohydr-

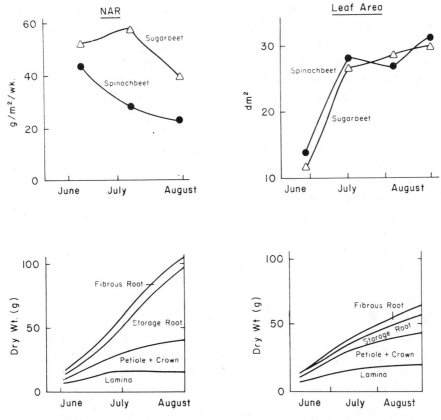

Fig. 2.2. Influence of "sink" size on net assimilation rate (NAR) of *Beta vulgaris*; lower left, sugar beet; lower right, spinach beet. (Redrawn by permission from G. N. Thorne and A. F. Evans, *Ann. Bot.*, n.s. 28: 499–508, © 1964, Oxford University Press)

ates and nitrogenous reserves was highest when the tops had attained a maximum growth and were at an advanced stage of maturity. Thus, frequent cutting of alfalfa at immature stages of growth caused a continuous reduction of organic reserves. Under this kind of cultivation, alfalfa not only showed decreased growth of roots and tops but suffered severe winter injury. Albert (1927) similarly observed that the roots of frequently cut alfalfa plants had low percentages of dry matter, total N, and carbohydrates. In addition, he found that in spring, alfalfa roots contained little starch and relatively high amount of sugars. In summer, reserves were in the form of starch, and in winter, there was a relatively high amount of soluble carbohydrates and a lower starch content.

The fourth type of evidence was obtained by the use of ^{14}C-labeled compounds. Rabideau and Mericle (1953) fed $^{14}CO_2$ to 8- to 10-day-old corn seedlings and found that a large proportion of the labeled compound was incorporated into the alcohol-soluble fraction containing sugars, organic acids, and amino acids of both root and shoot apices. The amount of $^{14}CO_2$ incorporated into the 3-mm root apices was about twice that taken into the 2-mm shoot apices. Hartt *et al.* (1963), using $^{14}CO_2$ fed to selected intact leaves of sugarcane (*Saccharum* sp.), have shown that labeled sucrose was the principal compound translocated throughout the plant. The distribution pattern was from the fed leaf to the stalk and downward. Upon arrival in the stalk, sucrose first moved to the center and then turned downward, but before getting as far as the roots, some sucrose found its way into an upward-moving system and was transported to the growing apex. The velocity of radioactive sucrose was calculated at 2.5 cm/minute. Additionally, Hartt *et al.* (1964) noted that defoliation of the upper leaves on the stalk decreased competition of assimilates and made available a larger quantity of sucrose moving into the stem and roots. Similarly, Williams (1964) observed that timothy (*Phleum pratense* L.) seedlings of different ages translocated ^{14}C-labeled assimilates more rapidly to young expanding leaves and growing roots than to older tissues.

In sweetpotatoes (*Ipomoea batatas*), Sekioka (1961, 1962, 1963) showed that ^{14}C-labeled sucrose applied to an intact leaf was rapidly translocated and stored in the roots. He observed that more foliar-applied sucrose accumulated in the roots as air and soil temperatures decreased; also, the plants grown under relatively low light intensity (Table 2.4) accumulated more of the foliar-applied sucrose in their roots than did plants grown under relatively high light intensity.

Table 2.4. Influence of environment on translocation of leaf-applied ^{14}C-sucrose to other parts of sweetpotatoes

Treatments	cpm/mg dry wt			
	Roots	Upper stem	Lower stem	Leaves
Soil temperature				
15C	101	125	61	32
20C	96	179	34	36
25C	56	104	18	39
Air temperature				
15C	207	99	43	26
20C	85	75	27	34
25C	130	84	24	89
Light intensity				
High	89	22	25	18
Low	179	12	20	11
Humidity				
High	78	20	17	17
Low	136	24	22	27

Sources: Sekioka, 1961, 1962, 1963.

B. Gas Exchange

Land plants exist in a gaseous milieu with their aerial portions completely bathed in the atmosphere but their subterranean portions only partially in contact with the gaseous phase located between the pore spaces of the soil medium. Often water displaces and occupies the air space of the soil, and it may create a stress condition detrimental to the growth of plants. However, although some plants succumb to poor aeration, others can thrive under similar conditions (Scholander *et al.*, 1955; Jensen *et al.*, 1964.) It is of interest, therefore, to consider some examples of gas exchange between roots and shoots and how it affects the metabolism of the whole plant.

Knight (1924) indicated that one of the problems in studying aeration was the difficulty in distinguishing between the negative effects due to lack of oxygen and the positive toxic effects due to high CO_2 levels. Leonard and Pinckard (1946) found by varying the O_2 concentration from 0% to 90% and maintaining a constant 10% CO_2 that cotton shoots and roots grew best between 5% to 21% O_2 concentrations. Oxygen concentrations above or below this range retarded both shoot and root development. In constrast, by maintaining the O_2 concentration at 21% and varying the CO_2 concentration from 0%, they obtained healthy shoot growth at CO_2 concentrations not exceeding 30% and healthy root growth at CO_2 concentrations not exceeding 15%. In addition, they noted that cotton plants could withstand an anaerobic soil condition quite well; they attributed ability to movement of O_2 from the tops to the roots. However, it is not known how long cotton can tolerate an anaerobic condition without injury.

Using pumpkin (*Cucurbita pepo*) seedlings, Brown (1947) demonstrated that when air was supplied to the root and hydrogen to the shoot, H was released from the root. When air was supplied to the shoot and N to the root, O was evolved from the root and N absorbed by it. He noted that the shoot was essential for gaseous movements and suggested that the transport of gases in solution took place through the cotyledons. Evans and Ebert (1960) used labeled ^{15}O on broad beans (*Vicia faba*) and observed the movement of ^{15}O down the primary root. Similarly, Barber *et al.* (1962) found that ^{15}O-labeled air moved from the shoot to the root in rice (*Oryza* spp.) and barley (*Hordeum vulgare* L.) plants by a simple gaseous diffusion process through the continuous intercellular spaces. They reported that gas spaces in rice roots constituted 5% to 30% of root tissue, whereas barley had less than 1%. Thus, under waterlogging conditions rice plants thrived (Valoras and Letey, 1966), but barley made very poor growth or died (Arnon, 1937). Livingston and Beall (1934) showed, on the other hand, that with a relatively "enriched CO_2"

in the soil solution, as much as 5% or more of CO_2 from the soil moved via the transpiration stream or the phleom to the shoot.

Aeration and nonaeration can affect both the growth and the composition of plants. Loehwing (1934) found greater dry weight and greater carbohydrate and mineral contents in the tops and roots of aerated soybean (*Glycine max* Merrill) plants than in nonaerated ones (Table 2.5). It may be noted that the decrease in dry weight and carbohydrate content of roots of nonaerated plants was greater than the decrease of similar components in the nonaerated shoot. However, with the exception of N, the mineral nutrients content of the roots changed less than that of the shoot. These observations are consistent with the view that the organs remote from the site of synthesis or absorption tend to be the most severely penalized under conditions unfavorable for these metabolic processes (Brouwer, 1962). Thus, when shoot-synthesized carbohydrates become deficient, the roots are the most affected organs; in the case of root-absorbed mineral nutrients, the shoot suffers more when these elements are limited.

Table 2.5. Response of 56-day-old soybean plants to soil aeration

Observations[†]	Shoot		Root	
	Aerated	Nonaerated	Aerated	Nonaerated
Dry wt (g)	0.70	0.48	0.20	0.10
Total sugars	21.0	12.6	21.8	3.9
Starch	95.6	61.7	47.8	15.0
Nitrogen	23.2	19.2	7.2	3.2
Phosphorus	4.5	2.6	0.7	0.9
Potassium	9.7	7.0	1.6	1.6
Calcium	13.1	10.4	2.3	2.7
Magnesium	2.9	2.1	1.6	1.1
Ash	64.0	57.0	1.8	13.4

Source: Loehwing, 1934.
† Except dry wt, all the measurements are in mg; the values given are on a per plant basis.

Hopkins *et al.* (1950) analyzed the content of both macronutrients and micronutrients in tomato plants grown with oxygen concentrations varying from 0.5% to 21% in the root zone. They found that iron, manganese, potassium, and phosphorus contents were highest at 21% oxygen concentration, while aluminum, copper, calcium, and magnesium contents remained unchanged over this range of oxygen concentration. The content of boron and sodium decreased significantly with increasing oxygen concentration. With soybeans, Shive (1941) found that low nitrate accumulation corresponded to a high rate of absorption at low O_2 tension (Fig. 2.3a). This seemingly contradictory observation, however, was explained on the basis of a high rate of NO_3 reduction. The rate of NO_3 reduction and absorption was low at high O_2 tension, and the best growth of soybean plants was observed at 6 ppm O_2. Since all the N supplied and absorbed was in the form of NO_3, the values of non-nitrate-N (Fig. 2.3b) indicated that a good proportion of the total NO_3-N absorbed had been reduced and transformed into other nitrogenous compounds within the plant (Fig. 2.3c).

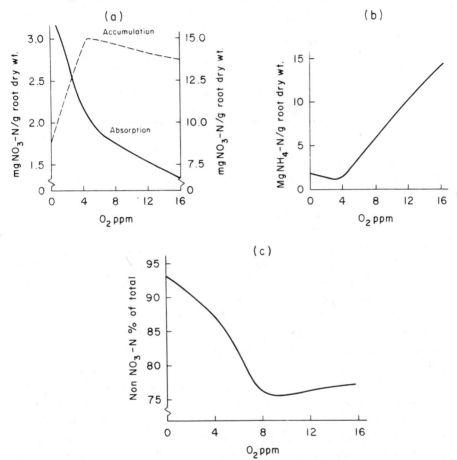

Fig. 2.3. Effects of oxygen concentrations on nitrate (*a*) accumulation and absorption, (*b*) reduction, and (*c*) utilization in soybeans (*Glycine max* Merrill). (Redrawn by permission from J. W. Shive, *Soil Sci.* 51 : 445–60, © 1941, The Williams & Wilkins Co., Baltimore)

Plant species differ in their responses to aeration (Vlamis and Davis, 1944; Leyton and Rousseau, 1957). Hopkins *et al.* (1950) found that tomato and tobacco (*Nicotiana tabacum* L.) plants responded to increasing concentrations of oxygen with an increase in fresh weight of shoots and roots, while soybeans showed relatively little response (Table

Table 2.6. Differential growth responses of three plant species to varying oxygen supply to the roots

O$_2$ content (%)	Fresh wt (g)					
	Tomato		Tobacco		Soybean	
	Shoot	Root	Shoot	Root	Shoot	Root
0.5	62	11	134	40	55	23
1.5	—	—	338	81	62	36
2.0	80	18	—	—	—	—
3.2	—	—	284	103	61	46
5.0	101	24	—	—	—	—
6.4	—	—	297	89	60	42
21.0 (Air)	123	30	338	96	68	43

Source: Hopkins et al., 1950.

2.6). Clark and Shive (1925) and Durrell (1941) also noted that aerated tomato plants had greater dry weight than nonaerated plants.

C. Location of Major Metabolic Activities

The specialized organs of plants require a wide variety of organic and inorganic compounds for their growth and nurture (Skoog, 1944; Torrey, 1954; Hannay et al., 1959; Street, 1959). By recognizing this nutritional interdependence, we can locate sites of synthesis, utilization, or accumulation. Thus, in section III A, we discussed how the leaf-synthesized carbohydrates were translocated and used or stored in such organs as the roots. We may note further some of the developmental changes in the distribution pattern of carbohydrates and their mode of dissimilation in the roots. For example, Rogozinska et al. (1965) found in 10-mm segments of corn root apices that starch and sucrose contents were high near the 0.5- to 2.5-mm regions while glucose and fructose were highest toward the 4.5- to 9.5-mm regions. In addition, the content of reducing sugars showed rapid changes with time. Beevers and Gibbs (1954a, 1954b), on the other hand, showed that the glucose in 1- to 2-mm corn root apices was metabolized via the Embden-Meyerhof-Parnas glycolytic pathway; however, in both the shoot and the root tissues of peas (*Pisum sativum* L.), carrots (*Daucus carota* L.), sunflowers (*Helianthus annuus* L.), and parsley (*Petroselinum crispum* Mill.), the glucose was respired via the direct oxidative glycolytic pathway (hexose monophosphate shunt). Other workers (Machlis, 1944a, 1944b; Berry and Brock, 1946; Berry, 1949) have also demonstrated in onion (*Allium cepa* L.) and barley root apices a well-defined pattern of respiration rates, which decreased with distance from the apex. Such respiratory activities exhibited by roots may be connected with mineral ion uptake or protein synthesis (Yemm and Willis, 1956; Steward and Sutcliffe, 1959; Beevers, 1961). Similarly, in the shoot the respiratory events may be connected with growth and the differentiation of reproductive organs (Kamerbeek, 1962).

Another important aspect of plant metabolism concerns the absorption of inorganic N and its assimilation into various organic compounds. In this section, we shall only deal with a few selected examples to emphasize some aspects of N metabolism. For a comprehensive review of the subject, the work by Chibnall (1939), Street (1949), Webster (1959), McKee (1962), and Hewitt and Cutting (1968) should be consulted.

Using ^{15}N, Cocking and Yemm (1961) showed that after a few hours of assimilation, glutamine and glutamic acid were highly labeled with the isotope in barley seedling roots. Subsequently, the labeled amino acids appeared in the root protein. Viets _et al._ (1946) found by chemical analysis of corn plants grown in ammonium nutrient culture that ammonia, glutamine, asparagine, and residual amino-N accumulated more rapidly and reached higher values in the roots than in the tops. They suggested that the amides and amino acids originated in the roots and then were translocated to the tops. Shishing-El (1955) has also shown with radish (_Raphanus sativus_ L.), sweetpotato, and carrot roots that inorganic ammonium or nitrate nitrogen was rapidly assimilated into amides. He noted that sweetpotato and carrot tissues were able to assimilate more NH_4 than NO_3, while the radish root tissues were able to assimilate more NO_3.

Vickery _et al._ (1936) found in sugar beets (_Beta vulgaris_ cv. 'Detroit Dark Red') that with increasing increments of $(NH_4)_2SO_4$ fertilization, the levels of glutamine increased correspondingly more in the roots than in the shoots. Clark (1936) made a thorough study of the various N fractions in the shoots and roots of tomato plants following NO_3 and NH_4 nutrition. He found that NH_4-treated tomato plants showed a greater content of protein-N, soluble organic-N, and amides in leaves, stems, and roots than NO_3-treated plants (Table 2.7). In addition, the dry weight and the organic acid fraction, consisting of oxalic, malic, and citric acids, were greater in the NO_3-treated plants.

By far the most complete study on root-shoot relationship was conducted by Pate (1968) and Pate and Wallace (1964) on the nutritional system of the field pea (_Pisum arvense_ L.). They showed that a considerable fraction of the inorganic N absorbed by the roots entered organic compounds such as amino acids and amides before being delivered to the aerial parts of the plant. In turn, the shoot provided the sugars that formed the C skeletons for the synthesis of organic compounds of N in

Table 2.7. Distribution of nitrogenous compounds in tomatoes under the influence of ammonium- and nitrate-nitrogen nutrition

| | % (dry wt basis) | | | | | |
| | NO_3 nutrition† | | | NH_4 nutrition† | | |
Fractions	Leaves	Stem	Root	Leaves	Stem	Root
Insoluble-N	3.89	0.93	2.65	5.13	1.27	3.09
Soluble-N	1.47	2.68	1.53	1.60	2.25	1.01
Total N	5.36	3.61	4.18	6.73	3.52	4.10
NH_3-N	0.04	0.03	0.04	0.16	0.14	0.08
NI_3-N	0.39	1.68	0.62	0.02	0.05	0.03
Soluble org.-N	1.04	0.98	0.88	1.42	2.06	0.90
Glutamine-N	0.04	0.11	0.04	0.28	1.13	0.21
Asparagine-N	0.05	0.04	0.03	0.28	0.32	0.11
Amino-N	0.24	0.13	0.10	0.23	0.09	0.17

Source: Clark, 1936.
† Nitrate-N was supplied as 0.0042 M $Ca(NO_3)_2$ and ammonium-N as 0.0042 M $(NH_4)_2 SO_4$.

the roots. A proportion of these amino acids were then transported to the aerial parts, where they supplemented the leaf-synthesized amino acids in protein synthesis.

Nitrate-reducing enzymes present in the roots and shoots of higher plants suggest that NO_3-N is first reduced to NH_3 before its assimilation into organic compounds. Street (1949) and Vaidyanathan and Street (1959) demonstrated nitrate reductase activity in tomato extracts. Sanderson and Cocking (1964a, 1964b) further showed the presence of nitrate reductase activity in the root and shoot of tomatoes. Similarly, Minotti and Jackson (1970) found nitrate reductase in both the root and shoot of wheat seedlings. Miflin (1968) noted nitrate and nitrite reductase activities in the supernatants of barley extract.

In addition to carbohydrates and nitrogenous compounds, plant organs need other factors such as vitamins for growth. Wilson and Withner (1946) found niacin, riboflavin, and thiamine in the fruit, leaves, and roots of five tomato cultivars. Bonner (1938, 1940, 1942, 1944) demonstrated that thiamine, pyridoxine, and pantothenic acid accumulated above the girdle of tomato stems and suggested that these substances were synthesized in the shoot and transported downward for root growth.

Roots can also serve as sites for the synthesis of metabolites of economic importance. An example is the production of the alkaloid nicotine in the root system of tobacco. Dawson (1942) showed in reciprocal grafts of tomato that when tobacco scions were grown upon tomato root stocks, little or no accumulation of nicotine occurred in the shoots of tobacco. However, when tomato scions were grown upon tobacco root stocks, a large amount of nicotine was found in the leaves, and smaller quantities were found in the stems and fruits. These results indicated the roots of tobacco as the site of nicotine synthesis.

D. Sites of Ion Uptake

The essentiality of mineral elements for the completion of the life cycle of plants has been discussed by Arnon (1961). Generally, the root system of plants constitutes the primary organ of ion absorption (Prevot and Steward, 1936; Broyer, 1961; Russell and Shorrocks, 1959). However, it has been shown that mineral nutrients can be absorbed through the foliage of plants. Bukovac and Wittwer (1957) showed that leaf-absorbed radioactive rubidium, sodium, and potassium were readily absorbed and translocated in beans (*Phaseolus vulgaris*), but calcium, strontium, and and barium were relatively immobile. The radioactive elements of magnesium, phosphorus, chlorine, sulfur, zinc, copper, manganese, iron, and molybdenum were absorbed and showed intermediate mobility in decreasing order of the elements listed. Although the foliar application of nutrients may be a satisfactory method of dealing with certain problems of mineral element deficiency that are not readily corrected by other means, it seems doubtful that such applications can be adopted profitably as a general means of continuous nutrient supply for crop growth in the field (Boynton, 1954; Wittwer and Teubner, 1959). Thus, despite the fact that the root system is not the only site of nutrient uptake, it still remains, from a physiological and practical standpoint, the primary site of ion uptake.

Steward *et al.* (1942) noted that the 1- to 3-cm apex of the growing barley root not only accumulated more bromine and rubidium ions from the external solution than the older segments back of the apex but also was the region from which more of the ions were removed for transfer to the shoot via the cortex and xylem. Kramer and Wiebe (1952) found that the accumulation of radioactive P in barley root apices was not restricted to the meristematic region of the roots but often occurred several centimeters behind the root apex (Table 2.8). Furthermore, ^{32}P also accumulated in the root hair region of both barley and tomato roots, and this ^{32}P was seen to be more readily translocated than that in the meristems. They indicated that among roots of similar past history there were much greater variations in respiration and salt uptake than in root structure. Additional study by Webe and Kramer (1954) has shown that although the meristematic regions of barley roots accumulated large amounts of mineral nutrients, relatively small amounts of these nutrients were translocated from this region to other organs of the plant. The

Table 2.8. Nitrogen and phosphorus distribution pattern in barley root apex

Distance from root apex (mm)	Total N/segment (μg)	Total P/segment (μg)
0–2	0.514	0.099
2–4	0.269	0.054
406	0.255	0.052
6–8	0.279	0.054
8–10	0.264	0.057
18–20	0.294	0.063
48–50	0.260	0.050

Source: Kramer and Wiebe, 1952.

greatest amount of translocation to the shoot occurred from a region several centimeters behind the root apex where the xylem was fully differentiated (Table 2.9). From this evidence, we see first that rapid accumulation of ions occurs only in actively metabolizing cells and second that the root system is not a homogeneous unit, for the tissues of the root display a graded activity in ion uptake influenced by the stage of root development and root configuration (Humphries, 1950; Olsen and Kemper, 1968; Barley, 1970).

E. Growth Regulators (Phytohormones)

Some three decades ago, several workers (Thimann, 1934; Went, 1938; Chibnall, 1939; Overbeek, 1939) indicated that the root system may be a site of production of phytohormones, in addition to its conventional roles of anchorage, storage, and water and salt absorption. The former role, however, was not seriously acknowledged until recently.

Table 2.9. Translocation of radioactive isotopes from different regions of barley seedling root apex to shoot

Distance from root apex at which isotope was supplied (mm)	% of isotope translocated to shoot			
	^{32}P	^{86}R	^{131}I	^{35}S
0–4	1.3	4.2	1.0	1.7
7–10	8.5	14.3	28.3	5.2
27–30	34.4	14.7	28.9	11.8
57–60	24.9	9.4	22.7	9.2

Source: Wiebe and Kramer, 1954.

1. Auxin

Thimann (1934) has shown in oat (*Avena sativa*) seedlings two forms of auxinlike substances. One form that readily diffused into dextrose agar blocks was designated as "free"; the second form, which was solvent extracted but did not diffuse out, was designated as "bound." Additionally, he found that these substances exhibited a basipetal gradient in the coleoptile but an acropetal gradient in the root (Fig. 2.4). He stated that

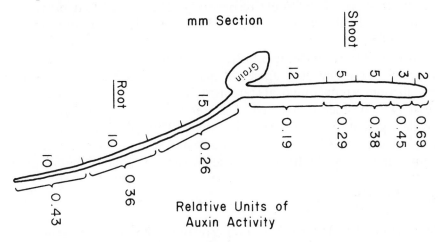

Fig. 2.4. Pattern of auxin distribution in the shoot and root of oat (*Avena sativa*) seedlings. (Redrawn by permission from K. V. Thimann, *J. Gen. Physiol.* 18: 23–24, copyright 1935, the Rockefeller Institute for Medical Research)

the presence of growth substances in the root may be due to (*a*) accumulation from the shoot apex or (*b*) production in the root apex but dependent upon a supply of precursor from the seed or plumule. Overbeek (1939) observed that the 3- to 4-mm pea (*Pisum sativum* cv. 'Alaska') root apices from 48-hour-old seedlings showed no more auxin diffusion into agar blocks after 6 to 9 hours. His conclusion was that the auxinlike substances diffusing from pea root apices were already present in the apices at the time they were excised from the seedlings and were not produced. He noted, however, that production may occur in intact roots since in vitro culture of excised pea roots could synthesize auxin. Britton *et al.* (1956) detected auxinlike compounds in excised 10-mm Sutton's Best-of-All tomato root apices from 4- to 5-week-old seedlings, cultured on a modified White's growth medium. The growth-promoting activities, however, were marginal. Subsequently, Thurman and Street (1960) found extracts of excised 20-mm tomato root apices stimulated the growth of oat coleoptiles. The greater part of the auxinlike activity was found in the aqueous fraction, although the acidic ethyl acetate fraction also showed some auxin activity.

Lahiri and Audus (1960) found auxinlike compounds in the neutral and acidic ether and water-soluble fractions of bean (*Vicia faba* cv. 'Green Leviathan') seedling roots. Further studies by Burnett *et al.* (1965) of 12-day-old bean (*Vicia faba*) roots failed to show the presence of indoleacetic acid in the acidic ether-soluble fraction following purification by diethylaminoethylether (DEAE) column chromatography. In the water-soluble ether-insoluble fraction, however, tryptophan and 3,4-dihydroxy phenylalanine (DOPA) were demonstrated at 12 and 20 mg/kg fresh weight of roots respectively. They noted that other substances, probably including some additional indole compounds, were unidentified.

In view of the evidence mentioned above and elsewhere (Aberg, 1957), it seems likely that root apices of plant species are capable of auxin production. An unequivocal statement, however, is made difficult by at least two considerations: first, the presence of high indoleacetic acid oxidase activity in root tissues (Janssen, 1970), which may destroy or mask the presence of auxin compounds, and second, the dependence of roots upon carbohydrates for metabolism and continual growth.

2. Gibberellins

Went (1938, 1943) noted that shoot growth in Alaska peas and tomatoes was dependent upon a factor, caulocaline, supplied by the root system. This root influence on shoot elongation was not due to water and salt uptake, since conditions adequately supplying these substances could not replace the effect of the root system. The nature of caulocaline, however, is unknown except that it is a factor(s) essential for shoot growth. In onions Abdalla and Mann (1963) have shown that continuous root removal from the bulbs presented shoot development. Similarly, Jackson (1955, 1956a, 1956b) observed in Marglobe tomato that adventitious roots partially prevented injury to the shoot when the original root system was flooded. These plants showed less epinasty and greater shoot growth than flooded plants without adventitious roots. Furthermore, shoot growth resumed if adventitious roots were allowed to develop following flooding but failed to do so if adventitious roots were removed. He also noted that plants without adventitious roots were dark-green and stocky while those with adventitious roots were pale-green and slender. The presence of gibberellinlike substances in excised tomato roots (Butcher, 1963) and the responses of these roots to exogenous gibberellins (Butcher and Street, 1960) suggested that the tomato system produced gibberellinlike substances.

Phillips (1964a, 1964b) indicated the possibility of a root-synthesized growth hormone which was transported to the shoot in common sunflowers. Subsequently, Phillips and Jones (1964) and Jones and Phillips

(1967) confirmed the presence of gibberellinlike substances in the root exudates of this species. The amount of gibberellinlike activity produced by the roots was about $0.05\mu g$ gibberellin A_3 equivalent plant^{-1} day^{-1}. Sitton *et al.* (1967b) also found that exudates from a decapitated sunflower root system continued to show gibberellinlike activity 4 days after decapitation. Additionally, they showed that incubation of these root apices in 2-^{14}C mevalonate yielded (-)-kauren 19-ol, an intermediate in gibberellin biosynthesis. Thus, the available evidence points to the fact that the root system is a site of production of gibberellinlike substances.

3. Cytokinins

Chibnall (1939) suggested that some hormonal influence from the root system was probably responsible for the regulation of leaf protein decomposition. Later, he showed that the rate of protein breakdown in bean leaves was indeed much reduced if the leaves were rooted (Chibnall, 1954), although the suspected hormonal factor was not identified. Went (1943) and Went and Bonner (1943) also noted that the tomato root system supplied a factor which prevented leaf chlorosis. The observation that 6-furfurylamino purine (kinetin) could delay chlorosis and protein breakdown implied that the hormonal root factor of Chibnall and Went resembles a cytokinin.

Kende (1965) have shown kinetinlike activity in the root exudates of sunflower plants. Itai and Vaadia (1965) further noted that water-stressed sunflower plants contained significantly less kinetinlike activity in the roots than nonstressed plants. This suggested that the alteration in shoot growth under water-stress conditions may be the result of a decrease of cytokininlike substances from the roots (Sitton *et al.*, 1967a). The origin of these substances was shown to be localized at the 1- to 3-mm meristematic root apices (Weiss and Vaadia, 1965). In addition to a reduction in cytokininlike activity due to water stress on the sunflower root system, Itai and Vaadia (1970) have also reported a diminution in the level of these substances in both the exudates and leaves of 6- to 8-week-old tobacco (*Nicotiana rustica*) plants under the influence of dry-air stress applied to the shoot.

Mullins (1967) observed that the inflorescences on woody grape (*Vitis vinifera* L.) cuttings failed to develop in the absence of roots. The retention of the inflorescences on the cuttings was influenced by the presence of roots. In the absence of roots, however, application of synthetic cytokinins such as 6-benzylamino purine (BAP) or 6-(benzylamino)-9-(2-tetrahydropyranyl)-9H-purine (SD8339) to the bases of unrooted cuttings in solution cultures or directly to emergent inflorescences promoted inflorescence growth. Skene and Kerridge (1967) and Skene (1970) have demonstrated the presence of cytokininlike substances in the exudates of grapes. These results substantiated Mullin's observation that cytokinins were essential for further development of grape inflorescences. Bui-Dang-Ha and Nitsch (1970) reported the isolation of 2 mg of ribosylzeatin from 250 kg of fresh roots of chicory (*Cichorium intybous* L.). This evidence strongly suggests that the root system of plants is capable of cytokinin production.

4. Abscisic Acid

In the plant extracts of tomato and bean roots, growth-inhibiting zones having R_f values on paper chromatograms corresponding to the so-called β-inhibitor complex have been recognized (Lahiri and Audus, 1960; Thurman and Street, 1960). It seems likely that a portion of this β-inhibitor is abscisic acid-like material (Milborrow, 1967).

Examination of the occurrence of abscisic acid-like substances in the various organs of 18 plant species revealed that only the roots of sycamore (*Acer pseudoplatanus*) showed abscisic acid-like activity (Milborrow, 1967). Lenton *et al.* (1968) also detected abscisic acid in the xylem sap of willow (*Salix viminalis* L.). Since abscisic acid-like substances occurred mainly in the shoot, it is possible that their presence in the roots was translocated from the shoot rather than by synthesis in the root system itself. Further work is needed to clarify this point.

IV. SHOOT: ROOT RATIO AND ITS ALTERATION

Shoots and roots together constitute the entire higher plant structure. In the preceding sections we have seen that various organic and inorganic substances serve important functions in the metabolism and growth of plants. But in the economy of the whole plant, specialized organs such as the shoot and root are in constant competition for available energy for their nurture and development. A measure of the resultant pattern of differential growth of the two organs, expressed as the shoot:root (S:R) ratio, can thus provide an index for the performance of each organ in a certain growth environment. For example, if food energy favors shoot growth at the expense of root growth, the plant in question will exhibit a relatively high S:R value. Conversely, if root growth is favored over shoot growth, the plant will have a lower S:R value. The S:R ratios may thus help to ascertain how environmental and chemical factors affect and modify the growth of the shoot and root. It should be noted, however, that the S:R ratios of plant species differ, and for a particular species the S:R value may vary with chronological age, stage of morphological development, and the kinds of growing environments.

A. Chronological Age

Le Clerc and Breazeale (1911) showed that the S:R ratio of wheat seedlings grown in nutrient culture increased with age from seeding (Fig. 2.5). During the first 3 days of growth, the radicle grew faster than the plumule; this relation was reflected in a low S:R value. Subsequently, the radicle grew slower than the plumule, and there was a corresponding increase in the S:R value. Pearsall (1923) also observed in pea seedlings initial low S:R values, followed by relatively higher S:R values as growth of the seedling progressed (Fig. 2.5).

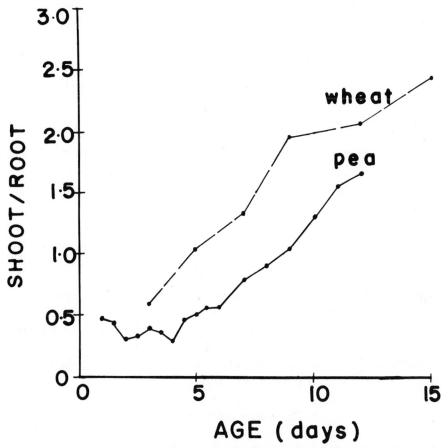

Fig. 2.5. Change in shoot: root ratio of wheat (*Triticum aestivum*) and pea (*Pisum sativum*) seedlings with advancing age. (Replotted from data of Le Clerc and Breazeale, 1911, and Pearsall, 1923)

B. Developmental Stage

The growth of the root is synchronized with the morphological stage of the shoot. In the vegetative phase, shoot and root growth proceeds con-

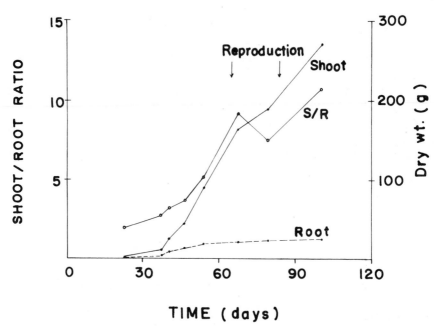

Fig. 2.6.   Influence of tasseling and silking in altering the shoot:root ratio of corn (*Zea mays*). (Foth, 1962)

currently in a linear fashion, but with the shoot growing at a faster rate. With the advent of flowering and fruiting, however, root growth slows or ceases abruptly due to a shortage of photosynthates from the shoot; the productive shoot diverts and monopolizes the available assimilates at the expense of the root. The change in the relationship is shown by a decrease in S:R values during the reproductive phase (Fig. 2.6). In tomatoes, Copper (1955) observed a sharp decline in root development coinciding with the time of fruit development. Also, during the maturation of modified storage structure such as the onion bulb, Kato (1963) found a marked decrease in both root weight and number as bulbing proceeded. Consequently, S:R ratios were higher at the postbulbing stage than at the prebulbing stage (Fig. 2.7).

C. Environmental Modification

Both physical and chemical factors can alter S:R ratios of plants. Yanada and Karimata (1969a, 1969b) have shown that soil types influenced the growth of the shoot and root of crop plants. They found that plants grown on a sandy soil had lower S:R ratios than similar plants grown on a loamy soil. It seems, however, that while the physical structure of the soil can affect S:R ratios of these plants, the chemical fertility factor was not clearly delineated. Ransom and Parija (1955) observed changes in S:R ratios in wheat, rice, and bean plants by exposing them to different

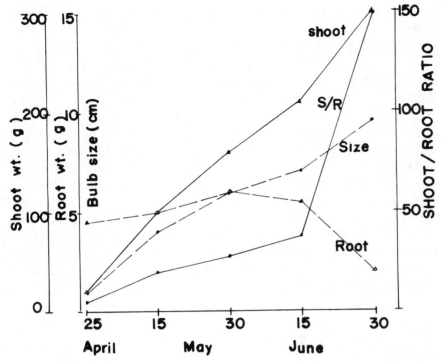

Fig. 2.7. Relationship between shoot: root ratio and bulk development of onion (*Allium cepa*). (Kato, 1963)

O_2 concentrations. In wheat they found that the S:R ratios decreased as O_2 concentrations dropped from 21% to 1.5%, whereas rice showed higher S:R ratios with decreasing O_2 concentrations.

Bora and Selman (1969) obtained a lower S:R ratio in tomato plants grown at 17C than in plants grown at 27C. Davis and Lingle (1961) similarly observed that a 15C root temperature caused a lower S:R ratio than a 25C temperature. They indicated that the decreased shoot growth at 15C was due not to a deficiency of mineral nutrients or water supply but rather to an endogenous mechanism. The nature of this mechanism is known; conceivably, it is hormonal (Went, 1943; Skene and Kerridge, 1967).

Other factors such as pruning, moisture, and light also modify the S:R ratio. Chandler (1919) found an increase in the S:R value of fruit trees following heavy pruning and attributed it to a reduction of root development. By increasing the soil moisture level from between 7.5% and 15.5% to 21%, Shank (1945) observed an increase in the S:R ratio of corn from 2.47 to 3.40. Crist and Stout (1929) showed that long photoperiods caused a decrease in S:R ratio of radishes by promoting the development of the taproot over that of the shoot. In addition to radishes, Weaver and Himmel (1929) found that the relative development of the shoots and roots of seven other plant species was altered by photo-periods.

Mineral nutrients generally have a decisive influence on the growth pattern of crop plants. Under natural growing conditions, some physical factors such as moisture, light, and temperature do not readily lend themselves to man's control. Chemical fertilization, however, is within his control, and he can manipulate nutrients to suit his needs. Shank (1945) found that application of increasing amounts of N and P to hybrid corn resulted in higher S:R ratios. Similarly, Turner (1922, 1926) showed higher S:R ratios in corn and barley with increased N concentrations. He attributed the increased S:R ratios to a greater carbohydrate utilization by the shoot at the expense of the root. In lettuce (*Lactuca sativa*), Crist and Stout (1929) found that S:R ratios increased with increasing amounts of $Ca(NO_3)_2$ application. Johnson and Ware (1948) obtained significant increases in both storage root and vine weights of sweet-potatoes grown under field conditions by applying increasing amounts of N fertilizers. The S:R ratio also increased correspondingly.

V. CONCLUDING REMARKS

First, I have considered the plant here from the standpoint of its compo-nent organs and their functions. This approach, however, is mainly one of convenience in order to gain a clearer view of plant behavior. There-fore, we must not lose sight of the fact of the plant is a unitary multicel-lular organism. Second, this topic is so vast that my survey undoubtedly fails to do justice on all points. I hope, however, that my review may serve to stimulate further work into the nature of plants, for further knowledge about root-shoot relationships can be an aid in maximizing crop plant productivity.

References

Abdalla, A. A., and L. K. Mann. 1963. Bulb development in the onion (*Allium cepa L.*) and the effect of storage temperature on bulb rest. *Hilgardia* 35: 85–112.

Aberg, B. 1957. Auxin relations in roots. *Ann. Rev. Plant Physiol.* 8: 153–80.

Albert, W. B. 1927. Studies on the growth of alfalfa and some perennial grasses. *J. Amer. Soc. Agron.* 19: 624–54.

Arnon, D. I. 1937. Ammonium and nitrate nitrogen nutrition of barley at different seasons in relation to hydrogen-ion concentration, manganese, copper and oxygen supply. *Soil Sci.* 44: 91–113.

——. 1961. Growth and function as criteria in determining the essential nature of inorganic nutrients. In *Mineral Nutrition of Plants*, ed. E. Truog, pp. 313–41. Madison: University of Wisconsin Press.

Aung, L. H., and W. C. Kelly. 1966. Influence of defoliation on vegetative, floral and fruit development in tomatoes (*Lycopersicon esculentum* Mill.). *Proc. Amer. Soc. Hort. Sci.* 89: 563–70.

Barber, D. A., M. Elbert, and N. T. S. Evans. 1962. The movement of ^{15}O through barley and rice plants. *J. Exp. Bot.* 13: 397–403.

Barley, K. P. 1970. The configuration of the root system in relation to nutrient uptake. In *Advances in Agronomy*, ed, N. C. Brady, New York: Academic Press.

Beevers, H. 1961. *Respiratory Metabolism in Plants*. New York: Harper and Row.

Beevers, H., and M. Gibbs. 1954a. Position of C^{14} in alcohol and carbon dioxide formed from labelled glucose by corn root tips. *Plant Physiol.* 29: 318–21.

——. 1954b. The direct oxidation pathway in plant respiration. *Plant Physiol.* 29: 322–24.

Berry, L. J. 1949. The influence of oxygen tension on the respiratory rate in different segments of onion root. *J. Cell Comp. Physiol.* 33: 41–66.

Berry, L. T., and M. J. Brock. 1946. Polar distribution of respiratory rate in the onion root tip. *Plant Physiol.* 21: 542–49.

Bonner, J. 1938. Thiamine (Vitamin B_1) and the growth of roots: The relation of chemical structure to physiological activity. *Amer. J. Bot.* 25: 543–49.

——. 1940. On the growth factor requirements of isolated roots. *Amer. J. Bot.* 27: 692–701.

——. 1942. Transport of thiamine in the tomato plant. *Amer. J. Bot.* 29: 136–42.

——. 1944. Accumulation of various substances in girdled stem of tomato plants. *Amer. J. Bot.* 31: 551–55.

Bonnett, O. T. 1966. Inflorescences of maize, wheat, rye, barley and oats: Their initiation and development. *Ill. Agr. Exp. Sta. Bull.* 721: 5–105.

Bora, B. C., and I. W. Selman. 1969. Growth and nitrogen accumulation in young tomato plants treated with gibberellic acid. *J. Exp. Bot.* 20: 288–301.

Boynton, D. 1954. Nutrition by foliage application. *Ann. Rev. Plant Physiol.* 5: 31–51.

Britton, G., S. Housley, and J. A. Bentley. 1956. Studies in plant growth hormones. V. Chromatography of hormones in excised and intact roots of tomato seedlings. *J. Exp. Bot.* 7: 239–51.

Brouwer, R. 1962. Distribution of dry matter in the plant. *Neth. J. Agr. Sci.* 10: 361–76, 399–408.

Brown, R. 1947. The gaseous exchange between the root and the shoot of the seedling of *Cucurbita pepo*. *Ann. Bot.*, n.s. 11: 417–37.

Broyer, T. C. 1961. The nature of the process of inorganic solute accumulation in roots. In *Mineral Nutrition of Plants*, ed. E. Truog, pp. 187–249. Madison: University of Wisconsin Press.

Bui-Dang-Ha, D., and J. P. Nitsch. 1970. Isolation of zeatin riboside from the chicory root. *Planta* (Berl.) 95: 119–26.

Bukovac, M. J., and S. H. Wittwer. 1957. Absorption and mobility of foliar applied nutrients. *Plant Physiol.* 32: 428–35.

Burnett, D., L. T. Audus, and H. D. Zinsmeister. 1965. Growth substances in the roots of *Vicia faba*. *Phytochemistry* 4: 891–904.

Butcher, D. N. 1963. The presence of gibberellins in excised tomato roots. *J. Exp. Bot.* 14: 272–80.

Butcher, D. N., and H. E. Street. 1960. The effects of gibberellins on the growth of excised tomato roots. *J. Exp. Bot.* 11: 206–16.

Chaloner, W. G. 1970. The rise of the first land plants. *Biol. Rev.* 45: 353–77.

Chandler, W. H. 1919. Some results as to the response of fruit trees to pruning. *Proc. Amer. Soc. Hort. Sci.* 16: 88–101.

Chibnall, A. C. 1939. *Protein Metabolism in the Plant*, p. 266. New Haven: Yale University Press.

———. 1954. Protein metabolism in rooted runner-bean leaves. *New Phytol.* 53: 31–37.

Clark, H. E. 1936. Effect of ammonium and of nitrate nitrogen on the composition of the tomato plant. *Plant Physiol.* 11: 5–24.

Clark, H. E., and J. W. Shive. 1925. Influence of continuous aeration upon the growth of tomato plants in solution culture. *Soil Sci.* 34: 37–41.

Clowes, F. A. L. 1961. *Apical Meristems*. Oxford: Blackwell Scientific Publications.

Cooper, A. J. 1955. Further observations on the growth of the root and the shoot of the tomato plant. *Proc. 14th Int. Hort. Congr.*, pp. 589–95.

Cocking, E. C., and E. W. Yemm. 1961. Synthesis of amino acids and proteins in barley seedlings. *New Phytol.* 60: 103–16.

Crist, J. W., and G. J. Stout. 1929. Relation between top and root size in herbaceous plants. *Plant Physiol.* 4: 63–85.

Crowther, F. 1934. Studies in growth analyses of the cotton plant under irrigation in the Sudan. I. The effects of different combinations of nitrogen applications and water supply. *Ann. Bot.* 48: 877–913.

———. 1941. Studies in growth analyses of the cotton plant under irrigation in the Sudan. II. Seasonal variations in development and yeild. *Ann. Bot.*, n.s. 5: 509–33.

Davis, R. M., and J. C. Lingle. 1961. Basis of shoot response to root temperature in tomato. *Plant Physiol.* 36: 153–62.

Dawson, R. F. 1942. Accumulation of nicotine in reciprocal grafts of tomato and tobacco. *Amer. J. Bot.* 29: 66–71.

De Ropp, R. S. 1947. Studies in the physiology of leaf growth. IV. The growth and behavior *in vitro* of dicotyledonous leaves and leaf fragments. *Ann. Bot.*, n.s. 11: 439–47.

Durrell, W. D. 1941. The effect of aeration on growth of the tomato in nutrient solution. *Plant Physiol.* 16: 327–41.

Eaton, F. M. 1931. Root development as related to character of growth and fruitfulness of the cotton plant. *J. Agr. Res.* 43: 875–83.

Eaton, F. M., and H. E. Johann. 1944. Sugar movement to roots, mineral uptake and the growth cycle of the cotton plant. *Plant Physiol.* 19: 507–18.

Esau, K. 1953a. *Plant Anatomy*. New York: John Wiley.

———. 1953b. Anatomical differentiation in shoot and root axes. In *Growth and Differentiation in Plants*, ed. W. E. Lommis, pp. 69–100. Ames, Iowa: Iowa State College Press.

Evans, L. T. 1969. *The Induction of Flowering*. Ithaca, N.Y.: Cornell University Press.

Evans, N. T. S., and M. Ebert. 1960. Radioactive oxygen in the study of gas transport down the root of *Vicia faba*. *J. Exp. Bot.* 11: 246–57.

Foth, H. D. 1962. Root and top growth of corn. *Agron. J.* 54: 49–52.

Gericke, W. F. 1922. Differences effected in the protein contents of grain by applications of nitrogen made at different growing periods of the plants. *Soil Sci.* 14: 103–9.

Hannay, J. W., B. L. Fletcher, and H. E. Street. 1959. Studies on the growth of excised roots. IX. The effects of other nutrient ions upon the growth of excised tomato roots supplied with various nitrogen sources. *New Phytol.* 58: 142–54.

Hartt, C. E., H. P. Kortschak, and G. O. Burr. 1964. Effects of defoliation, deradication and darkening the blade upon translocation of ^{14}C in sugarcane. *Plant Physiol.* 39: 15–22.

Hartt, C. E., H. P. Kortschak, A. J. Forbes, and G. O. Burr. 1963. Translocation of ^{14}C in sugarcane. *Plant Physiol.* 38: 305–18.

Hewitt, E. J., and C. V. Cutting. 1968. *Recent Aspects of Nitrogen Metabolism in Plants*. New York: Academic Press.

Hicks, P. A. 1928a. The carbon/nitrogen ratio in the wheat plant. *New Phytol.* 27: 1–46.

———. 1928b. Distribution of carbon/nitrogen ration in the various organs of the wheat plant at different periods of its life history. *New Phytol.* 27: 108–16.

Hopkins, H. T., A. W. Specht, and S. B. Hendricks. 1950. Growth and nutrient accumulation as controlled by oxygen supply to plant roots. *Plant Physiol.* 25: 193–209.

Humphries, E. C. 1950. The absorption of ions by excised root systems. I. Apparatus and preliminary experiments. *J. Exp. Bot.* 1: 282–300.

———. 1962. Dependence of net assimilation rate on root growth of isolated leaves. *Ann. Bot.*, n.s. 27: 175–82.

———. 1967. The effect of different root temperatures on dry matter and carbohydrate changes in rooted leaves of *Phaseolus* spp. *Ann Bot.*, n.s. 31: 59–69.

Itai, C., and Y. Vaadia. 1965. Kinetin-like activity in root exudate of water-stressed sunflower plants. *Physiol. Plant.* 18: 941–44.

———. 1970. Cytokinin activity in water stressed shoots. *Plant Physiol.* 47: 87–90.

Jackson, W. T. 1955. The role of adventitious roots in recovery of shoots following flooding of the original root system. *Amer. J. Bot.* 42: 816–19.

———. 1956a. Flooding injury studied by approach-graft and split root system techniques. *Amer. J. Bot.* 43: 496–502.

———. 1956b. The relative importance of factors causing injury to shoots of flooded tomato plants. *Amer. J. Bot.* 43: 637–39.

Janssen, M. G. H. 1970. Indoleacetic acid oxidase, peroxidase and polyphenoloxidase of pea roots. *Acta Bot. Neerl.* 19: 73–80.

Jensen, C. R., J. Letey, and L. H. Stolzy. 1964. Labeled oxygen: Transport through growing corn roots. *Science* 144: 550–52.

Jensen, C. R., L. H. Stolzy, and J. Letey. 1967. Tracer studies of oxygen diffusion through roots of barley, corn and rice. *Soil Sci.* 103: 23–29.

Johnson, E. S., and W. H. Dore. 1929. The influence of boron on the chemical composition and growth of the tomato plant. *Plant Physiol.* 4: 31–62.

Johnson, W. K., and L. M. Ware. 1948. Effects of rates of nitrogen on the relative yields of sweetpotato vines and roots. *Proc. Amer. Soc. Hort. Sci.* 52: 313–16.

Jones, R. L., and I. D. J. Phillips. 1967. Effect of CCC on the gibberellin content of excised sunflower organs. *Planta* (Berl.) 72: 53–59.

Kamerbeek, G. A. 1962. Respiration of the iris bulb in relation to the temperature and growth of the primordia. *Acta Bot. Neerl.* 11: 331–410.

Kato, T. 1963. Physiological studies on the bulbing and dormancy of onion plant. I. The process of bulb formation and development. *J. Jap. Soc. Hort. Sci.* 32: 81–89.

Keeble, F., M. G. Nelson, and R. Snow. 1930. The integration of plant behavior. II. The influence of the shoot on the growth of roots in seedlings. *Proc. Roy. Soc.* (London), ser. B, 106: 182–88.

Kende, H. 1965. Kinetin-like factors in the root exudate of sunflowers. *Proc. Nat. Acad. Sci.* (U.S.) 53: 1302–7.

Knight, R. C. 1924. The response of plants in soil and water-culture to aeration of the roots. *Ann. Bot.* 38: 305–25.

Kramer, P. J. 1956. Physical and physiological aspects of water absorption. In *Encyclopedia of Plant Physiology*, ed. W. Ruhland, 3: 124–59, 188–214.

Kramer, P. J., and H. H. Wiebe. 1952. Longitudinal gradients of ^{32}P absorption in roots. *Plant Physiol.* 27: 661–74.

Lahiri, A. N., and L. J. Audus. 1960. Growth substances in the roots of *Vicia faba. J. Exp. Bot.* 11: 341–50.

Le Clerc, J. A., and J. F. Breazeale. Translocation of plant food and elaboration of organic plant material in wheat seedlings. *U.S.D.A. Bull.* 138: 7–32.

Lenkel, W. A. 1927. Deposition and utilization of reserve foods in alfalfa plants. *J. Amer. Soc. Agron.* 19: 596–623.

Lenton, J. R., M. R. Bowen, and P. F. Saunders. 1968. Detection of abscistic acid in the xylem sap of willow (*Salix viminalis* L.) by gas-liquid chromatography. *Nature* 220: 86–87.

Leonard, O. A., and J. A. Pinckard. 1946. Effect of various oxygen and carbon dioxide concentrations on cotton root development. *Plant Physiol.* 21: 18–36.

Leyton, L., and L. Z. Rousseau. 1957. Root growth of tree seedlings in relation to aeration. In *The Physiology of Forest Trees*, ed. K. V. Thimann, pp. 467–75. New York: Ronald Press.

Livingston, B. E., and R. Beall. 1934. The soil as direct source of carbon dioxide for ordinary plants. *Plant Physiol.* 9: 237–59.

Loehwing, W. F. 1934. Physiological aspects of the effect of continuous soil aeration on plant growth. *Plant Physiol.* 9: 567–83.

Machlis, L. 1944a. The respiratory gradient in barley roots. *Amer. J. Bot.* 31: 281–82.

——. 1944b. The influence of some respiratory inhibitors and intermediates on respiration and salt accumulation of excised barley roots. *Amer. J. Bot.* 31: 183–92.

McKee, H. S. 1962. *Nitrogen Metabolism in Plants.* Oxford: Clarendon Press.

Maggs, D. H. 1964. Growth rates in relation to assimilate supply and demand. I. Leaves and roots as limiting regions. *J. Exp. Bot.* 15: 574–83.

——. 1965. Growth rates in relation to assimilate supply and demand. II. The effect of particular leaves and growing regions in determining the dry matter distribution in young apple trees. *J. Exp. Bot.* 16: 387–404.

Mason, T. G., and E. J. Maskell. 1928a. Studies on the transport of carbohydrates in the cotton plant. I. A study of diurnal variation in the carbohydrates of leaf, bark and wood and of the effects of ringing. *Ann. Bot.* 42: 189–253.

——. 1928b. Studies on the transport of carbohydrates in the cotton plant.

II. The factors determining the rate and the direction of movement of sugars. *Ann. Bot.* 42: 571–636.

Miflin, B. J. 1968. Nitrate reducing enzymes in barley. In *Recent Aspects in Nitrogen Metabolism*, ed. E. J. Hewitt and C. V. Cutting, pp. 85–88. New York: Academic Press.

Milborrow, B. V. 1967. The indentification of (+)-abscision II in plants and measurements of its concentrations. *Planta* (Berl.) 76: 93–113.

Minotti, P. L., and W. A. Jackson. 1970. Nitrate reduction in the roots and shoots of wheat seedlings. *Planta* (Berl.) 95: 36–44.

Mullins, M. G. 1967. Morphogenetic effects of roots and of some synthetic cytokinins in *Vitis vinifera* L. *J. Exp. Bot.* 18: 206–14.

Nightingale, G. T., L. G. Schermerhorn, and W. R. Robbins. 1928. The growth status of the tomato as correlated with organic nitrogen and carbohydrates in roots, stems and leaves. *N.J. Agr. Exp. Sta. Bull.* 461: 3–38.

Olsen, S. R., and W. D. Kemper. 1968. Movement of nutrients to plant roots. In *Advances in Agronomy*, ed. A. G. Norman, New York: Academic Press. 20: 91–151.

Overbeek, J. van. 1939. Is auxin produced in roots? *Proc. Nat. Acad. Sci.* (U.S.) 25: 245–48.

Pate, J. S. 1968. Physiological aspects of inorganic and intermediate nitrogen metabolism with special reference to the legume *Pisum arvense* L. In *Recent Aspects of Nitrogen Metabolism in Plants*, ed. E. J. Hewitt and C. V. Cutting, pp. 219–40. New York: Academic Press.

Pate, J. S., and W. Wallace. 1964. Movement of assimilated nitrogen from the root system of the field pea (*Pisum arvense* L.). *Ann. Bot.*, n.s. 28: 83–99.

Pearsall, W. H. 1923. Correlations in development. *Ann. Bot.* 37: 261–75.

Phillips, I. D. J. 1964a. Root-shoot hormone relations. I. The importance of an aerated root system in the regulation of growth hormone levels in the shoot of *Helianthus annuus*. *Ann. Bot.*, n.s. 28: 17–35.

——. 1964b. Root-shoot hormone relations. II. Changes in endogenous auxin concentration produced by flooding of the root system in *Helianthus annuus*. *Ann. Bot.*, n.s. 28: 37–45.

Phillips, I. D. J., and R. L. Jones. 1964. Gibberellin-like activity in bleeding-sap of root systems of *Helianthus annuus* detected by a new dwarf pea epicotyl assay and other methods. *Planta* (Berl.) 63: 269–78.

Popham, R. A. 1964. Developmental studies of flowering. *Brookhaven Symposia Biol.*, no. 16: 138–56.

Prevot, P., and F. C. Steward. 1936. Salient features of the root system relative to the problem of salt absorption. *Plant Physiol.* 11: 509–34.

Rabideau, G. S., and L. W. Mericle. 1953. The distribution of ^{14}C in the root and shoot apices of young corn plants. *Plant Physiol.* 28: 329–33.

Ransom, S. L., and B. Parija. 1955. Experiments on growth in length of plant organs. II. Some effects of depressed oxygen concentrations. *J. Exp. Bot.* 6: 80–93.

Rogozinska, J. H., P. A. Bryan, and W. G. Whaley. 1965. Developmental changes in the distribution of carbohydrates in the maize root apex. *Phytochemistry* 4: 919–24.

Russell, R. S., and V. M. Shorrocks. 1959. The relationship between transpiration and the absorption of inorganic ions by intact plant. *J. Exp. Bot.* 10: 301–16.

Sanderson, G. W., and E. C. Cocking. 1964a. Enzymic assimilation of nitrate in tomato plants. I. Reduction of nitrate to nitrite. *Plant Physiol.* 39: 416–22.

——. 1964b. Enzymic assimilation of nitrate in tomato plants. II. Reduction of nitrite to ammonia. *Plant Physiol.* 39: 425–31.

Scholander, P. F., L. van Dam, and S. I. Scholander. 1955. Gas exchange in the roots of mangroves. *Amer. J. Bot.* 42: 92–98.

Sekioka, H. 1961. Effect of temperature on the translocation of sucrose-C^{14} in the sweetpotato plant. *Crop Sci. Soc. Jap. Proc.* 30: 27–30.

——. 1962. The influence of light intensity on the translocation of sucrose-C^{14} in the sweetpotato plant. *Crop Sci. Soc. Jap. Proc.* 31: 159–62.

——. 1963. The effect of some environmental factors on the translocation and storage of carbohydrate in the sweetpotato, potato and sugar beet. (1) Relationships between the translocation of carbohydrate or radioisotopes and the soil and air temperatures. *Kyushu Univ. Sci. Bull.* 20: 107–18.

Seward, A. C. 1931. *Plant Life through the Ages*, pp. 60–154. New York: Macmillan.

Shank, D. B. 1945. Effects of phosphorus, nitrogen and soil moisture on top-root ratios of inbred and hybrid maize. *J. Agr. Res.* 70: 365–77.

Shishing-El, E. D. H. 1955. Absorption and assimilation of inorganic nitrogen from different sources by storage root tissue. *J. Exp. Bot.* 6: 6–16.

Shive, J. W. 1941. The balance of ions and oxygen tension in nutrient substrates for plants. *Soil Sci.* 51: 445–60.

Sitton, D., C. Itai, and H. Kende. 1967a. Decreased cytokinin production in the roots as a factor in shoot senescence. *Planta* (Berl.) 73: 296–300.

Sitton, D., A. Richmond, and Y. Vaadis. 1967b. On the synthesis of gibberellins in roots. *Phytochemistry* 6: 1101–5.

Skene, K. G. M. 1970. The relationship between the effects of CCC on root growth and cytokinin levels in the bleeding sap of *Vitis vinifera* L. *J. Exp. Bot.* 21: 418–31.

Skene, K. G. M., and G. H. Kerridge. 1967. Effect of root temperature on cytokinin activity in root exudate of *Vitis vinifera* L. *Plant Physiol.* 42: 1131–39.

Skoog, F. 1944. Growth and organ formation in tobacco tissue culture. *Amer. J. Bot.* 31: 19–24.

Steward, F. C., M. O. Mapes, and P. V. Ammirato. 1969. Growth and morphogenesis in tissue and free cell culture. In *Plant Physiology*, vol. VB, ed. F. C. Steward, pp. 329–76. New York: Academic Press.

Steward, F. C., P. Prevot, and J. A. Harrison. 1942. Absorption and accumulation of rubidium bromide by barley roots. Localization in the root of cation accumulation and of transfer to the shoot. *Plant Physiol.* 17: 411–21.

Steward, F. C., and J. F. Sutcliffe. 1959. Plants in relation to inorganic salts. In *Plant Physiology*, vol. 2, ed. F. C. Steward, pp. 253–478. New York: Academic Press.

Stocking, C. R. 1956. Histology and development of the root. In *Encyclopedia Plant Physiology*, ed. W. Ruhland, 3: 173–87.

Street, H. E. 1949. Nitrogen metabolism of higher plants. *Advance Enzymol.* 9: 391–454.

——. 1959. Special problems raised by organ and tissue culture. Correlations between organs of higher plants as a consequence of specific metabolic requirements. In *Encyclopedia Plant Physiology*, ed. W. Ruhland, 11: 153–78.

——. 1962. The physiology of roots. In *Viewpoints in Biology*, ed. J. D. Carthy and C. L. Duddington, pp. 1–49. London: Butterworths.

Thimann, K. V. 1934. Studies on the growth hormones of plants. VI. The distribution of the growth substance in plant tissues. *J. Gen. Physiol.* 18: 23–34.

Thorne, G. N., and A. F. Evans. 1964. Influence of tops and roots on net assimilation rate of sugar-beet and spinach beet and grafts between them. *Ann. Bot.*, n.s. 28: 499–508.

Thurman, D. A., and H. E. Street. 1960. The auxin activity extractable from excised tomato roots by cold 80% methanol. *J. Exp. Bot.* 11: 188–97.

Torrey, J. G. 1954. The role of vitamins and micronutrient elements in the nutrition of the apical meristem of pea roots. *Plant Physiol.* 29: 279–87.

Turner, J. W. 1922. Studies on the mechanism of the physiological effects of certain mineral salts in altering the ratio of top growth to root growth in seed plants. *Amer. J. Bot.* 9: 415–45.

——. 1926. The effect of varying the nitrogen supply on the ratios between the tops and roots in flax. *Soil Sci.* 21: 303–6.

Vaidyanathan, C. S., and H. E. Street. 1959. Nitrate reduction by aqueous extracts of excised tomato roots. *Nature* 184: 531–33.

Valoras, N., and J. Letey. 1966. Soil oxygen and water relationships to rice growth. *Soil Sci.* 101: 210–15.

Vickery, H. B., G. W. Pucher, and H. E. Clark. 1936. Glutamine metabolism of the beet. *Plant Physiol.* 11: 413–20.

Viets, F. G., Jr., A. L. Moxon, and E. I. Whitehead. 1946. Nitrogen metabolism of corn (*Zea mays*) as influenced by ammonium nitrition. *Plant Physiol.* 21: 271–89.

Vlamis, J., and A. R. Davis. 1944. Effects of oxygen tension on certain physological response of rice, barley and tomato. *Plant Physiol.* 19: 33–51.

Weaver, J. E. 1926. *Root Development of Field Crops.* New York: McGraw-Hill.

Weaver, J. E., and W. E. Bruner. 1927. *Root Development of Vegetable Crops.* New York: McGraw-Hill.

Weaver, J. E., and W. J. Himmel. 1929. Relation between the development of root system and shoot under long- and short-day illumination. *Plant Physiol.* 4: 435–57.

Webster, G. C. 1959. *Nitrogen Metabolism in Plants.* New York: Row, Peterson.

Weiss, C., and Y. Vaadia. 1965. Kinetin-like activity in root apices of sunflower plants. *Life Sci.* 4: 1323–26.

Went, F. W. 1938. Specific factors other than auxin affecting growth and root formation. *Plant Physiol.* 13: 55–80.

——. 1943. Effect of the root system on tomato stem growth. *Plant Physiol.* 18: 51–65.

Went, F. W., and D. M. Bonner. 1943. Growth factors controlling tomato stem growth in darkness. *Arch. Biochem.* 1: 439–52.

White, P. R. 1934. Potentially unlimited growth of excised tomato root tips in a liquid medium. *Plant Physiol.* 9: 585–600.

Whittington, W. J. 1969. *Root Growth*, pp. 361–76. London: Butterworths.

Wiebe, H., and P. J. Kramer. 1954. Translocation of radioactive isotopes from various regions of roots of barley seedlings. *Plant Physiol.* 29: 342–48.

Williams, R. D. 1964. Assimilation and translocation in perennial grasses. *Ann. Bot.*, n.s. 28: 419–25.

Wilson, K. S., and C. L. Withner. 1946. Stock-scion relationships in tomatoes. *Amer. J. Bot.* 33: 796–801.

Wittwer, S. H., and F. G. Teubner. 1959. Foliar absorption of mineral nutrients. *Ann. Rev. Plant Physiol.* 10: 13–32.

Yanada, K., and A. Karimata. 1969a. Ecological studies on the root system of vegetable crops in different soils. (2). Root system of radish, komatsuna and tomato plants. *J. Agr. Sci.* (Tokyo) 14: 49–70.

——. 1969b. Ecological studies on the root system of vegetable crops in different soils. (3). Root system of the cauliflower and squash plants. *J. Agr. Sci.* (Tokyo) 14: 71–84.

Yemm, E. W., and A. J. Willis. 1956. The respiration of barley plants. IX. The metabolism of roots during the assimilation of nitrogen. *New Phytol.* 55: 229–52.

3.   Plant Ionic Status

Konrad Mengel

FROM the time it became known that inorganic compounds play a major part in plant nutrition, numerous investigations have been made of the mineral contents of plants and special plant organs. Therefore, a vast amount of data is available, but the interpretation of these data often meets with difficulties.

The variations in the mineral contents of the different plant species and plant organs can be explained only if one knows the processes and the relationships responsible for the ionic status of living plants. In the last two decades, experimental results from various authors (Dijkshoorn, 1957, 1962; Dijkshoorn *et al.*, 1968; Coic, 1964; Coic *et al.*, 1962; Jackson and Stief, 1965; Jackson and Edwards, 1966; Jungk, 1967; Kirkby and Mengel, 1967; Hiatt, 1968) have given some insight into the processes affecting the ionic status of plants. In this respect there are four factors of major importance:

1. Assimilation of inorganic compounds
2. Active ion uptake
3. Long-distance transport
4. Specific ion requirements of various plant species

Antagonistic interactions can also exert a marked influence on the ionic status of plant tissues, but because these processes are discussed more thoroughly in Chapter 5, they are not described here in detail. All deductions presented below are based on the assumption that plant cells and cell organelles tend to reach electroneutrality.

A major factor influencing the ion content of plant tissues is metabolic activity. Because plant nutrients are inorganic compounds, the incorporation of these inorganic substances, chiefly ions, is a basic process of plant metabolism. Therefore, it is of interest to know to what extent assimilatory processes directly or indirectly affect the ionic status of plant tissues.

I. ASSIMILATION OF INORGANIC COMPOUNDS

The most important assimilatory process is the photosynthetic incorporation of CO_2. Actually, the compound being accepted by ribulosediphosphate is bicarbonate, and not CO_2. It is presumed that green plant cells are able to convert CO_2 very quickly into HCO_3^- and *vice versa* by means of a carboanhydrase (Hatch and Slack, 1970):

bicarbonate   ribulose-diphosphate   phosphoglyceric acid

In the assimilatory process, the inorganic anion bicarbonate is converted into the carboxylic group of phosphoglycerate. This intermediary product does not accumulate in large amounts in the cell but is rather quickly reduced to phosphoglyceric aldehyde. Thus no major accumulation of this organic anion will occur. The anions involved in this process are bicarbonate and glycerate. The bicarbonate concentration is also rather low; and, therefore, there is no major demand for cations to balance higher bicarbonate and phosphoglycerate concentrations.

The dark fixation of CO_2 with phosphoenol pyruvate or enolpyruvate as CO_2 acceptors results in the formation of new carboxylic groups:

phosphoenolpyruvate   oxalacetate

$$\begin{array}{c} CH_2 \\ \| \\ C-OH \end{array} + \ *CO_2 \quad \xrightarrow[\;NADPH+H^+ \qquad NADP+\;]{} \quad \begin{array}{c} *COOH \\ | \\ CH_2 \\ | \\ HCOH \\ | \\ COOH \end{array}$$

enolpyruvate   malate

The newly synthesized organic anions, oxalacetate and malate, can accumulate to higher concentrations, thus affecting the cation uptake or cation retention.

The overall reaction shows that the newly synthesized carboxylic group has a hydrogen atom which can be dissociated, but the anion equivalent of the carboxylic group is balanced by H^+. It can be assumed that when usual pH values in plant cells prevail, most of the carboxylic groups are present in a dissociated form.

The mobility of H^+ in the cell differs widely from that of organic anions. Supposing that there is a consistent synthesis of new carboxylic groups, it may be assumed that this would result in diffusion potentials (concentration gradients). Hydrogen ions, diffusing more quickly and not so impeded by cell membranes, can penetrate cell boundaries at a higher rate than the organic anion. These differences in diffusion rates can lead to the buildup of diffusion potentials. Therefore, a slight excess of anionic equivalents in the cell can occur; this might be the cause of the negative electrical potential of plant cells measured by various authors (Etherton and Higinbotham, 1961; Pallaghy and Scott, 1969; Poole, 1969; Macklon and Higinbotham, 1970). There is no doubt that other processes, too, could be responsible for the negative electrical potential of living plant cells, e.g., active ion uptake mechanisms, such as an inwardly directed anion pump or an outwardly directed sodium pump or carrier transport (Dainty, 1962; Pitman *et al.*, 1971).

Nevertheless, the fact that anion equivalents are directly or indirectly produced in the cell by metabolism plays an important part in the ionic status of the cell (Hiatt and Hendricks, 1967). Supposing that ions can penetrate through pores of biological membranes by diffusion or mass-flow, it can be assumed that a negative potential of the cell represents a physiological "sink" for cations.

A basic assumption in this respect is that anion equivalents are balanced almost completely by cation equivalents and *vice versa*. The assumption is supported by experimental results of various authors (Coic *et al.*, 1962; Dijkshoorn, 1957; Mengel, 1965; Jungk, 1967; Kirkby and Mengel, 1967). The fact that living cells have a negative electrical potential does not contradict this assumption, because at a difference in electrical potential of about 100–200 mV the surplus of anions in the cell is so small that it can hardly be measured by chemical methods. When it is assumed that the causes for the negative potential are metabolic processes producing anion equivalents, the important question is whether a physicochemical equilibrium between the ions of the outer medium and the ions of the cell can be reached. If all ions could diffuse through the various cell membranes, particularly through the plasmalemma, the electrical potential between the cell and the outer medium in the state of equilibrium would be zero. The assumption that all ions can diffuse through the cell membranes is, however, not justified. Rather, it is well known that indiffusible anion equivalents, such as carboxylic and phosphate groups of macromolecules (proteins, nucleic acids, phospholipids, pectic acids), are present in the different cell compartments. These indiffusible anionic groups can affect Donnan (1924) potentials because they induce an excess of anion equivalents in the cell.

It is not yet completely clarified whether the negative charges of living cells are mere Donnan potentials. Since the electrical potential difference between the cell and the outer medium varies widely, it seems that the prevailing negative electrical charges are not due to Donnan potentials alone but mainly can be ascribed to direct metabolic processes producing anion equivalents in the cell (Pitman *et al.*, 1971). Processes such as the synthesis of organic anions as described above are probably also active anion uptake. There is no doubt that such ion distribution patterns corresponding to Donnan systems affect the ionic status of plant tissues (Hiatt, 1968), but the cell is far from being a simple Donnan system. The ion relationships of the cell are further complicated by the fact that the entrance of ions into the cell is often a selective active uptake process and not a diffusion through the pores of the cell membranes. Otherwise, the distribution of cations would follow a pattern which favors the uptake of multivalent cations. But such a distribution pattern of cations does not accord with the cation contents actually found in plant tissues, which under normal conditions have higher contents of monovalent cations. Nevertheless, passive cation uptake and cation retention depend to a large extent on the diffusible and indiffusible anionic equivalents of the cell.

Another important process influencing the ionic status of plants is the uptake and incorporation of nitrate. Nitrogen is one major constituent of organic material. Therefore, the plant has a high demand for N, which for the most part is taken up by many higher plants in the form of nitrate-N (NO_3-N). In order to establish an equilibrium between anions and cations, the nitrate taken up in larger quantities has to be balanced either by an equivalent amount of cations or by the release of an equivalent amount of anions. The latter process plays a minor part because the amount of anions that can be released in exchange for nitrate is rather limited. It can hardly be assumed that inorganic anions, such as chloride, phosphate, or sulfate, are released by plant roots in exchange for nitrate to establish electro-neutrality. If there actually should exist such an anion exchange, it could be supposed for OH^- and/or HCO_3^- (Kirkby and Mengel, 1967). But this exchange covers only a small portion of total nitrate uptake. Otherwise, a rather high accumulation of H ions in the cell would occur, because the release of HCO_3^- resulting from CO_2 is balanced by H^+ according to the reaction

$$CO_2 + H_2O \rightleftharpoons H^+ + HCO_3^-.$$

The major part of nitrate being absorbed is balanced by a simultaneous uptake of cations. This does not imply that the active uptake of nitrate will necessarily produce a passive uptake of cations attracted by the anion equivalents. The cation uptake can be active and selective as well. Since in most cases potassium (K^+) accounts for the greatest part of total cation uptake, it is supposed that the active uptake of K^+ largely contributes to balancing the absorbed nitrate. The overall reaction of nitrate reduction is

$$NO_3^- + 8\,H \rightarrow NH_3 + 2\,H_2O + OH^-.$$

At the reduction step, nitrite $\rightarrow NH_3$, OH^- is released, representing the anion equivalent which was present before in the nitrate. Therefore, the cation/anion balance is not affected by nitrate reduction. The cation that at first balanced the nitrate will then balance OH^-. The OH-anion released by this process favors the solvation of CO_2 and increases the bicarbonate concentration, e.g.,

$$CO_2 + OH^- \rightarrow HCO_3^-.$$

It is feasible that this higher bicarbonate concentration directly affects the synthesis of organic anions. Experimental data obtained by Bedri *et al.* (1960) and also by Chouteau (1963) demonstrate that an increase in bicarbonate concentration stimulates the synthesis of organic anions. It can, therefore, be assumed that the anion equivalent of the nitrate is shifting from nitrate over OH^- and HCO_3^- to an organic anion. Some of the OH^- or HCO_3^- produced in the process of nitrate reduction will diffuse out of the plant and raise the pH value of the root environment (Kirkby and Mengel, 1967). This is the actual reason the pH value of the root medium is increased with nitrate nutrition. It is probable that nitrate reduction occurring in the roots in particular would be responsible for the release of OH^- or HCO_3^- ions to the outer medium.

The assimilation of inorganic sulfur (SO_4^{2-}) by the plant cell resembles in many respects the assimilation of nitrate-N (Kirkby, 1967). Although the different steps of sulfate-S assimilation have not yet been fully clarified, the overall reaction can be expressed as:

$$HO-CH_2-\overset{\overset{\displaystyle NH_2}{|}}{CH}-COOH + SO_4^= + 8H \longrightarrow$$

serine

$$HS-CH_2-\overset{\overset{\displaystyle NH_2}{|}}{CH_2}-COOH + 3H_2O + 2OH^-$$

cysteine

The reduction process leads to the release of hydroxyl ions, which, as indicated above, stimulates the synthesis of organic anions. This assumption is well supported by experiments of MacDonald and Laties (1964), who found that potato (*Solanum tuberosum* L.) slices treated with a K_2SO_4 solution incorporated more CO_2 than those supplied with a KCl solution.

Figure 3.1 shows the organic anion content of roots, stems, and leaves of sunflowers (*Helianthus annuus* L.) that were supplied in one treatment with KCl, in the other with K_2SO_4 (Mengel, 1965). Although the increment in the organic anion content of the SO_4-fed plants is not dramatic, a certain trend in this direction can be seen in the roots, the stems, and particularly in the leaves. The difference in the organic anion content of SO_4-fed and Cl-fed plants, however, cannot be expected to be considerable, for the quantity of sulfate being reduced in plant tissues is rather small compared with that of nitrate (see Table 3.1).

The deduction presented above implies that the total amount of organic N and S in plants fed by nitrate and sulfate is equivalent to the total amount of organic anions plus the amount of released HCO_3^- or OH^-. Kirkby (1969) demonstrated that the organic N contained in the plant tissue agreed fairly well with the content of organic anions. The ionic balance sheet presented in Table 3.1 demonstrates that the total anion uptake exceeded the total cation uptake by 10 meq. This means that 10 anion equivalents had been taken up in exchange for

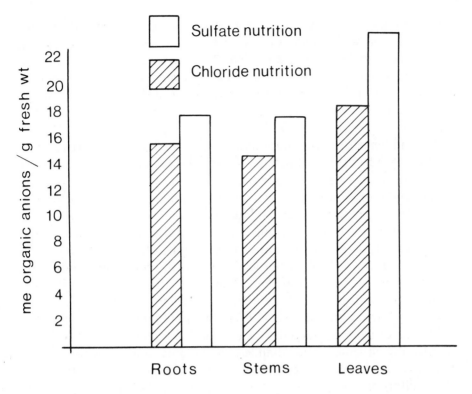

Fig. 3.1. Effect of a KCl and a K$_2$SO$_4$ nutrition upon the content of organic anions of sunflowers (*Helianthus annuus*)

Table 3.1. Ionic balance sheet of tomato plants grown with nitrate as the nitrogen source

	Uptake		Utilized anions	Accumulated anions		
	Anions	Cations	(meq/10 plants)	Inorganic	Organic	
	N 147	K 65	N$_o$† 138	NO$_3$ 9	TCA‡ and oxalic acids	97
	Cl 11	Na 11	S$_o$§ 4	Cl 11	uronic acids	33
	P 13	Mg 21		H$_2$PO$_4$ 13		
	S 13	Ca 77		SO$_4$ 9		
Total	184	174	142	42		130
			-10‖			
			$\overline{132}$			

Source: Reprinted by permission from Kirby, pp. 215–35, *Ecological Aspects of the Mineral Nutrition of Plants*, ed I. H. Rorison *et al.*, ⓒ 1969, Blackwell Scientific Publications Ltd.

† N$_o$ = organic N.
‡ TCA = tricarboxylic acid or citric acid cycle.
§ S$_o$ = organic S.
‖ Minus OH$^-$ released.

OH^- or HCO_3^-, whereas the major part of the OH^- originating from the nitrate and sulfate reduction contributed to the synthesis of organic anions. As can be seen from Table 3.1, the total of utilized anions minus the release of OH^- (132 meq) agrees very well with the total of organic anions (130 meq). Further, it can be seen that the total amount of anions (172 meq) is nearly identical with the total of cations (174 meq).

The influence of ammonium assimilation on the ionic status differs considerably from that of nitrate-N (NO_3-N) assimilation. The overall pattern is that NH_4-fed plants have very low contents of organic anions and inorganic cations, whereas the contents of inorganic anions, particularly of phosphate and sulfate, often increase (Coic *et al.*, 1962; Ehrendorfer, 1964, 1971; Jungk, 1967; Kirkby and Mengel, 1967). The interpretation of these relationships is difficult because N nutrition in the sole form of NH_4^+ often leads to growth retardation, indicating a disturbance of cell metabolism. The most striking sign of NH_4-fed plants is the extremely low content of diffusible organic ions. As reported by Kirkby and Mengel (1967), tomato (*Lycopersicon esculentum* Mill.) leaves from plants fed with nitrate had a tenfold higher content of diffusible organic anions than leaves of plants supplied with NH_4-N. On the other hand, the content of uronic acid, generally representing the indiffusible carboxylic groups of pectic substances, was hardly affected by the different forms of N nutrition.

Plants supplied with NH_4-N as the only N source have to take up rather large amounts of cations because they have to fill their total N requirements with NH_4^+. This might be the reason the uptake of other cation species is depressed. The presence of NH_4^+ in the nutrient medium may even induce a net loss of K^+ and Na^+ from the roots (Minotti *et al.*, 1969). The content of total anion equivalents in NH_4-fed plants is low because the assimilation of NH_4-N does not stimulate the synthesis of organic anions. In fact, it has an opposite effect. In the NH_4-N assimilation process, H^+ is split off from NH_4^+, resulting in a drop of pH which adversely affects the synthesis of organic anions. Therefore, the lack of diffusible organic anions may be directly due to processes connected with NH_4-N assimilation.

It has not yet been established whether the low content of organic anions in plants fed NH_4-N is the true reason for poor growth. Jungk (1967), when plotting the yield of tomato plants and of *Sinapis alba* vs. the $C - A$ values, found that good growth can only be obtained at high $C - A$ values. The term $C - A$ *value*, used by several authors, means the difference between the inorganic cation equivalent (C) and the inorganic anion equivalent (A) of a plant or a plant organ (de Wit *et al.*, 1963; Noggle, 1966; Jungk, 1967; Dijkshoorn *et al.*, 1968). This difference represents the total amount of organic anions. The relationship between the $C - A$ values (organic anion content) and growth does not answer the question whether poor growth is due to the low organic anion content.

The low cation content of NH_4-fed plants is probably not the actual reason for the poor growth. Even with extreme growth retardation due to NH_4^+, typical deficiency symptoms were not observed. This agrees well with Jungk's (1967) conclusion, based on experiments with tomatoes given varying nutrient supplies, that the poor growth of NH_4-fed plants cannot be explained by an insufficient content of K^+. It seems probable that the high levels of NH_4^+ often found in these plants have a direct negative and even toxic effect on metabolic processes (Puritch and Barker, 1967; Matsumoto et. al., 1968) and that the consistent H ion source resulting from the assimilation of NH_4-N affects cell metabolism in some way. In this respect, experimental data of Barker et al. (1966) are of interest, showing that the detrimental effect of NH_4^+ is lowered or even overcome by the addition of carbonate to the nutrient solution. NH_4-N assimilation, being favored by a high carbohydrate level of the plant (Michael et al., 1965), is not the limiting factor of growth of plants treated with NH_4-N because such plants show rather high contents of insoluble organic N and high contents of soluble amino acids.

In contrast to NH_4-N or NO_3-N taken up in ionic form, urea is a neutral compound and its absorption should not cause a change in the ion status of plants. As a matter of fact, it has an intermediate effect because it does not favor cation uptake as much as nitrate and, on the other hand, it does not depress cation uptake as much as ammonium. The content of organic anions in urea-fed plants is not as high as in plants supplied with NO_3-N (Kirkby and Mengel, 1967; Ruthsatz, 1967).

It is reasonable to assume that the uptake of phosphate is balanced by the uptake of cation equivalents or the release of anion equivalents. The actual amount of cation uptake or anion release per phosphorus atom depends on the dissociation of phosphate; at higher pH values HPO_4^{2-} would be available, whereas at lower pH values $H_2PO_4^-$ is taken up.[1] The first-mentioned compound needs two cations, the second only one cation equivalent. The degree of dissociation of organic and inorganic phosphates also plays an important part in the retention of cations because at a high degree of dissociation the counter H ions can be replaced by other cations.

[1] $H_2PO_4^-$ is the predominant form of phosphate absorbed. The $H_2PO_4^-$ form will dominate below pH 6.8, approximately equal amounts of $H_2PO_4^-$ and HPO_4^{2-} will exist at pH 6.8, and PO_4^{3-} will be the predominant form at high pH values. For brevity, $H_2PO_4^-$ will be used to refer to phosphate unless otherwise noted. One should keep in mind that the specific ionic form will be determined by the pH of the medium.—Ed.

The fundamental incorporation process, the aerobic or photosynthetic phosphorylation, is an esterification. Full knowledge of the mechanism of this process has not yet been obtained. If it is a true esterification with release of one H_2O molecule, as indicated in the following equation, it would not affect the cation retention capacity of the cell because the reacting hydroxyls would be present in an undissociated form:

$$Adenosine-O-\underset{\underset{O^-}{|}}{\overset{\overset{O}{\|}}{P}}-O \sim \underset{\underset{O^-}{|}}{\overset{\overset{O}{\|}}{P}}\underset{}{-}\boxed{OH + H}\,O-\underset{\underset{O^-}{|}}{\overset{\overset{O}{\|}}{P}}-OH \longrightarrow$$

$$Adenosine-O-\underset{\underset{O^-}{|}}{\overset{\overset{O}{\|}}{P}}-O \sim \underset{\underset{O^-}{|}}{\overset{\overset{O}{\|}}{P}}-O \sim \underset{\underset{O^-}{|}}{\overset{\overset{O}{\|}}{P}}-OH + H_2O$$

Not only does metabolism affect cation uptake and retention by the synthesis of new anion equivalents, but anion equivalents are continuously decomposed by decarboxylation processes. It seems unlikely that under normal metabolic conditions with sufficient O_2 supply these decarboxylation processes lead to a net loss of carboxylic groups (Egmond and Houba, 1970). But as soon as the roots are exposed to anaerobic conditions, losses of organic anion equivalents will occur that reduce the cation retention capacity of the roots considerably (Hiatt and Lowe, 1967).

Finally it should be stressed that carboxylation and decarboxylation processes play a major part in maintaining the cation/anion balance in the cell. This is well demonstrated by experimental results obtained by Hiatt (1967a, 1967b) showing that in roots which have taken up an excess of anions (e.g., Cl^-), the content of organic anions is lowered; whereas in roots with an excess of cation uptake (K^+ from K_2SO_4), the opposite effect occurs.

II. SELECTIVE ION UPTAKE

Assimilation of inorganic plant nutrients, particularly the synthesis of organic anions, is of great importance for cation uptake and cation retention. However, this process is only one of the major factors. The characteristics of ion contents and ion distribution cannot be explained by the synthesis of organic material and the resulting passive ion uptake and diffusion processes alone. The most important factors controlling ion contents and ion distribution patterns are active ion uptake mechanisms. The reasons for the assumption of such active and selective ion uptake processes and their hypothetical mechanisms will be discussed more thoroughly in Chapter 4. Here it should only be stressed that the uptake rates for the various ion species differ widely. This is particularly true when the uptake rates are related to equal concentrations of the respective ion species in the nutrient solution. In the soil solution, for example, the Ca^{2+} concentration usually is from 10 to 100 times higher than the K^+ concentration (Mengel *et al.*, 1969), whereas uptake rates are higher for K^+ than for Ca^{2+} This clearly demonstrates the very efficient and selective K^+ uptake mechanism of higher plants (Berry and Ulrich, 1968). Similar efficient uptake mechanisms are supposed to exist for nitrate and phosphate and probably also for ammonium and chloride. Plant cells need these active uptake mechanisms in order to pump enough inorganic material, which is indispensable for high growth rates, into the cell within a short time. Whereas nitrate, ammonium, and phosphate are needed for the synthesis of various organic compounds, high potassium quantities are a prerequisite for optimal activation of numerous enzymes (Evans and Sorger, 1966). This is the reason younger tissues, on a dry matter basis, show higher contents of N (inorganic and organic), P and K than older ones. Since young tissues are rich in water, the contents of these elements are not so high when expressed on a fresh weight basis. As has been shown by Jungk (1970), the actual concentrations of K^+ and NO_3^- in plant saps are rather constant and to a high degree independent of the age of the plant. These concentrations are important not only for the activation of enzymes but for their contribution to the osmotic potential of the plant cell.

III. LONG-DISTANCE TRANSPORT

The selective uptake mechanism also favors the transport of the respective ion species over long distances because high ion uptake rates often result in high releases of ions into the xylem of the roots. Therefore NO_3^-, $H_2PO_4^-$, and K^+ are transported to the tips of stems and leaves at higher rates than Ca^{2+}, Mg^{2+}, and SO_4^{2-}. According to experimental results of Haeder and Mengel (1969), the rate of K^+ transport to the tips of younger leaves is favored by a good nitrogen supply of these organs. Such tissues are indeed physiological sinks for N, K, and P, and their requirements for these elements are filled not only by the ions taken up by the root system but also by the mobilization of these elements in older plant parts. It is well known that older leaves export amino acids and K^+ to the younger leaves. This export is increased when the supply from the root system is reduced. In extreme cases, the older leaves are so greatly depleted of amino acids and K^+ or HPO_4^- that deficiency symptoms will occur. This is the reason that symptoms of N, P, and K deficiencies are first observed at the older leaves. The translocation of potassium, nitrogen compounds, and phosphates from the older to the younger leaves, fruits, or even roots is effected predominantly via the phloem. Therefore, the secretion of potassium, amino acids, and phosphates into the sieve elements of the phloem plays a major part in the ion distribution of higher plants and particularly in basipetal transport. The phloem sap contains only small amounts of inorganic ions. Its main inorganic component is K^+, contributing about 80%–90% to the total cation content. The contents of NO_3^- (Martin, 1969) and $H_2PO_4^-$ are very low, proving that N and P in the phloem are transported predominately in organic forms. Phosphate, taken up from the root medium, is quickly transported to the youngest leaves. From here a basipetal redistribution occurs which is believed to take place as organic phosphates in the phloem (Morard, 1970).

One may speculate whether the highly efficient K^+ uptake mechanism present in the root cells also exists in the various membranes of phloem cells, effecting there a high secretion of K^+ from the adjacent cells into the phloem sap, where K^+ balances the various anion equivalents (Dijkshoorn *et al.*, 1968). Since the phloem sap is rather rich in K^+, this results in high K^+ contents in plant organs toward which the phloem sap flows. Fruits, such as apples (*Malus pumila* Mill.), grapes (*Vitis* spp.), and tomatoes, and storage tissues, such as beet (*Beta* spp.) roots and potato tubers, contain relatively high amounts of K^+ compared with leaves or roots (Fig. 3.2).

In contrast to K^+, little Ca^{2+} is secreted into the phloem sap (Hoad and Peel, 1965). Therefore, Ca^{2+} has very poor basipetal movement and once transported into the leaf, will remain there. This is the reason older leaves have higher Ca^{2+} contents than younger ones, for they have been exposed for a longer time to the Ca^{2+} stream of the xylem sap. The most important anion equivalent of Ca^{2+} transported to the leaves is nitrate, the major part of which is reduced in the leaves and incorporated into organic compounds, such as amino acids. Nitrate reduction, as explained above, stimulates the synthesis of organic anions. Whereas the N of the imported nitrate will be exported in the form of amino acids, the organic anions remain in the older leaves and balance the rather immobile Ca^{2+}. In agreement with this deduction, Kirkby and de Kock (1965) report that there is a parallel

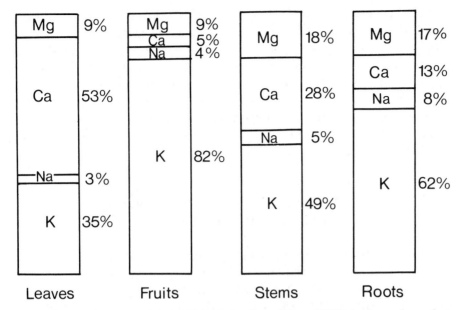

Fig. 3.2. Percentages of cations (K + Na + Ca + Mg = 100%) in the various plant parts of tomatoes (*Lycopersicon esculentum*)

increase of the Ca^{2+} and organic anion content with increasing age of the leaves.

The very poor Ca^{2+} transport from the leaf back to the stem and other plant organs is also the reason the Ca:K ratio of leaves usually is higher than the Ca:K ratio of stems. The acropetal Ca^{2+} transport largely depends on the transpiration intensity, which directly affects the Ca^{2+} supply of the tips of young leaves and stems. As the Ca^{2+} of older tissues can scarcely be mobilized, Ca^{2+} deficiencies start from the tips of stems and younger leaves. These deficiencies are also found in plants with older organs rich in Ca^{2+} (Loneragan and Snowball, 1969). Other cations, particularly Mg^{2+}, can favor the acropetal transport of Ca^{2+} at the xylem cell walls (Iserman, 1970).

The transport of Mg^{2+} in the phloem and its basipetal mobility were once believed to be very low. But recent experimental results of Steucek and Koontz (1970) show that Mg^{2+}, in contrast to Ca^{2+}, is transported in the phloem and can be translocated easily to basipetal plant parts. This agrees with the fact that fruits and storage organs can contain relatively high contents of Mg^{2+} compared with Ca^{2+}. Since Mg^{2+} deficiency in higher plants starts from the older leaves, Mg^{2+} is supposed to be translocated from the older leaves to the younger ones.

Heavy metals are usually present in plant tissues only in small amounts. It is presumed that they are transported mainly in a chelated

form which favors good basipetal and acropetal mobility. Plants
supplied with higher amount of heavy metals (manganese, copper,
zinc, and cobalt) show an accumulation of these elements, particularly
in the older leaves, often resulting in toxicity symptoms. The same
applies to boron, which under humid conditions can be excreted by
guttation (Oertli, 1962). The acropetal transport of B depends on
transpiration intensity. Low transpiration rates can induce B deficiency
in the youngest leaves even when the nutrient medium contains enough
available B (Michael *et al.*, 1969).

IV. SPECIFIC ION REQUIREMENTS OF VARIOUS PLANT SPECIES

The ionic pattern described above is characteristic of almost all higher
plants. However, remarkable variations can be found among plant
species. It is known that dicotyledons generally have higher contents
of bivalent cations than monocotyledons; the reverse holds true for
monovalent cations (Loneragan and Snowball, 1969). The reason for
this difference is not yet clarified. Coic (1964), when comparing tomato
plants with corn (*Zea mays*), came to the assumption that plant species
in which nitrate is reduced predominately in the roots preferentially
transport K^+ and not Ca^{2+} to the tops. He supposed that K^+ has a
higher affinity for amino anions than Ca^{2+}. In this case, the most
important cation balancing the N transported in amino anions from
the roots to the aboveground plant parts is K^+. In plants in which a
good part of the absorbed nitrate is reduced in the leaves, the transported
nitrate is balanced by Ca^{2+} and K^+. With this concept, Coic (1964)
explains the differences in the Ca and K contents of monocotyledons
and dicotyledons by assuming that in monocotyledons nitrate is predominately reduced in the roots and in dicotyledons in the roots and
tops. This concept needs further experimental support.

   The nitrate reduction capacities greatly depend on the plant species
(Pate, 1969). The ability of various herbaceous plants to reduce nitrate
in the roots increases in order for the following genera (Pate, 1969):
*Xanthium, Stellaria, Trifolium, Avena, Zea, Impatiens, Helianthus, Hor-
deum, Phaseolus, Vicia, Raphanus,* and *Lupinus. Xanthium* transports
nearly all nitrate into the aboveground plant parts, whereas *Rhaphanus*
and *Lupinus* reduce the nitrate predominately in the roots. The nitrate
content of aerial plant parts is also influenced considerably by the
nitrate supply from the soil. High nitrate availability in the soil results
in high nitrate content, and even plants that had been believed to
reduce the nitrate in the roots, such as apple trees, do transport nitrate
to the leaves when they are heavily fertilized with nitrate. Nitrate

induces the synthesis of nitrate reductase in the leaves (Klepper and Hageman, 1969).

Although most plant species have a similar uptake potential for K^+, considerable differences exist among the various plant species in the Na^+ uptake potential. *Beta* species are able to take up Na^+ at similar rates as K^+. These species also show high rates of Na^+ transport from the roots to the tops. As was observed by Marschner and Schafarczyk (1967) in *Beta vulgaris*, Na^+ is very mobile and can be transported from the roots to the tops and back to the roots, whereas in corn (*Zea mays*) the transport of Na^+ to the aerial plant parts is very weak. The same applies to the redistribution to the roots. As a rule, plant species with a low Na^+ acropetal transport potential, such as corn, beans (*Phaseolus vulgaris*), and sunflowers, may have fairly high Na^+ contents in the roots. It is therefore the transport mechanism rather than the uptake mechanism that is responsible for the low Na^+ content of the tops. The reason for this poor Na^+ transport in several plant species is not yet fully clarified. It can be supposed that these species secrete Na^+ into the xylem vessels at very low rates or even lack such a specific Na^+ secretion mechanism. Pitman (1965), on the basis of his experiments with beans, reports that Na^+ is accumulated in large amounts in the cells of the cortex tissue. Thus only small quantities of Na^+ are available for the xylem. After the cortex tissue has been saturated with Na^+, an appreciable amount of Na^+ can be transported to the upper plant parts. Experimental data of Shone *et al.* (1969) support the assumption that Na^+ is reabsorbed from the tissues adjacent to the xylem vessels. Na^+ uptake also depends on the salt status of the roots. When this status is low, Na^+ is taken up at higher rates (Pitman *et al.*, 1968). There is no doubt that the capability of Na^+ transport is a hereditary feature which has been developed to a various extent by the different plant species during evolution. It is amazing to see that even among Gramineae remarkable differences are to be found in the Na^+ content of the tops. *Lolium perenne* and *Holcus lanatus* have fairly high contents of Na^+ in the tops, whereas *Poa pratensis*, *Arrhenatherum elatius*, and *Festuca rubra* are low in Na^+ content (Griffith and Walters, 1966). Even the varieties of one individual species, e.g., the varieties of *Lolium perenne*, can differ considerably in their Na^+ contents (Werner and Todt, 1970).

References

Barker, A. V., R. J. Volk, and W. A. Jackson. 1966. Root environment acidity as a regulatory factor in ammonium assimilation by the bean plant. *Plant Physiol.* 41: 1193–99.

Bedri, A. A., A. Wallace, and W. A. Rhoads. 1960. Assimilation of bicarbonate by roots of different plant species. *Soil Sci.* 89: 257–63.

Berry, W. L., and A. Ulrich. 1968. Cation absorption from culture solution by sugar beets. *Soil Sci.* 106: 303–8.

Chouteau, J. 1963. Etude de la nutrition nitrique et ammoniacale de la plante de tabac en presence des doses croissantes de bicarbonate dans le milieu nutritif. *Ann. inst. exp. du tabac de Bergerac,* vol. 4, no. 2.

Coic, M. Y. 1964. Sur le determinisme de l'absorption des cations minéraux par les genres et espèces végétales: Influence de la localisation du métabolisme de l'azote. In Académie d'Agriculture de France, *Extrait du procésverbal de la Séance du 24 Juin 1964*, pp. 925–32.

Coic, M. Y., Ch. Lesaint, and F. Le Roux. 1962. Effects de la nature ammoniacale ou nitrique de l'alimentation azotée et du changement de la nature de cette alimentation sur le métabolisme des anions et cations chez la tomate. *Ann. Physiol. Veg.* 4: 117–25.

Dainty, J. 1962. Ion transport and electrical potentials in plant cells. *Ann. Rev. Plant Physiol.* 13: 379–402.

de Wit, C. T., W. Dijkshoorn, and J. C. Noggle. 1963. Ionic balance and growth of plants. *Verslagen van landbouwkundige onderzoekingen NR.* 69.15 Wageningen, The Netherlands.

Dijkshoorn, W. 1957. A note on the cation-anion relationships in perennial ryegrass. *Neth. J. Agr. Sci.* 5: 81–85.

——. 1962. Metabolic regulation of the alkaline effect of nitrate in utilization in plants. *Nature* 194: 165–67.

Dijkshoorn, W., D. J. Lathwell, and C. T. de Wit. 1968. Temporal changes in carboxylate content of ryegrass with stepwise change in nutrition. *Plant Soil* 29: 369–90.

Donnan, F. G. 1924. The theory of membrane equilibria. *Chem. Rev.* 1: 73–90.

Egmond, F. van, and V. J. G. Houba. 1970. Production of carboxylates (C-A) by young sugar-beet plants grown in nutrient solution. *Neth. J. Agr. Sci.* 18: 182–87.

Ehrendorfer, K. 1964. Einfluss der Stickstofform auf Mineralstoffaufnahme and Substanzbildung bei Spinat (*Spinacia oleracea* L.). *Bodenkultur* 15: 1–13.

——. 1971. Untersuchungen über die quantitative Abhängigkeit des Oxalsäuregehaltes von den Mineralstoffgehalten einschliesslich des Natriums in Spinatblättern (*Spinacea oleracea* L.). *Bodenkultur* 22: 34–48.

Etherton, B., and N. Higinbotham. 1961. Transmembrane potential measurements of cells of higher plants as related to salt uptake. *Science* 131: 409–10.

Evans, H. J., and G. J. Sorger. 1966. Role of mineral elements with emphasis on the univalent cations. *Ann. Rev. Plant Physiol.* 17: 47–77.

Griffith, G., and R. J. K. Walters. 1966. The sodium and potassium content of some grass genera, species, and varieties. *J. Agr. Sci.* 67: 81–89.

Hatch, M. D., and C. R. Slack. 1970. Photosynthetic CO_2-fixation pathways. *Ann. Rev. Plant Physiol.* 21: 141–62.

Haeder, H. E., and K. Mengel. 1969. Die Aufnahme von Kalium und Natrium in Abhängigkeit von Stickstoffernährungszustand der Pflanze. *Landw. Forsch.* 23: 79–91.

Hiatt, A. J. 1967a. Reactions *in vitro* of enzymes involved in CO_2 fixation accompanying salt uptake by barley roots. *Z. Pflanzenphysiol.* 56: 233–45.

——. 1967b. Relationship of cell sap pH to organic acid change during ion uptake. *Plant Physiol.* 42: 294–99.

——. 1968. Electrostatic association and Donnan phenomena as mechanisms of ion accumulation. *Plant Physiol.* 43: 893–901.

Hiatt, A. J., and S. B. Hendricks. 1967. The role of CO_2 fixation in accumulation of ions by barley roots. *Z. Pflanzenphysiol.* 56: 220–32.

Hiatt, A. J., and R. H. Lowe. 1967. Loss of organic acids, amino acids, K, and Cl from barley roots treated anaerobically and with metabolic inhibitors. *Plant Physiol.* 42: 1731–36.

Hoad, G. V., and A. J. Peel. 1965. Studies on the movement of solutes between the sieve tubes and surrounding tissues in willow. *J. Exp. Bot.* 16: 433–51.

Isermann, K. 1970. Der Einfluss von Adsorptionsvorgängen im Xylem auf die Calcium-Verteilung in der höheren Pflanze. *Z. Pflanzenernähr. Bodenk.* 126: 191–203.

Jackson, P. C., and D. G. Edwards. 1966. Cation effects on chloride fluxes and accumulation levels in barley roots. *J. Gen. Physiol.* 50: 225–41.

Jackson, P. C., and K. J. Stief. 1965. Equilibrium and ion exchange characteristics of potassium and sodium accumulation by barley roots. *J. Gen. Physiol.* 48: 601–16.

Jungk, A. 1967. Einfluss von Ammonium- und Nitrat-Stickstoff auf das Kationen-Anionen-Gleichgewicht in Pflanzen und seine Beziehung zum Ertrag. *Landw. Forsch.*, supplement 21: 50–63.

——. 1970. Mineralstoff- und Wassergehalt in Abhängigkeit von der Entwicklung von Pflanzen. *Z. Pflanzenernähr. Bodenk.* 125: 119–29.

Kirkby, E. A. 1967. A note on the utilization of nitrate, urea and ammonium nitrogen by *Chenopodium album*. *Z. Pflanzenernähr. Bodenk.* 117: 204–9.

——. 1969. Ion uptake and ionic balance in plants in relation to the form of nitrogen nutrition. In *Ecological Aspects of the Mineral Nutrition of Plants*, ed. I. H. Rorison *et al.*, pp. 215–35. Symp. Brit. Ecol. Soc. no. 9, Sheffield, 1968. Oxford and Edinburgh: Blackwell Scientific Publications.

Kirkby, E. A., and P. C. de Kock. 1965. The influence of age on the cation-anion balance in the leaves of brussels sprouts (*Brassica olerace var. gemmifera*). *Z. Pflanzenernähr., Düng., Bodenk.* 111: 197–203.

Kirkby, E. A., and K. Mengel. 1967. The ionic balance in different tissues of the tomato plant in relation to nitrate, urea or ammonium nutrition. *Plant Physiol.* 42: 6–14.

Klepper, L., and R. H. Hageman. 1969. The occurrence of nitrate reductase in apple leaves. *Plant Physiol.* 44: 110–14.

Loneragan, J. F., and K. Snowball. 1969. Calcium requirements of plants. *Aust. J. Agr. Res.* 20: 465–78.

MacDonald, J. R., and G. G. Laties. 1964. A comparative study of the influence of salt type and concentration on $^{14}CO_2$ fixation in potato slices at 25°C and 0°C. *J. Exp. Bot.* 15: 530–37.

Macklon, A. E. S., and N. Higinbotham. 1970. Active and passive transport of potassium in cells of excised pea epicotyls. *Plant Physiol.* 45: 133–38.

Marschner, H., and W. Schafarczyk. 1967. Influx und Efflux von Natrium und Kalium bei Mais- and Zuckerrübenpflanzen. *Z. Pflanzenernähr. Bodenk.* 118: 187–201.

Martin, P. 1969. Untersuchungen mit ^{15}N zur Wanderung von Stickstoff in der Pflanze. *Landw. Forsch.* 23/I. supplement, pp. 70–71.

Matsumoto, H., N. Wakiuchi, and E. Takahashi. 1968. Changes of sugar levels in cucumber leaves during ammonium toxicity. *Physiol. Plant.* 21: 1210–16.

Mengel, K. 1965. Das Kationen/Anionen-Gleichgewicht in Wurzel, Stengel and Blatt von *Helianthus annuus* bei K-Chlorid- und K-Sulfaternährung. *Planta* 65: 358–68.

Mengel, K., H. Grimme, and K. Németh. 1969. Potentielle und effektive Verfügbarkeit von Pflanzennährstoffen im Boden. *Landw. Forsch.* 23/I. supplement, pp. 79–91.

Michael, G., H. Schumacher, and H. Marschner. 1965. Aufnahme von Ammonium- und Nitratstickstoff aus markiertem Ammoniumnitrat und deren Verteilung in der Pflanze. *Z. Pflanzenernähr., Düng., Bodenk.* 110: 225–38.

Michael, G., E. Wilberg, and K. Kouhsiahi-Tork. 1969. Durch hohe Luftfeuchtigkeit induzierter Bormangel. *Z. Pflanzenernähr. Bodenk.* 122: 1–3.

Minotti, P. L., D. C. Williams, and W. A. Jackson. 1969. Nitrate uptake by wheat as influenced by ammonium and other cations. *Crop. Sci.* 9: 9–14.

Morard, 1970. Distribution de phosphore étudiée au moyen de l'isotope radioactif et par colorimétrie chez le sarrasin (*Fagopyrum esculentum*, var. La Harpe) cultivé sur solution nutritive. *C. R. Acad. Sci.*, ser. D, 270: 2075–77.

Noggle, J. C. 1966. Ionic balance and growth of sixteen plant species. *Soil Sci. Soc. Amer. Proc.* 30: 763–66.

Oertli, J. J. 1962. Loss of boron from plants through guttation. *Soil Sci.* 94: 214–19.

Pallaghy, C. K., and B. I. H. Scott. 1969. The electrochemical state of cells of broad bean roots. II. Potassium kinetics in excised root tissue. *Aust. J. Biol. Sci.* 22: 585–600.

Pate, J. S. 1969. The movement of nitrogenous solutes in plants. In *Coordination Meeting on Recent Developments in the Use of N-15 in Soil-Plant Studies*. Sofia, Bulgaria.

Pitman, M. G. 1965. Sodium and potassium uptake by seedlings of *Hordeum vulgare*. *Aust. J. Biol. Sci.* 18: 10–24.

Pitman, M. G., A. C. Courtice, and B. Lee. 1968. Comparison of potassium and sodium uptake by barley roots at high and low salt status. *Aust. J. Biol. Sci.* 21: 871–81.

Pitman, M. G., S. M. Mertz, Jr., J. S. Graves, W. S. Pierce, and N. Higinbotham. 1971. Electrical potential differences in cells of barley roots and their relation to ion uptake. *Plant Physiol.* 47: 76–80.

Poole, R. J. 1969. Carrier-mediated potassium efflux across the cell membrane of Red Beet. *Plant Physiol.* 44: 485–90.

Puritch, G. S., and A. V. Barker. 1967. Structure and function of tomato leaf chloroplasts during ammonium toxicity. *Plant Physiol.* 42: 1229–38.

Ruthsatz, B. 1967. Der Einfluss verschiedenartiger Stickstoffdüngung auf den Säurespiegel und einige Grössen des Wasserhaushaltes von *Fagopyrum esculentum* und *Rumex intermedius*. *Flora*, ser. B, 157: 36–55.

Shone, M. G. T., and D. T. Clarkson, and J. Sanderson. 1969. The absorption and translocation of sodium by maize seedlings. *Planta* 86: 301–14.

Steucek, C. G., and H. V. Koontz. 1970. Phloem mobility of magnesium. *Plant Physiol.* 46: 50–52.

Werner, W., and O. Todt. 1970. Gefäss- und Feldversuche über die Beeinflussung des Magnesium- und Natriumgehaltes verschiedener Gräser durch die Düngung. *Das wirtschaftseigene Futter*, no. 1, 48–58.

4. Ion Uptake and Translocation

Konrad Mengel

THE selective uptake of chemical compounds supplied by the environment is an outstanding characteristic of living systems. Not only uptake but also transport and separation of various compounds in the cell are essential processes of metabolism. It is well known that cells have various compartments in which metabolic cycles are recurring. Transport of the different chemical compounds into and out of these compartments is a prerequisite for metabolic activity. Also, ion uptake and translocation have to be considered as part of this activity. Cell compartments are separated by biological membranes representing barriers for chemical compounds. The transport through these barriers and processes mediating this transport are therefore of outstanding importance.

I. PASSIVE TRANSPORT THROUGH CELL MEMBRANES

Cell membranes consist of a protein fraction and a lipid fraction, each accounting for about 50% of the total membrane material. The bimolecular layer of the lipids is covered on both sides by a protein layer. Although the lipid fraction is the actual barrier for hydrophilic particles, the protein layers stabilize the whole system. This is only a very rough picture of a biological membrane; the various cell membranes differ considerably in details. But for the question discussed here, the principal characteristic of biological membranes, i.e., to form barriers for hydrophylic particles, is presumed to be a common property. If these barriers are completely impervious to ions, the ion transport through membranes can only be performed by special mechanisms.

The question whether biological membranes possess pores filled with water through which small hydrophylic particles, such as inorganic ions, can penetrate via diffusion or even mass-flow is not yet completely clarified. But several indirect arguments support the assumption that such pores exist (Passow, 1963). These pores can only exert a sieve effect, favoring the penetration of smaller particles more than that of larger ones, and the special ion requirements of the various cell compartments cannot be met in this way. The more important processes are selective ion uptake mechanisms, which will be discussed below in more detail.

When it is assumed that biological membranes are not completely impervious to passive penetration (diffusion, mass-flow) of small hydrophilic particles, it can also be supposed that this kind of passive transport obeys physical laws. The net ion flux in either direction will stop as soon as a state of equilibrium is reached. With particles that are not electrically charged, e.g., sucrose molecules, the equilibrium is attained when equal sucrose concentrations exist on either side of the membrane. Ions with an electrical charge, however, are affected not only by chemical but also by electrical potentials at either side of the membrane. Therefore the equilibrium is reached when equal electrochemical potentials exist at both sides of the membrane (Dainty, 1962). This state can be expressed by the Nernst equation:

$$E = \psi_i - \psi_o = RT/zF \cdot ln\,[a_o/a_i],$$

where E = electrical potential difference,

ψ_i, ψ_o = electrical potentials at the inner and outer sides of the membrane, respectively,

R = gas constant,

T = absolute temperature,

z = valence of the ion,

F = Faraday constant,

a_i, a_o = ion activity at the inner and outer sides of the membrane, respectively.

As long as this equilibrium is not attained, there will be a passive net ion movement down the electrochemical gradient from one side to the other of the membrane. This is also true for the whole cell and its environment, separated by the plasmalemma, as well as for the various cell compartments (vacuoles, chloroplasts, mitochondria, endoplasmic reticulum) and the symplasm, which are also separated by biological membranes.

It can be assumed that in most metabolically active cells no equilibrium will exist between the two sides of the membrane because the metabolic processes are continuously producing disequilibria. It has further been proved that plant cells have a negative electrical charge in relation to their outer medium. Therefore a consistent force must exist, drawing cations into the cell and inhibiting the passive entry of anions. As the concentrations of various anions, particularly of nitrate and chloride, in the cell or cell compartments are often considerably higher than the concentrations in the outer solution (Higinbotham *et al.*, 1962; Pierce and Higinbotham, 1970), it is supposed that these ion species are transported against an electrochemical gradient into the cell. Therefore, a kind of mechanism capable of performing such an "up-hill" transport must be presumed. Such a mechanism is designated as "active" transport, for the cell has to perform it actively (Ussing, 1961; Dainty, 1962).

The processes involved in cation transport are more complicated. The higher concentration of a specific cation species in the cell as compared to that in the outer medium is not a definite proof for the active uptake of this cation species. When an electrical potential difference of 58 mV is assumed (at which point the cell is negatively charged), an equilibrium is obtained for a monovalent cation when the cation activity in the cell is approximately ten times higher than the cation activity in the outer medium. As the electrical potential differences between plant cells and their environment are on the order of 100–200 mV, it is supposed that cells can even accumulate cations in the absence of an active transport mechanism.

Higinbotham *et al.* (1962), in evaluating ion activities and electrical potential differences in young plant tissues, came to the conclusion that chloride and nitrate are taken up actively; that for Ca^{2+}, Mg^{2+}, and Na^+ an outwardly directed active secretion mechanism exists; and that the K^+ activities of the cell and its outer medium obey, for the most part, the Nernst equation. If this is true, no active K^+ uptake mechanism exists, whereas an active secretion mechanism is supposed for the other cation species, Na^+, Mg^{2+}, and Ca^{2+}. Also, Blount and Levedahl (1960), on the basis of experiments with the marine alga *Halicystis ovalis*, postulated an active chloride and Na^+ transport, while they assumed K^+ to be transported only along the electrochemical gradient (passive transport).

The conclusions of Higinbotham *et al.* (1962) would be correct only if the values were measured in a state of equilibrium. When, for example, the Ca^{2+} concentration of the cell is lower than predicted by the Nernst equation, an active outwardly directed Ca^{2+} secretion mechanism is not indicated in any case, for the phenomenon can also be ascribed to the fact that the uptake of Ca^{2+} has not yet come to an equilibrium. Since the uptake rate for Ca^{2+} is usually low, particularly if this uptake is assumed to be a diffusion through membrane pores, it is technically difficult to ascertain the Ca^{2+} equilibrium. It should further be stressed that metabolic processes in living cells continuously produce imbalances. The assumption that Ca^{2+} and Mg^{2+} are actively transported out of the cell is therefore very problematic. Rather, it can be assumed that no active uptake mechanism exists for these two ion species.

As Na^+ can penetrate into plant cells at higher rates than Ca^{2+} and Mg^{2+}, the question arises whether the Na^+ imbalance is due to an outwardly directed Na^+ pump or to an active Na^+ secretion mechanism. It is highly probable that plants living in seawater possess such an active mechanism (Blount and Levedahl, 1960; Dainty, 1962), but for higher plants such a mechanism is doubtful. In experiments with roots of corn (*Zea mays*), Mengel and Pflüger (1972) found that the efflux rates for Na^+ and K^+ were similar at low temperatures (2C). At room temperature, however, the efflux rate of Na^+ did not change, whereas that of K^+ was considerably reduced. If the Na^+ release were due to an active outwardly directed Na^+ secretion, higher rates could be expected at room temperature. Also, under anaerobic conditions the Na^+ release was hardly affected. It therefore seems that no active Na^+ secretion exists in corn roots, and it is very unlikely that such a Na^+ secretion mechanism will be found in the roots of other crop plants. Pierce and Higinbotham (1970), in studying the fluxes of K^+, Na^+, and Cl^- in oat (*Avena sativa*) coleoptile tissues, found that the influx: efflux ratios through the plasmalemma did not agree with the activity ratios (ion activity of the outer solution : ion activity of the cytoplasm). The flux ratios of K^+ and Cl^- were higher than the corresponding activity ratios, indicating an active uptake, while the flux ratio for Na^+ was lower than the Na^+ activity ratio, which indicated in the authors' opinion an active outwardly directed Na^+ transport mechanism.

In many cases, the theoretical values derived from the Nernst equation do not agree very well with the measured ion activities and electrical potential differences. In this respect, it should be remembered that measuring electrical potentials in the tiny cells of higher plants is tricky (Ling, 1969). Besides, plant cells and their environments are not simple systems, and it can be assumed that several interactions exist between the various factors influencing the ion distribution and the electrical potentials. Without regard to these interactions, conclusions based on ion activities and electrical potentials may be wrong, as demonstrated by Shone (1968).

It seems certain that electrical potential differences also exist between the various cell compartments (Spanswick and Williams, 1964; Vorobiew, 1967; Etherton, 1968), which to some extent depend on the ion content of the different compartments.

With regard to K^+, it appears that, particularly with low concentrations in the outer solution, an active uptake mechanism is needed to bring about the rather high K^+ concentration found in living cells (Etherton, 1968; Spanswick and Williams, 1964; Vorobiew, 1967; Poole, 1969; Macklon and Higinbotham, 1970). Higher K^+ concentrations of the outer solution (greater than 1 mM) tend to decrease the negative potential in the cell (Etherton and Higinbotham, 1961; Higinbotham *et al.*, 1964), which can be ascribed to a higher passive entrance (diffusion) of K^+ into the cells (Pitman *et al.*, 1969).

Under normal field conditions with usual K^+ concentrations of less than 1 mM (Mengel *et al.*, 1969), the active K^+ uptake is the most important process. The passive uptake would be too weak to supply the plant with enough K^+ for high growth rates. This active uptake performed against an electrochemical gradient results in K^+ concentrations in the root cells that in the state of dynamic equilibrium are about 2×10^4 times higher than the K^+ concentration of the outer solution (Mengel and Haeder, 1971). Such a dynamic equilibrium is brought about by active K^+ uptake against passive K^+ diffusion out of the cell through the membrane pores. The K^+ concentration of the outer solution at which this equilibrium occurred ranged from $1\mu M$ to $3\mu M$, thus being much lower than the K^+ concentrations of the soil solution. This indicates that the K^+ release from normal intact plants under field conditions is negligible (Haeder, 1971).

II. INTERACTIONS BETWEEN ION INFLUX AND EFFLUX

Figure 4.1 gives a very simplified scheme of the relationships between active and passive ion uptake and influx and efflux processes. The active uptake mechanisms are inwardly directed, and they are rather efficient, which means that they pump in ions at fairly high rates even when the concentration of the respective ion species in the outer solution is very low. The passive fluxes, i.e., diffusion through the membrane pores, are controlled by the electrochemical gradient existing between either side of the membrane. High ion concentrations of the outer solution therefore seem to favor passive uptake processes (influx), whereas at low concentrations net efflux through the pores can occur. Therefore the ion potential of the cell depends largely on active uptake processes (Hiatt and Lowe, 1967; Harris *et al.*, 1967; Weigl, 1967; Mengel and Herwig, 1969). In addition, the ion retention by the cell is greatly influenced by the permeability of the membrane. A lack of Ca^{2+} or high H^+ concentrations in the outer solution increases the permeability of the membrane and the efflux rates (Handley *et al.*, 1965; Steveninck, 1965; Marschner and Mengel, 1966; Marschner and Schafarczyk, 1967; Mengel and Helal, 1967; Pitman, 1969).

As has been concluded from experiments with excised barley (*Hordeum vulgare*) roots (Jackson and Stief, 1965), the total cation content of the cell is controlled by the anion equivalent in the cell. At a high

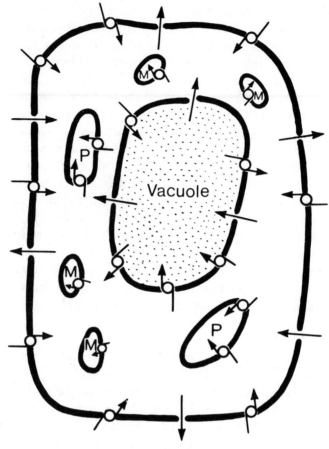

→   passive influx or efflux

⟋σ⃗   carrier transport or ion pump

P =   Plastids

M=   Mitochondria

Fig. 4.1.   Simplified scheme of active ion uptake and passive influx
and efflux processes of a plant cell

anion potential (diffusible + indiffusible anions), a high retention
capacity for cations is also observed. Hydrogen ions are also involved
in the establishment of a balance (Jackson and Adams, 1963). In
excised roots an equilibrium between influx (active uptake + passive
movement) and efflux (passive) often is obtained (Mengel and Schneider,
1965; Jackson and Edwards, 1966). For intact plants, this equilibrium
is not easily reached because the aerial plant parts represent a heavy
sink for the ions taken up by the roots (Johansen *et al.*, 1970). As already

mentioned, such an equilibrium between the influx and efflux of K^+ can only be reached at extremely low K^+ concentrations in the outer solution (about $2\mu M$).

The cations in the cell can compete for anionic counterparts, and a cation species taken up actively or passively at high rates can replace another cation species, resulting in an efflux of the latter (Maas and Leggett, 1968). By these processes cation contents in the cell are obtained that accord well with interactions denoted as antagonistic effects. However, the problem remains to be solved whether the antagonistic effects between cations can be ascribed to such an unspecific replacement in the cell. Of interest in this respect are the experimental results of Jaegere *et al.* (1963), showing that tomato (*Lycopersicon esculentum*) plants, after a change from nitrate to ammonium nutrition, lost a considerable portion of the cation content (K^+, Na^+, Mg^{2+}, Ca^{2+}), whereas the anion uptake was enhanced.

The system presented in Figure 4.1 is a very simplified one. The various compartments in the plant cell are assumed to have ion uptake mechanisms of their own, and ion distribution patterns often are directly connected with metabolic processes, such as respiration or photosynthetic activity (Dilley, 1964; Mitchell, 1967; Hecht-Buchholz and Marschner, 1970). Larkum (1968) in chloroplasts found chloride and K^+ concentrations of 340 mM, whereas the concentrations in the cytoplasm were in the range of only 20–100 mM. This demonstrates that the gross ion content of a cell or of plant tissues gives only a very rough insight into the actual ionic status.

III. ACTIVE UPTAKE MECHANISMS

As already mentioned above, there are good reasons for assuming active uptake mechanisms. Until now, however, no direct proof for the existence of such mechanisms could be given. The most common concept explaining active uptake processes in plant cells is the carrier theory (Jacobson and Overstreet, 1947; Epstein and Hagen, 1952). It furnishes an explanation for three basic processes of active ion uptake—selective uptake, transport through a lipid medium, or transport against an electrochemical gradient—and the close relationship between ion uptake and metabolism. The chemical properties of the various supposed carriers have not yet been elucidated. But it seems reasonable to state that molecules of lipid nature effect the transport through the lipid medium of biological membranes (Hokin and Hokin, 1963; Ward and Frantl, 1963; Jain *et al.*, 1969; Moore and Schechter, 1969; Dobler *et al.*, 1969; Wheeler and Whittam, 1970).

Konrad Mengel

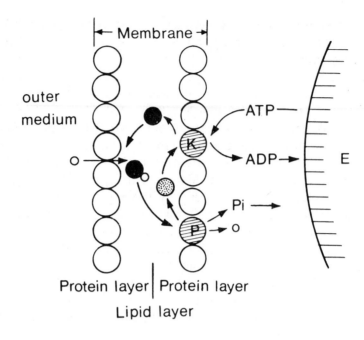

Fig. 4.2. Simplified scheme of ion-carrier transport through a membrane

In Figure 4.2 a very rough scheme of a carrier transport is given. The carrier, a lipid molecule, is diffusible in the lipid layer of the membrane. In the inwardly directed protein layer, a phosphokinase is located which phosphorylates the carrier by transferring a phosphoryl group from the ATP to the carrier. In this phosphorylated state the carrier is able to adsorb the specific ion; this is achieved at the outwardly directed side of the lipid layer. The ion-loaded carrier, diffusing in the lipid phase, meets, by chance, a phosphatase situated at the inner lipid layer which splits off the phosphate. By this process, the ion and phosphate are released into the cytoplasm, while the inactivated carrier remains in the lipid phase. It meets the phosphokinase again, and a new cycle can start. This concept is a mere assumption, particularly with regard to the enzymes' being involved, but it accords well with the characteristics of an active transport as outlined above and with experimental results. This agreement is especially true for ATP as an energy source for active ion transport through membranes. The inorganic phosphate released by phosphatase and the ADP split off by phosphokinase diffuse to the chloroplast or to the mitochondrion where a new ATP molecule can be synthesized. As various authors have proved, active ion uptake depends on photosynthetic or oxidative ATP synthesis (Weigl, 1963, 1967; Stoner *et al.*, 1964; Nobel and Packer, 1965; Kylin, 1966; Cockrell *et al.*, 1967; Barber, 1968; Jeschke, 1970; Nobel, 1970).

It is not yet completely clear whether each ion species has a carrier of its own, but it seems probable that only specific ion species, such as nitrate, phosphate, chloride, and potassium, taken up from dilute concentrations at high rates, possess carrier systems. Epstein (1966) postulated two different carrier systems for K^+; system I was highly selective, working at rather low concentrations (less than 0.2 mM K^+). System II was less selective, working in a concentration of more than 0.5 mM K^+. Also, Lüttge and Laties (1966), on the basis of experiments with young corn plants, took two K^+ uptake systems for granted, one working in a low concentration range (less than 0.5 mM K^+), the other at higher concentrations (greater than 1.0 mM K^+). Epstein (1966) supposed that both uptake systems were carrier systems; however, Lüttge and Laties (1966) conceded that system II was possibly a diffusion process. This latter assumption was supported by the fact that at low concentrations the K^+ uptake rate was not affected by the counter anion whereas at high concentrations K^+ was absorbed from a KCl solution at higher rates than from a K_2SO_4 solution. From KCl chloride is also taken up at high rates, thus improving the electrochemical gradient for an inwardly directed K^+ diffusion.

Epstein (1966) and Lüttge and Laties (1966) supposed that uptake mechanism I was located in the plasmalemma and mechanism II in the tonoplast. There is no doubt that different uptake mechanisms may occur in different cell membranes, but the assumption that the less selective K uptake mechanism, working at higher concentrations, is located in the tonoplast needs further experimental support.

One of the most fascinating phenomena in ion uptake by living cells is their capability to differentiate between K^+ and Na^+. This is particularly true for animal cells, e.g., cells of muscles, nerves, and erythrocytes, where the K^+ concentration is about 50 times higher than in extracellular liquids. For Na^+ the reverse is true. This reciprocal distribution of K^+ and Na^+ in extracellular and intracellular liquids is explained by the existence of a so-called Na^+ pump (Kuijpers and Vleuten, 1967). Its activity is directly ascribed to an ATPase (Post *et al.*, 1960) which can transport K^+ into and Na^+ out of the cell by splitting ATP into ADP and inorganic phosphate (Opit and Charnock, 1965; Lowe, 1968). The transport of cations is supposed to be mediated by a conformation change of the enzyme protein (Lowe, 1968; Somogyi, 1968). An outstanding feature of this ATPase is its high sensitivity to ouabain.

It is not known whether such a Na^+ pump also exists in plant cells. Experimental results obtained by Blount and Levedahl (1960) with *Halicystis ovalis* support the assumption that there is such an outwardly directing Na^+ pump in plants living in an environment rich in Na^+ (seawater). But according to experimental data of Kylin (1966) for *Scenedesmus* living in fresh water, the existence of a Na^+ pump does not seem likely. It appears reasonable to assume that plants living in seawater developed a Na^+ pump to protect the cells from too high a Na^+ concentration, whereas plants living in an environment poor in Na^+ lost this capability or never developed such a mechanism. The high sensitivity of the Na^+ pump in animal cells to ouabain affecting the Na^+ and K^+ fluxes could not be established for tissues of higher plants (Mengel, 1963; Cram, 1968).

The outstanding property of animal and plant cells of taking up K^+ with extreme selectivity is to a certain extent elucidated by new bio-chemical data. Müller and Rudin (1967) reported that antibiotics, such as valinomycin, dinactin, gramicidin, and enniatin B, affect the permeability of artificial lipid membranes considerably. The relative values of permeability affected by valinomycin for Li^+, Na^+, Cs^+, K^+, and Rb^+ correspond to 1, 1.4, 210, 395, 920, showing that the permeability for K^+ was about 300 times higher than that for Na^+. It is not yet known whether this permeability goes back to a carrier transport or whether the antibiotic molecules form pores by aggregation through which a selective alkaline cation transport takes place (Dobler *et al.*, 1969).

IV. ION TRANSPORT

In comparison with active uptake processes, ion transport in the plant over long distance is not so dramatic a problem. Some aspects of this problem have already been discussed in Chapter 3. Ion transport processes begin in the root-soil boundary layer. Under normal growth conditions, there exists a concentration gradient between the bulk of the soil solution and the solution of the root free space for several ion species (K$^+$, phosphate, nitrate) (Drew *et al.*, 1969; Drew and Nye, 1970). Therefore ions reach the free space by diffusion and also by mass-flow, brought about by the water consumption of the plant. The soil solution extends into the root free space, which consists of rather large pores and intercellular spaces of the cell wall. Thus ions reaching the plasmalemma can be taken up actively or passively as described above.

Exchange of cations between the clay minerals and root surface (contact exchange) is possible, but it is of minor importance for the proper ion uptake process. As can be seen in Figure 4.3, the exchange layer between the clay surface and the outer surface of the cell wall is about 15 Å deep. Cations that have shifted from the clay to the root surface by exchange adhere to the very outermost surface of the cell wall at a distance of about 10,000 Å from the proper ion uptake mechanism in the plasmalemma. The cations, in order to reach this mechanism, have to be exchanged again and must overcome the distance by diffusion or mass-flow. The assumption that ectodesmata are plasmatic structures extending into the cell walls seems incorrect (Franke, 1964, 1967).

Ions that have penetrated the plasmalemma by active or passive transport are translocated in the cytoplasm, probably by plasmatic streaming. They can be transported via the plasmodesmata from one cell to the other. Reliable data on the nature of this transport are scarce. But Arisz (1956) and Arisz and Sol (1956) have proved that this transport in the symplasm is influenced by metabolism. A further problem which needs more study is the ion transport through the Casparian strip. It

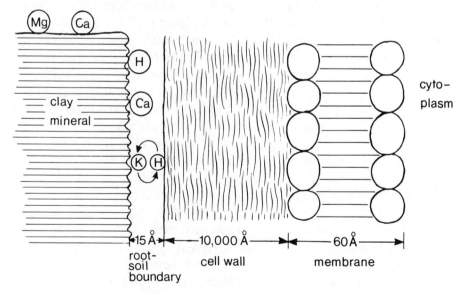

Fig. 4.3. Dimensions of the "contact exchange" and distances between the root-clay boundary zone and the plasmalemma

is supposed that ions in the symplasm pass this barrier by transport through the plasmatic fraction of the endodermis, whereas the transport of water and hydrophylic particles in the cell walls is considerably restrained by the lignin and lipid material of the Casparian strip.

Whether the secretion of ions into the xylem vessels is an active process still remains to be explained. As the xylem sap is negatively charged, relative to the solution, it seems possible that anions are in some way conveyed into the xylem sap by active transport. Thus an electrochemical gradient is built up which enables the passive entrance of cations (Bowling and Spanswick, 1964). Data of Lüttge and Laties (1966) agree well with this assumption.

The ion concentration of the xylem sap, usually much higher than that of the outer solution, represents an osmotic potential which attracts water from the surrounding tissues and thus results in root pressure (Mengel and Pflüger, 1969). This root pressure, pressing the xylem sap in an acropetal direction, brings about a mass-flow of the solutes of the xylem sap, which under conditions of high air moisture is an important mechanism translocating the ions from the roots to the upper plant parts (Locher and Brouwer, 1964). This is especially true for younger plants; in older plants with higher stems the root pressure is not strong enough to supply the plant tips with nutrients (Michael and Marschner, 1962).

The most important acropetal transport mechanism, translocating the ions over long distances at rather short intervals, is caused by the transpiration stream. Under normal transpiration conditions this transport is quicker than the lateral ion movement through the symplasm or through cell wall material (Rinne and Langston, 1960). The acropetal transport rates clearly depend on the transpiration intensity, which is influenced not only by the relative humidity of the air but also by the osmotic potential of the nutrient solution (Linser and Herwig, 1963). The ions thus transported to the different aboveground plant parts follow the veins of the xylem (Rinne and Langston, 1960). On their way, they are laterally distributed with the water over the various interspaces of the cell walls of neighboring tissues. Thus they reach again plasmatic boundaries (plasmalemma) through which they can be transported actively or passively by mechanisms already described above. The transpiration stream does not discriminate between the various ion species; but mechanisms exist that selectively pick up specific ion species from the xylem sap and transport them into the cytoplasm of the adjacent cells, thus effecting a typical ion distribution pattern.

The long-distance basipetal transport is effected by the phloem sap. It is not yet clarified whether active transport processes are involved in the translocation of organic and inorganic substances through the sieve plates of sieve vessels (Biddulph, 1969). The inorganic ion found in by far the largest amounts in the phloem sap is K^+, with concentrations of about 100 mM (Bowling, 1968). This is the reason K^+ is very mobile in the plant and can be translocated rather quickly in any direction.

References

Arisz, W. H. 1956. Significance of the symplasm theory for transport across the roots. *Protoplasma* 46: 1–62.

Arisz, W. H., and H. H. Sol. 1956. Influence of light and sucrose on the uptake and transport of chloride in *Vallisneria* leaves. *Acta Bot. Neerl.* 5: 218–46.

Barber, J. 1968. Light induced uptake of potassium and chloride by *Chlorella pyrenoidosa. Nature* 217: 876–78.

Biddulph, O. 1969. Mechanisms of translocation of plant metabolites. In *Physiological Aspects of Crop Yield*, pp. 143–66. Madison. Wis.: Amer. Soc. Agron.

Blount, R. W., and B. H. Levedahl. 1960. Active sodium and chloride transport in the single-celled marine alga *Halicystis ovalis. Acta Physiol. Scand.* 49: 1–9.

Bowling, D. J. F. 1968. Evidence for the electroosmosis theory of transport in the phloem. *Planta* 80: 21.

Bowling, D. J. F., and R. M. Spanswick. 1964. Active transport of ions across the root of *Ricinus communis. J. Exp. Bot.* 15: 422–27.

Cockrell, R. S., E. J. Harris, and B. C. Pressman. 1967. Synthesis of ATP driven by a potassium gradient in mitochondria. *Nature* 215: 1487–88.

Cram, W. J. 1968. The effects of ouabain on sodium and potassium fluxes in excised root tissue of carrot. *J. Exp. Bot.* 19: 611–16.

Dainty, J. 1962. Ion transport and electrical potentials in plant cells. *Ann. Rev. Plant Physiol.* 13: 379–402.

Dilley, R. A. 1964. Light-induced potassium efflux from spinach chloroplasts. *Biochem. Biophys. Res. Commun.* 17: 716–22.

Dobler, M., J. D. Dunitz, and J. Krajewski. 1969. Structure of the K^+ complex with Enniatin B, a macrocyclic antibiotic with K^+ transport properties. *J. Mol. Biol.* 42: 603–6.

Drew, M. C., and P. H. Nye. 1970. The supply of nutrient ions by diffusion to plant roots in soil. III. Uptake of phosphate by roots of onion, leek and rye-grass. *Plant Soil* 33: 545–63.

Drew, M. C., P. H. Nye, and L. V. Vaidyanathan. 1969. The supply of nutrient ions by diffusion to plant roots in soil. I. Absorption of potassium by cylindrical roots of onion and leek. *Plant Soil* 30: 252–70.

Epstein, E. 1966. Dual pattern of ion absorption by plant cells and by plants. *Nature* 212: 1324–27.

Epstein, E., and C. E. Hagen. 1952. A kinetic study of the absorption of alkali cations by barley roots. *Plant Physiol.* 27: 457–74.

Etherton, B. 1968. Vacuolar and cytoplasmic potassium concentrations in pea roots in relation to cell-to-medium electrical potentials. *Plant Physiol.* 43: 838–40.

Etherton, B., and N. Higinbotham. 1961. Transmembrane potential measurements of cells of higher plants as related to salt uptake. *Science* 131: 409–10.

Franke, W. 1964. Zur Frage der Struktur der Ektodesmen. A. Die Ektodesmen als plasmatische Strukturen. *Planta* 63: 279–300.

——. 1967. Mechanisms of foliar penetration of solutions. *Ann. Rev. Plant Physiol.* 18: 281–300.

Haeder, H. E. 1971. Kaliumabgabe reifender Gerste. *Z. Pflanzenernähr. Bodenk.* 29: 125–32.

Handley, R., A. Metwally, and R. Overstreet. 1965. Divalent cations and the permeability to Na of the root meristem of *Zea mays. Plant Soil* 22: 200–206.

Harris, E. J., K. van Dam, and B. C. Pressman. 1967. Dependence of uptake of succinate by mitochondria on energy and its relation to potassium retention. *Nature* 213: 1126–27.

Hecht-Buchholz, C., and H. Marschner. 1970. Veränderungen der Feinstruktur von Zellen der Maiswurzelspitze bei Entzug von Kalium. *Z. Pflanzenphysiol.* 63: 416–27.

Hiatt, A. J., and R. H. Lowe. 1967. Loss of organic acids, amino acids, K, and Cl from barley roots treated anaerobically and with metabolic inhibitors. *Plant Physiol.* 42: 1731–36.

Higinbotham, N., B. Etherton, and R. J. Foster. 1962. Concentration gradients of the mayor inorganic nutrient ions as related to cell electropotential gradients. *Plant Physiol.* 37: xlvii.

——. 1964. Effect of external K, NH_4, Na, Mg and H ions on the cell transmembrane electropotential of *Avena* coleoptiles. *Plant Physiol.* 39: 196–203.

Hokin, L. E., and M. R. Hokin. 1963. Biological transport. *Ann. Rev. Biochem.* 32: 553–78.

Jackson, P. C., and H. R. Adams. 1963. Cation-anion balance during potassium and sodium absorption by barley roots. *J. Gen. Physiol.* 46: 369–86.

Jackson, P. C., and D. G. Edwards. 1966. Cation effects on chloride fluxes and accumulation levels in barley roots. *J. Gen. Physiol.* 50: 225–41.

Jackson, P. C., and K. J. Stief. 1965. Equilibrium and ion exchange characteristics of potassium and sodium accumulation by barley roots. *J. Gen. Physiol.* 48: 601–16.

Jacobson, L., and R. Overstreet. 1947. A study of the mechanism of ion abosrption by plant roots using radioactive elements. *Amer. J. Bot.* 34: 415–20.

Jaegere, R. de, C. Lesaint, and Y. Coic. 1963. Sur l'excrétion d'ions minéraux: influence du changement de nature de l'alimentation azotée. *Ann. Physiol. Veg.* 5 (4): 263–76.

Jain, M. K., A. Strickholm, and E. H. Cordes. 1969. Reconstitution of an ATP-mediated active transport system across black lipid membranes. *Nature* 222: 871–72.

Jeschke, W. D. 1970. Der Influx von Kaliumionen bei Blättern von *Elodea densa*, Abhängigkeit vom Licht, von der Kaliumkonzentration und von der Temperatur. *Planta* 91: 111–28.

Johansen, C., D. G. Edwards, and J. F. Loneragen. 1970. Potassium fluxes during potassium absorption by intact barley plants of increasing potassium content. *Plant Physiol.* 45: 601–3.

Kuijpers, W., and A. C. van der Vleuten. 1967. Cochlear function and sodium and potassium activated adenosine triphosphatase. *Science* 157: 949–50.

Kylin, A. 1966. Uptake and loss of Na, Rb, and Cs in relation to an active mechanism for extrusion of Na in *Scenedesmus*. *Plant Physiol.* 41: 579–84.

Larkum, A. W. D. 1968. Ionic relations of chloroplasts in vivo. *Nature* 218: 477–49.

Ling, G. N. 1969. Measurements of potassium ion activity in the cytoplasm of living cells. *Nature* 221: 386–87.

Linser, H., and K. Herwig. 1963. Untersuchungen zur abhängigkeit der Nährstoffaufnahme vom osmotischen Druck der Aussenlösung. *Protoplasma* 57: 588–600.

Locher, J. T., and R. Brouwer. 1964. Preliminary data on the transport of water, potassium and nitrate in intact and bleeding maize plants. *Mededel. 238 I.B.S. Wageningen*, pp. 41–49.

Lowe, A. G. 1968. Enzyme mechanism for the active transport of sodium and potassium ions in animal cells. *Nature* 219: 934–36.

Lüttge, U., and G. G. Laties. 1966. Dual mechanisms of ion absorption in relation to long distance transport in plants. *Plant Physiol.* 41: 1531–39.

Macklon, A. E. S., and N. Higinbotham. 1970. Active and passive transport of potassium in cells of excised pea epicotyls. *Plant Physiol.* 45: 133–38.

Marschner, H., and K. Mengel. 1966. Der Einfluss von Ca - und H-Ionen bei unterschiedlichen Stoffwechselbedingungen auf die Membranpermeabilität junger Gerstenwurzeln. *Z. Pflanzenernähr., Düng., Bodenk.* 112: 39–49.

Marschner, H., and W. Schafarczyk. 1967. Influx and Efflux von Natrium und Kalium bei Mais- und Zuckerrübenpflanzen. *Z. Pflanzenernähr. Bodenk.* 118: 187–201.

Mass, E. V., and J. E. Leggett. 1968. Uptake of ^{86}Rb and K by excised maize roots. *Plant Physiol.* 43: 2054–56.

Mengel, K. 1963. Der Einfluss von ATP-Zugaben und weiterer Stoffwechselagenzien auf die Rb-Aufnahme abgeschnittener Gerstenwurzeln. *Physiol. Plant.* 16: 767–76.

Mengel, K., H. Grimme, and K. Németh. 1969. Potentielle und effektive Verfügbarkeit von Pflanzennährstoffen im Boden. *Landw. Forsch.* 23/I: 79–91.

Mengel, K., and H. E. Haeder. 1971. The effect of the nitrogen nutritional status of intact barley plants on the retention of potassium. *Z. Pflanzenernähr. Bodeak.* 128: 105–15.

Mengel, K., and M. Helal. 1967. Der Einfluss des austauschbaren Ca^{++} junger Gerstenwurzeln auf den Flux von K$^+$ und Phosphat-eine Interpretation des Viets-Effektes. *Z. Pflanzenphysiol.* 57: 223–34.

Mengel, K., and K. Herwig. 1969. Der Einfluss der Temperatur auf die K-

Retention, die Effluxrate und auf die Atmung junger abgeschnittener Getreidewurzeln. *Z. Pflanzenphysiol.* 60: 147–55.

Mengel, K., and R. Pflüger. 1969. Der Einfluss verschiedener Salze und verschiedener Inhibitoren auf den Wurzeldruck von *Zea mays*. *Physiol. Plant.* 22: 840–49.

——. 1972. The release of potassium and sodium from young excised roots of *Zea mays* under various efflux conditions. *Plant Physiol.* 49: 16–19.

Mengel, K., and B. Schneider. 1965. Die K-Aufnahme als Funktion der Influxrate und der Zellpermeabilität mathematisch und experimentell an der K-Aufnahme junger Gerstenwurzeln dargestellt. *Physiol. Plant.* 18: 1105–14.

Michael, G., and H. Marschner. 1962. Einfluss unterschiedlicher Luftfeuchtigkeit und Transpiration auf Mineralstoffaufnahme und-verteilung. *Z. Pflanzenernähr., Düng., Boden.* 96 (141): 200–212.

Mitchell, P. 1967. Proton-translocation phosphorylation in mitochondria, chloroplasts and bacteria: natural fuel cells and solar cells. *Federation Proc.* 26, no. 5.

Moore, J. H., and R. S. Schechter. 1969. Transfer of ions against their chemical potential gradient through oil membranes. *Nature* 222: 476–77.

Müller, P., and D. O. Rudin. 1967. Development of K^+-Na^+ discrimination in experimental bimolecular lipid membranes by macrocyclic antibiotics. *Biochem. Biophys. Res. Commun.* 26: 398–405.

Nobel, P. S. 1970. Relation of light-dependent potassium uptake by pea leaf fragments to the pK of the accompanying organic acid. *Plant Physiol.* 46: 491–93.

Nobel, P. S., and L. Packer. 1965. Light-dependent ion translocation in spinach chloroplasts. *Plant Physiol.* 40: 633–40.

Opit, L. J., and J. S. Charnock. 1965. A molecular model for a sodium pump. *Nature* 208: 471–74.

Passow, H. 1963. Passive Permeabilität von Zellmembranen. Zur Frage der Penetration durch Poren. *Verhandl. Ges. Naturforscher Ärzte*, pp. 40–48. Göttingen, Heidelberg: Springer Verlag Berlin.

Pierce, W. S., and N. Higinbotham. 1970. Compartments and fluxes of K^+, Na^+ and Cl^- in *Avena* coleoptile cells. *Plant Physiol.* 46: 666–73.

Pitman, M. G. 1969. Adaption of barley roots to low oxygen supply and its relation to potassium and sodium uptake. *Plant Physiol.* 44: 1233–40.

Pitman, M. G., N. Higinbotham, J. Graves, and S. Mertz, Jr. 1969. Electropotential differences in cells of barley roots and their relation to dual mechanisms of ion uptake. *Plant Physiol.* 44 (96): 20.

Poole, R. J. 1969. Carrier-mediated potassium efflux across the cell membrane of Red Beet. *Plant Physiol.* 44: 485–90.

Post, R. L., C. R. Merritt, C. R. Kinsolving, and C. D. Albricht. 1960. Membrane adenosine triphosphatase as a participant in the active transport of sodium and potassium in the human erythrocyte. *J. Biol. Chem.* 235: 1796–1802.

Rinne, R. W., and R. G. Langston. 1960. Studies on lateral movement of phosphorus 32 in peppermint. *Plant Physiol.* 35: 216–19.

Shone, M. G. T. 1968. Electrochemical relations in the transfer of ions to the xylem sap of maize roots. *J. Exp. Bot.* 19: 468–85.

Somogyi, J. 1968. The effect of proteases on the $(Na^+ + K^+)$-activated adenosine triphosphatase system of rat brain. *Biochim. Biophys. Acta* 151: 421–28.

Spanswick, R. M., and E. J. Williams. 1964. Electrical potentials and Na, K and Cl concentrations in the vacuole and cytoplasm of *Nitella translucens*. *J. Exp. Bot.* 55: 193–200.

Steveninck, R. F. M. van. 1965. The significance of calcium on the apparent permeability of cell membranes and the effects of substitution with other divalent ions. *Physiol. Plant.* 18: 54–69.

Stoner, C. D., T. K. Hodges, and J. B. Hanson. 1964. Chloramphenicol as an inhibitor of energy-linked processes in maize mitochondria. *Nature* 203: 258–69.

Ussing, H. H. 1961. Experimental evidence and biological significance of active transport. In *Biochemie des aktiven Transportes*, pp. 1–14. Göttingen, Heidelberg: Springer Verlag Berlin.

Vorobiev, L. N. 1967. Potassium ion activity in the cytoplasm and the vacuole of cells of *Chara* and *Griffithsia*. *Nature* 216: 1325–26.

Ward, H. A., and P. Frantl. 1963. Transfer of hydrophilic cations from an aqueous to a lipophilic phase by phosphatidie acids. *Arch. Biochem. Biophys.* 100: 338–39.

Weigl, J. 1963. Die Bedeutung der energiereichen Phosphate bei der Ionenaufnahme durch Wurzeln. *Planta* 60: 307–21.

——. 1967. Beweis für die Beteiligung von beweglichen Transportstrukturen (Trägern) beim Ionentransport durch pflanzliche Membranen und die Kinetik des Anionentransports bei *Elodea* im Licht und Dunkeln. *Planta* 75: 327–42.

Wheeler, K. P., and R. Whittam. 1970. ATPase activity of the sodium pump needs phosphatidylserine. *Nature* 225: 449–50.

5. Ionic Interactions and Antagonisms in Plants

A. J. Hiatt and James E. Leggett

NUTRIENT absorption by plants is usually referred to as ion uptake or ion absorption because it is the ionic form in which nutrients are absorbed. Cations and anions may be absorbed independently and may not be absorbed in equal quantities. Electroneutrality must be maintained, within reasonable limits, in the plant and in the growth medium. Therefore, ionic relationships achieve major importance in plant nutrition. The ionic form of a nutrient supplied to the plant root, as well as subsequent changes in its ionic form within the plant, has a very marked effect on metabolism and on absorption of other nutrients.

Soil fertility experiments often involve two or more plant nutrients varied in factorial design. These experiments usually confirm interactions among the varied nutrients but seldom give information that can be used to describe the nature of the interactions. In fact, it is usually difficult to distinguish between the interactions that occur in the soil and those that occur in the plant. The easiest means of avoiding these complications is to study the soil and plant systems independently. The complexities of studying plant nutrition with whole plants has led plant physiologists to refine their experiments by using single-salt solutions and small plant segments or single cells. This approach has been rewarding, but use of this information in understanding integrated plant systems has certain limitations.

Ionic interactions may occur as cation-cation interactions, anion-anion interactions, or cation-anion interactions. Cation-cation or anion-anion interactions are most prominent at the membrane level, i.e., in processes involved in transport across the cell membrane. These inhibitory interactions are mostly of a competitive nature. On the other hands, cation-anion interactions are prominent both at the membrane level and in cellular processes after absorption. We still do not completely understand the basic processes of ion transport across cellular membranes, but we know something about ion interactions at the membrane level. Probably less understood are interactions that occur after absorption. These have been less well investigated and may actually be much more important in a plant growing under normal field conditions. Growth, change in chemical nature of an absorbed ion, synthesis of organic ions, chemical or electrostatic binding of ions, and recycling of ions all influence ion interactions after absorption. These processes are removed from membrane transport phenomena but are as important in total nutritional processes as movement across the membrane.

I. IONIC STATUS OF THE PLANT

In comparison with major nutrient elements, the effect of micronutrient ions on ionic balance is negligible; therefore attention is given here only to the major nutrient elements. The ionic form of six major nutrient elements that may be absorbed by plant roots is shown in Table 5.1. Of these, nitrogen has the greatest impact upon interionic relationships because (1) it is utilized in large quantities, (2) it may be absorbed as either an anion or a cation, and (3) its ionic form is changed by metabolism.

Table 5.1. Ionic form of nutrients absorbed by plants

Nutrient	Ionic form
Nitrogen	NO_3^-, NH_4^+
Phosphorus	$H_2PO_4^-$, HPO_4^{2-}
Potassium	K^+
Calcium	Ca^{2+}
Magnesium	Mg^{2+}
Sulfur	SO_4^{2-}

In most situations nitrogen is absorbed predominantly in the NO_3^- form. Plants invariably contain higher concentrations of inorganic cations than inorganic anions, whether or not they are absorbed in equal quantities. This is caused by the conversion of NO_3^- to NH_4^+ and into organic forms. Numerous investigations have shown that ionic equilibrium is maintained in the plant by the synthesis of organic anions. Kirkby and Mengel (1967) made a detailed study of ionic balance in tomato (*Lycopersicon esculentum*) plants in relation to nitrogen nutrition. Table 5.2 contains data on the leaves of plants grown in complete nutrient solution with NO_3^- as the source of N. The leaves contained 280 meq N per 100 g dry wt; however, only 4 meq per 100 g were in the form of the anion, NO_3^-. Difference in inorganic cation and anion concentration was compensated by organic anions. These same general relationships hold in all plant tissues, although the levels of specific inorganic cations and anions vary with different species, with different tissues in the same plant, and with ionic composition of growth medium.

Table 5.2. Ionic balance in leaves of tomato plants grown on nitrate nitrogen

Cations (meq per g 100 dry wt)		Anions			
		Inorganic		Organic	
Ca^{2+}	161	SO_4^{2-}	22	Uronic acid	44
Mg^{2+}	30	$H_2PO_4^-$	13	Oxalic acid	41
K^+	58	Cl^-	12	Nonvolatile organic acids	117
Na^+	19	NO_3^-	4		
Total	268		51		202
Total cations	268			Total anions	253

Source: Kirby and Mengel, 1967.

II. ION INTERACTIONS AT THE MEMBRANE LEVEL

A. Short-Term Uptake

Ion interactions occur at any site where ions are involved in physiological processes. During absorption periods of several hours, interactions may come into play in transport across the membrane, in associating with organic ions within the cell, and at other levels of transport and metabolism. Therefore, short-term absorption periods of less than an hour give the most reliable indications of ion interactions at the membrane level.

Interactions of ions in short-term uptake studies may vary considerably depending upon the ion concentration and ion ratios of the nutrient medium. Epstein and Hagen (1952) indicated that the nature of inhibitory effect of Na^+ on Rb^+ absorption changed when the concentration of either ion was significantly changed. Subsequently, it became evident that dual systems existed for ion absorption and that the phenomenon observed by Epstein and Hagen probably occurred because the range of concentrations they used spanned both mechanisms. Hagen and Hopkins (1955) defined two systems for phosphate absorption, and Fried and Noggle (1958) demonstrated the existence of two distinct systems for the absorption of Rb^+, K^+, Na^+, and Sr^{2+}. Dual patterns of absorption have since been shown for many ions in many different plant materials, leading Epstein (1966, 1969) to conclude that dual systems are universal in mature tissues of higher plants.

The two systems of absorption are identified by the range of concentrations at which they are functional. Absorption by system I, the low concentration component, approaches maximal rates at ion concentrations of 0.1 to 1 mM. System II, the high concentration component, contributes to ion uptake when ion concentrations exceed approximately 1 mM. Therefore, when salt concentrations exceed 1 mM, both systems contribute to ion absorption, so that neither system can be studied independently of the other.

Although systems I and II differ in several properties, two major differences are pertinent to ionic interrelations. Absorption by system I functions relatively independent of the counter ion. On the other hand, system II is markedly influenced by the counter ion (Epstein, 1966; Hiatt, 1968). In fact, absorption by system II requires that the cation and the anion be absorbed in equivalent quantities (Hiatt, 1968). Another difference is the level to which the two systems can accumulate ions. System I may accumulate ions to very high accumulation ratios in plant tissues. However, when correction is made for the contribution of system I to absorption from high-salt solutions, the contribution of system II results in accumulation ratios of less than 1 (Hiatt, 1968). On the basis of these observations, it has been proposed that system II is a result of diffusion of neutral salts in accord with Donnan phenomena.

Reisenauer (1966), in a survey of ion concentrations in soil solutions, found that 45 % of the soil solutions contained less than 1 mM K^+ and 95 % contained less than 5 mM K^+. Likewise, 90 % of the soil solutions contained less than 3 mM NO_3^-. Thus the normal range of ion concentrations is below the concentrations at which system II makes significant contributions to ion absorption. Ion uptake via system II is negligible under normal physiological conditions and becomes significant only under unusual or artificial conditions. Nevertheless, studies of ion uptake in the range of system II have provided considerable insight toward the understanding of absorption processes. To minimize complications, the two systems will be considered separately.

1. Interactions in Absorption by System I

A hyperbolic isotherm is obtained when ion uptake is plotted against ion concentration in the range of system I, and at higher concentrations a limiting rate of absorption is approached asymptotically. Figure 5.1

shows results of such an experiment (Hiatt, 1970b). Many reactions in which one reactant binds or associates with another show the same type of concentration dependence. One example is the binding of a substrate by an enzyme as described by Michaelis-Menten enzyme kinetics. Another is the Langmuir adsoption isotherm, which describes the adsorbance of a substance by an adsorbent.

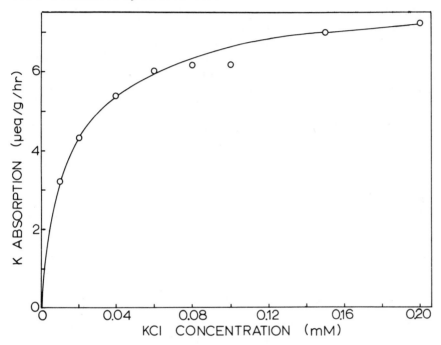

Fig. 5.1. K absorption by excised barley roots as a function of KCl concentration in the range of system I

Epstein and Hagen (1952) described ion uptake as a carrier-mediated process analogous to enzymatic reactions. The carriers are considered to combine with an ion at the outer membrane surface, moving as an ion complex to the inner surface, with subsequent internal release of the ion. Whether or not the absorption process involves the action of a mobile carrier, it seems evident that ions in some way interact with the cell membrane. Kinetics studies will not distinguish between mobile carriers and reaction or binding of ions with other membrane sites (Hiatt, 1968). The interaction between the membrane and the ion is of electrostatic nature regardless of the mechanism of transport across the cell membrane. Consequently, interactions would be expected for inorganic ions competing for ionic sites in or on the membrane.

Certain physical properties of ions obviously influence their interactions with other inorganic ions and with membrane sites. The most important of these properties are charge and hydrated radii. Univalent ions are absorbed faster than divalent or multivalent ions of similar hydrated radii, and ions of small hydrated radii are absorbed faster than ions of large hydrated radii. The hydrated radii of univalent cations in Angstroms are: Li^+, 10.03; Na^+, 7.90; K^+, 5.32; NH_4^+, 5.37; Rb^+, 5.09; and Cs^+, 5.05 (Jenny, 1935). When two of these ions are present in the same absorption medium, an increase in the concentration of one will decrease the absorption of the other. The mutual inhibition is more pronounced between ions of similar hydrated radii.

   Inhibition of uptake of one ion by another is usually treated kinetically in the same way as inhibition of enzyme reactions. Epstein and Hagen (1952) first used this technique in studying competition between univalent cations during absorption by excised barley (*Hordeum vulgare*) roots. They obtained straight lines when they plotted the reciprocal of the rate of uptake against the reciprocal of the concentration. They obtained the

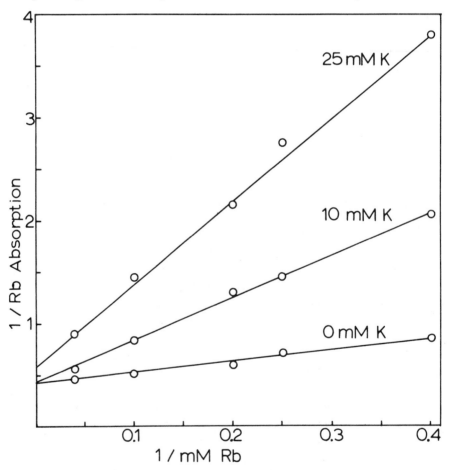

Fig. 5.2. Double-reciprocal plots of rate of absorption of Rb as affected by 0, 10, and 25 mM K

plot in Figure 5.2 when the effect of Rb^+ concentration on Rb^+ uptake was determined in the absence and presence of K^+. The straight lines obtained at different concentrations of the inhibitor intersect at the same point on the ordinate if the ions compete with one another. The results in Figure 5.2 indicate that K^+ and Rb^+ interact with the same absorption sites in transport across the membrane. While these data were obtained with Rb^+ concentrations in the range of system II, similar competitive effects were reported by Fried and Noggle (1958) in the range of system I. In fact, the literature is replete with examples which indicate that Rb^+ and K^+ act as analogues in ion uptake. Most K^+-activated enzymes are also activated by Rb^+ almost as effectively as by K^+ (Evans and Sorger, 1966). The analogous behavior of K^+ and Rb^+ would be expected since the ions have similar charge and similar hydrated radii.

Since Cs^+ and NH_4^+ have hydrated radii very similar to K^+ and Rb^+, they would be expected to compete effectively for the same absorption sites. Bange and Overstreet (1960) demonstrated that K^+ and Rb^+ competed strongly with Cs^+ during absorption by excised barley roots. Ammonium ions also effectively competed with Cs^+, but NH_4^+ did not inhibit Cs^+ uptake as much as did K^+ or Rb^+.

Evidence is convincing that univalent ions having similar hydrated radii, K^+, NH_4^+, Rb^+, and Cs^+, interact with the same absorption sites and each competitively inhibits the absorption of the other. Of these, only K^+ and NH_4^+ are present in growth media under normal physiological conditions. Rubidium has been used extensively in ion uptake studies as an analogue for K^+ because of the availability of a convenient radioisotope.

Interactions of Na^+ or Li^+ with the other univalent cations in the range of system I are not as clear-cut as interactions among Rb^+, K^+, NH_4^+, and Cs^+. Additionally, the concentration of Ca^{2+} in the absorption medium has a greater effect on Na^+ and Li^+ absorption than it does on the other cations. The hydrated radii of Na^+ and Li^+ are 7.9 and 10 Å, respectively. These radii are considerably larger than those of the other univalent cations, and consequently the behavior of these cations in ion uptake would be expected to differ from that of the smaller ions.

Numerous studies have reported on the interaction of Na^+ and K^+ or Rb^+. Interaction between these ions varies with concentration (system I vs. system II), plant species, and concentration of Ca^{2+}. Consequently, in considering Na^+ and K^+ interactions, the conditions of the experiment must be clearly defined.

Epstein (1961) has shown that in the absence of Ca^{2+}, Na^+ inhibits Rb^+ uptake in the range of system I. The inhibitory effect of Na^+ on Rb^+ uptake was partially reversed by the addition of Ca^{2+} to the medium. Conversely, the inhibitory effect of K^+ on Na^+ uptake was actually increased by the presence of Ca^{2+}. The inhibitory effect of Na^+

on K^+ uptake in the presence of Ca^{2+} was subsequently shown to be competitive (Epstein et al., 1963). An idea of relative selectivity for Na^+ and K^+ in absorption by system I can be obtained by comparing Na^+ and K^+ uptake from solutions containing equimolar concentrations of these ions. Excised barley roots absorbed approximately 3 times as much K^+ as Na^+ from solutions containing 0.05 mM Na^+ and K^+ (Hiatt, 1970a) and approximately 10 times more K^+ from solutions of 0.005 mM Na^+ and K^+ (Hiatt, 1969). These and numerous other reports lead to the conclusion that plant roots are selective in absorption of K^+ over Na^+ when the two ions are in the same solution together. Nevertheless, there is a definite interaction between the ions which appears to be of competitive nature.

Interactions involving divalent cations are much more complex than those involving univalent cations. In general, Ca^{2+} enhances the absorption of ions with small hydrated radii such as K^+ and Rb^+ and decreases the absorption of ions with large hydrated radii such as Li^+ and Na^+ (Bange and Schaminie-Dellaert, 1968; Epstein, 1961; Handley et al., 1965; Hooymans, 1964; Jacobson et al., 1961; Kahn and Hanson, 1957; Mengel and Helal, 1967; Rains et al., 1964; Waisel, 1962a, 1962b). However, there are reports of inhibition of K^+ uptake by Ca^{2+} (Elzam and Hodges, 1967; Kahn and Hanson, 1957; Hooymans, 1964). Epstein (1961) considered Ca^{2+} to be essential to the integrity of the selective ion transport mechanisms. Waisel (1962b) concluded that passive movement of ions across the outer cell membrane frequently constitutes the rate-limiting step of ion accumulation and that Ca^{2+} affects the selective permeability of the membrane. Handley et al. (1965) reached a similar conclusion and proposed that the Ca^{2+} effect may be related to membrane pore size. Estimates of the equivalent pore radius of different animal membranes based on osmotic pressure developed across membranes in the presence of diffusible salts ranged from 3.5 to 6.5 Å (Solomon, 1960). The equivalent pore radius of kidney slices from Necturus became significantly larger when Ca^{2+} was removed from the medium (Whittenbury et al., 1960). Handley et al. (1965) proposed that Ca^{2+} stabilizes the cell membrane with a consequent decline in permeability. Because of their large hydrated radii, passage of Na^+ and Li^+ should be restricted by the presence of Ca^{2+} while the passage of Rb^+ and K^+ should be restricted to a lesser extent. Handley et al. (1965) suggested that the elimination of Na^+ interference with K^+ by Ca^{2+} reported by Epstein (1961) could be explained on this basis.

Absorption rates for divalent cations are less than those for the monovalent cations. This reduced rate, in conjunction with difficulties in precise divalent cation analysis, is primarily responsible for the scarcity of absorption data. Calcium absorption has been investigated to a greater extent than the other divalent cations, a result of the interest aroused by the Ca^{2+} effect on ion absorption.

Calcium absorption can be analyzed in a similar manner as the monovalent ion absorption. Strontium has been used as an isotope for Ca^{2+} since they compete as equals in the absorption process. The work of Epstein and Leggett (1954) on strontium absorption in the range of system I gave evidence for Sr^{2+} inhibition by Ca^{2+}, Ba^{2+}, and Mg^{2+}. Calcium and Ba^{2+} inhibition was described as competitive, i.e., identical absorption site for Sr^{2+}, Ca^{2+}, and Ba^{2+}, while Mg^{2+} inhibition was uncompetitive, i.e., at a site removed from the Sr^{2+} absorption site.

Barley roots accumulate small amounts of Sr^{2+} or Ca^{2+}, $1\mu eq/g$ to $5\mu eq/g$ as compared with $100\mu eq/g$ of K^+. It is of interest that roots which do not accumulate Ca^{2+} (barley, soybeans, pinto beans, and mung beans) can be utilized as examples of ion influx activation by Ca^{2+}.

Calcium absorption can be readily observed in corn roots. In this tissue the Ca^{2+} content can approach the K^+ level under comparable conditions. Mass (1969) reported values at $35\mu eq/g$ Ca^{2+} and $50\mu eq/g$ K^+ from solutions containing 10 meq/l of these ions. Evidence for Ca^{2+} activation of K^+ absorption in corn roots was not clearly delineated. However, the data suggested a mutual inhibition rather than a Ca^{2+} activation of K^+.

Magnesium absorption by excised roots is generally equal to or greater than Ca^{2+} absorption. Differences among tissues are observed for Mg^{2+} absorption in the presence of Ca^{2+}. Calcium inhibits Mg^{2+} absorption by barley (Moore *et al.*, 1961a, 1961b) and corn (*Zea mays*) (Maas, 1969; Maas and Ogata, 1971) but only slightly in soybeans (*Glycine max*) (Leggett and Gilbert, 1969). These interactions are further complicated by the absence of Ca^{2+} absorption in barley and soybeans. Thus a clear distinction cannot be made between inhibition localized at the initial transport process and that localized at an internal site. Inhibition of Mg^{2+} absorption in soybeans can be significant in the presence of Ca^{2+} plus K^+. Such apparent differences make divalent cation absorption more difficult to understand than the monovalent cation absorption.

Cell membrane characteristics are an integral part of the absorption process. The divalent cations, particularly Ca^{2+}, greatly affect membrane phenomena, with further modification from the other cations. Therefore, interactions of divalent cations during transport would be expected to vary for different membranes. In a sense this divalent cation influences its own absorption by inducing changes in the rate-limiting membrane.

Absorption rates of cations are influenced somewhat by the counter ion. Cation absorption will usually be greater with a monovalent ion than with a divalent anion system. This can be further defined by absorption being anion-limited for system II. Potassium absorption with the concentration range of system I (Hiatt, 1968) was identical for Cl^- and SO_4^{2-}, whereas at higher concentrations (system II) K^+ absorption was greatly increased by Cl^-. Similar results were reported for Ca^{2+} absorption through the system II component of corn roots (Maas, 1969). The nitrate anion usually is associated with higher cation absorption rates because of its metabolic fate. Nitrate anion provides a mobile anion for transport, then is destroyed by reduction forming an organic anion. Destruction of NO_3^- maintains an anion gradient as well as a cation-binding site. Sulfate and phosphate are less effective than NO_3^- as a result of their slower mobilities and lower accumulation levels.

Anion absorption studies have not created a volume of literature comparable to cation absorption, probably because of the lack of interactions between the anions. If one excludes the hydroxyl ion, only similar anions exhibit a significant competition in short-time absorption periods. These have been restricted to Cl^--Br^- and SO_4^{2-}-SeO_4^{2-} interactions, which are examples of competitive inhibition (Epstein, 1953; Leggett and Epstein, 1955).

The NO_3^--Cl^- interaction is of interest since both are monovalent, usually found in the soil root environment, and relatively mobile. Nitrate represents a readily transportable anion which upon reduction can be immobilized. Although for short absorption periods NO_3^- and Cl^- are not mutually inhibitive, they do interact when presented to a growing plant. For example, barley plants grown in a complete nutrient solution with variable NO_3^- and constant Cl^+ absorb decreasing amounts of Cl^- with increasing NO_3^- levels (Table 5.3). Chloride levels in the tissue are reduced with low additions of NO_3^-. The nature of the inhibitory action has not been elucidated.

Table 5.3. Potassium, chloride, and nitrate uptake by plants from solutions of increasing nitrate concentrations

Treatment (meq/liter)			Tissue content (μeq/g fresh weight)			
			Cl		K	
NO_3	Cl	K	Shoot	Root	Shoot	Root
0	1	1.0	65	46	15	19
0.1	1	1.1	33	13	33	33
0.5	1	1.5	13	5	39	52
1.0	1	2.0	12	3	63	55
2.0	1	3.0	7	0	80	26
5.0	1	6.0	5	1	90	29

Note: Plants were grown 7 days in modified Hoagland's solution.

2. Interactions in Absorption by System II

System II participates in ion absorption from solutions with salt concentrations greater than 1 mM. Many studies of ion uptake have been conducted using salt concentrations of 1 to 50 mM. Since system I also functions at maximum capacity in this concentration range, ion uptake is the summation of uptake by both systems. Consequently, it is difficult to differentiate between ion interactions of system I and system II at ion concentrations within the range of system II.

Epstein and co-workers demonstrated that K^+ and Cs^+ competitively inhibited Rb^+ absorption in the concentration range of system II (Epstein and Hagen, 1952) and that Na^+ competitively inhibited K^+ absorption (Epstein *et al.*, 1963). System I exhibits much greater selectivity for K^+ over Na^+ than does system II.

3. Effect of Counter Ion

The initial rates of ion uptake in the range of system I are not influenced significantly by the rate of absorption of the counter ion. The absorption of Rb^+ or K^+ is the same whether the counter ion is a slowly absorbed anion (SO_4^{2-}) or a rapidly absorbed anion (Cl^-) (Epstein *et al.*, 1963; Hiatt, 1968). Likewise, the absorption of Cl^- is almost the same when the counter ion is Ca^{2+}, a slowly absorbed cation, or K^+, a rapidly absorbed cation. It is apparent that if conditions of electroneutrality are to be met, excess absorption of an ion must be countered by exchange of an ion of the same charge from the root. Excess cation absorption results in exchange of hydrogen ions from the root (Jackson and Adams, 1963; Hiatt, 1967a). Excess anions are absorbed in exchange for HCO_3^- (Jackson and Adams, 1963; Hurd and Sutcliffe, 1957). Consequently, the pH of the absorption solution changes according to the relative rates of cation and anion absorption. Salts may be classified as physiologically acid-forming or physiologically base-forming on the basis of the relative mobilities of the anion and cation.

In contrast to ion absorption by system I, absorption by system II is strikingly influenced by the rate of absorption of the counter ion (Epstein *et al.*, 1963; Hiatt, 1968). Figures 5.3 and 5.4 illustrate the influence of counter ions on ion uptake. Absorption from solutions of less than 10^{-4} to 10^{-3} M represents system I absorption, while absorption from solutions of greater ion concentrations is the sum of absorption by both systems.

Ion absorption to system II requires that cations and anions be absorbed in equivalent quantities (Hiatt, 1968). There is no system II absorption of Cl^- by excised barley roots in $CaCl_2$ because Ca^{2+} is absorbed in negligible quantities (Fig. 5.3). Likewise, K^+ absorption is much slower from K_2SO_4 because SO_4^{2-} is absorbed much slower than Cl^- (Fig. 5.4). Furthermore, uptake of ions by system II does not result in accumulation of ions against a concentration gradient. These properties are consistent with diffusion, and it has been proposed that ion uptake by system II is by diffusion of neutral salts (Hiatt, 1968).

B. Long-Term Uptake

Ion uptake by excised, low-salt roots is usually maintained at a constant rate for the first 3 to 6 hours. As the ion concentration in the tissue increases, the rate of uptake slows, and it becomes zero when the tissues are saturated. Depending on the ion concentration of the solution, the tissues will reach saturation within 8 to 24 hours. It is very likely that ion interactions at the membrane level observed in short-term experiments continue much in the same manner during long-term uptake by

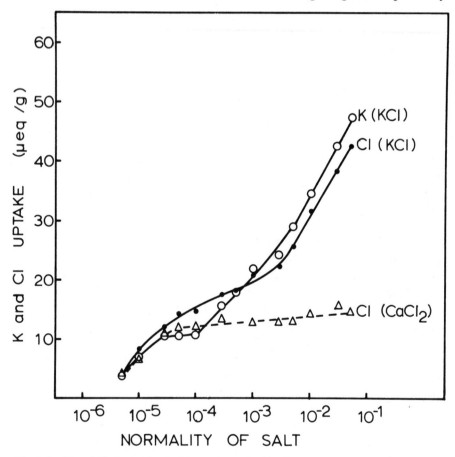

Fig. 5.3. K and Cl absorption in 4 hours by excised barley roots as a function of concentration of KCl or $CaCl_2$. (Hiatt, 1968)

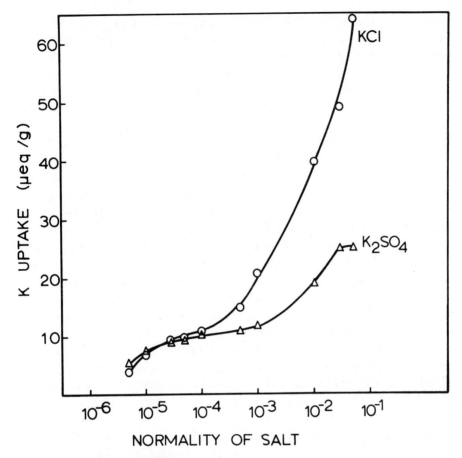

Fig. 5.4. K absorption in 4 hours by excised barley roots as a function of concentration of KCl or K_2SO_4. (Hiatt, 1968)

excised roots. However, as the concentrations of the absorbed ions begin to build up in the cell, other factors may come into play which override or mask interactions that occur in or on the membrane. Furthermore, unequal absorption of cations and anions induces metabolic changes, particularly in organic anion metabolism, which influence subsequent ion uptake. To distinguish from ion interaction at the membrane level, interactions resulting from phenomena occurring after absorption are referred to here as interactions at the cellular level.

III. INTERACTIONS AT THE CELLULAR LEVEL

A. Long-Term Uptake by Excised Roots

When cations are absorbed in excess of anions, the excess cations are absorbed in exchange for H^+ ions. Likewise, excess anions are absorbed in exchange for HCO_3^-. Ulrich (1941) demonstrated that when K^+ was absorbed in excess of the associated anion, the excess K^+ entering the cells was balanced by an increase of organic ions within the tissues. Subsequently, other investigators (Hiatt, 1967a; Hiatt and Hendricks, 1967; Jackson and Coleman, 1959; Jacobsen, 1955; Jacobson and Ordin, 1954) have similarly reported that organic acid content of roots increased when cations were absorbed in excess of anions and decreased when anions were absorbed in excess of cations. Jacobson and Ordin (1954) demonstrated that organic acid increase in tissues was stoichiometric with the excess absorption of cations over anions. In excised barley roots malic acid was the organic anion that changed in concentration.

Jacobson (1955) found that $^{14}CO_2$ was incorporated into malic acid much faster when roots were absorbing cations in excess of anions and suggested that malate was produced by a carboxylation reaction. Hiatt and Hendricks (1967) observed that $^{14}CO_2$ was incorporated into organic acids in quantities sufficient to account for observed increases in net organic acid concentration under conditions of excess cation absorption. Evidence is convincing that malate synthesis in response to excess cation absorption results from the carboxylation of phosphoenol pyruvate (PEP) to oxaloacetate (OAA) with subsequent reduction of oxaloacetate to malate by malic dehydrogenase (Jackson and Coleman, 1959; Hiatt, 1967b; Jacoby and Laties, 1971). This phenomenon is basic to cation–anion interrelations in plants, and its importance merits a detailed discussion.

When barley roots were incubated for 6 hours in solutions of K_2SO_4, $CaCl_2$, and KCl (Hiatt, 1967a), changes in organic acid content were proportional to the differences between cation and anion uptake (Table 5.4). Increase in organic acid content of roots in K_2SO_4 was approximately equivalent to excess cation uptake. The direction of organic acid change in KCl depended upon which ion was absorbed in excess. Although organic acid content of roots in $CaCl_2$ decreased markedly, the decrease in organic acid content was not equivalent to excess Cl^- uptake. The organic acid content ranged from 18 to 58 meq/g depending on the salt being absorbed.

If excess cations are absorbed in exchange for H^+ from the root, a change in cell sap pH might be expected. Ulrich (1941, 1942) noted slight increases in cell sap pH under these conditions. Table 5.5 shows the pH change of expressed sap of roots in 1 mM K_2SO_4. Absorption of SO_4^{2-} was negligible. There was a definite increase in pH of expressed cell sap within 15 minutes, and the pH continued to increase for 2 hours. These changes appear rather small; however, the measurement gave the average cell sap pH for the whole root, and it is likely that local pH changes may be considerably greater.

Table 5.4. Effect of concentration of K_2SO_4, $CaCl_2$, and KCl on cation uptake, anion uptake, and organic acid change of barley roots

Salt	Concentration (normality)	Substrate volume (liters)	Cation uptake ($\mu eq/g$)	Anion uptake ($\mu eq/g$)	Organic acid change ($\mu eq/g$)
K_2SO_4	10^{-5}	8	9	<1	8.7
	10^{-4}	4	12	<1	12.2
	10^{-3}	4	17	<1	15.1
	10^{-2}	4	22	2.1	18.0
	5×10^{-2}	4	25	6.4	20.6
$CaCl_2$	3×10^{-5}	8	<1	14	−9.0
	10^{-4}	8	<1	14	−10.7
	10^{-3}	4	<1	15	−9.7
KCl	5×10^{-5}	8	16	19	−2.9
	10^{-4}	4	14	20	−4.6
	5×10^{-4}	2	23	26	−0.4
	10^{-3}	2	28	29	−0.2
	5×10^{-3}	2	39	36	0.8
	10^{-2}	2	43	39	1.2

Source: Hiatt, 1967a.
Note: Roots were incubated for 6 hours in the indicated solutions. Initial levels were: K content = $18\mu eq/g$; Cl content = $4\mu eq/g$; organic acid content = $28.4\mu eq/g$.

Table 5.5. Change of expressed root sap pH with time of incubation in 10^{-3} M K_2SO_4.

Incubation period	Expressed sap pH
Initial roots	5.48
15 minutes	5.51
30 minutes	5.54
1 hour	5.56
2 hours	5.59
4 hours	5.59
6 hours	5.59

Source: Hiatt, 1967a.

Table 5.6 shows the pH of sap expressed from roots that were incubated in solutions of KCl, $CaCl_2$, and K_2SO_4. The shifts in pH of expressed sap corresponded to differences in cation and anion uptake

Table 5.6. Effect of concentration of K_2SO_4, $CaCl_2$, and KCl on pH of substrate solution and expressed cell sap

Salt	Concentration (normality)	Substrate		Expressed sap pH
		Initial pH	Final pH	
K_2SO_4	10^{-5}	5.70	—†	5.49
	10^{-4}	5.70	—	5.52
	10^{-3}	5.70	—	5.54
	10^{-2}	5.70	—	5.56
$CaCl_2$	10^{-4}	5.63	5.78	5.13
	10^{-3}	5.65	5.82	5.07
	10^{-2}	5.68	5.90	5.08
KCl	10^{-5}	5.70	5.77	5.38
	10^{-4}	5.70	6.02	5.21
	10^{-3}	5.70	5.86	5.27
	10^{-2}	5.70	5.63	5.47

Source: Hiatt, 1967a.
† Final pH levels of K_2SO_4 solutions are not given because pH was maintained at 5.5 to 5.7 with KOH.
Note: Expressed sap pH of initial roots = 5.45; incubation period = 2 hours.

as indicated by changes in the pH of the substrate solution and by the data in Table 5.4. The pH of cell sap responded to differences in cation and anion uptake even at salt concentrations of 10^{-5} M.

Since system II absorbs cations and anions in equivalent quantities (Hiatt, 1968), ion absorption related to organic acid changes should be a property of system I absorption. The data in Table 5.4 illustrate that the concentration of organic acids responds to imbalances in cation and anion uptake at all salt concentrations. Therefore, organic acid change in response to ion uptake is considered to be a property of system I (Hiatt, 1967a; Jacoby and Laties, 1971). Organic acid changes that occur with salt concentrations in the range of system II can be attributed to the action of system I, which is functioning at maximum capacity.

Malic acid synthesis in response to excess cation absorption is apparently through carboxylation of PEP to OAA, which is rapidly reduced to malate by malic dehydrogenase. Plant roots possess three carboxylating enzymes (Fig. 5.5). The first product of CO_2 fixation is OAA (Hiatt and Hendricks, 1967; Ting and Dugger, 1967); therefore, the reaction is catalyzed by either PEP carboxylase or PEP carboxykinase. PEP arises from glucose via glycolysis, as illustrated in Figure 5.6.

The reduction of glyceraldehyde-3 phosphate to 3-phosphoglyceric acid is markedly influenced by pH, so that malate synthesis would be

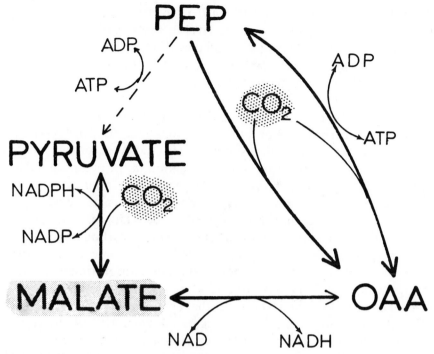

Fig. 5.5. Schematic illustration of reactions that yield organic acids by carboxylation of 3-C intermediates of the glycolytic pathway

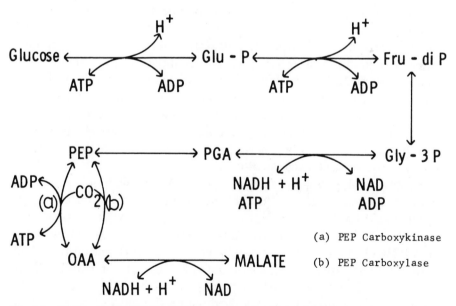

Fig. 5.6. Schematic illustration of the pathway of synthesis of organic acids. *Glu-P:* glucose phosphate; *Fru-diP:* fructose diphosphate; *PGA:* phosphoglyceric acid; *PEP:* phosphoenolpyruvic acid; *OAA:* oxaloacetic acid

favored by increase in pH (Hiatt, 1967b). Conversely, a decrease in cell pH would favor a reduction in the level of malate. Hiatt (1967b) has proposed that both carboxylation and decarboxylation are catalyzed by PEP carboxykinase, a reversible reaction. The control of tissue malate levels through this scheme is proposed to be by shifts in cellular pH induced by cation/H^+ exchange.

Jacoby and Laties (1971) have proposed that carboxylation is catalyzed by PEP carboxylase and that bicarbonate level of the cytoplasm controls organic acid synthesis. Cation/H^+ exchange or any H^+-requiring process leads to bicarbonate formation in the cytoplasm. Also, supplying K^+ as a bicarbonate salt results in increased organic acid synthesis and K^+ absorption. Regardless of whether carboxylation is catalyzed by PEP carboxylase or PEP carboxykinase, and whether control of malate levels is by cytoplasmic pH or bicarbonate levels, the net effect of ion absorption imbalance would be the same. In either case, cation/H^+ exchange resulting from excess cation absorption would lead to organic acid synthesis. As will be seen in subsequent discussions in this paper, any physiological process that leads to increased cytoplasmic pH or bicarbonate level results in increased malic acid synthesis.

During the first hour or two of excess cation absorption, organic acid synthesis lags behind cation uptake; however, stoichiometry is achieved after about three hours (Jacoby and Laties, 1971). Jacoby and Laties (1971) attribute stoichiometry to transfer of the salts of the organic acids to the vacuole, where they are unavailable metabolically. Organic acid synthesis continues for at least an hour after removal from salts producing excess cation absorption, indicating some residual of the stimulus for organic acid synthesis. The vacuole comprises greater than 95% of the cell volume, and consequently at equilibrium the major portion of organic acids and inorganic ions should be in the vacuole. No evidence has been presented which indicates that salts are accumulated from the cytoplasm into the vacuole against a gradient. Thus movement of salts into the vacuole may be a passive process, with salts existing predominantly in the vacuole because of its large volume.

Jacoby and Laties provided evidence that organic acid synthesis is stimulated when the HCO_3^- level of the treatment solution is increased. Organic acid levels in the tissue could be increased without significant cation uptake by exposing the roots to trisHCO_3^-. The absorption of K^+ from $KHCO_3^-$ was several times greater than K^+ absorption from KCl or K_2SO_4. They attributed the stimulated K^+ absorption by $KHCO_3$ to the stimulated synthesis of organic acids by HCO_3^-.

Organic acid anions are synthesized within the cell. At the pH of cell sap they are 80% to 90% dissociated and, consequently, must be salts of inorganic cations. Therefore, organic acids serve as nondiffusible anions within a semipermeable membrane. When net absorption is considered,

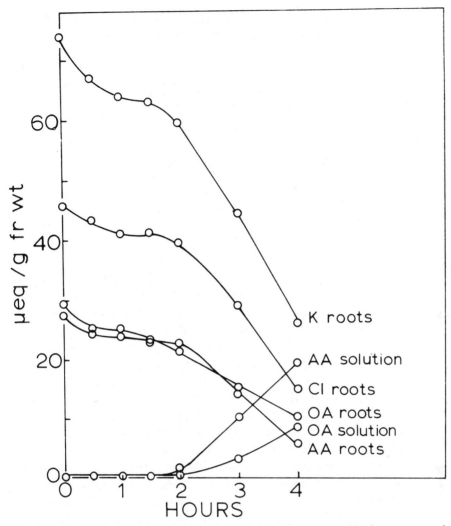

Fig. 5.7. Organic acid, amino acid, K^+, and Cl^- content for high-KCl barley roots treated anaerobically for various intervals of time. Roots were pretreated for 12 hours in an aerated solution containing 1 mM KCl and 0.2 mM $CaSO_4$ and then placed under N_2 in 0.2 mM $CaSO_4$.

in contrast to rate of uptake in short-term studies, organic ions must be considered an integral part of ion accumulation.

Comparison of changes in organic and inorganic ions in plant roots placed under anaerobic conditions (Hiatt and Lowe, 1967) provides useful information on the relationship of organic ions to inorganic ion accumulation. Excised barley roots were incubated for 12 hours in 1 mM KCl and 0.2 mM $CaSO_4$. The roots were then transferred to an anaerobic solution of 0.2 mM $CaSO_4$ for periods up to 4 hours. Changes in K^+, Cl^-, organic acids, and amino acids are shown in Figure 5.7. Organic

Table 5.7. Effect of duration of anaerobiosis on loss of K, Cl, organic acids, and amino acids from roots pretreated for 12 hours in 1 mM KCl

Duration of treatment (hr)	Loss (μeq/g roots)					
	K	OA	K − OA	Cl	AA	OA + AA
0.5	6	4	2	2	3	7
1.0	9	4	5	4	4	8
1.5	10	6	4	4	4	10
2.0	14	8	6	6	5	13
3.0	29	14	15	17	13	27
4.0	47	18	29	31	22	40

Source: Hiatt and Lowe, 1967.

acid and amino acid content of the roots decreased rapidly during the first 30 minutes but changed little between 30 minutes and 2 hours. After 2 hours under anaerobiosis, the organic acid decrease was a result of leakage to the solution. Amino acids also leaked to the solution after 2 hours. The loss of K^+ and Cl^- from the roots exhibited a similar pattern.

Table 5.7 shows the loss of K^+, Cl^-, organic acids, and amino acids from the roots. During the first 3 hours Cl^- loss was approximately equivalent to amino acid loss, and K^+ loss was equivalent to the loss of organic acids plus amino acids. Since cations other than K^+ made up only a small fraction of the total diffusible cation content, the organic acids were predominately in the form of K^+ salts. When organic acids are lost, either by decarboxylation or leakage, an equivalent amount of K^+ should be lost from the tissue. Subtraction of organic acid decrease from K^+ decrease gives an estimate of K^+ lost in excess of organic acid. This fraction of K^+ loss was equivalent to the loss of Cl^- and amino acids. The stoichiometry illustrated by these data suggests that in high KCl roots, organic acids exist as K^+ salts and amino acids exist as KCl salts. Roots treated with cyanide or 2,4-dinitrophenol exhibited similar stoichiometry in loss of organic acids and K^+.

Experiments with other organisms have also provided evidence of binding of inorganic ions by organic ions. Studies with yeast cells suggest that the cell behaves as an ion exchanger within a rate-limiting barrier through which ions move by simple diffusion (Leggett *et al.*, 1962, 1965; Leggett and Olsen, 1964; Olsen, 1968, 1969; Olsen and Tripp, 1969). Schaedle and Jacobson (1965) proposed that K^+ accumulation by *Chlorella* was limited by the ability of the cells to create organic acid anions. Both of these organisms accumulated cations but not inorganic anions.

Jackson and Stief (1965) allowed barley roots to equilibrate with 10 mM KCl plus NaCl solutions in which the Na:K ratios were 1:10, 1:1, and 10:1. Regardless of the size of the Na:K ratio in the treatment solutions, the total Na + K content of the roots was 78 to 79 meq/g fresh weight. Total Na + K content of roots was constant in excised barley roots incubated for 24 hours in a NaCl-KCl replacement series of 1 mM salt (Hiatt, 1968). Bear and Prince (1945) found the sum of the equivalent of Ca^{2+}, Mg^{2+}, and K^+ per unit weight of alfalfa (*Medicago sativa*) to be relatively constant although the ratio of the cations varied considerably in alfalfa grown on different soils. Thus, it appears that plant species have a characteristic upper limit of cation content. The preceding data suggest that the upper limit of ion accumulation is determined by the availability of charged sites supplied by organic ions. A logical extension of this statement is that inorganic ions must be associated with nondiffusible organic ions in order to be held in the cell at a concentration higher than that of the surrounding medium. In other words, ion accumulation requires that absorbed ions become associated with organic ions of opposite charge. Jacoby and Laties (1971) found that root tip cells (nonvacuolated) lost K^+ when malate was metabolized. For such a mechanism to operate, the cell must be capable of synthesizing new organic ions or otherwise possess a reservoir of organic ions which may associate with absorbed ions.

As already discussed, accumulation of cations without concurrent anion accumulation can be explained on the basis of synthesis of organic acid anions. However, when roots absorb cations and anions in equivalent amounts, as with KCl, neither organic nor amino acids change in concentration. It appears likely that in low-salt roots the positive and negative charges of amino acids are intramolecularly neutralized, or the amino acids may exist as dimers (Hiatt and Lowe, 1967; Hiatt, 1968). They might therefore serve as a reservoir of positive and negative charged sites if inorganic salts such as KCl or KNO_3 could compete with the intramolecular or intermolecular charge neutralization. When roots are placed in a salt solution, accumulation would continue until a new equilibrium with the salt solution is established; however, accumulation from a dilute salt solution would not exceed the concentration of available organic ions within the cell.

Six-day-old, low-salt barley roots contain $25\mu eq$ to $30\mu eq$ each of organic and amino acids per gram fresh weight. In addition, there are other ionic charges associated with membranes, cell walls, and protein and other organic molecules. The organic and amino acids alone provide a total of $50\mu eq$ to $60\mu eq$ of negative charges and $25\mu eq$ to $30\mu eq$ of positive charges per gram fresh weight. Thus they can potentially bind $50\mu eq$ to $60\mu eq$ of cations and $25\mu eq$ to $30\mu eq$ of anions. These quantities are of the same order of magnitude as the total K^+ and Cl^- accumulation from dilute KCl solutions (see Fig. 5.7). Although organic and amino acids are not the only organic ions that associate with in-

Table 5.8. Potassium accumulation by six-day-old barley roots from K_2SO_4, from KCl, or from KCl followed by K_2SO_4

Treatment		K uptake (μeq/g)		Total K uptake (μeq/g)
0–12 hr	12–24 hr	0–12 hr	12–24 hr	
K_2SO_4	K_2SO_4	35	16	51
KCl	KCl	54	13	67
KCl	K_2SO_4	54	21	75

Source: Hiatt, 1968.
Note: After incubation for 12 hours in 10^{-3} M K_2SO_4 or KCl, roots were transferred to new solutions as indicated for an additional 12 hours.

organic ions, they are the primary ones with the potential for changing in concentration or form.

Potassium absorption from K_2SO_4 occurs without concurrent absorption of the associated anion, and organic acids are synthesized in quantities equivalent to the K^+ absorbed. On the other hand, when roots absorb KCl, both K^+ and Cl^- are absorbed and there is little change in organic acids; it is postulated that the absorbed K^+ and Cl^- are primarily associated with amino acids. During K^+ absorption from single-salt solutions of KCl or K_2SO_4, K^+ becomes associated with amino acids and organic acids, respectively. Therefore, if both binding systems can be brought into play, the roots should be able to accumulate a larger amount of K^+. To test this hypothesis, roots were incubated for 24 hours in K_2SO_4 or KCl and for 12 hours in KCl followed by 12 hours in K_2SO_4. The roots were salt saturated within 24 hours (Hiatt, 1968). Table 5.8 shows that during the initial 12-hour period, a greater quantity of K^+ was absorbed from KCl than from K_2SO_4. During the 12- to 24-hour interval, K^+ uptake from K_2SO_4 exceeded that from KCl. When the roots were transferred from KCl to K_2SO_4 during the second 12-hour interval, K^+ uptake during this interval increased, and total K^+ accumulation increased. These data are consistent with the hypothesis that K^+ accumulation may be increased by utilizing both organic acids and other anions, presumably amino acids, to bind K^+ ionically.

Chloride absorption from $CaCl_2$ is unaccompanied by Ca^{2+} and presumably enters the cell in exchange for HCO_3^-. Absorption of Cl^- from $CaCl_2$ is initially as rapid as Cl^- absorption from KCl; however, the net accumulation of Cl^- from $CaCl_2$ at equilibrium is approximately one-half the net accumulation from KCl (Hiatt, 1968). Table 5.4 shows that organic acid levels of roots decrease markedly when roots absorb Cl^- from $CaCl_2$. If Cl^- is to become associated with the positive charge of amino acids, a cation must also be available to associate with the negative charge of the amino acid. In low-salt roots the organic acids

Table 5.9. Effect of preincubation of roots in K_2SO_4 on subsequent Cl accumulation from $CaCl_2$

Pretreatment	Absorption treatment	Cl uptake (μeq/g)
Experiment I		
$CaSO_4$ only	24 hr in 10^{-4} M $CaCl_2$	26.9
10^{-4} M K_2SO_4 + $CaSO_4$	24 hr in 10^{-4} M $CaCl_2$	34.8
Experiment II		
None	20 hr in 10^{-3} M $CaCl_2$	28.8
10^{-3} M K_2SO_4 + $CaSO_4$	20 hr in 10^{-3} M $CaCl_2$	34.3
None	24 hr in 10^{-3} M KCl	52.7

Source: Hiatt, 1968.
Note: Pretreated roots were placed for 6 hours in the indicated solutions and then rinsed for 10 minutes before being placed in the absorption solutions. All solutions contained 2×10^{-4} M $CaSO_4$.

exist primarily as K^+ salts. It is suggested that when Cl^- is absorbed unaccompanied by a cation, the decarboxylation of organic acids releases K^+, which can then associate with amino acids along with Cl^-. This suggests that Cl^- uptake unaccompanied by a cation is limited by the supply of K^+ that becomes available when organic acids are decarboxylated. Therefore, if organic acids and K^+ are increased by pretreatment of roots with K_2SO_4, there should be a subsequent increase in the ability of roots to accumulate Cl^- from $CaCl_2$. The results of such an experiment are shown in Table 5.9. Chloride accumulation was increased by preincubation of roots in K_2SO_4, suggesting that Cl^- accumulation from $CaCl_2$ may be increased by increasing the endogenous level of organic acids and K^+ in the tissue.

The preceding discussion presents strong evidence that inorganic ion accumulation is very closely associated with organic ion levels and changes in plant tissues. The quantitative relationships are strikingly close. Although much attention has been given to rates and kinetics of ion transport, until recent years little attention was paid to the ionic relations within the cell after absorption. We propose that diffusible inorganic ions in cells are retained by association or binding with organic ions that are synthesized in the cell and are retained there by a membrane. Free salts, such as KCl or KNO_3, would not be retained in the cell because of leakage through the membrane. Ion accumulation, therefore, would require electrostatic association between inorganic and organic ions. Organic and amino acids are the major organic ions involved in ion accumulation; however, other charged sites would necessarily contribute to the overall association of inorganic ions with organic ions.

In addition to providing a better understanding of overall ion relations in plants, the concept that ion retention and organic ions are closely interrelated has important implications for the mechanism of ion accumulation in plants. The concept that ion transport into cells is accomplished by mobile carriers is based primarily on the agreement between enzyme kinetics and ion uptake kinetics, and on the concept that ions are accumulated against a concentration gradient. This latter idea is very likely a misconception. Although ions are indeed accumulated in plants at much higher levels than in growth media, the association of inorganic ions with organic ions markedly reduces the activity of these ions within the cell. In other words, ions may not actually be accumulated against an activity gradient. Consequently, a mobile carrier within the membrane would not be essential for ion accumulation. The following mechanism is proposed as an alternative to the concept of carrier-mediated absorption.

Basically the proposed mechanism involves diffusion-exchange and Donnan phenomena. We suggest that roots may accumulate ions from dilute salt solutions (system I) in the following manner. When roots are placed in a salt solution of less than 1 mM concentration, the ions diffuse or exchange into the cell along an activity gradient. Inside the cell the ions become associated with organic ions of opposite charge. This maintains an activity gradient, and diffusion-exchange occurs until equilibrium is achieved. New negative charges may be created by synthesis of organic acids in response to changes in cell pH or bicarbonate levels (Hiatt, 1967a; Jacoby and Laties, 1971). In low-salt roots a reservoir of both negative and positive charges may exist in the form of intermolecularly or intramolecularly neutralized amino acids.

With solutions of greater than 1 mM salt concentration, total ion uptake consists of system I absorption, which is operating at near maximum rates, plus the additional contributions from system II. Ion uptake by system II is markedly dependent upon the nature of the counter ion, and an ion absorbed by this mechanism must be accompanied by an ion of opposite charge. Furthermore, absorption by system II does not result in significant accumulation in the plant tissues. Since plant cells possess a relatively high concentration of organic ions, which are limited by a semipermeable membrane, the cells behave like a Donnan system. It has been proposed that system II absorption is a result of diffusion of neutral salts according to Donnan phenomena (Hiatt, 1968).

The cell membrane would serve a twofold purpose in ion accumulation by this process: the selectivity or preferential exclusion of ions and the retention of organic ions within the cell. Kinetics observed in ion uptake are consistent with adsorption of ions onto fixed charged sites within the membrane or with the alternate adsorption and displacement of ions through a series of fixed ion exchange sites in the membrane. Certainly, membrane pore size contributes to ion selectivity, since rates of absorption decrease as hydrated ionic radii increase. A more detailed discussion of kinetics, selectivity, and respiratory requirements in ion uptake by this proposed mechanism is presented elsewhere (Hiatt and Lowe, 1967; Hiatt, 1968).

Regardless of whether the carrier concept or the exchange-diffusion concept more accurately describes ion transport across membranes, the interaction of organic and inorganic ions is a major determinant of accumulation levels and ion distribution in plants.

To this point discussion has been limited to ions that are not changed or metabolized in the cell. The two major nutrient elements, phosphorus and nitrogen, are metabolized and incorporated into organic compounds. As such, they are no longer a part of the inorganic ion pool in cells. As a result of phosphorylation, phosphate is incorporated into nucleic acid, sugar phosphates, and other phosphate esters. Thus, even though the plants contain relatively high level of phosphate, only a small fraction is in the free form. Organic compounds containing phosphate would provide negative sites in the cell.

Under most soil conditions N is absorbed as NO_3^-. Nitrate is reduced to NH_4^+ and incorporated into amino acids, proteins, and other nitrogenous compounds. As with phosphate, most of the N in plants exists in the organic form. The low-salt, excised roots used in most ion absorption studies initially do not have nitrate reductase. Consequently, NO_3^- is not reduced and may accumulate in cells in the same manner as does Cl^-. Nitrate is the major anion absorbed by intact growing plants. The metabolism of NO_3^- results in the conversion of an anion to a cation. Consequently, the ionic balance in the tissue is markedly affected. The effect of N nutrition on total ionic interactions in plants is discussed in the following section.

B. Long-Term Uptake by Intact Plants

The form and quantity of N supplied to plants have a marked influence on the accumulation of other ions and on the organic ion levels of plants. Kirkby and Mengel (1967) conducted an excellent study to characterize this phenomenon. They grew tomato plants for 20 days in nutrient solutions with N provided as NO_3^-, NH_4^+, and urea. Thus, N was supplied as an anion, a cation, and as an undissociated molecule. The samples were divided into leaves, petioles, stems, and roots and were analyzed for cations, inorganic anions, and organic anions. A summary of their results is presented in Table 5.10.

Compared with cation concentration, inorganic anion concentration varied little with different forms of N. Cation concentration of plants

Table 5.10. Influence of the form of nitrogen nutrition on the cation-anion balance in the tissues of tomato plants

Tissue	N source	Cations			Anions				
		Indiffusible	Diffusible	Total	Nonvolatile organic acids	Inorganic	Uronic	Oxalate	Total
				(meq/100 g dry wt)					
Leaves	NO$_3$	86	182	268	117	51	44	41	253
	Urea	79	131	210	41	69	40	25	175
	NH$_4$	52	79	131	11	64	46	8	129
Petioles	NO$_3$	136	222	358	147	73	69	61	350
	Urea	120	212	332	94	130	68	49	341
	NH$_4$	67	127	194	18	98	63	16	195
Stems	NO$_3$	99	202	301	114	67	54	58	293
	Urea	91	175	266	58	91	50	43	242
	NH$_4$	55	85	140	12	56	55	18	141
Roots	NO$_3$	59	134	193	45	85	56	22	208
	Urea	51	136	187	30	84	52	16	182
	NH$_4$	31	74	105	5	53	42	1	101

Source: Kirkby and Mengel, 1967.

with NO_3^- was approximately twice that of plants grown with NH_4^+. The difference between cation content and inorganic anion content was balanced by organic acid anions. Malic acid constituted the major change in organic acids; however, citric and oxalic acids changed considerably in concentration.

Plants receiving NO_3^- absorbed more NO_3^- than total cations. Absorbed NO_3^- does not remain in the ionic form but is converted to organic form. As indicated in Table 5.10, any imbalance between cation and anion content is compensated for by organic acid anions. So far, we have discussed only organic acid changes in response to differences in cation and anion uptake *per se*. In the situation here organic acid levels change in response to ion imbalances created by metabolism of nitrate. The stimulus for organic acid synthesis appears to be the same in both cases. Table 5.11 shows the pH of expressed sap of the various tissues. Hydrogen ions were utilized in the reduction of NO_3^-, leading to an increase in the pH of cell sap in plants receiving NO_3^-. On the other hand, the assimilation of NH_4^+ leads to the production of H ions, and consequently, cell pH becomes more acid in NH_4^+-treated plants. The direction of pH change is correlated with organic acid levels in the same manner as with imbalanced ion absorption of excised roots. Bicarbonate can be produced from H_2CO_3 by the utilization of H^+ in nitrate reduction. Organic acid levels can therefore be controlled by shifts in cell pH or in bicarbonate level.

It has generally been considered that cation absorption is greater when the cation is supplied with NO_3^- because the high mobility of NO_3^- favors more rapid absorption of the accompanying cation. By this explanation of the effect of NO_3^- on cation accumulation, the organic acid changes would simply be a means of maintaining electroneutrality in the cell when anion levels are decreased by the reduction of NO_3^-, and would not be causally related to stimulated cation accumulation.

Now that we have a better understanding of the stimulus for organic acid synthesis, we believe an alternate concept better explains the greater accumulation of cations when N is supplied as NO_3^-. The reduction of

Table 5.11. Influence of the form of nitrogen nutrition on the pH of macerated tissues of tomato plants

N source	Leaves	Petioles	Stems	Roots
NO_3	5.50	5.45	5.45	5.60
Urea	5.50	5.30	5.30	5.00
NH_4	5.00	4.70	4.80	4.70

Source: Kirkby and Mengel, 1967.

NO_3^- leads to organic acid synthesis in response to shifts in pH or bicarbonate levels in the tissue. The organic anions created serve as a source of nondiffusible negative charges for the accumulation of cations. In this context we see the synthesis of organic acids as a cause of stimulated cation accumulation. This is consistent with the hypothesis that ion accumulation depends upon association of inorganic ions with organic ions.

Jackson and Williams (1968) found that cation uptake by wheat (*Triticum aestivum*) seedlings in a 24-hour period was much greater from NO_3^- salts than from Cl^- salts. We have made similar observations with barley seedlings.

Observations of other tissues are consistent with the above explanation of the stimulation of cation accumulation by organic acid synthesis. Pentzer (1924) observed that peel tissue of apple (*Malus pumila*) fruits affected with Jonathan spot was higher in pH than unaffected tissue. Richmond *et al.* (1964) made a similar observation and also found that affected tissue contained approximately three times the cations and organic acids in normal tissue.

Knowledge of the interrelationships of organic and inorganic ions provides a better understanding of ion accumulation and retention as well as of nutritional relationship with respect to composition, growth, and morphology. De Wit *et al.* (1963) suggested that plants require a normal range of organic anion content for good growth and that yields are reduced when organic anion content falls below the normal range. It is apparent from the preceding discussion that increases in NO_3^- fertilization will increase organic acid levels in tissues and that yield response to N should correlate with organic acid levels.

One other physiological process related to ion transfer in plants merits consideration. Recent investigations have indicated that K^+ accumulation is responsible for development of osmotic pressure in guard cells (Fischer, 1968, 1971; Sawhney and Zelitch, 1969). K^+ accumulation in guard cells is unaccompanied by an anion, and we suggest that organic acids are synthesized in the guard cell to maintain electroneutrality. It is generally considered that K^+ uptake by the guard cell is an active process involving carrier mediation and that organic acid synthesis comes after K^+ uptake. It is tempting to speculate that the initial action is the synthesis of organic acids as a product of photosynthesis and that K^+ uptake is a passive process along an electropotential gradient.

It should be apparent from the foregoing discussions that not only the quantity of specific nutrients but also the form supplied will influence plant growth. Replacement of all or a portion of the NO_3-N in growth media with NH_4-N may have a significant impact upon both cation and NO_3^- absorption, as well as up on organic acid levels. The effect of NH_4^+ is twofold. The NH_4 ion competes in absorption with other cations;

however, the more important influence may be on the organic acid levels. The lower organic acid levels induced by replacing NO_3^- with NH_4^+ will lower the capacity of the plant to accumulate cations.

The predominant anions absorbed by plants are NO_3^-, $H_2PO_4^-$, SO_4^{2-}, and Cl^-. Except when Cl^- is added by fertilization, this anion will be present in relatively low levels. In the total ionic balance, NO_3^- has the largest influence because of the large quantities utilized in plant growth. Usually uptake of NO_3^- exceeds the total absorption of cations. Thus NO_3^- fertilizers are physiologically basic. However, the reduction of NO_3^- to NH_4^+ with subsequent incorporation into organic N compounds leaves the plant with an excess of cations that is balanced by organic acids.

The anions NO_3^-, $H_2PO_4^-$, and SO_4^{2-} are all metabolized in the plant and converted to organic forms. In contrast to cations, there is no need for the plant to accumulate high levels of these anions. Generally the level of free inorganic anions in plants is considerably lower than the level of cations. Crooke and Knight (1962) compiled data on the cation and anion content of a number of plants. In the majority of the plants the inorganic cation content was three to six times that of the inorganic anion content. The implication of these observations is that the requirement for organic cations to retain anions is only a fraction of the requirement for organic anions to retain cations.

We do not imply that all inorganic ions in leaves of plants exist as salts of organic ions. Undoubtedly free salts move from the soil solution to the leaves in the transpiration stream where they remain after loss of water by transpiration. Cation accumulation in leaves, particularly that fraction in excess of anions, is conclusively associated electrostatically with organic acid anions. Although free amino acids have been implicated in anion accumulation by roots, the extent to which the level of inorganic ions in leaves is correlated with organic cations has not been experimentally determined and must remain a matter of speculation.

IV. SUMMARY

Most plants contain 50 to 100 meq of negative charges per kilogram of fresh weight of tissue. Equal cation concentrations are required to meet conditions of electroneutrality. The majority of cations in plant tissues are in the inorganic form, predominantly K^+, Ca^{2+}, and Mg^{2+}, and the majority of anions are in the organic form. The organic ions are synthesized within the tissue, while inorganic ions are absorbed from the growth medium.

Ionic interactions occur on and within all plant cells. In short-term studies with low-salt plants, the most evident interactions occur at the limiting membrane of root cells. These interactions are expressed as competitive inhibition, uncompetitive inhibition, and stimulation. Kinetic studies of ion absorption over short-term intervals are useful in revealing the nature of the interactions, particularly among ions that react with the similar sites in or on the membrane. In general, ions with similar properties compete in absorption. Interactions involving divalent cations are complex because ionic binding of these cations by the membrane modifies the physical nature of the membrane. Ion absorption selectivity is imparted by a rate-limiting membrane, apparently the plasmalemma.

When absorption periods exceed a few hours, ionic interactions within the cell influence the properties and capacity of ion accumulation. Changes in organic ion composition induced by ion uptake and ion metabolism in turn influence the capacity of the tissue to accumulate inorganic cations and anions. For example, the capacity to accumulate cations is modified by factors that induce changes in organic acid levels. The capacity for inorganic cation accumulation is apparently determined by the availability of organic anions in tissue; however, the specific cations that satisfy the potential for accumulation are determined by the nutrient composition of the media and the selectivity imposed in transfer across the limiting membrane.

Contrasting viewpoints exist about the significance of inorganic-organic ion relationships in cells. A popular viewpoint is that organic ion levels change in response to ion imbalance, simply maintaining charge balance within the tissues. On the other hand, we favor the idea that changes in organic ion levels have a causal relationship to inorganic ion accumulation.

References

Bange, G. G. J., and R. Overstreet. 1960. Some observations on absorption of cesium by excised barley roots. *Plant Physiol.* 35: 605–8.

Bange, G. G. J., and A. Schaminie-Dellaert. 1968. A comparison of the effect of calcium and magnesium on the separate components of alkali cation uptake by excised barley roots. *Plant Soil* 28: 177–81.

Bear, F. E., and A. L. Prince. 1945. Cation-equivalent constancy in alfalfa. *J. Amer. Soc. Agron.* 37: 217–22.

Crooke, W. M., and A. H. Knight. 1962. An evaluation of published data on the mineral composition of plants in the light of cation-exchange capacities of their roots. *Soil Sci.* 93: 365–73.

de Wit, C. T. W. Dijkshoorn, and J. C. Noggle. 1963. Ionic balance and growth of plants. *Verslagen van landbouwkundige onderzoekingen NR.* 69.15 Wageningen, The Netherlands.

Elzam, O. E., and T. K. Hodges. 1967. Calcium inhibition of potassium absorption in corn roots. *Plant Physiol.* 42: 1483–88.

Epstein, E. 1953. Mechanism of ion absorption by roots. *Nature* 171: 83.

——. 1961. Essential role of calcium in selective cation transport by plant cells. *Plant Physiol.* 36: 437–44.

——. 1966. Dual pattern of ion absorption by plant cells and by plants. *Nature* 212: 1324–27.

——. 1969. Mineral metabolism of halophytes. In *Ecological Aspects of the Mineral Nutrition in Plants*, ed. I. H. Rorison, pp. 345–55. Philadelphia: Blackwell.

Epstein, E., and C. E. Hagen. 1952. A kinetic study of the absorption of alkali cations by barley roots. *Plant Physiol.* 27: 457–74.

Epstein, E., and J. E. Leggett. 1954. The absorption of alkaline earth cations by barley roots: kinetics and mechanism. *Amer. J. Bot.* 41: 785–91.

Epstein, E., D. W. Rains, and O. E. Elzam. 1963. Resolution of dual mechanisms of potassium absorption by barley roots. *Proc. Nat. Acad. Sci.* 49: 685–92.

Evans, H. J., and G. J. Sorger. 1966. Role of mineral elements with emphasis on the univalent cations. *Annu. Rev. Plant Physiol.* 17: 47–76.

Fischer, R. A. 1968. Stomatal opening: role of potassium uptake by guard cells. *Science* 160: 784–85.

——. 1971. Role of potassium in stomatal opening in the leaf of *Vicia faba*. *Plant Physiol,* 47: 555–58.

Fried, M. and J. C. Noggle. 1958. Multiple site uptake of individual cations by roots as affected by hydrogen ion. *Plant Physiol.* 33: 139–44.

Hagen, C. E., and H. T. Hopkins. 1955. Ionic species in orthophosphate absorption by barley roots. *Plant Physiol.* 30: 193–99.

Handley, R., A. Metwally, and R. Overstreet. 1965. Effects of Ca upon metabolic and nonmetabolic uptake of Na and Rb by root segments of *Zea mays*. *Plant Physiol*. 40: 513–20.

Hiatt, A. J. 1967a. Relationship of cell sap pH to organic acid change during ion uptake. *Plant Physiol*. 42: 294–98.

——. 1967b. Reactions *in vitro* of enzymes involved in CO_2 fixation accompanying salt uptake by barley roots. *Z. Pflanzenphysiol*. 56: 233–45.

——. 1968. Electrostatic association and Donnan phenomena as mechanisms of ion accumulation. *Plant Physiol*. 43: 893–901.

——. 1969. Accumulation of potassium and sodium by barley roots in a K-Na replacement series. *Plant Physiol*. 44: 1528–32.

——. 1970a. An anomaly in potassium accumulation by barley roots. I. Effect of anions, Na concentration, and length of absorption period. *Plant Physiol*. 45: 408–10.

——. 1970b. An anomaly in potassium accumulation by barley roots. II. Effect of calcium concentration and ^{86}Rb concentration. *Plant Physiol*. 45: 411–14.

Hiatt, A. J., and S. B. Hendricks. 1967. The role of CO_2 fixation in accumulation of ions by barley roots. *Z. Pflanzenphysiol*. 56: 220–32.

Hiatt, A. J., and R. H. Lowe, 1967. Loss of organic acids, amino acids, K, and Cl from barley roots treated anaerobically and with metabolic inhibitors. *Plant Physiol*. 42: 1731–36.

Hooymans, J. J. M. 1964. The role of calcium in the absorption of anions and cations by excised barley roots. *Acta Bot. Neerl*. 13: 507–40.

Hurd, R. G., and J. F. Sutcliffe. 1957. An effect of pH on uptake of salt by plant tissue. *Nature* 180: 233.

Jackson, P. C., and H. R. Adams. 1963. Cation-anion balance during potassium and sodium absorption by plant roots. *J. Gen. Physiol*. 46: 369–86.

Jackson, P. C., and K. J. Stief. 1965. Equilibrium and ion exchange characteristics of potassium and sodium accumulation by barley roots. *J. Gen. Physiol*. 48: 601–16.

Jackson, W. A., and N. T. Coleman. 1959. Ion absorption by bean roots and organic acid changes brought about through CO_2 fixation. *Soil Sci*. 87: 311–19.

Jackson, W. A., and Doris Craig Williams. 1968. Nitrate-stimulated uptake and transport of strontium and other cations. *Soil Sci. Soc. Amer. Proc*. 32: 698–704.

Jacobson, L. 1955. Carbon dioxide fixation and ion absorption in barley roots. *Plant Physiol*. 30: 264–68.

Jacobson, L., R. J. Hannapel, D. P. Moore, and M. Schaedle. 1961. Influence of Ca on selectivity of ion absorption process. *Plant Physiol*. 36: 58–61.

Jacobson, L., and L. Ordin. 1954. Organic acid metabolism and ion absorption in roots. *Plant Physiol*. 29: 70–75.

Jacoby, B., and G. Laties. 1971. Bicarbonate fixation and malate compartmentation in relation to salt-induced stoichiometric synthesis of organic acid. *Plant Physiol*. 47: 525–31.

Jenny, H. 1935. *J. Phys. Chem.* 36: 2217–58.

Kahn, J. S., and J. B. Hanson. 1957. The effect of calcium on potassium accumulation in corn and soybean roots. *Plant Physiol.* 32: 312–16.

Kirkby, E. A., and K. Mengel. 1967. Ionic balance in different tissues of the tomato plant in relation to nitrate, urea, or ammonium nutrition. *Plant Physiol.* 42: 6–14.

Leggett, J. E., and E. Epstein. 1955. Kinetics of sulfate absorption by barley roots. *Plant Physiol.* 31: 222–26.

Leggett, J. E., and W. A. Gilbert. 1969. Magnesium uptake by soybeans. *Plant Physiol.* 44: 1182–86.

Leggett, J. E., W. R. Heald, and S. B. Hendricks. 1965. Cation binding by baker's yeast and resins. *Plant Physiol.* 40: 665–71.

Leggett, J. E., and R. A. Olsen. 1964. Anion absorption by baker's yeast. *Plant Physiol.* 39: 387–90.

Leggett, J. E., R. A. Olsen and B. D. Spangler. 1962. Cation absorption by baker's yeast as a passive process. *Proc. Nat. Acad. Sci.* 11: 1949–56.

Maas, E. V. 1969. Calcium uptake by excised maize roots and interactions with alkali cations. *Plant Physiol.* 44: 985–89.

Maas, E. V., and Gen Ogata. 1971. Absorption of magnesium and chloride by excised corn roots. *Plant Physiol.* 47: 357–60.

Mengel, K., and M. Helal. 1967. Der Einfluss des austauschbaren Ca^{2+} junger Gerstenwurzeln auf den Flux von K^+ und Phosphat—eine Interpretation des Viets-Effektes. *Z. Pflanzenphysiol.* 57: 223–34.

Moore, D. P., L. Jacobson, and R. Overstreet. 1961a. Uptake of calcium by excised barley roots. *Plant Physiol.* 36: 53–57.

——. 1961b. Uptake of magnesium and its interaction with calcium in excised barley roots. *Plant Physiol.* 36: 290–95.

Olsen, R. A. 1968. The driving force of an ion in the absorption process. *Soil Sci. Soc. Amer. Proc.* 32: 660–64.

——. 1969. Cation equilibration in yeast. *Soil Sci. Soc. Amer. Proc.* 33: 983–84.

Olsen, R. A., and S. Tripp. 1969. Ion behavior in yeast. I. The cell as an exchanger system. *Soil Sci. Soc. Amer. Proc.* 33: 410–12.

Pentzer, W. T. 1924. Color pigment in relation to development of Jonathan spot. *Proc. Amer. Soc. Hort. Sci.* 22: 66–69.

Rains, D. W., W. E. Schmid, and E. Epstein. 1964. Absorption of cations by roots. Effect by hydrogen ions and essential role of calcium. *Plant Physiol.* 39: 274–78.

Reisenauer, H. M. 1966. Mineral nutrients in soil solution. In *Environmental Biology,* ed. P. L. Altman and D. S. Dittmer, p. 507. Bethesda, Md.: Fed. Amer. Soc. Exp. Biol.

Richmond, A. E., D. R. Dilley, and D. H. Dewey. 1964. Cation, organic acid, and pH relationships in peel tissue of apple fruits affected with Jonathan spot. *Plant Physiol.* 39: 1056–60.

Sawhney, B. L., and I. Zelitch. 1969. Direct determination of potassium ion accumulation in guard cells in relation to stomatal opening in light. *Plant Physiol.* 44: 1350–54.

Schaedle, M., and L. Jacobson. 1965. Ion absorption and retention by *Chlorella pyrenoidosa*. I. Absorption of potassium. *Plant Physiol.* 40: 214–20.

Solomon, A. K. 1960. Measurements of equivalent pore radius in cell membranes. In *Membrane Transport and Metabolism,* ed. A. Kleinzeller and Z. Katyk, pp. 94–99. New York: Academic Press.

Ting, I. P., and W. M. Dugger, Jr. 1967. CO_2 metabolism in corn roots, I. Kinetics of carboxylation and decarboxylation. *Plant Physiol.* 42: 712–18.

Ulrich, A. 1941. Metabolism of non-volatile organic acids in excised barley roots as related to cation-anion balance during salt accumulation. *Amer. J. Bot.* 28: 526–37.

——. 1942. Metabolism of organic acids in excised barley roots as influenced by temperature, oxygen tension and salt concentration. *Amer. J. Bot.* 29: 220–27.

Waisel, Y. 1962a. The absorption of Li and Ca by barley roots. *Acta Bot. Neer.* 11: 56–68.

——. 1962b. The effect of Ca on the uptake of monovalent ions by excised barley roots. *Physiol. Plant.* 15: 709–24.

Whittenbury, G., N. Sugino, and A. K. Solomon. 1960. Effect of antidiuretic hormone and calcium on the equivalent pore radius of kidney slices from *Necturus*. *Nature* 187: 699–701.

6. Physiological Effects of pH on Roots

David P. Moore

PLANT roots are subjected to a wide variation in the pH of the medium within the normal physiological pH range. Inherent soil pH is a result of the factors that determine soil development, and therefore the effects of pH on plant roots are often confounded with other chemical properties of the soil. For this reason, most studies of pH effects have been done in nutrient solutions where pH and other variables can be more precisely established and maintained.

I. EFFECT OF pH ON ROOT GROWTH

A direct effect of pH on root growth was illustrated in experiments conducted by Arnon and Johnson (1942). Three species, bermudagrass (*Cynodon dactylon*), tomato (*Lycopersicon esculentum* Mill.), and lettuce (*Lactuca sativa*), were grown in nutrient solutions in which the pH was varied from 3 to 9. The pH of each treatment was maintained within $\pm$ 0.2 of a unit by daily adjustment, frequent solution changes, and use of large volumes of solution per plant. At pH 3 all the species showed a complete lack of root growth. The roots of the seedlings were severely damaged and collapsed soon after their exposure to this pH. Substantial root growth occurred at pH 4, and in the case of bermudagrass the maximum root growth occurred at pH 4. Root growth of tomatoes and lettuce was about half of that obtained at higher pH values, with a marked reduction in root growth occurring at pH 9; nevertheless, there was still a modest amount of growth. It is not clear to what extent reduced availability of the metal micronutrients may have been responsible for the reduced growth at high pH values since there was a steady decline in growth at pH values above 5.

At pH values of 4 and 5 additional calcium in the nutrient solution resulted in substantial improvement in the growth. This enhancement by Ca was not obtained at pH 6, suggesting that Ca may offset the harmful effects of H^+ (see section III). Sutton and Hallsworth (1958) also reported that Ca decreased the toxicity of H^+ in nutrient solution experiments. Their results showed that H^+ was much more toxic when there was a rapid renewal of the solution immediately adjacent to the roots.

Calcium was not as effective in preventing the above H^+ damage as it was when the roots were grown in sand culture or in agar at a comparable pH. Jackson (1967) suggested that this may be due to an increase in the pH immediately adjacent to the roots as a result of greater anion than cation uptake. In contrast to a nutrient solution, this layer of higher pH would not be readily dissipated in sand culture or agar. Greater anion uptake as compared to cation uptake in acid solutions has been reported by Jacobson *et al.* (1957) for excised barley (*Hordeum vulgare*) roots.

Ekdahl (1957) reported that growth of root hairs was more sensitive to pH changes than was root elongation. Increasing the pH from 5.5 up to 7.2 resulted in only a 10% increase in the rate of elongation of wheat (*Triticum aestivum*) roots. However, it should be pointed out that pH 5.5 is not an especially harsh H^+ environment. In contrast, root hair length was increased by over 40% over the same pH range. Root hair diameter was not affected. The distance back of the root tip to the first root hairs was also affected by the pH of the solution. At pH 5.5 the length of the hairless tip zone was 2.6 mm; at pH 7.2 it was 3.5 mm long.

Kerridge (1969) has reported small reductions in root elongation and root yield due to H^+. He grew wheat plants in nutrient solution and rigidly maintained the pH at 4.0 and 5.0. The plants showed only a negligible difference in dry weight of roots produced in 26 days. Root length was slightly longer at pH 5 than at pH 4. The difference in root length was somewhat greater at 14 days, suggesting that the plants may have acquired some tolerance to acidity with further age. Burstrom (1952) has reported a small reduction in root elongation due to increased H^+.

Although the data show that root growth is affected by extremes of pH, root growth is only negligibly affected by pH in the range from 4 to 8 if sufficient Ca is available and if excess toxic ions such as aluminum and manganese are not present. It is widely recognized that excessive levels of Al and Mn are major contributing factors in poor plant growth on acid soils (Jackson, 1967).

II. EFFECT OF pH ON ION TRANSPORT

The pH of the ambient solution markedly affects both cation and anion absorption by roots. In short-term experiments with excised roots, the maximum absorption rate occurs at pH values from 5 to 7. Below pH 5 cation absorption is sharply reduced by H^+. This same behavior has been noted for potassium uptake (Fawzy *et al.*, 1954; Jacobson *et al.*, 1960; Jacobson *et al.*, 1957; Murphy, 1959; Nielsen and Over-street, 1955), for lithium, sodium, rubidium, and cesium uptake (Jacobson *et al.*, 1960), for magnesium uptake (Moore *et al.*, 1961b), for calcium uptake (Maas, 1969), and for manganese uptake (Maas *et al.*, 1968). Similar behavior for cation uptake by intact plants was reported by Arnon *et al.* (1942) and by Olsen (1953).

Although maximum cation absorption rates occurred in the range from pH 5 to pH 7, higher pH values did not materially depress cation uptake in short-term experiments. Apparently the cation absorption mechanism *per se* is not affected by OH^-. Jacobson *et al.* (1957) reported that K uptake by barley roots maintained near maximum rates from pH 6 up to a pH above 10. Similar results were obtained for Li, Na, Rb, and Cs (Jacobson *et al.*, 1960) and for Mg (Moore *et al.*, 1961b).

In contrast to cation absorption, anion absorption is relatively less affected by H^+ but more strongly affected by OH^- in short-term experiments. Jacobson *et al.* (1957) reported that Br^- uptake by barley roots was maximum at about pH 5 and declined steadily as the pH was increased to about 10.5. Bromide uptake decreased at pH values below 5, but not as markedly as K uptake. Similar results were reported for Cl^- uptake by corn (*Zea mays*) roots in comparison to Ca uptake (Maas, 1969). Stout *et al.* (1951) showed that molybdenum uptake by tomatoes was sharply decreased by increasing OH^- concentration. Hagen and Hopkins (1955) reported a similar effect of OH^- on uptake of both $H_2PO_4^-$ and HPO_4^{2-} by barley roots. Although the effects of alkaline pH values on anion absorption are usually attributed to OH^- competition, at least part of the effects may be due to HCO_3^- (Hurd, 1958) since the amount of HCO_3^- found in solution also increases with pH.

One of the early formulations of the carrier theory took into consideration the effects of H^+ and OH^- on ion absorption (Jacobson *et al.*, 1950). Jacobson *et al.* expressed their model in the form of generalized equations analogous to chemical reactions:

$$HR + M^+ \rightleftharpoons MR + H^+ \text{ (for cations)}$$

and

$$R'OH + A^- \rightleftharpoons R'A + OH^- \text{ (for anions)},$$

where M^+ is the cation, A^- is the anion, and R and R' are hypothetical carrier molecules. This model adequately accounts for the effects of pH on absorption since it predicts that H^+ should compete with cations and OH^- should compete with anions. Indeed, the competitive nature of H^+ and OH^- has been demonstrated by kinetic analysis of absorption isotherms (Rains *et al.*, 1964; Hagen and Hopkins, 1955).

Calcium and other polyvalent cations apparently play a crucial role in maintaining the integrity of the absorption process, especially in the acid pH range. At pH values below about 6 these polyvalent cations strongly stimulated K absorption by excised barley roots (Fawzy *et al.*, 1954; Viets, 1944). Jacobson *et al.* (1960) showed that this stimulatory effect of Ca occurred only for certain cations and was limited to acid solutions. For instance, Ca stimulated K, Rb, and Cs absorption at pH 6 and below but actually inhibited absorption of these ions at pH values above neutrality. Thus for these cations Ca appeared to decrease the competitive effects of H^+ on absorption. Calcium did not stimulate Na absorption except in very acid solutions (pH 3), and it completely inhibited Li absorption over the entire pH range. On the other hand, Ca had neither a stimulatory nor an inhibitory effect on Mn absorption in acid solutions (Maas *et al.*, 1969).

In addition to its interaction with H^+, Ca has a general regulatory effect on ion absorption. In the absence of Ca, Li and Na both reduced the absorption of K; whereas in the presence of Ca, Li had no effect on K absorption and the effect of Na was greatly reduced (Jacobson *et al.*, 1960, 1961a, 1961b). Apparently Ca reduced the competition of Li and Na with K and thereby greatly increased the selectivity of the tissue for K. Other polyvalent cations had a similar effect on selectivity (Jacobson *et al.*, 1961a). Calcium markedly reduced the absorption of Mg by excised roots, but the effect could not be explained as competition for the same transport site and was therefore probably an effect on selectivity (Moore *et al.*, 1961b). Calcium by itself had no effect on Mn absorption by barley roots; but in the presence of Mg, Ca depressed the uptake of Mn (Maas *et al.*, 1969). In the same system Mg depressed Mn absorption to a greater extent in the presence of Ca than in its absence. Maas *et al.* concluded that this complex interaction was noncompetitive.

Although a number of polyvalent cations exhibit regulatory effects on ion transport, Ca is probably the single most important one in maintaining the integrity of the absorption mechanism (Epstein, 1961, 1962; Jacobson *et al.*, 1961a; Rains *et al.*, 1964). Calcium is unique among the polyvalent cations in that it is generally nontoxic. Wide variation in Ca level of the medium is common in nature without adverse effects *provided* excess levels of toxic ions are not present. Calcium is the only cation that can occupy the bulk of the exchange complex without toxic effects, especially in moderate to high exchange capacity soils. Many plants can grow quite well in calcareous soils containing free $CaCO_3$ or gypsum or in soils containing 20–30 meq of exchangeable Ca. Thus exchangeable Ca can be used over a wide range of exchange capacities to replace other exchangeable cations that are toxic when present in excessive amounts.

III. H$^+$ INJURY

In addition to its competitive effects in ion absorption, H$^+$ can be damaging to roots. At pH values below about 4, H$^+$ causes a loss of previously absorbed ions from root tissue. Sizable losses of K from roots exposed to low pH in short-term experiments have been reported (Fawzy *et al.*, 1954; Jacobson *et al.*, 1950, 1957, 1960; Nielsen and Overstreet, 1955). Similar results were reported for Mg (Moore *et al.*, 1961b) and Ca (Jacobson *et al.*, 1950; Moore *et al.* 1961a). Low pH also caused a loss of inorganic phosphorus, organic phosphorus, and soluble nitrogen from barley roots, which suggests that H$^+$ generally increased the permeability of the cell membranes and allowed cell constituents to leak out. At high temperatures H$^+$ was much more damaging to the tissue (Jacobson *et al.*, 1957).

Calcium and other polyvalent cations have been shown to protect the root tissue somewhat from the injurious effects of low pH (Fawzy *et al.*, 1954; Jacobson *et al.*, 1960). Apparently H$^+$ damage to the root in the absence of Ca is partially reversible (Rains *et al.*, 1964). At pH 3.9 without Ca, the rate of Rb absorption of barley roots decreased progressively with time until it was practically zero at the end of an hour. Adding Ca after 30 minutes partially restored the absorptive capacity, but the rate was somewhat less than where Ca was present from the beginning.

Considerable loss of K from corn root tips at normal physiological pH values was reported by Marschner *et al.* (1966). The K loss decreased as the pH was increased from 5.5 to 8, and Ca greatly reduced the loss of K. The effect of Ca was greatest at the lower pH and essentially disappeared as the pH increased to about 8. At pH 4.4 and below considerable damage to the tissue occurred in the absence of Ca. There was severe derangement of the fine structure of the cytoplasm due to H$^+$ injury, but this could largely be prevented by a low concentration of Ca (0.01 meq/1).

Thus it appears that Ca is probably indispensable not only as a regulator of selective ion transport but indeed in maintaining the integrity of the membranes. Hydrogen ions adversely affect both the ion transport mechanism and the permeability of cell membranes, and the strong interaction between Ca and H$^+$ suggests a common site of action.

IV. EFFECTS OF ROOT ACTIVITY ON THE pH OF
THE MEDIUM

Although the effects of pH on root processes are well known, the reverse, i.e., the effect of root activity on the pH of the medium, is often ignored. As will be seen, the change in pH around roots may have profound effects on the roots themselves.

The necessity for maintaining electrical neutrality and cation-anion balance in plant tissues and the attendant biochemical processes that control these phenomena have been well documented in this volume (Chapters 3 and 5). The net result of excess cation absorption is the net release of H^+ from the root, while the result of net excess of anion absorption is the release of OH^- or HCO_3^-. Hoagland and Broyer (1940) measured the pH changes that occurred when plants were grown in a complete nutrient solution. Initially the pH was slightly above 5, and it decreased within 5 hours to below 4. Then it rose, until after 20 hours it was approximately 7. They concluded that the pH drop was due to absorption of more cations than anions; the subsequent increase in pH resulted from absorption of more anions than cations. As the pH drops, anion absorption is less affected than cation absorption, and therefore the pH changes due to unbalanced uptake would be somewhat muted. A similar situation would exist as the pH rose; i.e., cation absorption would be less affected than anion absorption.

The so-called physiological acidity or alkalinity of a salt depends upon which ion of the salt, the cation or the anion, is most rapidly absorbed. Generally the monovalent cations are rapidly absorbed (Jacobson *et al.*, 1960), whereas the divalent cations, especially Ca, are more slowly absorbed (Maas, 1969; Moore *et al.*, 1961a, 1961b). The monovalent anions are generally more rapidly absorbed than the polyvalent anions (Hagen and Hopkins, 1955; Jacobson *et al.*, 1957; Leggett and Epstein, 1956; Stenlid, 1957). Thus a salt like K_2SO_4 would be physiologically acid since the K would enter roots more rapidly than the SO_4^{2-}. By the same token, $CaCl_2$ should be physiologically alkaline since the Ca enters slowly and the Cl^- rapidly.

Pitman (1970) reported that H^+ efflux from barley roots in K_2SO_4 solutions was about twice as rapid as that from roots in KCl solutions. These changes correlated well with the imbalance in cation-anion absorption rates. Although K and Cl^- are both classed as rapidly absorbed, apparently KCl is physiologically acid. Similar results were obtained for KBr (Jacobson *et al.*, 1961b). At high root:solution ratios, the pH of a 0.005 N KBr solution dropped from 6 to 4.3 in 3 hours. However, the physiological acidity of KCl was found to be concentration dependent. Pitman (1970) reported that net H^+ efflux was progressively higher as the KCl concentration increased from 0.1 mM to 10 mM. Hiatt (1967) reported that KCl could be either physiologically acid or alkaline depending on the concentration. At 10^{-5} M, KCl was almost neutral, at 10^{-4} and 10^{-3} M it was alkaline, and at 10^{-2} it was slightly acidic. Whether KCl is physiologically acid or alkaline depended entirely upon the balance between the K and the Cl$^-$ absorption (Hiatt, 1967; Pitman, 1970).

Calcium chloride was found to be consistently physiologically alkaline over a 100-fold concentration range (Hiatt, 1967). On the other hand, $CaSO_4$ was physiologically neutral (Pitman, 1970); this would be expected since both Ca and SO_4^{2-} are very slowly absorbed.

Aside from their significance to the ion transport mechanism and to the biochemical processes involved in cation-anion balance, changes in pH can also have profound effects on the external root environment. The gross pH changes noted in the bulk nutrient solution are a reflection of greater changes occurring immediately adjacent to the root surface or within the interstitial regions of the root cortex. Normally H$^+$ and OH$^-$ (or HCO$_3^-$) released from the root would be rapidly dissipated in the bulk nutrient solution. Depending on the rate of release and the rate of solution agitation, the gradient from the root surface could be substantial.

The pH changes in the root environment due to root activity are probably of considerably greater significance in soil systems than in nutrient culture. In nutrient solution the potential quantity of any ionic species is fixed, whereas in the soil system additional quantities of ions may be brought into solution (see Chapters 15 and 16). The possibilities for dissipation of H$^+$ or OH$^-$ are greatly reduced because the solution immediately adjacent to the root is relatively static. Mass-flow in the soil is usually toward the root of a transpiring plant, and therefore substantial pH changes could occur and persist adjacent to the root when there is a net release of H$^+$ or OH$^-$ (Sutton and Hallsworth, 1958).

Not only is the solubility of ions affected by pH, but the specific ionic form is also pH dependent. For instance, in the physiological pH range (4 to 9), two species of phosphate ions dominate, and their proportions are determined by pH. At pH 4, 98.6% of the phosphate is in the $H_2PO_4^-$ form and only 0.06% is in the divalent form, HPO_4^{2-}. At pH 9 only 1.5% of the phosphate is $H_2PO_4^-$ and 98.5% is HPO_4^{2-}. Furthermore, the monovalent form is absorbed more rapidly than the divalent form (Hagen and Hopkins, 1955). The two forms are absorbed via separate sites, both of which are competitively inhibited by OH$^-$. Thus as the pH is increased, the rate of P absorption is decreased through the direct action of OH$^-$ on ion absorption as well as by a change in the ionic species. A change in ionic species due to pH changes in the root zone has been proposed to account for the stimulation in P uptake by ammonium ions (see Chapter 21).

The relationship between pH and the toxicity of ammoniacal nitrogen is reviewed in Chapter 22 below. The pH of the system controls the distribution of ammoniacal N between the NH_4^+ form and NH_3 form. The latter is quite toxic to roots (Colliver and Welch, 1970; Warren, 1962), apparently because it is a neutral molecule and can readily penetrate the cell membranes. The ability of the roots to release a net excess of H^+ in the root zone would obviously lessen the toxicity from NH_3. Since it is a monovalent cation, NH_4^+ absorption should be relatively rapid and should lead to a reduction in pH around the root.

A striking example of root response to an ionic species controlled by pH is the toxicity of Al. The many effects that soluble Al has on plant growth are reviewed in Chapter 20; it appears that root growth is the most sensitive indicator of Al toxicity (Kerridge *et al.*, 1971). The first sign of Al toxicity is severe stunting and thickening of the roots, probably due to an inhibition of cell division (Clarkson, 1965). This characteristic has been used as a criterion for the identification of sensitive and tolerant varieties of cereals (Fleming and Foy, 1968; Kerridge *et al.*, 1971; Reid, 1971). Using root growth as an indicator of toxicity, Kerridge (1969) showed that a concentration of 2 ppm Al was considerably more toxic to wheat roots at pH 4.5 than at pH 4.0. He suggested that one of the hydrolysis products of Al^{3+} was responsible for the toxicity. To test this possibility, I have conducted experiments in which both pH and Al concentration were variables.

The basic technique described by Kerridge *et al.* (1971) was followed except for one essential modification. In their experiments the wheat roots were exposed continuously to a constant Al concentration, and the total length of root was used as a measure of root growth. In my experiments, wheat plants were started in an Al-free solution until the root length was 3–5 cm (about 48 hours after germination). The plants were then transferred to nutrient solutions containing Al for 48 hours. The Al solutions were of the same nutrient composition as the initial nutrient solutions except that P was omitted and Fe was added as $FeCl_3$ instead of as the chelate. After the 48-hour exposure to Al, the length of the primary root was measured, and the plants were returned to the original Al-free solutions. The plants were allowed to recover for 72 hours, and root length was again measured. The elongation of the primary root during the recovery period was used as the indicator of Al toxicity to roots. This was a very sensitive indicator since irreversible damage could be readily evaluated. Clarkson (1965) and Fleming and Foy (1968) had shown that primary roots did not recover when exposed to toxic levels of Al.

The experiments were conducted in the greenhouse at 21C to 27C with supplemental lighting to lengthen the day to 16 hours. The pH was rigorously maintained at the desired value by adjustment twice a day. With the large volume of solution used and the repeated adjustment, pH values were maintained within ± 0.02 units.

Two varieties of wheat were used, Brevor and Druchamp, which differed in their tolerance to Al (Kerridge *et al.*, 1971). Brevor is a sensitive variety, and Druchamp is a moderately tolerant variety. A single gene has been shown to control the response to Al of these two varieties (Kerridge and Kronstad, 1968). In a cross between these two varieties, the F_1 plants were moderately tolerant (like Druchamp), and the F_2 plants segregated on a 3:1 ratio of moderately tolerant to sensitive plants. Therefore, moderate tolerance was dominant over sensitivity, and a single major gene was probably involved.

Table 6.1 shows the results for the moderately tolerant variety Druchamp. When Al was omitted from the treatment solutions, subsequent root elongation was essentially independent of the pH. By contrast, when the plants were exposed to Al for 48 hours, subsequent root recovery was strongly dependent on the pH of the system. For any particular treatment the pH of the Al and Al-free solutions was maintained at the same value. At 2 ppm Al the pH 4.5 treatments showed a significant reduction in root recovery as compared to the pH 4.0 to 4.3 treatments. There was a slightly better recovery at pH 4.7 than at pH 4.5, and calculations showed that the solubility of Al had probably been exceeded. At 4 ppm a reduction in root elongation was evident in the pH 4.3 treatment, and a severe reduction was noted at pH 4.5 Again, the pH 4.7 treatment showed better recovery, and a visible precipitate was noted in this treatment solution. A similar pattern

Table 6.1. Primary root elongation within 72 hours after removal of wheat (*Triticum aestivum*, cv. 'Druchamp') plants from aluminum treatment solutions

pH	Root length (mm)			
	0 ppm Al	2 ppm Al	4 ppm Al	6 ppm Al
4.0	77	89	83	79
4.2	80	95	86	66
4.3	83	80	50	17
4.5	79	55	25	8†
4.7	85	62†	(66)†	(53)†

† Treatments where calculation predicts solubility of Al exceeded. Parentheses indicate treatments where visible precipitation was noted.

was noted at 6 ppm except that the effects of pH were more striking. Very little root elongation occurred in the pH 4.5 treatment, and some effect was noted at pH 4.2. An obvious precipitate was present in the pH 4.7 Al treatments.

The results for the variety Brevor are given in Table 6.2. In the absence of an Al treatment, pH had no effect on root elongation. The effect of pH of the Al treatment on subsequent root elongation was even more striking for this variety than for Druchamp. This was expected since Brevor is classed as an Al-sensitive variety. Very little root recovery was noted at 2 ppm and pH 4.5. At 4 ppm the roots showed absolutely no recovery in the pH 4.3 and 4.5 treatments, whereas at 6 ppm the results were similar except that pH 4.2 showed very little subsequent root growth. Again, root recovery was substantial where the solubility of Al was exceeded, i.e., at pH 4.7.

From these data it is obvious that the toxicity of a given concentration of Al was increased by increasing the pH of the system from 4.0 up until the solubility of Al was exceeded. Furthermore, the relative effects on the sensitive and the moderately tolerant varieties were similar. These results suggest that it was indeed a hydrolysis product of Al rather than the Al^{3+} form that was responsible for inhibiting root growth.

Table 6.2. Primary root elongation within 72 hours after removal of wheat (*Triticum aestivum*, cv. 'Brevor') plants from aluminum treatment solutions

	Root length (mm)			
pH	0 ppm Al	2 ppm Al	4 ppm Al	6 ppm Al
4.0	85	41	20	7
4.2	89	36	7	2
4.3	88	31	0	0
4.5	81	2	0	0†
4.7	87	38†	(31)†	(25)†

† Treatments where calculation predicts solubility of Al exceeded. Parentheses indicate treatments where visible precipitation was noted.

Figure 6.1 shows the theoretical relationship between pH and concentration for the two major ionic species of Al in the pH range from 4.0 to 5.0. These curves were obtained by taking a fixed concentration of Al at pH 4.0 and calculating the change in composition as the pH was increased to 5.0. The assumption of ideal behavior was made to simplify the calculations.

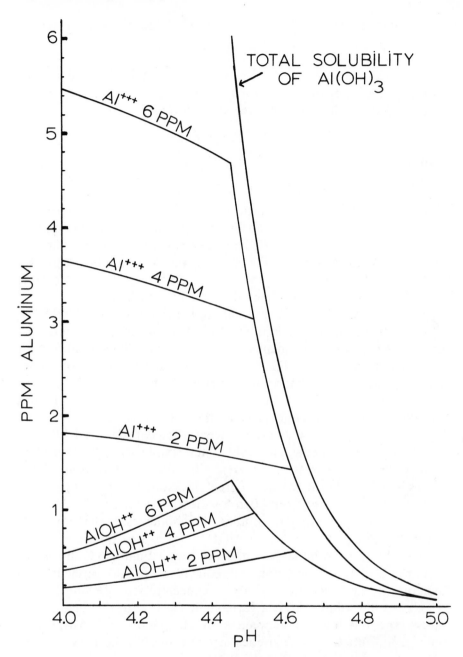

Fig. 6.1. Calculated relationship between pH and concentration of Al^{3+} and $AlOH^{2+}$

The total solubility of Al is controlled by the following reaction:

$$Al(OH)_3 \rightleftharpoons Al^{3+} + 3\ OH^-.$$

For these calculations, a pK of 32.3 was used (Raupach, 1963a). The distribution of the soluble Al between Al^{3+} and the $AlOH^{2+}$ forms is controlled by the following reaction (Raupach, 1963b):

$$Al^{3+} \rightleftharpoons AlOH^{2+} + H^+, \ pK = 5.0.$$

Therefore,

$$[Al^{3+}]/[AlOH^{2+}] = [H^+] \times 10^5.$$

This equation shows that the ratio of Al^{3+} to $AlOH^{2+}$ is determined solely by the pH of the system. Using this equation, the amounts of these two ions can be calculated for any particular concentration of soluble Al. Thus, at 6 ppm (Fig. 6.1), all the Al is soluble at pH 4.0 but there is 10 times as much Al^{3+} as $AlOH^{2+}$. As the pH increases above 4.0, the Al^{3+} form declines along the line marked "Al^{3+} 6 ppm," while the $AlOH^{2+}$ form increases along the line marked "$AlOH^{2+}$ 6 ppm." Until the solubility limit is reached (between pH 4.4 and 4.5), the total Al in solution remains at 6 ppm. When the solubility limit is reached, both forms decline in concentration. The curves for 4 ppm and 2 ppm are similar, but a higher pH is required before the limit of solubility is reached. Once the limit of solubility is reached, all concentrations follow the same curve as the pH is increased. At pH 5.0 the solubility is quite low, but the concentrations of Al^{3+} and $AlOH^{2+}$ have become equivalent. The only other simple hydrolysis product possible in these systems is $Al(OH)_2^+$, but the concentration of this species is so low that it can be ignored (Jackson, 1963).

As the pH is increased from 4.0 to 4.5, the Al^{3+} decreases, the $AlOH^{2+}$ increases, and the toxicity of soluble Al also increases; therefore, it is reasonable to conclude that the $AlOH^{2+}$ form is responsible for the adverse effects of soluble Al on plant roots. Figure 6.2 shows the relationship of root elongation (subsequent to the Al treatment) as a function of the calculated $AlOH^{2+}$ concentration. Obviously pH, as well as total soluble Al, is a variable in this plot of the data. Nonetheless, the relationship between root elongation and $AlOH^{2+}$ is quite good, and the two varieties separate nicely according to their tolerance to Al (Kerridge *et al.*, 1971).

The existence of a single major gene controlling the difference in response between Brevor and Druchamp varieties suggests a highly specific yet simple mechanism of Al tolerance. The concept of a single gene controlling the formation of a single polypeptide or protein is well recognized. Furthermore, the fact that tolerance was dominant over sensitivity (Kerridge and Kronstad, 1968; Reid, 1971) suggests

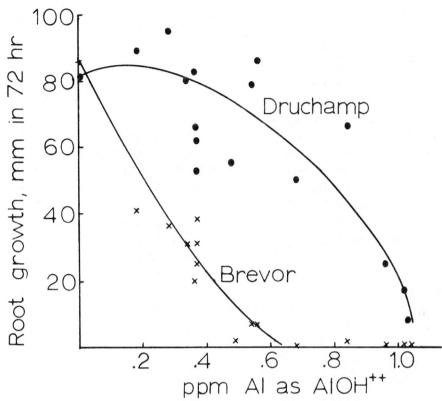

Fig. 6.2. Relationship between calculated concentration of AlOH^{2+} and root elongation (subsequent to Al treatment) for a sensitive (Brevor) and a moderately tolerant (Druchamp) variety of wheat (*Triticum aestivum*)

that an exclusion phenomenon may be responsible for tolerance to Al. The expression of the recessive allel is usually attributed to the failure to form the complete or active protein in the homozygous individual. If the ability to exclude Al depended on formation of a complete protein in the plasmalemma membrane, for instance, then tolerance would be dominant over sensitivity. In other words, sensitivity could be due to the failure to form a complete membrane. Exclusion of Al is also suggested by the fact that even the tolerant varieties can be affected if the Al concentration is high enough. Increasing the Al concentration apparently overrides the mechanism of tolerance. An alternative to the exclusion hypothesis is some sort of repair mechanism or means to prevent damage by Al once it is inside the cell. In this case sensitivity should be recessive if the protein responsible for repair or for preventing damage is incomplete or nonfunctional. It is unlikely, however, that Al affects only one system inside the cell, as would be demanded by the one-gene relationship. It seems more likely that once Al is inside the cell, any number of reactions may be inhibited. Therefore, the one-gene relationship most logically fits the exclusion hypothesis. Clarkson (1969) has proposed that, among other things, Al inhibits cell division by interfering with DNA synthesis in the S period of the mitotic cycle. Such a mechanism could account for the irreversible damage done to root meristems by toxic levels of Al. Certainly an exclusion phenomenon would be consistent with Clarkson's proposal, as would the repair and/or protection concept.

Since the solubility and the ionic form of Al are both determined by the pH, it is not at all certain which of these would be the most important in soil systems. If the soil solution remains saturated, then the concentration of $AlOH^{2+}$ would be determined by the solubility and could not change except as the total soluble Al changed. However, if the soil solution were not saturated, then $AlOH^{2+}$ could change independently of the total soluble Al as the pH changes. In any event, pH changes adjacent to the root could have profound effects on root growth where sufficient soluble Al was present.

The Al^{3+}-$AlOH^{2+}$, NH_4^+-NH_3, and $H_2PO_4^-$-HPO_4^{2-} systems are probably the most important ionic forms affected by pH in the physiological pH range. Other elements for which changes in ionic form may occur on the extremes of the physiological pH range are boron, sulfur, molybdenum, and zinc, although practically nothing is known about the physiological effects of such changes.

V. SUMMARY

The physiological effects of pH on plant roots are diverse and are often confounded with other changes in the system associated with pH. Hydrogen ions have a direct effect on growth of roots and root hairs, but this effect is not large in the physiological pH range if sufficient Ca is present in the root medium. Hydrogen ions also inhibit cation absorption by competition for absorption sites and through damage to the integrity of cell membranes. Calcium, again, exerts a protective action against the harmful effects of H^+. Cation absorption is not very sensitive to OH^-, but anion absorption appears to be competitively inhibited by OH^-. There is a net efflux of H^+ from the root when more cations than anions are absorbed. When anion absorption exceeds cation absorption, there is a net release of OH^- (or HCO_3^-) from the roots. Thus the root can bring about significant pH changes in the root environment. In nutrient solution the H^+ and OH^- released may be dissipated rapidly, but in soils the pH changes probably persist. The pH of the root environment can have significant effects on the ionic species formed, and in turn these can have significant physiological effects. The Al^{3+}-$AlOH^{2+}$ system is an example of the indirect effects of pH on root behavior.

References

Arnon, D. I., W. E. Fratzke, and C. M. Johnson. 1942. Hydrogen ion concentration in relation to absorption of inorganic nutrients by higher plants. *Plant Physiol.* 17: 515–24.

Arnon, D. I., and C. M. Johnson. 1942. Influence of hydrogen ion concentration on the growth of higher plants under controlled conditions. *Plant Physiol.* 17: 525–39.

Burstrom, H. 1952. Studies on growth metabolism of roots VIII. Calcium as a growth factor. *Physiol. Plant.* 5: 391–402.

Clarkson, D. T. 1965. The effect of aluminum and some other trivalent metal cations on cell division in the root apices of *Allium cepa. Ann. Bot.*, n.s., 29: 309–15.

——. 1969. Metabolic aspects of aluminum toxicity and some possible mechanisms for resistance. In *Ecological Aspects of the Mineral Nutrition of Plants*, ed. I. H. Rorison, pp. 381–97. Oxford: Blackwell Scientific Publications.

Colliver, G. W., and L. F. Welch. 1970. Toxicity of preplant anhydrous ammonia to germination and early growth of corn: II. Laboratory studies. *Agron. J.* 62: 346–48.

Ekdahl, I. 1957. The growth of root hairs and roots in nutrient media and bidistilled water and the effects of oxalate. *K. Lantbrukshogskol. Amn.* 23: 497–518.

Epstein, E. 1961. The essential role of calcium in selective cation transport by plant cells. *Plant Physiol.* 36: 437–44.

——. 1962. Mutual effects of ions in their absorption by plants. *Agrochimica* 4: 293–322.

Fawzy, H., R. Overstreet, and L. Jacobson. 1954. Influence of hydrogen ion concentration on cation absorption by barley roots. *Plant Physiol.* 29: 234–37.

Fleming, A. L., and C. D. Foy, 1968. Root structure reflects differential aluminum tolerance in wheat varieties. *Agron. J.* 60: 172–76.

Hagen, C. E., and H. T. Hopkins. 1955. Ionic species in orthophosphate absorption by barley roots. *Plant Physiol.* 30: 193–99.

Hiatt, A. J. 1967. Relationship of cell-sap pH to organic acid change during ion uptake. *Plant Physiol.* 42: 294–98.

Hoagland, D. R., and T. C. Broyer. 1940. Hydrogen ion effects and the accumulation of salt by barley roots as influenced by metabolism. *Amer. J. Bot.* 27: 173–85.

Hurd, R. G. 1958. The effect of pH and bicarbonate ions on the uptake of salts by disks of red beet. *J. Exp. Bot.* 9: 159–74.

Jackson, M. L. 1963. Aluminum bonding in soils: A unifying principle in soil science. *Soil Sci. Soc. Amer. Proc.* 27: 1–10.

Jackson, W. A. 1967. Physiological effects of soil acidity. In *Soil Acidity and Liming*, ed. R. W. Pearson, pp. 43–124. Agronomy 12. Madison, Wis.: American Society of Agronomy.

Jacobson, L., R. J. Hannapel, D. P. Moore, and M. Schaedle. 1961a. Influence of calcium on selectivity of ion absorption process. *Plant Physiol.* 36: 58–61.

Jacobson, L., R. J. Hannapel, M. Schaedle, and D. P. Moore. 1961b. Effect of root to solution ratio in ion absorption experiments. *Plant Physiol* 36: 62–65.

Jacobson, L., D. P. Moore, and R. J. Hannapel. 1960. Role of calcium in absorption of monovalent cations. *Plant Physiol.* 35: 352–58.

Jacobson, L., R. Overstreet, R. M. Carlson, and J. A. Chastain. 1957. The effect of pH and temperature on the absorption of potassium and bromide by barley roots. *Plant Physiol.* 32: 658–62.

Jacobson, L., R. Overstreet, H. M. King, and R. Handley. 1950. A study of potassium absorption by barley roots. *Plant Physiol.* 25: 639–47.

Kerridge, P. C. 1969. Aluminum toxicity in wheat (*Triticum aestivum* Vill. Host). Ph.D. Thesis. Oregon State Univ. 170 pp. University Microfilms, Ann Arbor, Mich., *Diss. Abstr.*, B, 29: 3159.

Kerridge, P. C., M. D. Dawson, and D. P. Moore. 1971. Separation of degrees of aluminum tolerance in wheat. *Agron. J.* 63: 586–91.

Kerridge, P. C., and W. E. Kronstad. 1968. Evidence of genetic resistance to aluminum toxicity in wheat (*Triticum aestivum* Vill. Host). *Agron. J.* 60: 710–11.

Leggett, J. E., and E. Epstein. 1956. Kinetics of sulfate absorption by barley roots. *Plant Physiol.* 31: 222–26.

Maas, E. V. 1969. Calcium uptake by excised maize roots and interactions with alkali cations. *Plant Physiol.* 44: 985–89.

Maas, E. V., D. P. Moore, and B. J. Mason. 1968. Manganese absorption by excised barley roots. *Plant Physiol.* 43: 527–30.

———. 1969. Influence of calcium and magnesium on manganese absorption. *Plant Physiol.* 44: 796–800.

Marschner, H. R., R. Handley, and R. Overstreet. 1966. Potassium loss and changes in the fine structure of corn root tips induced by H-ion. *Plant Physiol.* 41: 1725–35.

Moore, D. P., L. Jacobson, and R. Overstreet. 1961a. Uptake of calcium by excised barley roots. *Plant Physiol.* 36: 53–57.

Moore, D. P., R. Overstreet, and L. Jacobson. 1961b. Uptake of magnesium and its interaction with calcium in excised barley roots. *Plant Physiol.* 36: 290–95.

Murphy, R. P. 1959. Some factors influencing cation uptake by excised roots of perennial rye grass (*Lolium perenne*). *Plant Soil* 10: 242–49.

Nielsen, T. R., and R. Overstreet. 1955. A study of the role of hydrogen ion in the mechanism of potassium absorption by excised barley roots. *Plant Physiol.* 30: 303–9.

Olsen, C. 1953. The significance of concentration for the rate of ion absorption by higher plants in water culture. IV. The influence of hydrogen ion concentration. *Physiol. Plant.* 6: 848–58.

Pitman, M. G. 1970. Active H^+ efflux from cells of low-salt barley roots during salt accumulation. *Plant Physiol.* 45: 787–90.

Rains, D. W., W. E. Schmid and E. Epstein. 1964. Absorption of cations by roots. Effects of hydrogen ions and essential role of calcium. *Plant Physiol.* 39: 274–78.

Raupach, M. 1963a. Solubility of simple aluminum compounds expected in soils. II. Hydrolysis and conductance of Al^{3+}. *Aust. J. Soil Sci.* 1: 36–45.

——. 1963b. Solubility of simple aluminum compounds expected in soils. IV. Reactions of aluminum hydroxide under acid conditions. *Aust. J. Soil Sci.* 1: 55–62.

Reid, D. A. 1971. Genetic control of reaction to aluminum in winter barley. *Barley Genetics*, 2: 409–13. Pullman, Wash.: Washington State University Press.

Stenlid, G. 1957. Some differences between the accumulation of chloride and nitrate ions in excised wheat roots. *Physiol. Plant.* 10: 922–36.

Stout, P. R., W. R. Meagher, G. A. Pearson, and C. M. Johnson. 1951. Molybdenum nutrition of crop plants. I. The influence of phosphate and sulfate on the absorption of molybdenum from soils and solution cultures. *Plant Soil* 3: 51–87.

Sutton, C. D., and E. G. Hallsworth. 1958. Studies on the nutrition of forage legumes. I. The toxicity of low pH and high manganese supply to lucerne as affected by climatic factors and calcium supply. *Plant Soil* 9: 305–17.

Viets, F. G. 1944. Calcium and other polyvalent cations as accelerators of ion accumulation by excised barley roots. *Plant Physiol.* 19: 466–80.

Warren, K. S. 1962. Ammonia toxicity and pH. *Nature* 195: 47–49.

7. Biology of the Rhizosphere

A. D. Rovira and C. B. Davey

IN THIS volume in which the physical, chemical and biological aspects of the plant root and its environment are discussed, serious consideration should be given to the biology of the root surface and the surrounding soil of the rhizosphere. All nutrients that plants obtain from soil must pass through the rhizosphere; hence the biological and chemical changes created in this zone of soil by the plant root and the resident microbes could affect the welfare of the plant.

For many years soil microbiologists were concerned mainly with the changes in the microorganisms when roots grew through the soil. Detailed studies were made of the numbers and types of bacteria and fungi in the rhizosphere compared with unplanted soil. These studies, although of considerable interest to microbiologists, aroused little interest from plant physiologists or agronomists. However, over the last decade the questions have been: What do these millions of organisms on and around plant roots do to the host plant? Are they simply scavengers living off waste products from roots, or do they have beneficial or detrimental effects on plants? There is now considerable research by microbiologists and plant physiologists directed toward answering these questions. Another aspect of the biology of the rhizosphere that has been studied intensively over the past 15 years has been the nature and amounts of organic materials originating from healthy growing roots. These materials provide the foodstuffs for the microorganisms in the rhizosphere and also have important effects on soil properties and nutrient availability.

In this chapter, we shall stress these three aspects of the biology of the rhizosphere, allowing other contributors to cover the mycorrhizal associations between plants and fungi and the effects of root pathogens on plant growth in depth. Brief mention of mycorrhizae and pathogens is unavoidable because of the important ecological niches they occupy in the rhizosphere.

I. MICROBIOLOGY OF THE RHIZOSPHERE

It is now nearly 70 years since Hiltner (1904) observed that micro-organisms were more abundant in the soil surrounding plant roots than in soil remote from the root. He called this zone of soil in which the microflora was influenced by the plant root the *rhizosphere*. Many workers have subsequently shown that the microflora of the rhizosphere differs both quantitatively and qualitatively from that in the soil beyond the influence of the root.

Consideration of the nature of the rhizosphere leads to the conclusion that it is not a uniform, well-defined volume of soil. In fact, the rhizosphere represents a poorly defined zone of soil with a microbiological gradient in which the maximum changes to the microflora occur in the soil adjacent to the root and decline with distance away from the roots. There arises, then, the question whether the microflora that colonizes the root surface is really a part of the soil population and should be considered as part of the rhizosphere.

Attempts to better define the zones of influence have led to such terms as *outer rhizosphere, inner rhizosphere, root surface*, and *rhizoplane*. The heterogeneous nature of soil makes it impossible to define precisely where each of these zones begins and ends. Even the term *rhizoplane* is not technically correct because the organisms are not in a single plane over the surface of the roots. Direct microscopy of roots grown in soil reveals that over many parts of a root the bacteria can be from 10 to 40 cells deep while other parts appear to be relatively free of micro-organisms. Often the bacteria and fungi are found arranged in lines along the axis of the root. These linear concentrations of microorganisms are often associated with the functions of the epidermal cells. Finally, some organisms, notably the mycorrhizal fungi, may traverse all of these defined zones from the outer rhizosphere (or soil proper) to several cells deep in the root cortex. Thus all the zones are interrelated and overlapping.

A. Quantitative Changes in the Rhizosphere Microflora

1. Methods of Measuring Rhizosphere Population

a. Plate Counts. The traditional method of sampling the rhizosphere is to free the roots of much of the adhering soil by vigorous shaking, suspend the roots plus "firmly adhering soil" in a given volume of diluent, and prepare a dilution series from which aliquots are taken for counting. A comparison is made of the counts per gram of firmly adhering soil (R) with the counts per gram of soil taken some distance from the root (S). The ratio of these two counts was termed the *R:S ratio* by Katznelson (1946) and is the most widely accepted expression of the extent of the rhizosphere effect. Considering the greater concentration of organisms at the root surface, it is obvious that the vigor with which the sample is shaken will markedly affect the R:S ratio. When a small amount of soil is included with the roots in the rhizosphere sample, a higher R:S ratio is obtained than when a large amount of soil remains on the roots. Clark (1947) found that the R:S ratio of identical plants could be made to vary considerably by simply varying the amount of soil adhering on the roots at the time of suspension in diluent. Thus, although there is little doubt that the R:S ratio expresses the degree to which roots influence the soil microflora, caution is required in its use because of the ease with which simple manipulation of the sample can alter the ratio. It would certainly be unwise to make comparisons between results of different workers on the basis of R:S ratios, and comparisons between different plant species need to be confined to those with reasonably similar root systems with approximately similar root-to-soil weight ratios.

Because of the heterogeneous nature of the rhizosphere it is unlikely that any precise and simple sampling method will be found. The most satisfactory methods so far described are those of Ishizawa *et al.* (1957) and Louw and Webley (1959), who defined the rhizosphere as that soil which is removed from roots when these and the firmly adhering soil are gently shaken in sterile water, while the rhizoplane microflora is obtained by transferring these gently washed roots to a second flask and shaking vigorously. No doubt even with this method organisms will be washed from the roots' surface into the rhizosphere sample, but even so the results obtained for the rhizosphere population will be reasonably accurate and should form the basis of comparison with control soil. Also, the rhizoplane microflora estimated by this method will include many rhizosphere organisms since the roots cannot be washed free of all soil before the final shaking for the rhizosphere count.

The serial washing techniques used by Harley and Waid (1955), Parkinson *et al.* (1963), and Brown *et al.* (1962) to assess the rhizoplane populations of fungi and *Azotobacter* provide valuable information on the organisms that are tightly held to the root surface. Rouatt and Katznelson (1961) distinguished between the rhizosphere and rhizoplane microflora by shaking roots and soil to provide the rhizosphere sample and then, after several washings, macerating and suspending the roots to provide the rhizoplane sample. In the interpretation of these results consideration should be given to the likelihood that, at least for bacteria, the successive vigorous washings of the roots not only will remove the outer layers of the rhizoplane population but also will wash off colonies and cells intimately associated with a root but less adherent than those persisting through all the treatments. The ability of an organism to penetrate the cortical cells of the host or produce gum would enhance its chances of being rated a rhizoplane inhabitant.

Although the gentle washing technique used to separate the rhizosphere and rhizoplane microflora is considered to be the most satisfactory, the majority of the rhizosphere results are based on counts obtained on the combined sample of roots and firmly adhering soil without distinction between the rhizosphere and rhizoplane. In view of the problems involved in expressing results on the R:S ratio basis, what then is the best method of recording them? Expression both of numbers of organisms per gram of soil in the sample and of numbers per gram of root is probably the most satisfactory. On the soil weight basis, comparisons may be made between rhizosphere and nonrhizosphere soil (R:S ratio); while on a root weight basis, comparisons between different portions of the root system and between roots of different plants are more valid. Such comparisons on a root weight basis can be made only between root systems of similar morphology. Rovira and Stern (1961) found less variation when results were expressed on a root weight basis than on a soil weight basis. Probably the most satisfactory expression would be in terms of numbers of organisms per unit area of root surface. Unfortunately, the methodology for this determination and expression do not appear to have been developed yet.

Papavizas and Davey (1961) used a multiple-tube device for sampling 3-mm cores at distance of 0 to 22 mm from blue lupine (*Lupinus angustifolius* L.) roots. Blue lupine was selected because of its sturdy taproot and the predictable directions taken by its secondary and lateral roots during the seedling stages. Thus, these workers sampled the rhizosphere zone at known distances from the main root without interference from lateral roots. Their results (Table 7.1) indicate that the rhizosphere zone extended for 18 mm from the root, but although there is little doubt of the significance of the differences between populations of control soil and those of the rhizoplane and inner rhizosphere (0–3 mm), the small increase between the other zones (farther from root surfaces) and the control soil is within the variation obtained for replicate soil samples. An examination of some of the qualitative differences in fungi shows that the extent of the rhizosphere varied with species. For example, *Cylindrocarpon radicicola* occurred only in the rhizoplane; *Paecilomyces marquandii* had the greatest numbers in the rhizoplane and was not found 12 mm beyond the root; and *Aspergillus ustus* was most numerous in the rhizoplane, falling off with distance from the root but even at 18 mm twice as abundant as in the control soil. The *Fusarium oxysporum* and *Trichoderma viride* counts were not affected by the root. It is obvious that general bacterial and fungal counts may not be sufficiently sensitive

Table 7.1. Extent of the rhizosphere of 18-day-old blue lupine (*Lupinus angustifolius*) seedlings

Distance from root (mm)	Microorganisms (1000's per g oven-dried soil)			Fungi (per g oven-dried soil)		
	Bacteria	Strepto- mycetes	Fungi	*Aspergillus ustus*	*Cylindro- carpon radicicola*	*Paecilo- myces marquandii*
0†	159,000	46,700	355	5,650	4,940	9,000
0–3	49,000	15,500	176	3,360	0	2,800
3–6	38,000	11,400	170	2,920	0	1,600
9–12	37,400	11,800	130	2,880	0	1,500
15–18	34,170	10,100	117	2,270	0	0
80‡	27,300	9,100	91	1,000	0	0

† Rhizoplane.
‡ Control soil.

to define the zone of root influence, but the investigations of certain species may yield better indications of the extent of the zone of soil influenced by roots. Individual organisms appear to range from strongly rhizophilic to moderately rhizophobic.

b. *Direct Observation.* Direct examination of roots with a light micro-scope can give valuable information on the actual distribution of micro-organisms on the root surface. Although such information is difficult to quantify, greater use of direct observation is needed to gain the informa-tion necessary for our understanding of the ecology of the rhizoplane microorganisms.

By staining washed roots with a mixture of aniline blue, phenol, and acetic acid (Jones and Mollison, 1948), it is possible to observe the bacteria, fungi, and actinomycetes. Rovira (1956) used this method to show that roots of oats (*Avena sativa*) and tomatoes (*Lycopersicon esculentum*) are colonized by bacteria within six hours of emergence from the seed. The root tip was almost always found to be devoid of bacteria, which first appeared as isolated cells or small clusters in the zone of elongation. The older root portions, where the root surface was densely colonized, often had bacteria many cells deep. Different patterns of development in the rhizoplane microflora have been observed by this method. On roots of large canarygrass (*Phalaris tuberosa* L.), bacteria were clustered in colonies containing 8 to 100 or more cells; and on tomato roots, bacteria and often fungi aligned themselves along the root at junctions of cell walls. Light microscopy has also revealed that there are often many layers of bacteria overlying the root.

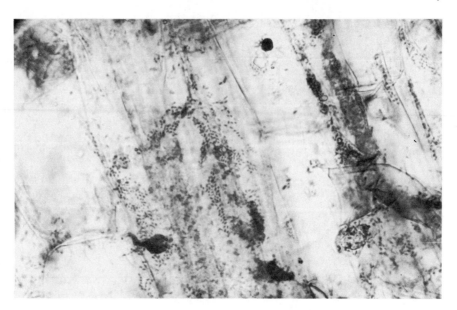

Fig. 7.1.   Bacteria on the surface of wheat (*Triticum aestivum*) grown in sand with plant nutrient solution for 7 days

Figure 7.1 illustrates the manner in which bacteria and fungi are distributed over the roots of wheat (*Triticum aestivum*) and clover (*Trifolium* sp.).

Parkinson *et al.* (1963) found by direct microscopy of roots of corn (*Zea mays*), cabbage (*Brassica oleracea*), and beans (*Phaseolus* sp.) that fungal hyphae are seldom observed in the root tip region but become more abundant toward the crown, although even there they cover only part of the root surface. In corn, only a small fraction of the root surface in the crown region was actually colonized. The authors suggested that mutual inhibition between hyphae may explain the empty spaces, but many of these spaces are colonized by bacteria that may be responsible for the sparse colonization by the fungi, either through antagonism or competition for nutrients.

During observations of stained roots of French bean (*Phaseolus* sp.) seedlings grown in University of California potting mixture, one of us (A. D. R.) has observed that the rootcap cells, each of which is surrounded by mucilaginous material, are virtually sterile for the first day after being sloughed off. After this time, bacteria progressively colonize the mucilaginous sheath until by the third day the discarded rootcap cells are completely surrounded by bacteria 3 to 4 cells deep. Similarly, the root tip region was free of bacteria, and it was not until the root was one to two days old that bacteria extensively colonized the root surface.

Jenny and Grossenbacher (1963) and Dart and Mercer (1964) used the electron microscope to observe the presence of a mucilaginous sheath several microns thick around roots. Jenny and Grossenbacher grew barley (*Hordeum vulgare*) seedlings in nonsterile synthetic soils (Ca-bentonite clay and Ca-permutite sand) and found the "mucigel" between the roots and soil to be sparsely colonized by bacteria. Dart and Mercer established that sterile medic roots are surrounded by a coarse, granular, and rather electron-dense material contained within a membrane layer of greater electron density; so it can be now concluded that this material is of plant rather than of microbial origin. After inoculation, rhizobia penetrated the outer membrane or "cuticle" and formed an almost continuous layer 5 to 10 cells deep between the bulk of the mucilaginous material and the cuticle. In general, the bacteria did not mix intimately with the mucilaginous material itself, but when this did occur, lysis of the rhizobia followed. Foster and Marks (1967) demonstrated by electron microscopy of sections of mycorrhizae of Monterey pine (*Pinus radiata*) that there is a greater number of bacteria in soil up to 20μ from the mantle than in soil beyond this zone. In Figure 7.2 the bacteria in the rhizosphere of the mycorrhiza can be seen occurring singly with wide spaces between them.

The stereoscan electron microscope can contribute much to the study of the orientation and distribution of microorganisms on the root surface. Gray (1967) and Dart (1971) have used this microscope to demonstrate the distribution of bacteria, including rhizobia, on the surface of roots and root hairs. The three-dimensional effect obtained with this technique illustrates in great detail the associations between the microflora and root cells, but as with all direct microscopy, care is needed to ensure that the fields illustrated represent the situation of a root in its natural soil environment. For example, our experience with normal light microscopy indicates that the rhizoplane populations of roots grown in agar or water culture are many times greater than the populations on soil-grown roots. Hence, future stereoscan electron microscopy of the root surface microflora should use soil-grown plants.

Selective demonstrations of bacteria on or in roots can be made using infrared color photography with unstained preparations. Using this photographic technique Casida (1968) observed bacteria as red objects while all other inorganic and organic materials appeared as contrasting colors. He observed bacteria on the root hairs of legumes and also *Rhizobium* cells in infection threads in root hairs.

Some of the most spectacular illustrations of microorganisms on the surfaces of roots and root hairs were produced by Trolldenier (1965a, 1965b) using roots stained with acridine orange and examining them with ultraviolet light, by which bacteria and fungi show up as bright green or red cells against a pale red-brown background.

The fluorescent antibody technique developed in medical microbiology for the identification of specific bacteria, protozoa, and fungi

Fig. 7.2. Ultrathin section of mycorrhiza of Monterey pine (*Pinus radiata*) grown in forest soil. *R:* root cells; *M:* mycorrhizal mantle; *Rh:* rhizosphere; *S:* soil. Arrows point to some of the bacterial cells. × 3,200

among a background of other organisms (Beutner, 1961) has been applied by Trinick (1970) in his studies of the ecology of *Rhizobium* species in the rhizoplane of clover. Trinick (1969) used this technique to identify the strains of *Rhizobium* responsible for nodule formation by treating replicate smears of nodule contents with a fluorescent antibody serum prepared against a range of *Rhizobium* strains. The technique

a

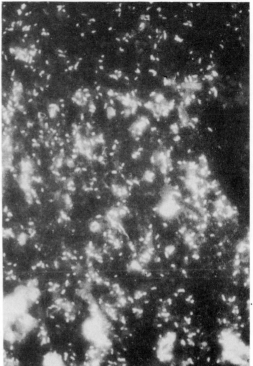

b

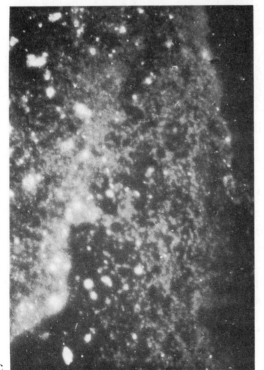

c

Fig. 7.3. Application of the fluorescent antibody technique in a study of the ecology of *Rhizobium* on the roots of subterranean clover (*Trifolium subterraneum*) grown in sand. The *Rhizobium* cells fluoresce a bright yellow-green while the other organisms fluoresce a faint orange-brown due to the gelatine-rhodamine counterstain. Part *a: R. trifolii* (strain WU290) in the absence of other organisms; *b: R. trifolii* (strain WU290) in the presence of 10 inhibitory organisms; *c: R. trifolii* (strain TA1) in the presence of 10 inhibitory organisms

apparently cannot be applied to identify bacteria on roots because of the high background fluorescence of plant material, but Trinick (1970) did find that impressions on glass slides could be satisfactorily treated with fluorescent antiserums. In this way he traced the development of strains of *Rhizobium* along roots of legumes growing in sand in the presence of other bacteria. His photographs in Figure 7.3 demonstrate how neutral and antagonistic bacteria isolated from soil influence the abundance of *Rhizobium* cells on the root surface of clover. In this way he showed that the presence of antagonistic and competitive micro-organisms did not reduce the population on roots of the WU290 strain of *Rhizobium trifolii* to the same extent as it did the population of strain TA1. Thus by fluorescence antibody staining and direct observation, Trinick found an explanation of why, in certain soils in field trials, strain WU290 was more successful as a seed inoculant than strain TA1.

2. Numbers of Microorganisms

There are numerous publications reporting that all of the plants which have been studied under a wide variety of conditions from arctic to tropic and wet to arid environments stimulate certain groups of soil organisms. Many of these data are presented on the R:S ratio basis, but it is difficult to make valid comparisons among the results of different workers because of differences in their sampling procedures, methods of cultivation, media, and so on.

a. Bacteria. It is not uncommon to find reports in the literature of R:S ratios as high as 100 for bacteria. More commonly they range from 5 to 20, but in any application it should be remembered that the R:S ratio is not an absolute figure, for it is dependent upon the soil type, root system, and manipulation of the sample. However, it is valid to take the results from one laboratory, where all the methods are constant, for a comparison of the effects of a range of plants on soil bacteria. Such a set of results was published by Rouatt and Katznelson (1961), in which they showed that of six crop plants studied, red clover (*Trifolium pratense*) exerted the greatest effect on the rhizosphere soil. Their results (Table 7.2) show the R:S ratio for each crop at six weeks.

In comparing the relative abilities of plants to stimulate bacteria, one should consider the results of Rivière (1960). He showed quite clearly that the R:S ratio for wheat was affected by the plant's stage of development. Using the entire root systems, the R:S ratios were: germination, 3.1; tillering, 27.7; heading, 16.8; and maturity, 5.4. The value at maturity may be somewhat inflated by the activity of saprophytic

Table 7.2. Comparison of the colony counts of bacteria in the rhizosphere of crop plants and in root-free soil

Crop	Colony count (10^6/g of soil)		R:S ratio
	Root-free	Rhizosphere	
Red clover (*Trifolium pratense*)	134	3,255	24
Oats (*Avena sativa*)	184	1,090	6
Flax (*Linum usitatissimum*)	184	1,015	5
Wheat (*Triticum aestivum*)	120	710	6
Corn (*Zea mays*)	184	614	3
Barley (*Hordeum vulgare*)	140	505	3

organisms on the increasing amount of senescent root tissue as the roots reach old age.

Martin (1971a) has used a different technique to assess the effects of roots on the soil microflora. In his experiments, wheat, Wimmera ryegrass (*Lolium rigidum*), and subterranean clover (*Trifolium subterraneum*) were grown in pots containing free-draining sandy soil through which a standard volume of water was passed each week from seeding to maturity. The number of bacteria in each leachate was counted by the pour-plate method. In each species of plant, maximum development of bacteria occurred at flowering.

These results of Rivière and Martin illustrate that, quite apart from problems of technique and sampling, it can be misleading to compare the different rhizosphere effects of different plant species on the basis of R:S ratios unless samples are taken at various stages of development of the plants. Any sample, therefore, must be considered a sample in time as well as in space.

The results of Louw and Webley (1959) demonstrate a further problem in assessing quantitative changes in the bacterial populations of the rhizosphere. Comparisons made by direct counting and plate counting with a nonselective medium showed that the ratio of direct count to plate count was lowest in the rhizoplane soil and highest in the control soil. Table 7.3, prepared from their data, shows that this ratio became smaller as the plants aged. This decrease in the ratio was probably caused by the selective stimulation by the roots of those bacteria capable of utilizing the nutrients of root exudates and by the easy adaptation of these organisms to artificial media. On the other hand, the majority of soil organisms will not develop on artificial agar media, although they may be viable and capable of growth when given appropriate nutrients.

Table 7.3. Ratios of direct count to plate count of control soil, rhizo-sphere, and rhizoplane of oats (*Avena sativa*)

Plant age (weeks)	Control	Rhizosphere	Rhizoplane
16	11.3	5.1	5.2
17	9.2	3.1	2.6
20	7.9	1.4	1.2

b. Fungi. The filamentous nature of fungi and the prolific sporulation of certain species make this group more difficult to assess quantitatively than the bacteria. Katznelson (1960) quotes R:S ratios of 10 and 19 for fungi from wheat and mangels (*Beta vulgaris*), respectively, but these figures were obtained by the standard dilution procedure, which estimates the numbers of fungal spores rather than the numbers or quantity of active mycelia. Rivière (1960) reported that the R:S ratio for fungi in the inner rhizosphere of wheat ranged from 150 to 274 at tillering and from 11 to 22 at maturity. These ratios are considerably higher than those reported by others and may be due to active sporulation by fungi at this stage of plant development and also to the fact that the sample was from the inner rhizosphere. At tillering the R:S ratio of the outer rhizosphere was 11.5.

The general conclusion that fungi are less stimulated by plant roots than bacteria is based upon a comparison of the numbers of cells or propagules, but in view of the much greater size of fungal hyphae it is possible that consideration of the rhizosphere effect on the basis of cell volume or cell mass might favor fungi. The filamentous growth of the fungi gives them certain advantages over bacteria in that they move considerable distances from a substrate base through soil (Burges, 1960). Fungi also can grow through and colonize material in dry soils (Griffin, 1963) and can move toward roots by chemotaxis and chemotrophy (Zentmyer, 1961). On the other hand, the higher growth rates of bacteria supplied with soluble organic compounds give them advantages over the fungi in colonizing roots. Not only do bacteria have a much shorter generation time than fungi or actinomycetes, but their exponential (2^n) growth habit far exceeds the cubic (n^3) growth habit of the other two groups of organisms in tissue production.

The serial root-washing technique was developed to distinguish between those fungi which exist mainly as spores in the rhizosphere and those which actually colonize the root surface (Harley and Waid, 1955). They showed that the initial root washings yielded colonies similar to those given by the dilution plate method but different from those developing from the root fragments. This technique showed the contrast between fungi on the root surface and those in soil, revealing many slow-growing sterile mycelia not found when unwashed roots were used. On the average, 4.3 types of viable mycelia were recorded per centimeter of root. The frequency of isolation of fungi from root segments increased with the age of the root, segments near the apex often appearing to be sterile. Different species of fungi were isolated from the various parts of the root system, indicating a succession of colonizers on the root.

The difficulty inherent in referring to the rhizosphere effect on fungi, as indicated by the R:S ratio, was amply disclosed in the widely varying R:S values which Papavizas and Davey (1961) found for fungi in general, the genus *Penicillium* in general, and for two species of the genus in particular. They reported R:S ratios for 18-day-old blue lupine as follows: fungi, 3.9; *Penicillium* spp., 9.0; *P. piscarium*, 46.7; and *P. lilacinum*, 3.3.

These same authors also showed that the rhizosphere effect on fungi could be completely negated when the soil was amended with crop residues, with or without mineral nitrogen.

When Parkinson and Clarke (1961) isolated fungi from 2-mm segments of onion (*Allium cepa*) roots, they found that 36% to 56% of the segments gave fungal colonies, while by direct observations they found fungi on 46% to 58% of the segments.

The vagueness of our understanding of the inner boundary of the rhizosphere was demonstrated in a study of the microbial recolonization of a fumigated nursery soil (Danielson and Davey, 1969). The species of fungi isolated from the rhizosphere were the same as those isolated from the soil, but seven of the nineteen species isolated from surface-sterilized roots were not recorded in either the soil or the rhizosphere. These seven species included six from roots grown in nonfumigated soil and one from fumigated soil. None of the seven was a known mycorrhiza-former or a known pathogen; yet they were definitely root inhabitants (*sensu* Garrett, 1956) and thus very much under the influence of the root. It might be argued that these seven species do not qualify to be considered as any part of the rhizosphere, but the fact that they were not seed-borne organisms indicates that they entered the root from the soil, and thus at some time in the development of the root they were "properly" rhizospheric and apparently also highly rhizophilic. Finally, it should be remembered that the remaining twelve species isolated from the surface-sterilized roots were all found in either or both the soil and the rhizosphere.

c. Actinomycetes. The problems of estimating the numbers of actinomycetes in the rhizosphere are similar to those involved in counting sporing fungi. It is generally accepted that roots stimulate actinomycetes less than they do bacteria. The actinomycetes merit special attention, because many of them produce antibiotics and are antagonistic to bacteria and fungi. Rouatt *et al.* (1951) found that rhizospheres of wheat, oats, lucerne (*Medicago sativa*), soybeans (*Glycine max*), mangels, and potatoes (*Solanum tuberosum*) supported proportionally more of the actinomycetes antagonistic to bacteria than did control soil. Strzelczyk (1961) reported similar results—a larger proportion of the actinomycetes isolated from wheat, radish (*Raphanus sativus*), and onion rhizospheres were antagonistic to *Azotobacter* than of the actinomycetes isolated from soil beyond the rhizosphere.

Rivière (1960) reported for wheat at tillering that the R:S ratio for actinomycetes of the inner rhizosphere ranged from 26.3 to 35.1, whereas for the outer rhizosphere the R:S range was from 2.1 to 11.5.

If some of the actinomycetes produce significant amounts of antibiotics at the nutritional levels found in the rhizosphere, then they could play a significant part in modifying the bacterial populations of the rhizosphere. Davey and Papavizas (1959) reported a relationship between the ratio of actinomycetes and bacteria in bean rhizospheres and the severity of disease caused by *Rhizoctonia solani*. These workers studied the effects of various crop residues on soil and rhizosphere populations and on *Rhizoctonia* disease severity. In general those crop residues associated with a decreasing ratio of bacteria to streptomycetes in the rhizosphere were also associated with a reduction in disease.

d. *Algae and Protozoa.* There are conflicting reports on the influence of plant roots on algae and protozoa. Algae are known as poor competitive saprophytes, but a few apparently can live in close association with roots in the dark. In the case of protozoa there would be an indirect effect in that extra foodstuff would be provided for them by the numerous bacteria of the rhizosphere.

Darbyshire (1966) and Darbyshire and Greaves (1967) reported that no more species of protozoa occurred in the rhizosphere than in interrow soil but at midsummer the cyst populations of flagellates and amoebae were higher in rhizosphere soil. These higher numbers of protozoa in the rhizospheres of ryegrass (*Lolium perenne*) persisted despite desiccation.

Small numbers of algae have been found in both the rhizosphere and root cortex cells. These have included blue-green algae in cycads and green algae in pines.

e. *Nematodes.* Free-living nematodes are more abundant in the rhizosphere than in soil beyond it (Varga, 1958), the R:S ratios ranging from 13 to 71 in the rhizospheres of various plant species (Henderson and Katznelson, 1961). The dependence of plant-parasitic nematodes upon the presence of host plants is well established, and one method of reducing such populations is crop rotation with nonsusceptible crop plants. A second control method is the use of trap crops such as *Cassia tora* (sickle senna). The roots of these crops are both attractive and poisonous to nematodes.

B. Qualitative Changes in the Rhizosphere Microflora

There is by now a considerable body of evidence which demonstrates that the microorganisms in the rhizosphere differ not only in number but also in type from the dominant types in soil away from roots.

1. Bacteria

The roots selectively stimulate the Gram-negative species of bacteria, which are characterized by their fast growth rates, growth response to amino acids and glucose, production of acid from glucose, and resistance to certain antibiotics.

One of the most consistent differences between bacteria isolated from the rhizosphere and those from other soil is in nutritional requirement. Lochhead, Katznelson, and co-workers have shown that a larger proportion of rhizosphere isolates than of soil cultures requires amino acids for maximal growth; some of their results are given in Table 7.4. Lochhead and Rouatt (1955) summarized 15 years of experiments covering nine plant species (several of different ages) in 23 trials to show that the proportions of bacteria requiring amino acids were consistently higher in the rhizosphere than in the control soil.

Table 7.4. Percentages of soil, rhizosphere, and rhizoplane bacteria from barley (*Hordeum vulgare*) roots in three nutritional groups

Nutritional group†	Control soil	Rhizosphere	Rhizoplane
I Inorganic salts, glucose	3.8	4.0	6.0
II Inorganic salts, glucose, amino acids	12.6	24.0	32.9
VII Inorganic salts, glucose, yeast, soil extracts	83.6	70.0	59.7

† For the definition of *nutritional group*, see Lochhead and Chase, 1943.

A further difference in the nutritional requirements of soil and rhizosphere bacteria was revealed in the ratios of direct counts to plate counts reported by Louw and Webley (1959). Only about 10% of soil bacteria grew on a nonselective agar while up to 72% of rhizosphere bacteria grew on the same medium. This probably indicates more fastidious nutritional requirements of the soil bacteria. Further, it means that, if the majority of soil bacteria are viable, the nutritional requirements of soil bacteria actually differ more from those of rhizosphere bacteria than is indicated by tests of cultures isolated on agar media. This is because we can only study those which can grow on laboratory media.

Besides the differences in nutritional requirement, there are differences in physiological activity. In general, rhizosphere bacteria are more responsive to readily available nutrients. Zagallo and Katznelson (1957) have shown that the average oxygen uptake of rhizosphere isolates supplied with various substrates, such as sucrose, glucose, acetate, succinate, and alanine, was greater than that of soil isolates supplied with similar substrates. However, there was considerable overlap in the activities of the individual members of the two groups. This indicates that both soil and rhizosphere populations isolated on agar media cover a similar range of organisms but that the rhizosphere sample is richer in bacteria capable of responding to nutrients such as glucose and alanine.

These results are but a few which demonstrate the qualitative differences between soil and rhizosphere bacteria. A further point to be examined is whether different plant species stimulate different groups of bacteria. The results of Krasil'nikov (1958), summarized in Table 7.5, demonstrate that, with model systems at least, plant species can stimulate different bacteria.

Further evidence of the specificity in the reactions between plants and some bacteria comes from studies by Robinson (1967) on the interactions between legumes and *Rhizobium* species. He found that *R. meliloti* was stimulated only in the rhizosphere of its host plant, lucerne, whereas *R. trifolii* showed a nonspecific behavior by proliferating in the rhizospheres of not only its host, clover, but also lucerne and oats.

2. Fungi

Parkinson *et al.* (1963) have shown a succession of fungi with age of root; not only do the fungi become more abundant in the older parts of the root, but the species change. They found that as the root grows through soil there is a successive colonization of the root from the soil; the root tip is almost devoid of fungi, the zone behind the root tip

Table 7.5. Growth of bacteria in Knop's solution in the presence of plant roots

Bacterium	Number of bacteria per ml after 20 days					
	Wheat	Cotton	Sugar beets	Clover	Lucerne	Peas
Rhizobium trifollii	1×10^4	1×10^7		1×10^8	1×10^6	1×10^7
Rhizobium meliloti	1×10^5	1×10^8	1×10^6	1×10^7	1×10^8	1×10^6
Azotobacter chroococcum	0	3×10^3	1×10^3	6×10^6	5×10^6	3×10^6
Pseudomonas fluorescens						
Strain 1	2×10^8	1×10^8	1×10^3	1×10^6	1×10^6	1×10^9
Strain 2	1×10^6	3×10^7	1×10^8	1×10^9	1×10^8	1×10^6

Table 7.6. Percentage frequency of occurrence of some fungi from washed bean roots, bean rhizosphere, and control soil

Fungus	Roots		Rhizosphere soil†		Control soil	
	40 days	190 days‡	Soil-washing techn.	Dilution plate techn.	Soil-washing techn.	Dilution plate techn.
Mortierella spp.	9	16	34	18	35	20
Mucor spp.	0	0	20	1	2	1
Cylindrocarpon radicola	5	39	0		2	1
Fusarium spp.	55	16	21	5	2	2
Gliocladium spp.	11	7		6	0	7
Penicillium spp.	16	2	7	55	7	44
Trichoderma viride	4	0	8	1	9	3
Sterile mycelia	0	9	4	2	3	1
Other species	0	11	6	12	40	21

† Rhizosphere soil from mature dwarf bean plants.
‡ Moribund roots.

contains the casual colonizers, and the older part of the root supports more specialized fungi, including *Fusarium oxysporum, Cylindrocarpon radicicola, Gliocladium* spp., *Penicillium* spp., *Trichoderma viride*, and sterile dark forms.

The results of Parkinson (1965) given in Table 7.6 demonstrate the influence of isolation technique on the range of fungal species from control soil and bean rhizosphere. The dominant genus found in both samples by the dilution plate technique was *Penicillium*, but the dominant genera found by the soil-washing technique (for isolating hyphae) were *Mortierella* in the soil and *Mortierella, Mucor,* and *Fusarium* in the rhizosphere. On the washed roots of 40-day-old plants *Fusarium* constituted more than half the isolates, whereas on moribund roots *Cylindrocarpon radicicola* was dominant.

There is ample evidence that the proximity of plant roots stimulates the germination of the spores of many nonpathogenic fungal species lying dormant in soil. Schroth and Hildebrand (1964) have cited several reports that the germination of fungal spores in the rhizosphere is non-specific, whereas Buxton (1957) reported that exudate from wilt-resistant peas (*Pisum sativum*) inhibited the germination of spores of *Fusarium oxysporum* Fr. f. *pisi* but that exudate from susceptible peas promoted it. Schroth and Hildebrand (1964) considered that a shortage of carbon and nitrogen limited the germination of spores in soil and that the exu-dation of compounds rich in C and N by roots of both host and nonhost plants resulted in spore germination. This explanation fits one of the two existing hypotheses concerning the mode of action of fungistasis; namely, that the failure of spores to germinate in soil is the result of an impoverishment of nutrients rather than the presence of inhibitory substances. However, Buxton's (1957) interpretation of his data would tend to indicate that both mechanisms (lack of stimulator and presence of inhibitor) are possible. Most indications are that fungistasis is less prevalent in the rhizosphere than it is in the soil proper.

Although the stimulation of fungal spore germination by plant roots may in many instances be nonspecific, the growth of fungi in the rhizosphere may often be selective; for example, Rangaswami and Balasubramanian (1963) found that the exudation of hydrocyanic acid from sorghum (*Sorghum bicolor*) roots inhibited the growth of *Helmin-thosporium turcicum* Pass. and *Fusarium moniliforme* Scheld but not other fungi such as *Aspergillus* spp.

A strong host specificity in the attraction of motile zoospores to the tip region of roots of susceptible plants was demonstrated by Zentmyer (1961) when he found that the zoospores of *Phytophthora cinnamomi* were attracted to the zone of elongation of avocado (*Persea americana*) roots whereas no such attraction occurred with mandarin orange (*Citrus sinensis*).

II. ORGANIC MATERIALS IN THE RHIZOSPHERE

The stimulation of microorganisms in the rhizosphere is undoubtedly due to the greater supply of readily available organic material in this·

zone than in soil away from the root. This organic material ranges from decaying and moribund roots and root cells to materials exuding from healthy intact roots. The nature of these exudates may be grouped according to their mobility through soil, that is, (*a*) diffusible-volatile, (*b*) diffusible–water-soluble, and (*c*) nondiffusible compounds. Most techniques used to study root exudates yield information only on the diffusible–water-soluble compounds, but evidence from the C.S.I.R.O. laboratories using ^{14}C-labeling techniques shows that for every unit of carbon exuded as water-soluble material, some 3 to 5 units are released as non-water-soluble mucilaginous material and rootcap cells and some 8 to 10 units as materials that are volatile under acidic conditions.

A. Methods of Study

In studies that aim to define the composition of exudates as they leave the root, care must be taken to avoid alteration of compounds by microorganisms or contamination by leaf washings and decomposing plant material. Hence, most precise studies have been conducted under well-defined conditions involving the axenic culture of plants in solution or sand culture. With solution culture, it is possible to estimate exudation progressively by changing the solution, but such growing conditions are artificial and care must be exercised in extrapolation of results to soil. Sand-culture conditions approximate the soil more closely and enable progressive sampling by leaching or by solution-recycling in a soil perfusion apparatus, but sand still lacks the chemical and physical complexity of soil. Further use should be made of the technique of Harmsen and Jager (1962, 1963). They used a synthetic soil of sand, feldspar, and kaolinite in measuring exudation of carbon from roots.

Martin (1971b) has studied the water-soluble compounds coming from the roots of wheat, clover, and ryegrass growing under more natural conditions. He grew the plants in a sandy soil in pots with free drainage, and after labeling the tops with $^{14}CO_2$, he studied the compounds present in leachates. He found differences in the amounts of ^{14}C-labeled material among plant species, but in this near-natural nonsterile environment, the exudates were largely compounds of high molecular weight. Using ultrafiltration techniques, Martin found that 45% of the material had a molecular weight above 10,000 and 70% was above 1,000; the material was also highly negatively charged. Work of others using sterile conditions indicates that compounds of low molecular weight make up the bulk of the soluble exudates released by roots. Thus Martin's results indicate that in natural soil the rhizosphere microorganisms rapidly utilize the low molecular weight material and subsequently release the carbon in more complex forms.

These results have major implications for the possibilities of decreasing the chemotactic attractiveness of a root to a pathogen. A reduction in infection may be possible through the encouragement of appropriate rhizosphere inhabitants.

Only little use has been made of radioisotopes in root exudate studies despite the fact that they provide a powerful tool, considering the small amounts of materials involved. For studies on the exudation of endogenous compounds from plant roots, the use of $^{14}CO_2$ has the advantage that the photosynthate is labeled and travels mainly as sucrose to the roots where, over the next 24 hours, the labeled carbon is incorporated into several compounds whose exudation can be studied (McDougall, 1970). This pulse-labeling technique is sensitive enough to record sites of exudation from single roots placed between sheets of moist filter paper (McDougall, 1968; McDougall and Rovira, 1970). McDougall (1968) has also studied interplant variability in exudation, the effects of medium-bathing the roots, and the influence of metabolic inhibitors.

The "sandwich" technique used by McDougall (1968) and Bowen (1968) for short-term exudation studies has now been adopted both to record the sites of exudation of ^{14}C-labeled compounds and to distingush between the diffusible and nondiffusible components (Rovira, 1969). In this technique plants are grown for one to two weeks with the roots between four strips of Whatman No. 1 chromatography paper wet with plant nutrient solution. The tops of the plants are enclosed in inflated thin polyethylene tubing into which air containing $^{14}CO_2$ is injected, each plant receiving from $10\mu c$ to $50\mu c$ as a single dose or as successive doses over several days. The radioactivity that accumulates in the paper strips represents the amount of the ^{14}C from the pulse label exuded from the roots. The level of radioactivity along the roots or along the paper strips is measured and recorded automatically by passing them through a radiochromatogram scanner (Bowen and Rovira, 1967). This technique has revealed that the apical regions of the clover and wheat roots exude a large amount of labeled material, most of which is nondiffusible. By making autoradiograms of the filter paper strips that are scanned, one can also obtain information on the lateral spread of the ^{14}C-labeled material. The relationship between ^{14}C inside the root and exudation can be seen from the scans in Figure 7.4 (*a* and *b*), while the autoradiograph (*c*) shows that besides the concentration of ^{14}C along the zone of elongation of the root, there is considerable radioactivity in discrete spots some distance away. Subsequent work has demonstrated that these spots correspond to fungal colonies that had utilized volatile ^{14}C compounds, probably mostly CO_2.

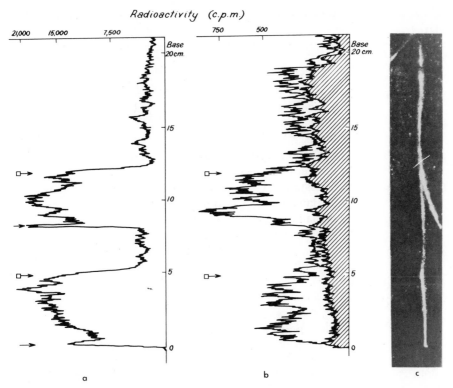

Fig. 7.4. Exudation of ^{14}C-labelled material from wheat (*Triticum aestivum*) roots. Roots were grown between sheets of Whatman No. 1 chromatography paper saturated in plant nutrient solution. After 6 days the tops received 50μcuries[14]CO_2 in 10 ml air, and 24 hours later the roots were moved from the paper. Part *a:* radioactivity inside two roots grown together between sheets of chromatography paper; *b:* radioactivity exuded into the paper by the roots scanned in *a*; the clear area represents the total exudate, and the cross-hatched area represents the diffusible exudate—beyond 1 mm from root; *c:* autoradiograph of paper used for scan *b*

Subba-Rao *et al.* (1962) treated tomato seedlings growing in soil with [14]CO_2 and measured the radioactivity that could be leached from the soil by water and alcohol and released after acid hydrolysis. They also attempted to measure the distances that exudates diffuse through. soil by measuring the radioactivity of soil cores taken at different distances from the roots. Their results indicate that variety Bonney Best released more radioactivity than did variety Geneva II, but no conclusion could be reached on any relationship between exudation and differences in resistance to *Verticillium albo-atrum*.

Slankis *et al.* (1964) labeled 9-month-old white pine (*Pinus strobus*) seedlings with [14]CO_2 for 8 days and found that the roots (grown under axenic conditions) released over 35 radioactive organic compounds.

Table 7.7. Compounds reported in wheat (*Triticum aestivum*) root exudate

Sugars	Amino acids	Organic acids	Nucleotides, flavonones	Enzymes
Glucose	Leucine and isoleucine	Oxalic	Flavonone	Invertase
Fructose	Valine	Malic	Adenine	Amylase
Maltose	γ-Amino butyric acid	Acetic	Guanine	Protease
Galactose	Glutamine	Propionic		
Ribose	α-Alanine	Butyric		
Xylose	β-Alanine	Valeric		
Rhamnose	Asparagine	Citric		
Arabinose	Serine	Succinic		
Raffinose	Glutamic acid	Fumaric		
Oligosaccharides	Aspartic acid	Glycolic		
	Glycine			
	Phenylalanine			
	Threonine			
	Tyrosine			
	Lysine			
	Proline			
	Methionine			
	Cystathionine			

B. Nature of Compounds

An indication of the range of compounds exuded by a single plant species, wheat (*Triticum aestivum* L.), is given in Table 7.7, compiled from reports by ten different workers. Since all of these compounds have been reported to exude from wheat roots and since wheat is not unique in this ability, it can be assumed that most plants will give off a wide range of compounds. Composite tables are of little value in any attempt to assess either the exudation from a particular plant species under a given set of conditions or the differences between the plant species. Comparisons between plants are valid only if conducted in identical environments and at comparable stages of plant development.

However, two points must be stressed when considering the nature of compounds in root exudate. The first point is that under natural, nonsterile conditions many of these simple compounds will not diffuse any great distance before being absorbed or modified by the microflora. Secondly, the techniques used largely ignore the volatile materials and the water-insoluble materials that, under natural conditions, may far exceed the soluble compounds.

The role of volatile substances in the soil is only poorly understood at present, but it has been shown (Cholodny, 1951; Davey, 1954) that the action of saprophytic soil organisms on organic residues results in the formation of volatile substances that have a profound effect on root behavior. These volatiles are quite separate from the volatiles released by the root. This entire subject is in need of much more research.

Current research by Krupa and Fries[1] and Melin and Krupa[2] is providing the first extensive, quantitative information on volatile exudates. These workers have studied volatiles produced by roots of Scotch pine (*Pinus sylvestris* L.) in axenic culture, in monoxenic culture with a mycorrhizal symbiont (*Boletus variegatus* Fr.), and in natural forest nursery soil. They also determined the volatiles produced by *B. variegatus* in pure culture. The major volatile compounds produced by the fungus were ethanol, isobutanol, isoamyl alcohol, acetoin, and isobutyric acid. The pine roots produced primarily monoterpenes and sesquiterpenes. The presence of the mycorrhizal fungus increased the production of the terpenes two to eight times but did not appreciably alter the compounds produced. Of the monoterpenes, α-pinene, 3-carene, and terpinolene were the chief products. In the mycorrhizal condition, the plant-produced volatiles were abundant while the fungus-produced volatiles were detected only in small amounts. The volatiles isolated from nursery-grown trees were essentially the same as those produced in pure culture.

On the basis of this work, Krupa and co-workers have proposed a hypothesis to explain a possible role of the host plant in the resistance of mycorrhizal root systems to pathogens and in the establishment and maintenance of the symbiotic (mycorrhizal) state.

C. Amounts of Soluble Exudates

Meshkov (1953) found that peas growing in nutrient solution for 20 days exuded from 2.9 to 4.3 mg of reducing sugar per plant (0.14% and 0.23% of dry weight) and corn exuded 8.4 and 8.2 mg per plant (0.23% and 0.35% of dry weight). Rivière (1960) reported that a single wheat plant growing axenically in nutrient solution produced 13 mg

[1] Krupa, S., and N. Fries, Studies on ectomycorrhizae of pine. I. Production of volatile organic compounds, in press in *Can. J. Bot.* Manuscript kindly provided for our use by S. Krupa.

[2] Melin, E., and S. Krupa, Studies on ectomycorrhizae of pine. II. Growth inhibition of mycorrhizal fungi by volatile organic constituents of *Pinus sylvestris* L. (Scotch pine) roots, in press in *Physiol. Plant.* Manuscript kindly provided for our use by S. Krupa.

acetic, 3.5 mg propionic, 2 mg butyric, and 1.5 mg valeric acid up to the tillering stage. Using a synthetic soil, which would provide a more natural environment than solution culture, Harmsen and Jager (1963) calculated that vetch (*Vicia* sp.) exuded carbon compounds equivalent to 1.6% to 2.9% of the carbon in their roots and wheat exuded 2.6 to 22.5 mg carbon per plant during a two-month period of growth.

The exudation of compounds with specific biological activity (stimulatory or inhibitory) is often so low in quantity that such compounds are barely detectable by chemical or chromatographic techniques; in such instances bioassay procedures must be used, e.g., nematode cyst-hatching factor and fungal zoospore attractants.

The chemical characterization of such active factors is only possible with the large-scale production of root exudate. For example, when Calum *et al.* (1949) studied the active factor responsible for the hatching of cysts of the potato root eelworm (*Heterodera rostochiensis*), they grew 150,000 tomato plants in sand culture and leached the sand on alternate days for one month to produce 12 g of the crude factor that stimulates hatching of the cysts.

D. Sites of Exudation

The zone of root immediately behind the tip has been considered the major source of exudates. This is based upon the finding of Pearson and Parkinson (1961) that when bean roots were grown between sheets of moist filter paper, exudation of ninhydrin-reacting compounds was confined to the zone behind the root tip. Other evidence of the importance of this zone is that it strongly attracts nematodes (Bird, 1959) and fungal zoospores (Zentmyer, 1961) and that germination of *Striga lutea* seed occurred mainly in the root tip region of sorghum with no germination after the root tip had passed (Brown and Edwards, 1944).

However, older parts of roots may also exude significant quantities of organic compounds. Frenzel (1960) used colonization of roots by mutants of *Neurospora* with specific nutrient requirements to show that, with sunflower (*Helianthus annuus*), threonine and asparagine came from the root tip whereas leucine, valine, phenylalanine, and glutamic acid were exuded from the root hair zone. Schroth and Snyder (1961) found that sugars and amino acids were exuded where the adventitious roots emerged as well as from behind the root tip.

Although McDougall (1968) reported that exudation of ^{14}C-labeled compounds occurred mainly from the zone of lateral roots after these were fully emerged, subsequent experiments have shown that a high proportion of the radioactivity comes from the tips of the lateral roots. In culture solutions with plants pulse-labeled with ^{14}C-carbon dioxide, radioactive material accumulated on the outside of the root tips, and when the roots were placed between sheets of moist filter paper to measure the sites of exudation, spots of high radioactivity occurred on the paper in as short a contact time as one minute (McDougall and Rovira, 1970).

Root hairs are also involved in exudation, as shown by Head (1964), who used time-lapse cinephotomicrography to record the buildup of droplets on apple (*Malus pumila*) root hairs. No analyses were made on this material to indicate the presence of organic compounds. Where plants do not have root hairs because the roots are mycorrhizal, the fungi involved and the type of mycorrhiza formed will influence root exudates.

Bowen (1968) measured the loss of chloride along roots by placing roots of Monterey pine labeled with ^{36}Cl between sheets of moist filter paper for different time periods, drying the papers, and measuring the radioactivity along the papers by passing them through a radio-chromatogram scanner. He showed that chloride exudation occurred along the entire length of these roots, which are unbranched and free of root hairs, but a higher proportion of the chloride was exuded from the apical region. Similar techniques have shown that phosphate is lost along the whole root in amounts proportional to the amount of ^{32}P-phosphate accumulated in the roots (Rovira and Bowen, 1970).

E. Factors Affecting Exudation

1. Plant Species

The amount, range, and balance of compounds in root exudates differ for different plant species. Vancura (1964) found differences between root exudates of wheat and barley with respect to certain sugars (galactose, glucose, and rhamnose), whereas other sugars occurred in similar amounts in exudates of both plants.

Smith (1969) has presented comprehensive analyses of the soluble exudates from roots of five tree species (Table 7.8), These results demonstrate the differences among species in the nature of their exudates and also that the organic acids, acetic and oxalic, constitute a major fraction of the exudate.

2. Age of Plant

More amino acids and sugars exuded from peas and oats during the first 10 days of growth than during the second 10 days (Rovira, 1956).

Table 7.8. Release of organic materials per seedling during 10-day growth period

| | Pine | | | | Black |
Material	Monterey	Sugar	Jack	Pitch	locust
Organic acid (μg)					
Acetic	113	332	146	42	65
Fumaric			7		
Glycolic	6				
Malonic		9			6
Oxalic	201	202	113	20	
Succinic	0.7	3	4	4	6
Carbohydrate (μg)					
Fructose		1			
Glucose	13	32	0.1	0.6	4
Rhamnose		0.1		—†	2
Sucrose		0.7		0.4	3
Amino acid or amide (μg)					
Alanine	24			2	
γ-aminobutryric acid	36	56	21	6	
Asparagine					3
Aspartic acid	14	42	7	2	21
Glycine	14	26	6	2	11
Leucine/isoleucine	42		11	3	6
Methionine					1
Phenylalanine	19	48	9	5	
Serine		32	10	4	2
Threonine		97		5	
Tyrosine					0.3
Root dry weight (mg)‡	3.6	9.6	1.9	1.9	4.2
Root length (cm)‡	4.0	6.0	1.8	2.0	5.0§

Note: Average of 3 runs, each run combining the exudates of 196 seedlings for each species.
† Less than 0.1μg.
‡ Average of 588 seedlings.
§ Fibrous.

Vancura and Hovadik (1965) found β-pyrazolylalanine in root exudate of cucumbers (*Cucumis sativus*) only at the early seedling stage. With tomatoes and red peppers they found that tyrosine occurred in the exudate only at fruiting and not at any other stage of growth.

Smith (1970), by developing a technique of collecting exudate aseptically from roots of mature trees, has compared the composition of exudate from roots of 3-week-old sugar maple (*Acer saccharum*) seedlings with that from roots of a 55-year-old mature tree. His results showed that per unit weight of root, seedlings exuded more sugars but considerably less organic acids than did mature trees.

3. Light and Temperature

The light intensity at which plants are growing affects the amounts and balance of compounds exuded into nutrient solution. Clover grown at full daylight intensity exuded more serine, glutamic acid, and α-alanine than plants grown under 60% shade. With tomatoes, the levels of aspartic acid, glutamic acid, phenylalanine, and leucine in exudate were reduced by shading (Rovira, 1959).

The release of amino acids and, especially, asparagine from roots of tomatoes and subterranean clover increased with rising temperature (Rovira, 1959). This effect is not universal, as Husain and McKeen (1963) found more amino acids in exudates from strawberry (*Fragaria* spp.) plants at 5C to 10C than at 20C to 30C; this was thought to account for the greater attack of strawberries at low soil temperatures by *Rhizoctonia fragariae*.

4. Plant Nutrition

Although it is obviously of considerable importance, the influence of plant nutrition on the exudation of organic compounds by roots has not been studied in much detail. The study of Bowen (1969) with Monterey pine seedlings demonstrates that nutrition affects exudation of amino acids; the total moles of amino nitrogen exuded per plant into culture solution over 2 to 4 weeks was 104×10^{-9} in complete nutrient solution, 25×10^{-9} with N-deficient solutions, and 248×10^{-9} with phosphate-deficient solutions. Bowen suggested that the higher exudation from phosphate-deficient plants is due to the doubling of amido and amino nitrogen in the roots of these plants rather than an increase in leakiness or a phosphate effect upon reabsorption of amino acids and amides. This was the first conclusive demonstration that nutrition affects exudation of organic compounds from roots and is important in predicting the significance of root exudates in plant-plant and plant-microbe relationships in soil.

5. Soil Moisture

Katznelson *et al.* (1955) found that temporary wilting of plants greatly increased the release of amino acids from roots. This factor would be important under field conditions. Similarly, Vancura and Garcia (1969)

found that millet plants which had been reversibly wilted exuded 190%
more α-amino nitrogen and 113% more reducing sugars than did un-
wilted plants.

Even when plants have adequate water for their own needs, the
water content of soil will influence the distance that exudates will
diffuse. Kerr (1964) demonstrated that more weight (that is, sugars)
was lost from germinating pea seeds in loam with high water content
than in drier soil. This loss of sugar was thought to be causally related
to the higher incidence of infection by *Pythium ultimum* in wet soil.
He also found that more material was lost from the pea seeds in sand
than in soil, probably due to greater ease of diffusion in sand.

Ivanov *et al.* (1964) showed that underground transfer of ^{14}C-labeled
corn root exudate to neighboring bean plants was greatest in wet soils;
while Wallace (1958) showed that the potato root eelworm hatching
factor diffused further in wet than in dry sand. In soil the diffusion
rate rose with increasing water content and reached 5 mm per day at
saturation (Wallace, 1961).

6. Medium Supporting the Roots

Most root exudate studies have been conducted in solution culture
for ease of sampling analysis, but to ascertain the true significance of
exudation under natural conditions it is necessary to perform experiments on roots growing in sand and soil. In one of the few comparisons
of exudation from solution-grown plants and sand-grown plants,
Boulter *et al.* (1966) found up to a sevenfold increase in the release
of certain amino acids when roots were grown in quartz sand. These
results need to be accepted cautiously because some of the additional
exudates in sand may have originated from root cells and root hairs
damaged in removal from the sand before leaching out the exudate.
Also, the root washings included in the exudate may have contained
leakage products from damaged cells.

Exudation from roots and seeds into soil may be quite different
from exudation into solution or sand (Kerr, 1964) because of the
different physical, chemical, and biological environments. It is difficult
to distinguish between organic substances originating from roots and
those present in soil, but the use of radioactive isotopes may overcome
this problem (Martin, 1971b).

7. Microorganisms

Although root exudate studies must be conducted initially under sterile
conditions to determine precisely the nature and amounts of substances

leaking from roots, we should know how the rhizosphere population modified the exudation process and the nature of the exudate. Microorganisms may affect exudation in several ways, the main ones being: (*a*) affecting the permeability of root cells, (*b*) affecting the metabolism of roots, (*c*) absorbing certain compounds in root exudates and excreting other compounds, and (*d*) altering nutrient availability to the plant.

Martin (1958) reported that filtrates of cultures of some bacteria and fungi, and also some antibiotics (e.g., penicillin), increased the exudation of scopoletin (6-methoxy-7-hydroxycoumarin) by oat roots. Norman (1955, 1961) found that certain polypeptide antibiotics, e.g., polymyxin that is formed by *Bacillus polymyxa* from soil, altered cell permeability and increased leakage. Interpretation of the significance of these results in field terms is difficult because the conditions under which the organisms are grown are quite different both physically and nutritionally from those under which a rhizosphere population grows, and also because the concentrations of biologically active substances under test may be quite different from the concentrations under natural conditions. In addition, any consideration of the significance of the rhizosphere population in modifying exudation must involve a concept of microecology, in which a wide variety of organisms occupy different niches on the roots and only those plant cells in the immediate vicinity of exudation-promoting organisms are likely to be affected.

The results of Martin (1971b) quoted earlier showing that under nonsterile conditions over 70% of the water-soluble ^{14}C leached from sand-supported roots has a molecular weight over 1,000 illustrate the difficulty of extrapolating laboratory-controlled experiments to the field.

8. Foliar Sprays

By the use of radioactively labeled materials it has now been shown that when certain compounds are applied to leaves they are absorbed, transported to the roots, and exuded into the surrounding soil; exploitation of this finding may have profound effects in the fields of plant-microbe and plant-plant interactions.

The application of compounds to leaves does not necessarily mean that they will be translocated to roots and exuded, for several steps are involved: adsorption, penetration, absorption, downward translocation, and exudation. Upward transport of applied compounds appears to be more common than downward.

The importance of molecular configurations was demonstrated by Linder *et al.* (1964) when they found that only minor changes in molecular structure could affect the translocation of compounds from leaves to roots. For example, 2,3,6-trichlorobenzoic acid applied to leaves was exuded unaltered by the roots, whereas 2,5-dichlorobenzoic acid moved from leaves to stem but could not be detected in roots or root exudate. The possibility exists that the configuration of part of the molecule may be involved in translocation while configuration of other parts is involved in chemical activity. Thus certain "carriers" might be developed upon which units with certain desired activities could be substituted. This would provide a powerful tool for rhizosphere investigations.

Hurtt and Foy (1965) and Reid and Hurtt (1970) demonstrated that the herbicides Dicamba (2-methoxy-3,6-dichlorobenzoic acid), Picloram (4-amino-3,5,6-trichloropicolinic acid), and 2,4,5-T (2,4,5-trichlorophenoxy acetic acid) are exuded from the roots of sprayed plants in amounts high enough to affect neighboring plants. This has important implications in plant ecology and weed control, especially considering the persistence of Picloram under some field conditions and the very low amounts of this material required for biological activity.

Attempts have been made to control the rhizosphere microflora by applying antibiotics to the leaves. Davey and Papavizas (1961) found a suppression of Gram-negative bacteria in the rhizosphere of coleus (*Coleus blumei* Benth.) which lasted from 8 to 12 days when streptomycin was applied to leaves. Experiments with ^{14}C-labeled streptomycin showed that the ^{14}C was translocated from the treated leaves laterally and down to the root tips within 24 hours. Vrany *et al.* (1962) found that applications of chloramphenicol to wheat leaves greatly increased exudation of amino acids and sugars from roots and led to fewer bacteria and more fungi in the rhizosphere. They also showed that the roots of chloramphenicol-treated plants contained antibiotics and that a relatively high proportion of the bacteria were resistant to chloramphenicol.

In studies of the control of nematodes by foliar sprays, Peacock (1966) showed that the compound 1,3,5-tricyano-3-phenyl pentane (TCPP) moved from the leaves of tomatoes to the roots and reduced the number of galls caused by the nematode *Meloidogyne incognita*. In some experiments it killed the nematodes in the sand surrounding the roots. Peacock found an interaction between the effectiveness of foliar application of TCPP, humidity, and soil moisture. Greater control of nematodes occurred when the tops were in a humid atmosphere and roots in drier soil. Rohde and Jenkins (1958) showed that the natural nematocide extracted from asparagus (*Asparagus plumosus*) roots, when applied to tomato leaves as a 0.1 % solution, protected the tomato roots for up to 10 days from attack by stubby root nematode, *Trichodorus christiei*.

Foliar applications of plant nutrients may influence the rhizosphere populations indirectly; e.g., foliar sprays of urea caused marked increases in the exudation of glucose, fructose, glutamine, and α-alanine and decreases in the amounts of organic acids (Agnihotri, 1964). Such changes in exudates can modify the rhizosphere microflora (Ramachandra-Reddy, 1959; Venkata-Ram, 1960; Horst and Kerr, 1962; Vrany *et al.*, 1962; Vrany, 1965).

Modification of root exudates by the foliar sprays may be one of the more feasible ways of modifying the rhizosphere to protect roots from pathogens and to improve plant nutrition, but more basic information is required before this can be applied on a practical level.

F. Significance of Exudates in Interactions between Plants

Decomposing plant residues from both native vegetation and crops often produce enough phytotoxins to reduce the productivity and at times cause the complete failure of succeeding crops (Holland, 1962; McCalla and Haskins, 1964; Patrick *et al.*, 1964; Patrick and Toussoun, 1965). There is now evidence that root exudates are involved in interactions between plants (Borner, 1960; Woods, 1960; Garb, 1961), but often the results are confused by toxins resulting from root and top residues.

Bonner and Galston (1944) and Bonner (1946) showed that guayule (*Parthenium argentatum* A.) roots exuded trans-cinnamic acid, which is highly toxic to the growth of these roots but has no effect upon tomatoes at the concentrations tested. The amount exuded was 1.6 mg of pure trans-cinnamic acid from 20,000 guayule roots; although this may appear to be small, it is enough to prevent the mixing of roots from neighboring plants. Trans-cinnamic acid was decomposed within 14 days in soil; so the authors concluded that it must be continually released by the roots and also that it would not persist in soil after removal of the guayule plants.

Examples of root washings (note: not root extracts) that inhibited other plants come from South Africa and the Soviet Union. Roux (1953) found that the washings from sand supporting *Trachypogon plumosus* inhibited the germination of *Tagetes minuta* and adversely affected the growth of plants from seeds that germinated. Mishustin and Naumova (1955) demonstrated that the roots of 3- to 4-year-old lucerne exuded saponins that retarded the growth of cotton but not of wheat.

The possible significance of root exudates in plant succession on abandoned fields has been demonstrated by Rice (1964). He found that leachates of sand that was supporting plants making up the second stage of plant succession were inhibitory to *Azotobacter* and *Rhizobium*. Rice also found that the roots of inoculated bean seeds planted in sand in which *Ambrosia elatior*, *Bromus japonicus*, and *Digitaria sanguinalis* were growing had fewer nodules than control roots; the nodules were smaller and gray in color compared with the large, bright pink nodules on the roots of control plants (E.L. Rice, personal communication). The plant species that are toxic to *Azotobacter* and *Rhizobium* will grow on soils low in N, and so their antagonism toward the N-fixing bacteria gives them a competitive advantage over species that require a higher level of soil N. It is not possible from Rice's results to say unequivocally that root exudates alone cause toxicity of some plants toward N-fixing bacteria. He conducted the experiments under non-sterile conditions, so that the soil would have contained the products of decomposition of sloughed-off and moribund root cells as well as modified exudates. These inhibitors are resistant to microbial decomposition in soil and may be similar to isochlorogenic and chlorogenic acids isolated from the leaves of *Ambrosia elatior* (Rice, 1965).

The underground transfer of natural plant constituents between neighboring plants has been demonstrated by the use of radioisotopes. Rahteenko (1958) found that ^{32}P-phosphate applied to the leaves of one tree species was transferred to trees 0.25 to 2 m apart. Some of this transfer may have been due to direct root contact and grafting and some to exudates.

After treating stumps of red maple (*Acer rubrum*) with ^{32}P and ^{45}Ca, Woods and Brock (1964) detected these radioisotopes in the leaves of 19 other species for distances up to 24 feet from the donor trees. Eight days after treatment, 33 % of all species around the donor trees contained ^{45}Ca, while 75 % contained ^{32}P, proving that leakage from roots of intact trees is similar to that of roots of stumps. These results indicate that a significant exchange of nutrients may occur in a forest. This exchange occurs through root exudation or through mutually shared mycorrhizal fungi with root grafting being of minor significance, especially in a mixed stand. The forest may be regarded as comprising many stems with but one or a few root systems. Ivanov (1962) demonstrated that ^{14}C applied as carbon dioxide to the top of either corn or beans appeared in the roots of neighboring untreated plants.

III. EFFECTS OF RHIZOSPHERE MICROORGANISMS

The bacteria, actinomycetes, and fungi of the rhizosphere and rhizoplane affect their host plants through their influence on such factors as the availability of nutrients, the growth and morphology of roots, the nutrient uptake processes, and the physiology and development of the plants. It is difficult to establish the magnitude of these effects because all plants growing under natural conditions live in association with microorganisms, and hence experimenters are faced with the difficulty of an adequate control treatment. The many reports of stimulation of plants by microorganisms where plant growth in sterilized soil is used as a control may represent no more than a destruction by the microflora of toxins formed during sterilization (Rovira and Bowen, 1966a). Such problems may be overcome in soil by using gamma sterilization instead of heat sterilization (Bowen and Rovira, 1961b) or alternative substrates such as solutions or sands that can be heat-sterilized without producing phytotoxins. In some studies the effects of the microorganisms under investigation have been observed in natural, unsterilized soil; these have provided further evidence of the effect of microorganisms on plants.

A. Availability of Nutrients

Root exudates and the rhizosphere microflora can affect plant growth indirectly, either for the benefit or detriment of the plant, through their effects on the availability of nutrients.

1. Decomposition of Organic Matter

Oxygen consumption and carbon dioxide evolution, which are indicative of microbial decomposition of organic compounds, are higher in rhizosphere soil than in control soil. Katznelson and Rouatt (1957) measured the respiration of soil from field crops and found that the soils from corn and soybean rhizosphere respired four times as fast as the control; those from barley, rye, and wheat rhizospheres, three times as fast; and those from oat rhizosphere, two times as fast. When substrates such as amino acids were added to these samples, respiration of the rhizosphere soil over an 8-hour period was increased 100%, but that of the control soil only 50%. Thus it appears that the rhizosphere not only is more active as a result of the exudate and root debris but also can be more active in decomposing soil organic matter. This high activity may be due to both a greater inoculum density and a more active metabolic state of the rhizosphere organisms.

2. Ammonification

R:S ratios of ammonifying bacteria often exceed 50 (Rouatt *et al.*, 1960), and as a consequence, the rhizosphere soil has a high potential for the release of ammonia from soil organic matter. However, studies with synthetic rhizosphere soil (soil treated with root exudate) showed that although such soils released more ammonia from peptides and amino acids than did control soils, they did not release more ammonia from soil organic matter (Rovira, 1956).

3. Nitrification

There are conflicting results on the question of nitrification in the rhizosphere. Goring and Clark (1948) found that the N was immobilized in the rhizosphere, giving an apparent inhibition of nitrification, but after removal of the roots there was a rapid accumulation of nitrate. Molina and Rovira (1964) and Rivière (1959) found that exudates did not inhibit nitrification of pure cultures of nitrifying bacteria and that the numbers of *Nitrosomonas* and *Nitrobacter* increased in the rhizospheres. By contrast, Rice (1964) found that extracts of roots of a wide range of plant species inhibited nitrification of pure cultures. He has also shown that nitrification in soil is also inhibited by root extracts. Recently, Moore and Waid (1971), using a leaching technique, demonstrated that the washings from ryegrass roots could reduce nitrification by up to 84% while the root washings from wheat, rape (*Brassica napus*), lettuce (*Lactuca sativa*), and onion also reduced nitrification but to a lesser extent. The inhibition by washings of ryegrass roots persisted for some 30 days following cessation of treatment of the soil with washings. In the case of rape and lettuce the inhibition was only temporary, and nitrification recurred after 5 days even in the presence of root washings. These experiments of Moore and Waid eliminated immobilization and denitrification as possible causes for the low content of nitrate and confirmed the report by Theron (1951) that inhibition of nitrification by grass roots was responsible for low levels of nitrate in permanent grassland soils.

It is obvious from these reports that the nitrification process in soil is affected differently by roots of various plant species. This will account, in part, for the conflicting reports in the literature.

4. Denitrification

With the high numbers of denitrifying bacteria in the rhizosphere (Katznelson *et al.*, 1956), it would be surprising if the losses of N from this zone were not greater than from control soil. Woldendorp (1963) demonstrated that 15% to 37% of fertilizer N was volatilized when applied to permanent grassland below field capacity, the losses from nitrate fertilizer being double those from ammonium fertilizer. He attributed this greater denitrification in the rhizospheres of living plants to three factors: greater numbers of denitrifying bacteria, greater oxygen consumption, and exudation of hydrogen donors by root systems.

5. Phosphate Availability

Microorganisms in the rhizosphere could increase the phosphate available to plants by dissolving water-insoluble mineral phosphate in soil and by mineralizing phosphate from organic compounds in soil. Studies with bacteria and fungi isolated from roots leave no doubt that there are many organisms living on and around roots which are capable of releasing inorganic phosphate from organic forms. However, apart from the paper by Gerretson (1948), who demonstrated that nonsterile plants absorbed more phosphate from insoluble calcium phosphate than did sterile plants, there is no conclusive evidence to prove that the rhizosphere organisms affect availability of phosphate. In the Soviet Union, *Bacillus megaterium* var. *phosphaticum* has been used extensively as a seed inoculum to increase the availability of soil phosphate. The evidence now indicates, however, that growth responses with this organism are caused by mechanisms other than phosphate release (Mishustin and Naumova, 1962). With regard to organic phosphates, Ridge and Rovira (1971) found less phosphatase activity associated with nonsterile than with sterile roots. The enzymes were confined to the root surface; so the zone of soil in which organic phosphates are hydrolyzed would be restricted.

6. Availability of Other Elements

In view of the wide range of organic substances with chelating properties that are released from roots, the availability of microelements may be increased in the rhizosphere. The uptake of zinc in particular could be modified greatly by the exudation of chelating compounds. With two varieties of oats (one susceptible and one resistant to manganese deficiency), Timonin (1946) showed that the susceptible variety harbored many more Mn-oxidizing bacteria in its rhizosphere than did the nonsusceptible variety. Fumigation of the soil reduced the population of Mn-oxidizing bacteria and eliminated Mn-deficiency symptoms in the oat plants. He did not, however, determine the reasons for the variation in the native difference in the rhizospheric populations of Mn-oxidizing bacteria. Deb and Scheffer (1970) compared the amino acid composition of exudates from eight varieties of oats with a range of susceptibilities to Mn deficiency. They found differences between varieties in the composition of exudates under nonsterile conditions, but they could not explain the differential absorption of Mn on the basis of amino acids in the exudates. Other compounds in the exudate or additional rhizosphere microorganisms may be involved in the differential uptake of Mn.

The differences between plant species in the uptake and solubility of calcium in the root zone (Wilkinson *et al.*, 1968; Barber and Ozanne, 1970) may well be associated with exudates from roots. Not all the activity need be due to organic compounds in the exudate. Riley and Barber (1969) demonstrated, with plants that received nitrate, that sufficient bicarbonate ions were exuded to raise the pH of the rhizosphere by one unit.

B. Growth and Morphology of Roots

The significance of root growth and morphology in the uptake of nutrients, especially nutrients of low mobility, e.g., phosphate, is undisputed (Nye, 1966), and hence if the rhizosphere microflora affects root growth, then the nutrition of the plant will be affected.

During an investigation of the carbon dioxide evolution from sterile and nonsterile sunflower roots growing in nutrient solution, Reuszer (1949) found that the presence of microorganisms affected both the roots and the tops of the plants. The root weights of the sterile and nonsterile plants were 7.1 and 2.5 g, and the top weights, 21.8 and 11.3 g, respectively. The nonsterile roots were yellow-brown and limp and tended to adhere together when removed from the flask, whereas the sterile roots were turgid and white.

Bowen and Rovira (1961a) found that the roots of tomatoes, subterranean clover, phalaris, and Monterey pine were stunted by the presence of soil microorganisms in both agar and sand culture (Table 7.9). The degree of stunting was affected by environmental conditions and also by the nature of the soil from which the inoculum was prepared. The 20% to 53% reduction in the number of lateral roots of plants grown in sand caused by some microorganisms could be important in soil, for Bowen and Rovira (1967) showed that lateral roots account for 27%, 79%, and 81% of the uptake of chloride, sulfate, and phosphate, respectively.

Welte and Trolldenier (1965) also observed the stunting of red clover roots by soil inoculum in plant-nutrient solution, but the nonsterile plants grew significantly longer roots when the solution was made up with tricalcium phosphate and calcium sulfate of low solubility; these

Table 7.9. Effects of rhizosphere microorganisms on root growth in sand

Plant	Primary-root length (cm)		Secondary-root length (cm)	
	Sterile	Nonsterile	Sterile	Nonsterile
Subterranean clover	8.3	5.7†	8.5	5.6
Tomato	8.0	3.6‡	15.4	10.0
Phalaris	24.4	14.6†	21.5	15.0†
Monterey pine	7.1	5.5§	7.1	4.3

† Significant at 1% level.
‡ Significant at 0.1% level.
§ Significant at 5% level.

authors, too, concluded that inhibition was caused by components of the soil population. Domsch (1963) and Otto (1965), however, have reported both stimulation and depression of root growth by soil fungi. Certainly the interactions among the many organisms and the roots are very complex, and hence either root stimulation or inhibition can be expected under natural conditions.

Barley and Rovira (1970) showed that uptake of phosphate from soil was increased by over 70% when pea roots developed hairs. Thus the finding that microorganisms can delay root hair formation and reduce numbers and lengths of root hairs in subterranean clover growing in sand (Bowen and Rovira, 1961a) and in peas growing in nutrient solution (Darbyshire and Greaves, 1970) could be important in the nutrition of plants growing in soils of low phosphate. However, the presence of mycorrhizae may complicate this relationship.

C. Nutrient Uptake Processes

Until lately little was known of the influence of microorganisms on plants, but recent work has shown conclusively that microorganisms external to the plant can compete with the plant for nutrients and affect the uptake and translocation of nutrients (see reviews by Barber [1968, 1969]).

1. Uptake, Translocation, and Incorporation of Phosphate

Krasil'nikov (1958) found that 17-day nonsterile barley plants absorbed twice as much phosphate as did sterile plants, and the reverse occurred with woody plants; in neither instance was uptake studied in relation to root growth. Welte and Trolldenier (1962) and Trolldenier and

Marckwordt (1962) found lower total ash contents in nonsterile than in sterile red clover and sunflower plants, but a recalculation of their data shows that nonsterile plants with smaller root systems absorbed up to 64% more nutrients per gram of roots than did sterile plants.

Bowen and Rovira (1966) and Rovira and Bowen (1966b), using short uptake times and levels of phosphate similar to those found in fertile soil, found not only that uptake of phosphate by nonsterile clover and tomatoes was 55% to 77% greater than by sterile plants but also that transport of phosphate in tomatoes was 4.4 times greater in nonsterile plants after 20 minutes. Greater transport of phosphate also occurred in nonsterile clover at 60 minutes.

Incorporation into the nucleic acid fractions of nonsterile roots is significantly larger than that of sterile roots, and concomitantly a smaller percentage is incorporated into the fraction that contains inorganic phosphate and soluble esters (Barber, 1966; Rovira and Bowen, 1966b; Barber and Loughman, 1967). Barber and Loughman also showed an increased incorporation of phosphorus into phosphoprotein and phospholipid fractions of the nonsterile barley roots. Longberg-Holm (1967) has demonstrated differences in the RNA fractions of nonsterile and sterile roots. In Barber and Loughman's experiments with phosphate-deficient plants, microbial effects on incorporation were extremely great at 3×10^{-8} M phosphate and still large at 3×10^{-5} M phosphate. Rovira and Bowen (1966a) found that incorporation into nucleic acid fractions was significantly increased in the presence of microorganisms. The root data are consistent with a rapid incorporation of phosphate into nucleic acids of multiplying bacteria in the rhizosphere, but there was a similar effect in the chemical distribution of radioactive phosphate in the tops of wheat, demonstrating a physiological effect of microorganisms on plant metabolism.

2. Competition between Roots and Microorganisms

Barber (1966) found translocation to the tops of nonsterile barley was 2.3% of the total compared with 20.4% for sterile plants. The plants had been grown in the absence of phosphate, and uptake was for 24 hours from a 3×10^{-8} M phosphate solution. Under these conditions the rhizoplane microflora was competing with the roots for phosphate. The percentage of translocation to the tops of nonsterile barley was lower than that for sterile barley until phosphate concentration in the uptake solution was 3×10^{-5} M. Such trapping of phosphate by microorganisms is indicated by photographs of Crossett (1967) and Barber *et al.* (1968).

It is difficult to predict the importance of microbial trapping of phosphate in the field, for much will depend on the rate of arrival of phosphate at the root surface, the continuity of the films of microbes around the root, and their capacity to absorb and store phosphate. It is important to note, however, that when Barber and Loughman (1967) used phosphate-deficient plants, significant trapping apparently occurred up to 3×10^{-5} M phosphate. It is probable, therefore, that microbial capture of phosphate may be of ecological significance, especially under conditions of P deficiency.

D. Physiology and Development

1. Sap Composition

Krasil'nikov (1961) found that sap from nonsterile cereals and legumes had a higher content of free amino acids than did sap from sterile plants, but there were no differences in the patterns of amino acids. The general levels of amino acids were higher in the presence of organisms, especially when either *Pseudomonas fluorescens* or a mixture of rhizosphere bacteria was the inoculum. This finding has been confirmed by Rempe and Kaltagova (1965), who also showed that the activities of some root enzymes and the chlorophyll contents of the tops were increased when rhizosphere microorganisms were present.

2. Physiological Development

The widespread use of *Azotobacter chroococcum* as a seed inoculant in the Soviet Union (Cooper, 1959; Mishustin and Naumova, 1962) has led to considerable study of its effects on the higher plants. Investigations with nonsterilized soils have shown that the organism causes wheat to come into ear earlier (Rovira, 1963) and causes the first truss of tomatoes to flower and produce ripe fruit earlier (Jackson *et al.*, 1964). A similar advancement in plant development has been obtained with seeds inoculated with *Clostridum pasteurianum* and *Bacillus polymyxa*.

Although sometimes marginal in statistical significance, yield increases are obtained often enough by inoculating seed with *Azotobacter* and other organisms to indicate that microorganisms in the rhizosphere can influence crop yields (Smith *et al.*, 1961; Mishustin and Naumova, 1962; Brown *et al.*, 1962; Ridge and Rovira, 1968).

The mechanisms by which the organisms affect plant growth are not yet fully elucidated, but Jackson *et al.* (1964) have correlative evidence that the response of tomatoes to *Azotobacter* inoculation is due to gibberellinlike substances. Krasil'nikov (1958) has also suggested that substances promoting plant growth may be responsible, but he considers that competition with inhibitory soil organisms and plant pathogens may be another important factor.

Increased yields following seed inoculation have been obtained with a wide variety of organisms that originate in control soil and rhizosphere, indicating that in noninoculated soils the natural rhizosphere microflora may well be affecting plant development and yields. After studying pure cultures of 25 saprophytic fungi and 5 plant species, Domsch (1963) put forward the hypothesis that there are stimulatory and inhibitory fungi and that in natural soil the two groups balance each other. Seed inoculation may reduce the effects of inhibitory organisms by competing with them, or it may increase the number of stimulatory organisms in the rhizosphere. Rivière (1963) reported similar inhibitory and stimulatory effects on wheat roots by bacteria from the rhizosphere of wheat and attributed the stimulation to the production of indole acetic acid.

IV. MODIFICATION OF THE RHIZOSPHERE MICROFLORA

The studies on the effects of *Azotobacter* and other bacteria on plant growth in natural soil and the finding by Chang and Kommedahl (1968) that corn seedling blight can be controlled by inoculating seed with *Bacillus subtilis* illustrate that the rhizosphere population can be modified to a significant degree by seed and root inoculation.

Brown *et al.* (1962) have shown that *Azotobacter* can be established in the rhizosphere of a variety of plants by inoculation of the seeds or roots. In field trials in South Australia similar conditions were found after the inoculation of wheat seed with *Azotobacter* and *Clostridium*, and although the numbers are not high compared with other bacterial species in rhizosphere population, they are considerably higher than in the rhizosphere of noninoculated wheat (Rovira, unpublished results). However, much more basic study is required on the ecology of the rhizosphere population and the movement of organisms along roots before we know whether it is possible to introduce cultures that will form the dominant component of the population in a highly competitive environment.

The soil microflora can be modified drastically by sterilization, and it is generally found that plant growth increases after partial or complete sterilization (Warcup, 1957; Ingestad and Molin, 1960; Martin, 1963;

Table 7.10. Effects of soil fumigation on growth, nutrient uptake, and yield of wheat (*Triticum aestivum*) in the field

(*a*) Total plant analysis at ear emergence

		Nutrient content								
		%						ppm		
Fumigation	Dry wt (g/plant)	N	P	S	K	Ca	Si	Cl	Zn	Mn
No	4.8	1.0	0.19	0.12	1.8	0.15	1.1	0.57	5	17
Yes	8.1	1.6	0.35	0.24	3.5	0.21	1.6	0.90	11	42

(*b*) Final harvest

Fumigation	Total plant dry wt (g/plot)	Grain (g/plot)	% N in grain
No	973	202	2.0
Yes	2,064	418	2.4

Source: Rovira and Hutton, C.S.I.R.O. Adelaide, personal communication.
Note: Fumigation treatment: Methyl bromide-chloropicrin (98:2) at 400 lb/acre 17 days before seeding.

Wilhelm, 1966; Danielson and Davey, 1969). The marked increase in growth and uptake of nutrients following fumigation of soil (Table 7.10) indicates that in a natural soil the uptake and translocation of nutrients is limited by the soil microflora. Some of this effect may be attributed to the destruction of root pathogens and release of nutrients from the biomass, but other significant factors may also be the destruction of nonpathogenic organisms that restrict root growth and nutrient uptake and the provision of N in the available ammonium form.

The results (Table 7.10) demonstrate that in soil large reserves of nutrients exist which by proper modification of the microflora and fauna of the soil could improve both the yield and quality of crops.

V. CONCLUSIONS

This chapter demonstrates the complexity of biology of the rhizosphere. Clearly, the microflora of the rhizosphere may influence plant growth in many ways. Although the most obvious effect of the rhizosphere population on plants is the damage caused by root pathogens, there are other effects that may be as great but by virtue of their more subtle effects upon the plant are not as obvious as effects upon the growth, morphology, and nutrient uptake of roots.

Despite the great deal of work already devoted to this zone of soil, we have not yet learned to control it so as to improve plant growth. True, there have been successes in controlling some root pathogens and in stimulating plant growth by seed inoculation, but our overall ignorance of the total ecology of the rhizospheres of plants growing in the field still precludes the immediate prospects of manipulating this important volume of the soil to increase plant growth. If we are to control the rhizosphere microflora consistently, then new techniques must be developed to study the population dynamics of the micro-organisms. These studies must be conducted at a fundamental level with relevance to particular field problems, e.g., root pathogens, nutrient uptake, and plant competition.

We cannot visualize a quick and easy path toward this goal of controlling the biology of the rhizosphere, but the rewards of success should be high in terms of the ultimate effects upon food production.

References

Agnihotri, V. P. 1964. Studies on Aspergilli. XIV. Effect of foliar spray of urea on the Aspergilli of the rhizosphere of *Triticum vulgare* L. *Plant Soil* 20: 364–70.

Barber, D. A. 1966. Effect of micro-organisms on nutrient absorption by plants. *Nature* 212: 638–40.

——. 1968. Micro-organisms and the inorganic nutrition of higher plants. *Ann. Rev. Plant Physiol.* 19: 71–88.

——. 1969. The influence of the microflora on the accumulation of ions by plants. In *Ecological Aspects of the Mineral Nutrition of Plants*, ed. I. H. Rorison, pp. 191–99. Symp. Brit. Ecol. Soc. no. 9, Sheffield, 1968. Oxford and Edinburgh: Blackwell Scientific Publications.

Barber, D. A., and B. C. Loughman. 1967. The effect of micro-organisms on the absorption of inorganic nutrients by plants. II. Uptake and utilization of phosphate by barley plants grown under sterile and non-sterile conditions. *J. Exp. Bot.* 18: 170–76.

Barber, D. A., J. Sanderson, and R. S. Russell. 1968. The influence of micro-organisms on the distribution of ^{32}P-labelled phosphate in roots. *Nature* 217: 644.

Barber, S. A., and P. G. Ozanne 1970. Autoradiographic evidence for the differential effect of four plant species in altering the calcium content of the rhizosphere soil. *Soil Sci. Soc. Amer. Proc.* 34: 635–37.

Barley, K. P., and A. D. Rovira. 1970. The influence of root hairs on the uptake of phosphate. *Soil Sci. Plant Anal.* 1: 287–92.

Beutner, H. E. 1961. Immuno-fluorescent staining—the fluorescent antibody method. *Bacteriol. Rev.* 25: 49–75.

Bird, A. F. 1959. The attractiveness of roots to the plant parasitic nematodes *Meloidogyne javanica* and *M. hapla. Nematologica* 4: 322–35.

Bonner, J. 1946. Relation of toxic substances to growth of guayule in soil. *Bot. Gaz.* 107: 343–51.

Bonner, J., and A. W. Galston. 1944. Toxic substances from the culture media of guayule which may inhibit growth. *Bot. Gaz.* 106: 185–98.

Borner, H. 1960. Liberation of organic substances from higher plants and their role in the soil sickness problem. *Bot. Rev.* 26: 393–424.

Boulter, D., J. J. Jeremy, and M. Wilding. 1966. Amino acids liberated into the culture medium by pea seedling roots. *Plant Soil* 24: 121–27.

Bowen, G. D. 1968. Chloride efflux along *Pinus radiata* roots. *Nature* 218: 686–87.

——. 1969. Nutrient status effects on loss of amides and amino acids from pine roots. *Plant Soil* 30: 139–42.

Bowen, G. D., and A. D. Rovira. 1961a. Effects of micro-organisms on plant growth. I. Development of roots and root hairs in sand and agar. *Plant Soil* 15: 166–88.

——. 1961b. Plant growth in irradiated soil. *Nature* 191: 936–37.

——. 1966. Microbial factor in short term phosphate uptake studies with plant roots. *Nature* 211: 665–66.

——. 1967. Phosphate uptake along attached and excised roots measured by an automatic scanning method. *Aust. J. Biol. Sci.* 20: 369–78.

——. 1969. The influence of micro-organisms on growth and metabolism of plant roots. In *Root Growth*, ed. W. J. Whittington, pp. 170–201. London: Butterworths.

Brown, M. E., S. K. Burlingham, and R. M. Jackson. 1962. Studies on *Azotobacter* species in soil. II. Populations of *Azotobacter* in the rhizosphere and effects of artificial inoculation. *Plant Soil* 18: 320–32.

Brown, R., and M. Edwards. 1944. The germination of the seed of *Striga lutea*. 1. Host influence and the progress of germination. *Ann. Bot.* (London), n.s. 8: 131–48.

Burges, N. A. 1960. Dynamic equilibria in the soil. In *The Ecology of Soil Fungi*, ed. D. Parkinson and J. S. Waid, pp. 185–91. Liverpool: Liverpool University Press.

Buxton, E. W. 1957. Some effects of pea root exudates on physiologic races of *Fusarium oxysporum* Fr. f. *pisi* (Linf.) Snyder and Hansen. *Trans. Brit. Mycol. Soc.* 40: 145–54.

Calum, C. T., H. Raistrick, and A. R. Todd. 1949. The potato eelworm hatching factor. I. The preparation of concentrates of the hatching factor and a method of bioassay. *Biochem. J.* 45: 513–19.

Casida, L. E. 1968. Infra-red color photography: selective demonstration of bacteria. *Science* 159: 199–200.

Chang, I., and T. Kommedahl. 1968. Biological control of seedling blight of corn by coating kernels with antagonistic micro-organisms. *Phytopathology* 58: 1395–1401.

Cholodny, N. G. 1951. Soil atmosphere as a source of organic nutrient substances for plants. *Pedology* 1: 16–19.

Clark, F. E. 1947. Rhizosphere microflora as affected by soil moisture. *Soil Sci. Soc. Amer. Proc.* 12: 239–42.

Cooper, R. 1959. Bacterial fertilizers in the Soviet Union. *Soils Ferts.* 22: 327–33.

Crossett, R. H. 1967. Autoradiography of ^{32}P in maize roots. *Nature* 213: 312–13.

Danielson, R. M., and C. B. Davey. 1969. Microbial recolonization of a fumigated nursery soil. *Forest Sci.* 15: 368–80.

Darbyshire, J. F. 1966. Protozoa in the rhizosphere of *Lolium perenne* L. *Can. J. Microbiol.* 12: 1287–89.

Darbyshire, J. F., and M. P. Greaves. 1967. Protozoa and bacteria in the rhizosphere of *Sinapis alba* L., *Trifolium repens* L., and *Lolium perenne* L. *Can. J. Microbiol.* 13: 1057–68.

——. 1970. An improved method for the study of the interrelationships of soil micro-organisms and plant roots. *Soil Biol. Biochem.* 2: 63–71.

Dart, P. J. 1971. Scanning electron microscopy of plant roots. *J. Exp. Bot.* 22: 163–65.

Dart, P. J., and F. V. Mercer. 1964. The legume rhizosphere. *Arch. Mikrobiol.* 44: 344–78.

Davey, C. B. 1954. Evaluation of composted fertilizers by microbiological methods of analysis. *Wis. Acad. Sci., Arts, and Letters* 43: 93–96.

Davey, C. B., and G. C. Papavizas. 1959. Effect of organic soil amendments on the Rhizoctonia disease of snap beans. *Agron. J.* 51: 493–96.

——. 1961. Translocation of streptomycin from Coleus leaves and its effect on rhizosphere bacteria. *Science* 134: 1368–69.

Deb, D. L., and F. Scheffer. 1970. Effect of the amino acid fraction of root exudate on the absorption of manganese by eight varieties of oat (*Avena sativa*) in sterile and non-sterile media. *Agrochimica* 15: 74–84.

Domsch, K. H. 1963. Der einfluss saprophytischer bodenpilze auf die jugendentwicklung höherer pflanzen. *Pflanzenkrankh. Pflanzenschutz.* 70: 470–75.

Foster, R. C., and G. C. Marks. 1967. Observations on the mycorrhizas of forest trees. II. The rhizosphere of *Pinus radiata* D. Don. *Aust. J. Biol. Sci.* 20: 915–26.

Frenzel, B. 1960. Zur Ätiologie der Anreicherung von Aminosäuren und Amiden im Wurzelraum von *Helianthus annus* L.: ein Beitrag zur Klärung der Probleme der Rhizosphäre. *Planta* 55: 169–207.

Garb, S. 1961. Differential growth-inhibitors produced by plants. *Bot. Rev.* 27: 422–43.

Garrett, S. D. 1956. *Biology of Root-Infecting Fungi.* Cambridge: Cambridge University Press. 293 pp.

Gerretson, F. C. 1948. The influence of micro-organisms on the phosphate intake by plants. *Plant Soil* 1: 51–81.

Goring, C. A. I., and F. E. Clark. 1948. Influence of crop growth on mineralization of nitrogen in the soil. *Soil Sci. Soc. Amer. Proc.* 13: 261–66.

Gray, T. R. G. 1967. Stereoscan electron microscopy of soil micro-organisms. *Science* 155: 1668–70.

Griffin, D. M. 1963. Soil moisture and the ecology of soil fungi. *Biol. Rev.* (Cambridge Phil. Soc.) 38: 141–66.

Harley, J. L., and J. S. Waid. 1955. A method of studying active mycelia on living roots and other surfaces in the soil. *Trans. Brit. Mycol. Soc.* 38: 104–18.

Harsmen, G. W., and G. Jager. 1962. Determination of the quantity of carbon and nitrogen in the rhizosphere of young plants. *Nature* 195: 1119–20.

——. 1963. Determination of the quantity of carbon and nitrogen in the rhizosphere of young plants. In *Soil Organisms*, ed. J. Doeksen and J. van der Drift, pp. 245–51. Amsterdam: North-Holland.

Head, G. C. 1964. A study of "exudation" from root hairs of apple roots by time-lapse cine-photomicrography. *Ann. Bot.* (London), n.s. 28: 495–98.

Henderson, V. E., and H. Katznelson. 1961. The effect of plant roots on the nematode population of the soil. *Can. J. Microbiol.* 7: 163–67.

Hiltner, L. 1904. Ueber neuere Erfahrungen und Probleme auf dem Gebiet der Boden-bakteriologie und unter besonderer Berücksichtigung der Gründingung und Brache. *Arb. Deut. Landw. Ges.* 98: 59–78.

Holland, A. A. 1962. The effect of indigenous saprophytic fungi upon nodulation and establishment of subterranean clover. In *Antibiotics in Agriculture*, ed. M. Woodbine, pp. 147–64. Proc. Univ. Nottingham, 9th Easter School. London: Butterworths.

Horst, R. K., and L. I. Kerr. 1962. Effects of foliar urea treatment on numbers of actinomycetes antagonistic to *Fusarium roseum* f. *cerealis* in the rhizosphere of corn seedlings. *Phytopathology* 52: 423–27.

Hurtt, W., and C. L. Foy. 1965. Some factors influencing the excretion of foliarly-applied dicamba and picloram from roots of Black Valentine Beans. *Plant Physiol.* (supp.) 40: 58.

Husain, S. S., and W. E. McKeen. 1963. Interactions between strawberry roots and *Rhizoctonia fragariae*. *Phytopathology* 53: 541–45.

Ingestad, T., and N. Molin. 1960. Soil disinfection and nutrient status of spruce seedlings. *Physiol. Plant.* 13: 90–103.

Ishizawa, S., T. Susuki, O. Sato, and H. Toyoda. 1957. Studies on microbial population in the rhizosphere of higher plants with special reference to the method of study. *Soil Plant Food* (Tokyo) 3: 85–94.

Ivanov, V. P. 1962. Mutual effect through the root system of a mixed crop of corn and broad beans. *Fiziol. Rast.* 9: 179–88.

Ivanov, V. P., G. A. Yacobsen, and B. S. Fomenko. 1964. Effect of soil moisture on metabolism of root excretions. *Fiziol. Rast.* 11: 630–37.

Jackson, R. M., M. E. Brown, and S. K. Burlingham. 1964. Similar effects on tomato plants of *Azotobacter* inoculation and application of gibberellins. *Nature* 203: 851–52.

Jenny, H., and K. Grossenbacher. 1963. Root-soil boundary zones as seen in the electron microscope. *Soil Sci. Soc. Amer. Proc.* 27: 273–77.

Jones, P. C. T., and J. E. Mollison. 1948. A technique for the quantitative estimation of soil micro-organisms. *J. Gen. Microbiol.* 2: 54–69.

Katznelson, H. 1946. The rhizosphere effect of mangels on certain groups of micro-organisms. *Soil Sci.* 62: 343–54.

———. 1960. Observations on the rhizosphere effect. In *The Ecology of Soil Fungi*, ed. D. Parkinson and J. S. Waid. Liverpool: Liverpool University Press.

Katznelson, H., and J. W. Rouatt. 1957. Studies on the incidence of certain

physiological groups of bacteria in the rhizosphere. *Can. J. Microbiol.*
3: 265–69.

Katznelson, H., J. W. Rouatt, and T. M. B. Payne. 1955. The liberation of
amino acids and reducing compounds by plant roots. *Plant Soil* 7: 35–48.

——. 1956. Recent studies on the microflora of the rhizosphere. *Trans. 6th
Int. Congr. Soil Sci.* (Paris), pp. 151–56.

Kerr, A. 1964. The influence of soil moisture on infection of peas by *Pythium
ultimum. Aust. J. Biol. Sci.* 17: 676–85.

Krasil'nikov, N. A. 1958. *Soil Micro-organisms and Higher Plants.* Moscow:
Academy of Sciences, U.S.S.R.. Eng. translation.

——. 1961. On the role of soil bacteria in plant nutrition. *J. Gen. Appl.
Microbiol.* (Tokyo) 7: 128–44.

Linder, P. J., J. W. Mitchell, and G. D. Freeman. 1964. Persistence and translocation of exogenous regulating compounds that exude from roots. *J. Agr.
Food Chem.* 12: 437–38.

Lochhead, A.G., and F. E. Chase. 1943. Qualitative studies of soil micro-
organisms. V. Nutritional requirements of predominant bacterial flora. *Soil
Sci.* 55: 185–95.

Lochhead, A. G., and J. W. Rouatt. 1955. The "rhizosphere effect" on the
nutritional groups of soil bacteria. *Soil Sci. Soc. Amer. Proc.* 19: 48–49.

Longberg-Holm, K. K. 1967. Nucleic acid synthesis in seedlings. *Nature* 213:
454–57.

Louw, H. A., and D. M. Webley. 1959. The bacteriology of the root region
of the oat plant grown under controlled pot culture conditions. *J. Appl.
Bacteriol.* 22: 216–26.

McCalla, T. M., and F. A. Haskins. 1964. Phytotoxic substance from soil
micro-organisms and crop residues. *Bacteriol. Rev.* 28: 181–207.

McDougall, B. M. 1968. The exudation of C^{14}-labelled substances from roots
of wheat seedlings. *Trans. 9th Int. Congr. Soil Sci.* (Adelaide) 3: 647–55.

——. 1970. Movement of C^{14}-photosynthate into roots of wheat seedlings
and exudation of C^{14} from intact roots. *New Phytol.* 69: 37–46.

McDougall, B. M., A. D. Rovira. 1970. Sites of exudation of C^{14}-substances
from wheat roots. *New Phytol.* 69: 999–1003.

Martin, J. K. 1971a. The influence of plant species and plant age on the rhizosphere microflora. *Aust. J. Biol. Sci.* 24: 1143–50.

——. 1971b. ^{14}C-labeled material leached from the rhizosphere of plants
supplied with $^{14}CO_2$. *Aust. J. Biol. Sci.* 24: 1131–42.

Martin, J. P. 1963. Influence of pesticide residues on soil microbiological and
chemical properties. *Residue Rev.* 4: 96–129.

Martin, P. 1958. Einfluss der Kulturfiltrate von Microorganismen auf die
Abgabe von Skopoletin aus den Keimwurzeln des Hafers (*Avena sativa* L.)
Arch. Mikrobiol. 29: 154–86.

Meshkov, M. V. 1953. Works of symposium on soil microbiology. In *Soil*

Micro-organisms and Higher Plants, ed. Krasil'nikov, p. 285. Translated and published for the N.S.F. and U.S.D.A. by the Israel Program for Scientific Translations, 1961.

Mishustin, E. N., and A. N. Naumova. 1955. Secretion of toxic substances by alfalfa and their effects on cotton and soil microflora. *Izvest. Akad. Nauk S.S.S.R., Ser. Biol.* 6: 3–9.

——. 1962. Bacterial fertilizers; their effectiveness and mode of action. *Mikrobiologiya* 31: 543–55.

Molina, J. A. E., and A. D. Rovira. 1964. The influence of roots on autotrophic nitrifying bacteria. *Can. J. Microbiol.* 10: 249–57.

Moore, D. R. E., and J. S. Waid. 1971. The influence of washings of living roots on nitrification. *Soil. Biol. Biochem.* 3: 69–83.

Norman, A. G. 1955. The effect of polymyxin on plant roots. *Arch. Biochem. Biophys.* 58: 461–77.

——. 1961. Microbial products affecting root development. *Trans. 7th Int. Congr. Soil Sci.* (Madison) 2: 531–36.

Nye, P. H. 1966. The effect of the nutrient intensity and buffering power of a soil and the absorbing power, size and root hairs of a root, on nutrient absorption by diffusion. *Plant Soil* 25: 81–105.

Otto, G. 1965. The influence exerted by fungi in the rhizosphere on the density and ramification of cucumber roots. In *Plant-Microbes Relationships*, ed. J. Macura and V. Vancura, pp. 204–19. Prague: Czech. Acad. Sci.

Papavizas, G. C., and C. B. Davey. 1961. Extent and nature of the rhizosphere of Lupinus. *Plant Soil* 14: 215–36.

Parkinson, D. 1965. The development of fungi in the root region of crop plants. In *Plant-Microbes Relationships*, ed. J. Macura and V. Vancura, pp. 69–75. Prague: Czech. Acad. Sci.

Parkinson, D., and J. H. Clarke. 1961. Fungi associated with seedling roots of *Allium porrum* L. *Plant Soil* 13: 384–90.

Parkinson, D., G. S. Taylor, and R. Pearson. 1963. Studies on the fungi in the root region. I. The development of fungi on young roots. *Plant Soil* 19: 332–49.

Patrick, Z. A., and T. A. Toussoun. 1965. Plant residues and organic amendments in relation to biological control. In *Ecology of Soil-Borne Plant Pathogens—Prelude to Biological Control*, ed. K. F. Baker and W. C. Snyder, pp. 440–59. Berkeley and Los Angeles: University of California Press.

Patrick, Z. A., T. A. Toussoun, and L. W. Koch. 1964. Effect of crop-residue decomposition products on plant roots. *Ann. Rev. Phytopathol.* 2: 267–92.

Peacock, F. C. 1966. Nematode control by plant chemotherapy. *Nematologica* 12: 70–86.

Pearson, R., and D. Parkinson. 1961. The sites of excretion of ninhydrin positive substances by broad bean seedlings. *Plant Soil* 13: 391–96.

Rahteenko, I. N. 1958. On the transfer of mineral nutrients from one plant to another owing to interactions between the root systems. *Bot. Zh.* 43: 695–701.

Ramachandra-Reddy, T. K. 1959. Foliar spray of urea and rhizosphere microflora of rice (*Oryza sativa* L.). *Phytopathol. Z.* 36: 286–89.

Rangaswami, G., and H. Balasubramanian. 1963. Release of hydrocyanic acid by sorghum roots and its influence on the rhizosphere microflora and plant pathogenic fungi. *Indian J. Exp. Biol.* 1: 215–17.

Reid, C. R. P., and W. Hurtt. 1970. Root exudation of herbicides by woody plants: allelopathic implications. *Nature* 225: 291.

Rempe, J. K., and O. G. Kaltagova. 1965. Influence of root microflora on the increase, development and activity of physiological processes in plants. In *Plant-Microbes Relationships*, ed. J. Macura and V. Vancura, pp. 178–85. Prague: Czech. Acad. Sci.

Reuszer, H. W. 1949. A method for determining the carbon dioxide production of sterile and non-sterile root systems. *Soil Sci. Soc. Amer. Proc.* 14: 175–79.

Rice, E. L. 1964. Inhibition of nitrogen-fixing and nitrifying bacteria by seed plants (I). *Ecology* 45: 824–37.

——. 1965. Inhibition of nitrogen-fixing and nitrifying bacteria by seed plants. II. Characterization and identification of inhibitors. *Physiol. Plant.* 18: 255–68.

Ridge, E. H., and A. D. Rovira. 1968. Microbial inoculation of wheat. *Trans. 9th Int. Congr. Soil Sci.* (Adelaide), 3: 473–81.

——. 1971. Phosphatase activity of intact young wheat roots under sterile and non-sterile conditions. *New Phytol.* 70: 1017–26.

Riley, D., and S. A. Barber. 1969. Bicarbonate accumulation and pH changes at the soybean (*Glycine max* (L.) Mere.) root-soil interface. *Soil Sci. Soc. Amer. Proc.* 33: 905–8.

Rivière, J. 1959. Contribution a l'étude de la rhizosphère du blé. D.Sc. Thesis. University of Paris.

——. 1960. Étude de la rhizoshpère du blé. *Ann. Agron.* 11: 397–440.

——. 1963. Rhizosphère et croissance du blé. *Ann. Agron.* 14: 619–53.

Robinson, A. C. 1967. The influence of host on soil and rhizosphere populations of clover and lucerne root nodule bacteria in the field. *J. Aust. Inst. Agr. Sci.* 33: 207–9.

Rohde, R. A., and W. R. Jenkins. 1958. *Basis of Resistence of* Asparagus officinalis *var*. altius L. *to the Stubby Root Nematode* Trichodorus christiei. Univ. Maryland Agr. Exp. Sta. Bull. A97, 19 pp.

Rouatt, J. W., and H. Katznelson. 1961. A study of the bacteria on the root surface and in the rhizosphere soil of crop plants. *J. Appl. Bacteriol.* 24: 164–71.

Rouatt, J. W., H. Katznelson, and T. M. B. Payne. 1960. Statistical evaluation of the rhizosphere effect. *Soil Sci. Soc. Amer. Proc.* 24: 271–73.

Rouatt, J. W., M. Lechevalier, and S. A. Waksman. 1951. Distribution of antagonistic properties among *Actinomycetes* isolated from different soils. *Antibiot. and Chemother.* 1: 185–92.

Roux, E. R. 1953. The effect of antibiotics produced by *Trachypogon plumosus* on the germination of seeds of the kakiebos (*Tage, minuta*). *South Afr. J. Sci.* 49: 334.

Rovira, A. D. 1956. Plant root excretions in relation to the rhizosphere effect. I. The nature of root exudate from oats and peas. *Plant Soil* 7: 178–94.

——. 1959. Root excretions in relation to the rhizosphere effect. IV. Influence of plant species, age of plants, light, temperature and calcium nutrition on exudation. *Plant Soil* 11: 53–64.

——. 1963. Microbial inoculation of plants. I. Establishment of free-living nitrogen-fixing bacteria in the rhizosphere and their effects on maize, tomato and wheat. *Plant Soil* 19: 304–14.

——. 1969. Diffusion of carbon compounds away from wheat roots. *Aust. J. Biol. Sci.* 22: 1287–90.

Rovira, A. D., and G. D. Bowen. 1966a. The effects of micro-organisms upon plant growth. 2. Detoxication of heat-sterilized soils by fungi and bacteria. *Plant Soil* 25: 129–42.

——. 1966b. Phosphate incorporation by sterile and non-sterile plant roots. *Aust. J. Biol. Sci.* 19: 1167–69.

——. 1970. Translocation and loss of phosphate along roots of wheat seedlings. *Planta* 93: 15–25.

Rovira, A. D., and W. R. Stern. 1961. Rhizosphere bacteria in grass-clover associations. *Aust. J. Agr. Res.* 12: 1108–18.

Schroth, M. N., and D. C. Hildebrand. 1964. Influence of plant exudates on root-infecting fungi. *Ann. Rev. Phytopathol.* 2: 101–32.

Schroth, M. N., and W. C. Snyder. 1961. Effect of host exudates on chlamydo-spore germination of the bean root rot fungus *Fusarium solani* f. *phaseoli*. *Phytopathology* 51: 389–93.

Slankis, V., V. C., Runeckles, and G. Krotkov. 1964. Metabolites liberated by roots of white pine (*Pinus strobus* L.) seedlings. *Physiol. Plant.* 17: 301–13.

Smith, J. H., F. E. Allison, and D. A. Soulides. 1961. Evaluation of phospho-bacterin as a soil inoculant. *Soil Sci. Soc. Amer. Proc.* 25: 109–11.

Smith, W. H. 1969. Release of organic materials from the roots of tree seedlings. *Forest Sci.* 15: 138–43.

——. 1970. Root exudates of seedling and mature sugar maple. *Phytopathology* 60: 701–3.

Strzelcyk, E. 1961. Studies on the interaction of plants and free-living nitrogen-fixing micro-organisms. II. Development of antagonists of *Azotobacter* in the rhizosphere of plants at different stages of growth in two soils. *Can. J. Microbiol.* 7: 507–13.

Subba-Rao, M. S., G. S. Bidwell, and D. L. Bailey. 1962. Studies of rhizosphere activity by the use of isotopically labeled carbon. *Can. J. Bot.* 40: 203–12.

Theron, J. J. 1951. The influence of plants on the mineralization of nitrogen and the maintenance of organic matter in the soil. *J. Agr. Sci.* 41: 289–96.

Timonin, M. I. 1946. Microflora of the rhizosphere in relation to the manganese-deficiency disease in oats. *Soil Sci. Soc. Amer. Proc.* 11: 284–92.

Trinick, M. J. 1969. Identification of legume nodule bacteroids by the fluorescent antibody reaction. *J. Appl. Bacteriol.* 32: 181–86.

——. 1970. Rhizobium interactions with soil micro-organisms. Ph.D. Thesis. University of Western Australia.

Trolldenier, G. 1965a. Fluoreszenmikroskopische untersuchung von mikroorganismenreinkulturen in der Rhizosphäre. *Z. Bakteriol. Parasitenk. Infekt. Hyg.* II. 119: 256–59.

——. 1965b. Fluoreszenmikroskopische untersuchung der Rhizosphäre. *Z. Landwirts. Forsch.* supplement 19: 1–7.

Trolldenier, G., and V. Marckwordt. 1962. Untersuchungen über den einfluss der bodenmikroorganismen auf rubidium- und calcium-aufnahme in nährlösung wachsender pflanzen. *Árch. Mikrobiol.* 43: 148–51.

Vancura, V. 1964. Root exudates of plants. 1. Analysis of root exudates of barley and wheat in their initial phases of growth. *Plant Soil* 21: 231–48.

Vancura, V., and J. L. Garcia. 1969. Root exudates of reversibly wilted millet plants (*Panicum miliaceum* L.). *Oecol. Plant.* (Gauthier Villars) 4: 93–98.

Vancura, V., and A. Hovadik. 1965. Root exudates of plants. II. Composition of root exudates of some vegetables. *Plant Soil* 22: 21–32.

Varga, L. 1958. Protozoa in the root zone of sugar beet. *Agrokem Talajtan.* 7: 393–400 (Eng. trans.).

Venkata-Ram, C. S. 1960. Foliar application of nutrients and rhizosphere microflora of *Camellia sinensis. Nature* 187: 621–22.

Vrany, J. 1965. Effect of foliar applications on rhizosphere microflora. In *Plant-Microbes Relationships*, ed. J. Macura and V. Vancura, pp. 21–25. Prague: Czech. Acad. Sci.

Vrany, J., V. Vancura, and J. Macura. 1962. The effects of foliar application of some readily metabolized substances, growth regulators and antibiotics on rhizosphere microflora. *Folia Microbiol.* 7: 61–70.

Wallace, H. R. 1958. Observations on the emergence from cysts and the orientation of larvae of three species of the genus *Heterodera* in the presence of host plant roots. *Nematologica* 3: 236–43.

——. 1961. Factors influencing the ability of *Heterodera* larvae to reach host plant roots. In *Recent Advances in Botany*, pp. 407–10. Toronto: University of Toronto Press.

Warcup, J. H. 1957. Chemical and biological aspects of soil sterilization. *Soils Ferts.* 20: 1–5.

Welte, E., and G. Trolldenier. 1962. Der einfluss der bodenmikroorganismen auf trockensubstanzbildung und aschegehalt in nährlösung wachsender pflanzen. *Arch. Mikrobiol.* 43: 138–47.

——. 1965. Effect of soil microflora and seed microflora on the growth of plants. In *Plant-Microbes Relationships*, ed. J. Macura and V. Vancura, pp. 186–92. Prague: Czech. Acad. Sci.

Wilhelm, S. 1966. Chemical treatments and inoculum potential of soil. *Ann. Rev. Phytopathol.* 4: 53–78.

Wilkinson, H. F., J. F. Loneragan, and J. P. Quirk. 1968. Calcium supply to plant roots. *Science* 161: 1245–46.

Woldendorp, J. W. 1963. The influence of living plants on denitrification. *Mededel. Landbouwh. Wageningen* 63: 1–100.

Woods, F. W. 1960. Biological antagonisms due to phytotoxic root exudates. *Bot. Rev.* 26: 546–69.

Woods, F. W., and K. Brock. 1964. Interspecific transfer of Ca^{45} and P^{32} by root systems. *Ecology* 45: 886–89.

Zagallo, A. C., and H. Katznelson. 1957. Metabolic activity of bacterial isolates from wheat rhizospheres and control soil. *J. Bacteriol.* 73: 760–64.

Zentmyer, G. A. 1961. Chemotaxis of zoospores for root exudates. *Science* 133: 1595–96.

8. Mycorrhizae

J. W. Gerdemann

IN NATURE the roots of most plants are invaded by fungi and transformed into mycorrhizae or "fungus-roots." The host and fungus live together in an intimate balanced relationship, and generally the association is beneficial to both organisms. The morphology of the root is modified; however, as long as a balanced relationship is maintained there are no pathological symptoms. Nutrients are absorbed from the soil by the fungus and released to host cells, and the mycorrhizal fungus in turn obtains food from the host.

The degree of benefit that each partner receives varies. Plants that lack chlorophyll may depend completely upon mycorrhizal fungi to supply them with both organic and inorganic nutrients. At the other extreme, it seems likely that some plant species are capable of absorbing sufficient nutrients from a highly fertile soil to achieve maximum growth without mycorrhizal fungi. Most mycorrhizal associations probably fall somewhere between these extremes; however, there is very little information on the degree of benefit that any plant species receives when grown at various well-defined nutrient levels. A better understanding of plant nutrition depends upon more knowledge of the role that mycorrhizal fungi play in nutrient uptake and of the degree of dependence that each host species has for a fungal symbiont.

There are several distinct kinds of mycorrhizae: (*a*) ectomycorrhiza, in which the fungus forms a compact mantle over the root surface and produces intercellular hyphae in the cortex; (*b*) ectendomycorrhiza, which is similar to ectomycorrhiza but has both intercellular and intracellular hyphae; and (*c*) endomycorrhiza, which has a loose network of hyphae in the soil and extensive growth within the cortex. Endomycorrhizae are subdivided into two groups, (1) those formed by septate fungi and (2) those formed by nonseptate fungi. The latter type is often called *phycomycetous* or *vesicular-arbuscular*

(VA) mycorrhiza. There have been a number of recent reviews of this subject. For comprehensive reviews, see Harley (1965, 1969), Bowen (1969), and Hacskaylo (1971). For literature reviews concerned with VA mycorrhizae, see Mosse (1963), Nicolson (1967), and Gerdemann (1968, 1970).

I. ECTOMYCORRHIZAE

Ectomycorrhizae are found on species in the Pinaceae, Betulaceae, Fagaceae, and a few other plant families. The hosts with few exceptions are trees or shrubs. The morphology of ectomycorrhizae varies depending on the host species and the species of mycorrhizal fungus. However, certain generalizations are possible. Typically ectomycorrhizae are thicker and more branched than nonmycorrhizal roots. The entire structure is enclosed in a layer of fungal tissue called the *fungal sheath* or *mantle*. Hyphae from the sheath penetrate between the epidermal and cortical cells to form a network of intercellular hyphae called the *Hartig net*. Strands of hyphae also grow from the sheath out into the soil, and these hyphae and the outer surface of the sheath constitute the absorbing surface of a mycorrhizal plant. Exchange of materials occurs primarily in the Hartig net, where there is extensive contact between fungus and host cells.

Ectomycorrhizal fungi produce auxins that are responsible for some of the morphological differences between mycorrhizal and nonmycorrhizal roots (Slankis, 1958; Moser, 1959). More recently it has been shown that an ectomycorrhizal fungus produces cytokinins when grown in culture (Miller, 1967). Whether or not cytokinins are produced in mycorrhizae is as yet unknown.

A modification of ectomycorrhiza, termed *ectendomycorrhiza*, sometimes occurs on ectomycorrhizal species. It is characterized by a thin fungal sheath and intracellular invasion of host cells by hyphae from the Hartig net. It occurs most frequently in forest tree nurseries and is believed to be caused by a specific unidentified fungus. It is usually replaced by typical ectomycorrhizae when seedlings are transplanted to forest soils (Mikola, 1965).

There are many species of fungi that produce ectomycorrhizae. Most of them are Basidiomycetes in the Agaricales and Gasteromycetes (Trappe, 1962). A number of Ascomycetes and a few Fungi Imperfecti and Phycomycetes also produce ectomycorrhizae. Some ectomycorrhizal fungi are highly specialized and are associated with only a single host species. Others, such as *Cenococcum graniforme* (Sow.) Ferd. & Winge, have extremely wide host ranges, and it is likely that *C. graniforme* can form mycorrhizae with most ectomycorrhizal hosts (Trappe, 1964).

Ectomycorrhizal fungi as a rule utilize only relatively simple carbo-hydrates. The majority of these fungi are deficient in thiamine, other vitamins, and organic growth factors that are obtained from the host. They cannot decompose lignin, and only a few have been shown to utilize cellulose. Therefore, it is unlikely that these fungi are capable of much decompostion of plant debris or of making appreciable growth in the soil apart from living roots (Harley, 1969).

The intensity of ectomycorrhizal infection tends to be highest when plants are grown in soils with a moderately low or unbalanced nutrient status. The intensity of infection may be reduced either in extremely infertile soil or under conditions of high balanced mineral nutrition (Hatch, 1937; Björkman, 1942; Harley, 1969). Higher levels of infection also occur in raw humus forest soils than in agricultural soil high in mineral nutrients but low in humus. Mycorrhizal development is also influenced by light intensity; increases in light intensity have been reported to increase the degree of mycorrhizal infection (Björkman, 1942).

There is abundant evidence that ectomycorrhizal infection increases nutrient absorption and improves plant growth (Harley, 1969). Various theories have been proposed to account for the increased capacity of ectomycorrhizal roots to absorb minerals, and it is probable that many factors are involved. Mycorrhizal infection increases the amount of absorbing surface in contact with soil. Mycorrhizal roots are stimulated to branch, and the diameters of individual mycorrhizae are greater than those of comparable nonmycorrhizal roots (Hatch, 1937). Ectomy-corrhizal infection also prolongs the life of roots, thus greatly increasing the total amount of actively absorbing surface area.

The amount of actively absorbing surface is further increased by the external hyphae that grow from the fungal sheath into the soil. In a series of papers Melin and Nilsson (Harley, 1969) have shown that such hyphae function in a similar manner to root hairs. Various nutrients were absorbed by hyphae and translocated to the host plant. Ecto-mycorrhizae remain active in the same position in the soil for relatively long periods of time while the effective absorbing portion of the un-infected root tip "moves" through the soil as the root elongates. Because of the time required for ion movement through soil, an organ that remains active in the same position for a longer time can obtain nutrients from a larger volume of soil (Bowen and Theodorou, 1967).

In addition to increasing the surface area through which absorption can take place, mycorrhizae absorb nutrients more rapidly per unit area than do nonmycorrhizal roots (Kramer and Wilbur, 1949; Harley and McCready, 1950).

Various nitrogen compounds are absorbed by mycorrhizae. Also, claims have been made that ectomycorrhizae or ectomycorrhizal fungi can fix atmospheric N, but there is little experimental evidence that supports this view. It is possible that ectomycorrhizal fungi increase the availability of sparingly soluble compounds, but the evidence for this is somewhat conflicting (Harley, 1969).

There is good evidence that a part of the benefit the plant derives from ectomycorrhizal infection is due to protection of the root tissue from attack by pathogens (Zak, 1964; Marx and Davey, 1969a, 1969b; Marx, 1970). Both the fungal sheath and Hartig net apparently form a mechanical barrier to infection of pine roots by *Phytophthora cinnamomi* Rands, a common root pathogen (Marx, 1970). Also, a number of mycorrhizal fungi produce antibiotics that inhibit *P. cinnamomi* (Marx, 1969a, 1969b).

Trees that are normally ectomycorrhizal can be grown without mycorrhizae in liquid media in which all of the required nutrients are readily available. It has also been reported that they can be grown without mycorrhizae in highly fertile soil, although some investigators have failed to obtain normal growth even when the trees were well fertilized (Briscoe, 1959; Harley, 1969; Zak, 1964). For all practical purposes, under natural conditions, mycorrhizal infection is necessary if trees are to survive and grow satisfactorily.

Ectomycorrhizal fungi generally are not native in soils where their hosts do not occur, and there are numerous reports of failure of pines to grow when introduced into regions where pines are not native (Zak, 1964). Recent attempts to grow pines in Puerto Rico provide a good example (Briscoe, 1959; Vozzo, 1971). Here pines failed to grow even when they were provided with fertilizers. However, when they were inoculated with mycorrhizal fungi by importing and planting seedlings bearing mycorrhizae or by adding soil collected from native pine stands in the southeastern United States, most of the seedlings survived and made rapid growth.

Inoculations made by using infected seedlings or soil from native stands are somewhat dangerous. There is always a good posibility that pathogenic microorganisms will be introduced along with the mycorrhizal fungi. Inoculation using pure cultures is a much safer procedure; this also has been successfully done in Puerto Rico (Vozzo, 1971).

Mikola (1969, 1970) has discussed in detail the importance and techniques of inoculation with ectomycorrhizal fungi.

II. ENDOMYCORRHIZAE

Endomycorrhizae can be subdivided into two groups on the basis of the septation of the hyphae of the fungi. Septate endophytes occur in the Ericales, Orchidaceae, and a few other plant groups. Mycorrhizae formed by nonseptate fungi, the phycomycetous or VA mycorrhizae, occur on more plant species than any other kind. Most cultivated plants as well as many forest and shade trees, shrubs, and wild herbaceous plants have VA mycorrhizae.

A. Septate Endophytes: The Ericales

The mycorrhizae on some Ericales are somewhat transitional between ectomycorrhizae and typical endomycorrhizae. In *Arbutus, Arctostaphylos*, and the Pyrolaceae and Monotropaceae the fungi form hyphal sheaths quite similar to those formed in an ectomycorrhizal type. However, the fungi penetrate the cortex and form coils within the cells that are digested as in a typical endomycorrhiza. In most Ericaceae a hyphal sheath is not formed, and the fungi produce a loose weft of hyphae in the soil surrounding the roots. The fungi penetrate the cortex and produce complex coils within the cells. These coils soon disintegrate and are believed to be digested by the host. The host cells survive both the infection and digestion processes. During infection the host nucleus enlarges, and when digestion is complete, it returns to its normal size.

The fungi have been isolated in pure culture, but fruiting was not obtained and their taxonomic position is unknown. There have been several studies on function in which growth increases of host species were obtained.

Monotropa lacks chlorophyll, and it depends upon mycorrhizal fungi for a source of organic compounds. Its roots are completely enclosed in a fungal mantle, and the same fungus forms an ectomycorrhiza with nearby tree roots. When d-glucose-^{14}C or ^{32}P-phosphate was injected into such trees, these substances were translocated to the *Monotropa* plants. Thus, *Monotropa* is an epiparasite which depends upon mycorrhizal fungi to obtain food from nearby trees. The fungus may obtain a growth-promoting factor from *Monotropa*, as its growth in vitro is greatly stimulated by *Monotropa* extracts (Björkman, 1960).

B. Septate Endophytes: The Orchidaceae

The orchids are highly dependent upon mycorrhizal infection for survival and normal growth. Orchid seeds are extremely small and contain little stored food. Germination and growth of seedlings can be obtained in the laboratory if the seeds are supplied with sugars, minerals, nutrients, and vitamins, but under natural conditions they can grow only if the embryo becomes infected. Orchid species that develop chlorophyll become partially self-supporting, and some terrestrial species with large green leaves and well-developed roots may be nearly free of mycorrhizae as adults. On the other hand, those species that lack chlorphyll remain completely dependent upon mycorrhizal fungi throughout their entire life.

The fungi that form mycorrhizae with orchids are Basidiomycetes, or Fungi Imperfecti which are likely to be of basidiomycetous affinity. The orchid fungi can easily be isolated and grown in pure culture, and in nature they are able to grow saprophytically. They can utilize cellulose and transport sugars and mineral nutrients (Smith, 1966). Some orchid fungi are virulent pathogens on other plants. *Rhizoctonia* isolates that are capable of forming mycorrhizae on orchids can cause disease on various other plant species (Downie, 1957). *Armillaria mellea* (Vahl.) Karst., which is pathogenic on many woody plant species, forms mycorrhizae with orchids. The same mycelium that is pathogenic on a tree may at the same time form mycorrhizae with nearby orchids.

Unless a balanced relationship is maintained at all times between the fungus and the orchid, the fungus can become pathogenic and kill its host. The endophytes are controlled in at least two ways: 1) The fungi produce hyphal coils within the cells that are digested by the host, and it is generally believed that during this digestion process nutrients are transferred to the host. 2) Orchid plants also produce phytoalexins that are toxic to the endophytes and apparently restrict the fungal invasion of host tissue (Gäumann *et al.*, 1960). These phytoalexins form in higher concentrations in plant parts, such as tubers, that normally are not infected than in roots that become mycorrhizal.

C. Nonseptate Endophytes: Vesicular-Arbuscular Mycorrhiza

The vesicular-arbuscular mycorrhiza is found on more plant species than any other type of mycorrhiza. It occurs in the bryophytes, pteridophytes, and spermatophytes (gymnosperms and angiosperms) and is of particular interest because it is found on so many economically important species. Most agronomic and horticultural crop plants as well as many forest trees have this type. It occurs throughout the world from the tropics to the Arctic, and there are very few plant associations that do not contain some species with VA mycorrhizae.

The VA mycorrhizal infections produce very little change in external root morphology, and because of this they are frequently overlooked. They can be recognized on relatively thin unsuberized roots by their bright yellow color, but this color disappears rapidly when exposed to bright light. The VA mycorrhizae are also surrounded by an extensive growth of hyphae; however, external hyphae are generally lost when the mycorrhizae are removed from soil and washed. The simplest way to diagnose and observe VA mycorrhizae is by clearing and staining (Gerdemann, 1955). This method permits the rapid examination of large numbers of roots for infection. After processing, the coarse irregular hyphae, vesicles, and arbuscules are easily visible under relatively low magnification.

The VA fungi produce appressoria on the root surface and penetrate the epidermis. They are confined to the epidermis and cortex, and they do not invade the endodermis, stele, or root meristem. Growth within the cortex may either be entirely intracellular or intracellular and intercellular depending on the host species. The hyphae form complex coils and loops within cells. Although hyphae are nonseptate when young and healthy, some septa form when growing conditions are unfavorable or the fungus is dying. Arbuscules, complex, highly branched structures similar to haustoria, are produced within the cells, and as they form the nuclei within the cells become greatly enlarged. The arbuscules are generally assumed to be the major site of nutrient exchange. Soon after the arbuscules form they are digested, and the nuclei in invaded cells return to their normal size. Vesicles are terminal ovate to globose structures that contain drops of yellow oil. In some hosts they are intercellular and in others they are intracellular. If they remain thin-walled, they function as temporary food storage organs, but if the vesicle wall becomes thickened, they can function as resting spores. Occasionally vesicles form in such abundance that the cortex is ruptured. It is uncertain whether this occurs only in old cortical tissue that is no longer functional or if healthy young cortical tissue is sometimes destroyed. Ordinarily, the fungus does not kill host cells; there are usually no pathological symptoms.

The VA mycorrhizae are produced by species of Endogonaceae (Mosse, 1956; Nicolson and Gerdemann, 1968; Gerdemann, 1971). These fungi are apparently obligate symbionts, and it is uncertain if they can be obtained in pure culture on nutrient media (Gerdemann, 1968). They produce very large spores that can be extracted from soil by wet-sieving and decanting (Gerdemann, 1955; Gerdemann and Nicolson, 1963). This method has been used to demonstrate the abundance of *Endogone* species in many soils (Gerdemann and Nicolson, 1963; Mosse and Bowen, 1968; Hayman, 1970). Spores obtained by sieving have also been used to establish pot cultures of *Endogone* species. Inoculum, consisting of spores, sporocarps, or mycorrhizae from such cultures, has been used to investigate the fungi and determine their effect on plant growth.

Many mycorrhizal *Endogone* species have been found, and each has an extremely wide host range (Gerdemann, 1971). The same species can form VA mycorrhizae on such diverse plants as maize (*Zea mays*), red clover (*Trifolium pratense*), soybeans (*Glycine max*), onions (*Allium cepa*), and strawberries (*Fragaria* sp.) (Gerdemann, 1961) as well as on maize and tuliptrees (*Liriodendron tulipfera*) (Gerdemann, 1965). It is unusual for fungi that behave as obligate parasites to have such wide host ranges. Apparently there has not been natural selection within most plants for defense mechanisms against these fungi.

Many investigators have tested the effect of VA mycorrhizal infection on plant growth in greenhouse experiments under a wide variety of conditions. Most of them have obtained growth stimulation, particularly at low phosphorus levels (Gerdemann, 1968, 1970). At high P levels experiments are generally not sensitive enough to determine if the much smaller growth differences are significant (Daft and Nicolson, 1966, 1969; Holevas, 1966; Baylis, 1967; Murdoch *et al.*, 1967). High levels of P also depress the formation of mycorrhizae and sporulation of the fungus (Daft and Nicolson, 1969). Under field conditions addition of N or complete fertilizer reduced the amount of infection and the number of *Endogone* spores (Hayman, 1970).

Ross and Harper (1970) obtained increased growth and yield of soybeans grown in fumigated soil in field plots that had been inoculated with an *Endogone* species. Much more information is needed on the effect of VA mycorrhiza on plants grown in fertile soil in the field. In Illinois soybeans are highly mycorrhizal even in the most fertile soils. Even a slight increase or decrease in yield resulting from mycorrhizal infection would be of considerable economic importance.

The VA mycorrhizae increase the ability of plants to absorb nutrients. Mycorrhizal root portions of roots accumulated more ^{32}P than comparable nonmycorrhizal portions of root tips (Gray and Gerdemann, 1969). Mycorrhizal plants remove more nutrients from soil (Gerdemann, 1964, 1965), and the plants contain larger quantities of nutrients (Mosse, 1957; Baylis, 1959, 1967; Gerdemann, 1964; Holevas, 1966; Ross and Harper, 1970; Gilmore, 1971). Plants with VA mycorrhizae can better utilize the less available forms of P (Daft and Nicolson, 1966; Murdoch *et al.*, 1967). Baylis (1970) has presented evidence that woody species which have poorly developed root hairs are obligatorily mycorrhizal in soils low in available P.

The VA mycorrhizae also decrease the resistance to water transport in soybeans (Safir *et al.*, 1971). The lower resistance was associated with the root mycorrhizal system rather than the stem, but it could not be related to increased root growth. It appeared to be related to the increased uptake of nutrients.

Since nearly all soils contain mycorrhizal *Endogone* species, large increases in growth would not be expected to result from field inoculation. There are, however, several possibilities for practical applications. There are many *Endogone* species, and it is likely that some are more efficient in nutrient absorption than others. It is possible that increased yields might be obtained by manipulation of species or increasing population levels. Seedlings started in sterilized soil are stimulated by inoculation, and this increased growth may be maintained when they are transplanted into nonsterile soil (Mosse *et al.*, 1969). If increased growth rates could be maintained for an appreciable time under field conditions, then inoculation of crops that are started in sterilized soil and later transplanted to the field may be of value.

In certain situations it may be desirable to inoculate soils after fumigation or heat sterilization. As a matter of fact, reduced growth from "soil toxicity" following fumigation of field soil may, in some instances, be due to the killing of *Endogone* species. Such toxins have not been identified, and in some cases effects have been partially overcome with applications of P (Gerdemann, 1968). Inoculation with an *Endogone* species prevented P deficiency symptoms on maize grown in steamed soil (Gerdemann, 1964). Also, inoculation of steamed soil with *Endogone* prevented zinc deficiency symptoms on peach (*Prunus persica*) seedlings (Gilmore, 1971). Whether or not such nutrient deficiencies can be prevented more economically by application of fertilizers requires further study. It is highly probable that plant species vary greatly in their degree of dependence on VA mycorrhizae. It is possible that many herbaceous plants with well-developed root hairs, when grown in highly fertile soil, receive little or no net benefit from VA mycorrhizae. On the other hand, many woody plant species that lack or have only poorly developed root hairs will probably prove much more dependent upon VA mycorrhizae. With such species it may not be practical to maintain nutrient levels high enough for the plant to grow at the maximum rate without VA mycorrhizal infection.

III. NONMYCORRHIZAL PLANTS

Aquatics and other plants growing in very wet habitats are usually reported to be nonmycorrhizal (Gerdemann, 1968). However, such plants may become mycorrhizal if the water table is lowered or if they are planted in well-drained soil. Rice grown in flooded soil is nonmycorrhizal, but rice planted in nonflooded soil has VA mycorrhizae.

A number of investigators have reported that some plant families appear to be nonmycorrhizal (Gerdemann, 1968). The Cruciferae and Chenopodiaceae are the only such families that contain a number of important crops species. Several other reportedly nonmycorrhizal families contain a few common ornamentals. Further observation will probably shorten the list of nonmycorrhizal families.

References

Baylis, G. T. S. 1959. The effect of vesicular-arbuscular mycorrhizas on growth of *Griselinia littoralis* (Cornaceae). *New Phytol.* 58: 274–80.

——. 1967. Experiments on the ecological significance of phycomycetous mycorrhizas. *New Phytol.* 66: 231–43.

——. 1970. Root hairs and phycomycetous mycorrhizas in phosphorus-deficient soil. *Plant Soil* 33: 713–16.

Björkman, E. 1942. Über die Bedingungen der Mykorrhizabildung bei Kiefer und Fichte. *Symbolae Bot. Upsalienses* 6: 1–190.

——. 1960. *Monotropa hypopitys* L.—an epiparasite on tree roots. *Physiol. Plant.* 13: 308–27.

Bowen, G. D. 1969. *The Roles of Mycorrhizae and Root Nodules in Tree Nutrition.* C.S.I.R.O. Div. of Soils. Reprint no. 708.

Bowen, G. D., and C. Theodorou. 1967. Studies on phosphate uptake by mycorrhizas. *Proc. 14th IUFRO* (Munich) 5: 116–38.

Briscoe, C. B. 1959. Early results of mycorrhizal inoculation of pine in Puerto Rico. *Caribbean Forest.* 20: 73–77.

Daft, M. J., and T. H. Nicolson. 1966. Effect of *Endogone* mycorrhiza on plant growth. *New Phytol.* 65: 343–50.

——. 1969. Effects of *Endogone* mycorrhiza on plant growth. II. Influence of soluble phosphate on endophyte and host in maize. *New Phytol.* 68: 945–52.

Downie, D. G. 1957. *Corticium solani*—an orchid endophyte. *Nature* (London) 179: 160.

Gäumann, E., J. Nüesch, and R. H. Rimpan. 1960. Weitere Untersuchungen über die chemischen Abwehrreaktionen der Orchideen. *Phytopathol. Z.* 38: 274–308.

Gerdemann, J.W. 1955. Relation of a large soil-borne spore to phycomycetous mycorrhizal infections. *Mycologia* 47: 619–32.

——. 1961. A species of *Endogone* from corn causing vesicular-arbuscular mycorrhiza. *Mycologia* 53: 254–61.

——. 1964. The effect of mycorrhiza on the growth of maize. *Mycologia* 56: 342–49.

——. 1965. Vesicular-arbuscular mycorrhizae formed on maize and tuliptree by *Endogone fasciculata. Mycologia* 57: 562–75.

——. 1968. Vesicular-arbuscular mycorrhiza and plant growth. *Ann. Rev. Phytopathol.* 6: 394–418.

——. 1970. The significance of vesicular-arbuscular mycorrhizae in plant nutrition. In *Root Diseases and Soil-borne Pathogens*, ed. T. A. Toussoun,

R. V. Bega, and P. E. Nelson, pp. 125–29. Berkeley, Los Angeles, London: University of California Press.

——. 1971. Fungi that form the vesicular-arbuscular type of endomycorrhiza. In *Mycorrhizae: Proceedings of the First North American Conference on Mycorrhizae*, ed. E. Hacskaylo, pp. 9–18. U.S.D.A. Forest Service, Misc. Publication 1189.

Gerdemann, J. W., and T. H. Nicolson. 1963. Spores of mycorrhizal *Endogone* species extracted from soil by wet sieving and decanting. *Trans. Brit. Mycol. Soc.* 46: 235–44.

Gilmore, A. E. 1971. The influence of endotrophic mycorrhizae on the growth of peach seedlings. *J. Amer. Soc. Hort. Sci.* 96: 35–38.

Gray, L. E., and J. W. Gerdemann. 1969. Uptake of phosphorus-32 by vesicular-arbuscular mycorrhizae. *Plant Soil* 30: 415–22.

Hacskaylo, E. (ed.). 1971. *Mycorrhizae: Proceedings of the First North American Conference on Mycorrhizae.* U.S.D.A. Forest Service, Misc. Publication 1189.

Harley, J. L. 1965. Mycorrhiza. In *Ecology of Soil-borne Plant Pathogens*, ed. K. F. Baker and W. C. Snyder, pp. 218–30. Berkeley and Los Angeles: University of California Press.

——. 1969. *The Biology of Mycorrhiza.* London: Leonard Hill.

Harley, J. L., and C. C. McCready. 1950. The uptake of phosphate by excised mycorrhizal roots of the beech. *New Phytol.* 49: 388–97.

Hatch, A. B. 1937. The physical basis of mycotrophy in *Pinus. Black Rock Forest Bull.*, no. 6: pp. 1–168.

Hayman, D. S. 1970. *Endogone* spore numbers in soil and vesicular-arbuscular mycorrhiza in wheat as influenced by season and soil treatment. *Trans. Brit. Mycol. Soc.* 54: 53–63.

Holevas, C. D. 1966. The effect of a vesicular-arbuscular mycorrhiza on the uptake of soil phosphorus by strawberry (*Fragaria* sp. var. Cambridge Favorite). *J. Hort. Sci.* 41: 57–64.

Kramer, P. J., and K. M. Wilbur. 1949. Absorption of radioactive phosphorus by mycorrhizal roots of pine. *Science* 110: 8–9.

Marx, D. H. 1969a. The influence of ectotrophic mycorrhizal fungi on the resistance of pine roots to pathogenic infections. I. Antagonism of mycorrhizal fungi to root pathogenic fungi and soil bacteria. *Phytopathology* 59: 153–63.

——. 1969b. The influence of ectotrophic mycorrhizal fungi on the resistance of pine roots to pathogenic infection. II. Production, identification, and biological activity of antibiotics produced by *Leucopaxillis cerealis* var. *piceina. Phytopathology* 59: 411–17.

——. 1970. The influence of ectotrophic mycorrhizal fungi on the resistance of pine roots to pathogenic infections. V. Resistance of mycorrhizae to infection by vegetative mycelium of *Phytophthora cinnamomi. Phytopathology* 60: 1472–73.

Marx, D. H., and C. B. Davey. 1969a. The influence of ectotrophic mycorrhizal fungi on the resistance of pine roots to pathogenic infections. III. Resistance of aseptically formed mycorrhizae to infection by *Phytophthora cinnamomi*. *Phytopathology* 59: 549–58.

——. 1969b. The influence of ectotrophic mycorrhizal fungi on the resistance of pine roots to pathogenic infections. IV. Resistance of naturally occurring mycorrhizae to infections by *Phytophthora cinnamomi*. *Phytopathology* 59: 559–65.

Mikola, P. 1965. Studies on the ectendotrophic mycorrhiza of pine. *Acta Forest. Fennica* 79: 1–56.

——. 1969. Afforestation of treeless areas. *Unasylva*. 22: 35–48.

——. 1970. Mycorrhizal inoculation in afforestation. *Int. Rev. of Forest Res.* 3: 123–96.

Miller, C. O. 1967. Zeatin and zeatin riboside from a mycorrhizal fungus. *Science* 157: 1055–56.

Moser, M. 1959. Beitrage zur Kenntnis der Wuchsstoffbeziehungen im Bereich ectotroper Mycorrhizen I. *Arch. Mikrobiol.* 34: 251–69.

Mosse, B. 1956. Fructifications of an *Endogone* species causing endotrophic mycorrhiza on fruit plants. *Ann. Bot.* (London), n.s. 20: 349–62.

——. 1957. Growth and chemical composition of mycorrhizal and nonmycorrhizal apples. *Nature* 179: 922–24.

——. 1963. Vesicular-arbuscular mycorrhiza: An extreme form of fungal adaptation. In *Symbiotic Associations: 13th Symposium of the Society for General Microbiology*, ed. P. S. Nutman and B. Mosse, pp. 146–70. London: Cambridge University Press.

Mosse, B., and G. D. Bowen. 1968. The distribution of *Endogone* spores in some Australian and New Zealand soils and in an experimental field soil at Rothamsted. *Trans. Brit. Mycol. Soc.* 51: 485–92.

Mosse, B., D. S. Hayman, and G. J. Ide. 1969. Growth responses of plants in unsterilized soil to inoculation with vesicular-arbuscular mycorrhiza. *Nature* 224: 1031–32.

Murdoch, C. L., J. A. Jackobs, and J. W. Gerdemann. 1967. Utilization of phosphorus sources of different availability by mycorrhizal and nonmycorrhizal maize. *Plant Soil* 27: 329–34.

Nicolson, T. H. 1967. Vesicular-arbuscular mycorrhiza—a universal plant symbiosis. *Sci. Progr.* (Oxford) 55: 561–81.

Nicolson, T. H., and J. W. Gerdemann. 1968. Mycorrhizal *Endogone* species. *Mycologia* 60: 313–25.

Ross, J. P., and J. A. Harper. 1970. Effect of *Endogone* mycorrhiza on soybean yields. *Phytopathology* 60: 1552–56.

Safir, G. R., J. S. Boyer, and J. W. Gerdemann. 1971. Mycorrhizal enhancement of water transport in soybean. *Science* 172: 581–83.

Slankis, V. 1958. The role of auxin and other exudates in mycorrhizal symbiosis of forest trees. In *The Physiology of Forest Trees*, ed. K. V. Thimann, pp. 427–43. New York: Ronald Press.

Smith, S. E. 1966. Physiology and ecology of orchid mycorrhizal fungi with reference to seedling nutrition. *New Phytol.* 65: 488–99.

Trappe, J. M. 1962. Fungus associates of ectotrophic mycorrhizae. *Bot. Rev.* 28: 538–606.

——. 1964. Mycorrhizal hosts and distribution of *Cenococcum graniforme*. *Lloydia* 27: 100–106.

Vozzo, J. A. 1971. Field inoculations with mycorrhizal fungi. In *Mycorrhizae: Proceedings of the First North American Conference on Mycorrhizae*, ed. E. Hacskaylo, pp. 187–96. U.S.D.A. Forest Service, Misc. Publication 1189.

Zak, B. 1964. The role of mycorrhizae in root disease. *Ann. Rev. Phytopathol.* 2: 377–92.

9. Diseases of Feeder Roots

W. A. Campbell and F. F. Hendrix, Jr.

THE previous chapters have dealt with various phases of root growth and development, with emphasis on the biochemistry, physiology, and various rhizosphere interrelationships of the biological environment. Rather than attempting to enlarge on these subjects, we wish to detail some of our experiences in the practical investigation of several specific diseases. These diseases are basically infections of the feeder or absorbing roots with symptom complexes suggesting nutrient starvation usually seen as deficiencies of a specific element such as nitrogen or iron. Investigation of these diseases, especially their earlier phases, was frustrating, but exhaustive research finally elucidated the succession of events leading to a type of disease now termed *feeder root necrosis.*

Diseases that affect the large roots and kill the cambial areas by direct mycelial invasion by the causal organism are well known in forestry and horticulture. *Fomes annosus* (Fr.) Karst not only girdles the large roots but causes decay in many conifers. *Clitocybe tabescens* (Fr.) Bres. attacks many woody species in the southeast (Rhoads, 1950, 1954; Ross, 1970). The closely related *Armillaria mellea* (Vahl. ex. Fr.) Kummer has been particularly troublesome as a root pathogen on fruit trees planted on old oak or orchard sites in California (Childs and Zeller, 1929; Heald, 1933). *Phymatotrichum omnivorum* (Shear) Dug., which causes Texas root rot, is an outright destroyer of the root systems of a large number of plants when they are planted on infested land (Garrett, 1956). Diseases of this nature are root rots whose destructive results are associated with massive lesions on the main roots or the root crown, and hence they can be clearly separated from those classified as fine root or feeder root necrosis.

I. FEEDER ROOTS DEFINED

By definition we are limiting the term *feeder roots* to the newly formed rootlets or the terminal end of extending roots. These are the final order of roots on plants. Typically each root terminates in a protective rootcap behind which is the meristematic region of high physiological activity where cell synthesis takes place. Behind this area is a zone of cell elongation which may vary in length according to species and rapidity of root growth. This area gives way to the region of cell maturation and tissue differentiation, which may remain succulent and nonlignified for a relatively long period. In many species root hairs are commonly present (Scott, 1963). They are evanescent structures in many plants and are sloughed off as the epidermis thickens in the maturation zone. Lateral roots develop from the pericyclic cells within the vascular tissues and push through the cortical layers and epidermis at regular intervals leading to the characteristic branching of a normal root system. As the lateral roots push through the cortical layer, the epidermal cells near the margin of the lateral root are ruptured and displaced, providing infection courts for pathogenic fungi and nematodes (Marx, 1969a; Marx and Davey, 1969b; Reuhle and Marx, 1971).

In the seedling stage most of the root system is an absorbing organ. As the roots thicken with age and become specialized for strength, storage, and conduction, the area for specialized and differential absorption is restricted to the morphologically simple tissue at the root tips (Scott, 1965). These tissues are relatively unprotected and susceptible to damage by pathogenic organisms. Since plants may have thousands of actively growing root tips, the loss of even a relatively large number at any one time will only temporarily reduce nutrient uptake in soils of moderate fertility, especially if conditions are favorable for root regeneration. Stored food reserves may serve as a substitute system should there be a temporary reduction in nutrient uptake. Prolonged feeder root destruction, however, or massive feeder root losses at one time may severely affect a plant's ability to absorb nutrients needed to produce food reserves. These reserves are needed for periodic seasonal demands such as leaf and fruit formation in the spring and bud formation later in the growing season. Reduction of these reserves below a minimum level causes starvation and quick or slow decline.

The conditions under which root growth takes place may determine the number and nature of the feeder roots. With rapid root extension, under optimum soil moisture and temperature conditions, the extent of succulent feeder root tissue may be much greater than that developing under less favorable conditions. The immediate rhizosphere area may vary greatly in the quantity and quality of root exudates and hence in the rhizosphere microbial population, which may inhibit or restrict by various mechanisms the growth and development of organisms (Rovira, 1965).

Root distribution has been studied for many horticultural and agronomic plants. Generally the fine roots are most abundant in the upper 10 cm of soil, while anchor and tap roots go deeper where moisture is retained and available for longer periods. The surface root layer develops in response to greater soil aeration, the presence of soil nutrients resulting from biological activity, and soil moisture interception during the growing season when rainfall is light or intermittent and penetrates only a few centimeters.

In summary, the feeder roots of many plants consist mainly of incompletely differentiated tissue composed of cells in a relatively high physiologic state, covered by a thin epidermal layer. Because of the accompanying root hairs and thin-walled cell membranes, selective absorption of mineral nutrients is possible at optimum rates for plant growth. At the same time root exudates contribute energy sources to the associated rhizosphere organisms, which may or may not compete with pathogens adapted for infection and growth in living root cells.

II. ACTION OF PATHOGENIC ORGANISMS

Early investigators of diseases of the kind now classified as feeder root necrosis were baffled by their failure to find normal numbers of feeder roots in the root zone of affected plants. Zentmyer *et al.* (1962), reporting on avocado decline, stated that they found few if any dead or live roots; just no roots. Decline diseases such as peach decline in the Southeast and citrus decline in Florida and Texas all had one thing in common—the lack of feeder roots (Chandler *et al.*, 1962; Stolzy *et al.*, 1965). Shortleaf pines affected by littleleaf disease are deficient in feeder roots, particularly those normally mycorrhizal (Copeland, 1952). When decline diseases progress to the point of causing host mortality, feeder roots are lacking, and the few that can be found are often in poor condition (Smalley and Scheer, 1963; Lorio, 1966; Oxenham and Winks, 1963). Various pathogenic fungi and nematodes are regularly associated with root deficiencies in decline diseases (Hine, 1961; McIntosh, 1964).

The actively growing root tips of young feeder roots are very vulnerable to infection. The active physiological processes in the meristematic tissue and incompletely differentiated tissues provide chemical attractants and stimulation for the zoospores of *Phytophthora* and *Pythium* species (Zentmyer, 1970; Biesbrock and Hendrix, 1970a,1970b). Nematodes are also attracted to these succulent tissues where the feeding of parasitic forms results in reduced root functions and feeder root necrosis (Ruehle, 1962, 1968, 1969a, 1969b; Hendrix and Powell, 1968).

Damping-off of seedlings is often related to feeder root necrosis. Here most of the seedling root system is in the same juvenile state as the feeder roots of older plants. Infection can take place at the ground line, causing affected plants to topple over and die. As the seedlings age and the larger roots become lignified and covered with a well-defined epidermis, the susceptible root tissue is restricted to the feeder roots. In nurseries, where infection opportunities closely follow excess moisture, overfertilization, or other ecological imbalances, feeder root damage may occur at any time, resulting in poor growth, chlorosis, and low survival following transplanting.

III. BIOLOGIC PROTECTION OF FEEDER ROOTS

Because of the abundance of pathogenic organisms in the soil, surviving plant species have developed specific root adaptations. One adaptation is the ability to replace lost or damaged feeder roots quickly. This is done by regenerating new roots basipetal to the dead root tips, so that their loss causes only a temporary setback in root extension necessary for absorption of water and nutrients. The capacity to generate new roots may be genetic in nature as well as a response to soil conditions. Poor soil aeration and low fertility drastically reduces a plant's ability to replace roots lost to feeder root necrosis (Zak, 1961).

With decline diseases, especially littleleaf of pines, feeder roots are destroyed almost as fast as they are replaced (Campbell and Copeland, 1954). On well-drained, fertile soils where root infection periods are relatively short, littleleaf symptoms do not develop even though the organism causing this disease, *Phytophthora cinnamomi* Rands, is present in the root zone. When physical conditions of the soil are poor and fertility is low, root penetration is restricted, root regeneration lags, symptoms of littleleaf appear when the trees are no older than 20 years. Potential littleleaf–resistant trees have been selected on severely diseased sites as a basis for developing strains of shortleaf pine that will either tolerate the causal organism or be better adapted to the soil and site conditions associated with littleleaf. Pathogenicity tests with progeny from these selections indicate resistance to infection by zoospores of the fungus (Bryan, 1965). Potentially resistant trees may show greater root growth during the season when soil temperatures are below 18C, needed for infection by *P. cinnamomi*. Hence these trees may produce a significant amount of root growth at a time when the feeder roots are relatively free from infection.

Most forest trees form complex mycorrhizal structures. Mycorrhizae enormously increase the root area active in the selective absorption of nutrients from the soil. The symbiont fungi stimulate the formation of short lateral roots and their proliferation into complex masses of various kinds (Marx and Zak, 1965). Each mycorrhizal root is encased in a fungus mantle which may vary in thickness depending upon environment and the specific symbiont. Mycelium penetrates between the cortical cells, forming the familiar Hartig net without apparent injury to the cells and generally stimulating them to become larger than those in nonmycorrhizal cortex tissue.

Recent studies have shown that these ectotrophic mycorrhizal structures are resistant to infection by zoospores or direct mycelial penetration by *P. cinnamomi*. The encasing mycelium acts as a physical barrier reinforced by antibiotics produced by the fungal symbionts. Even when the fungus mantle does not completely cover the tip of the short root, the tip resists infection, presumably because of the antibiotics diffused from the fungus mass. Nonmycorrhizal short roots close to mycorrhizal ones are also resistant if the symbiont forming the mycorrhizae produces antibiotics (Marx, 1969a, 1969b; Marx and Davey, 1969a, 1969b) The number of short roots that can become mycorrhizal depends upon the extent and vigor of the entire root system. Hence any massive or continuing feeder root infections reduce the extent of the root system and lead to a condition where the tree becomes deficient in total active absorbing area.

IV. *PHYTOPHTHORA* AND *PYTHIUM* SPECIES AS FEEDER ROOT PARASITIES

In the late 1930's a decline disease of shortleaf pine in the South attracted the attention of forest pathologists. The needles of affected trees turned yellow and were shorter than on healthy trees, and the crowns appeared tufted because of decreased shoot growth. Radial growth declined sharply with the onset of symptoms; diseased trees lived an average of only seven years after becoming chlorotic (Campbell and Copeland, 1954). Foliar analysis showed that the needles of diseased trees contained about half the nitrogen generally found in green needles (Roth *et al.*, 1948). Excavation of the root systems of diseased trees showed little difference in the number, size, and distribution of the major roots. Large patches of brown and inactive phloem tissue found in the large roots, indicating root starvation, were termed *brown patch*, and proved to be cortical areas excised by the cork cambium (Jackson, 1945). The trees were walling off the normal root storage areas because of a lack of food reserves. The only noticeable difference between diseased and healthy root systems was in the number and condition of the feeder roots and in the paucity of mycorrhizal short roots (Copeland, 1952).

After intensive efforts investigators failed to find causal agents in the aboveground parts of diseased trees. A number of fungi were isolated from necrotic areas and lesion margins on the larger roots of diseased plants, but none were known pathogens (Siggers and Doak, 1940). Later the root isolation work concentrated on the small roots 1 to 3 mm and smaller in diameter. These were often difficult to separate from soil and after plating on water agar yielded numerous nonpathogenic fungi and an occasional *Fusarium* species. Finally *P. cinnamomi* was isolated from 2% of the root pieces (Campbell, 1951b.). The fungus was known to occur in the littleleaf area from the history of root disease of chestnut, from which *P. cinnamomi* had been isolated (Crandall *et al.*, 1945). The fungus also attacked hardwood seedlings in several forest tree nurseries and was isolated from rhododendron in New Jersey (White, 1937).

Isolations of *P. cinnamomi* on agar were slow and time-consuming. Apples had been used by various investigators to recover *Phytophthora* species from diseased plant tissue, but stuffing small roots in puncture holes in apples did not increase recovery of *P. cinnamomi* over the 2% figure. But when soil from around the roots of littleleaf trees was stuffed into holes in apples, *P. cinnamomi* was readily isolated (Campbell, 1949). Further refinement in sampling technique proved the constant association of *P. cinnamomi* with littleleaf trees. Under healthy trees, *P. cinnamomi* was either present in small amounts or absent (Campbell and Copeland, 1954). By using the apple technique, researchers found the fungus to be distributed in the soil in greatest abundance in the feeder roots region or in the top 10 cm of soil (Campbell, 1951a). *Phytophthora cinnamomi* was also found in soil of the Southeast outside of the range of littleleaf (Campbell, 1951c). The apple technique proved effective for many *Phytophthora* species and for a number of *Pythium* species (Sutherland *et al.*, 1966).

The problem still remained to discover how the fungus was affecting the trees since it appeared to be restricted only to young tissue near the apices of the roots. Some greenhouse pathogenicity experiments helped to clarify the picture. Many of the early inoculation attempts gave inconclusive results because after the seedlings had grown for a season, even the poorest were pot-bound, and top symptoms were inconclusive. Only six weeks after a few potted shortleaf seedlings were planted in infested soil, the root tips were found to be discolored. These yielded pure cultures of *P. cinnamomi*. The rest of the pots were left for a longer period, and the infected root tips disappeared. There was no new root growth, and consequently little evidence of root damage. Root elongation had ceased, but regeneration had started with new root initials, appearing basipetal to the root tips.

The addition of *Phytophthora*-infested soil to water in quart jars in which the roots of avocado (*Persea americana*), pineapple (*Ananas sativa*), and pine were suspended showed that the liberated zoospores attacked only the area immediately behind the root tips, causing a brown discoloration and a sunken lesion. Zoosporangia formed in abundance on the lesion and occasionally some distance back of the lesion on succulent roots. Eventually the root tip died, generally with very little further advance of the mycelium into the older cells. Regardless of the abundance of zoospores or the number of healthy root tips present, not all the root tips were infected at any one infection period. In more critical studies shortleaf (*Pinus echinata*) and loblolly (*Pinus taeda*) seedlings were grown for two months in a nutrient solution at temperatures favorable for root growth. When these seedlings were transferred to jars containing tap water and a small cheesecloth bag containing wheat grains permeated with *P. cinnamomi*, root lesions appeared in three days. All the root tips were excised and plated on water agar. *Phytophthora cinnamomi* was isolated from 38% of 457 loblolly pine root pieces and from 58% of 386 shortleaf pine pieces (Campbell and Copeland, 1954). These experiments and others demonstrated that even under ideal laboratory conditions, zoospores of *P. cinnamomi* failed to colonize all of the available infection courts.

Feeder root necrosis caused by *P. cinnamomi* affects a variety of woody plants. Distribution of the fungus in the soil, the ecological conditions needed for its development and pathogenicity, and the susceptibility of root tissues have been studied on avocado in California and in South and Central America (Zentmyer *et al.*, 1962), conifers in New Zealand (Sutherland *et al.*, 1960), and azalea and rhododendron in New Jersey (White, 1937). This fungus was also isolated from soil and roots of peach trees in Georgia, Maryland, and New Zealand (Hendrix *et al.*, 1966; Mircetish and Keil, 1970). *Phytophthora cactorum* (Leb. & Cohn) Schroet. and P. *parasitica* Dastur. were also isolated from the soil in peach heeling-in beds in Tennessee (Hendrix *et al.*, 1966). Citrus decline was associated with a rich soil flora including a number of *Phytophthora* and *Pythium* species, all root pathogens with well-documented histories as the cause of root disease (Hendrix and Campbell, unpublished data).

Disease caused by water molds are influenced by rainfall distribution and soil drainage as well as soil temperature (Campbell and Presley, 1946; Biesbrock and Hendrix, 1970a, 1970b). Free water needed for zoospore formation and release may be supplied by rainfall or irrigation. *Phytophthora cinnamomi* is sometimes a root parasite capable of killing more than feeder roots. This fungus causes jarrah dieback in

Southwest Australia where it quickly kills *Eucalyptus marginata* Smith and associated species of trees and understory shrubs (Poder *et al.*, 1965). These plant species are very susceptible to this introduced pathogen, especially in areas with abundant moisture. The fungus is so virulent that almost all of the absorbing roots are quickly killed. The fungus meets with little competition from other soil organisms and once introduced into an area quickly spreads and becomes widely distributed.

Pythium species require the same conditions for development and pathogenicity as the closely related *Phytophthora* species. Most *Pythium* species produce zoospores, although a few depend upon germinating oospores or chlamydospores to furnish infective hyphae. *Pythium* species are particularly abundant in forest and horticultural nurseries, and many infected ornamental plants are delivered to the homeowner (Hendrix and Campbell, 1966, 1968; Vaartaja and Salisbury, 1961; Lumsden and Haasis, 1964). Again, investigators have noticed that these fungi attack juvenile tissue at the root crown but later, as the seedlings age, infect only the succulent roots (Vaartaja and Salisbury, 1961). Only the very young roots of large orchard trees are susceptible to infection (Miller *et al.*, 1966; Mulder, 1969). Soil temperature is very important, and certain *Pythium* species are favored by either high or low soil temperatures. Thus some species are more pathogenic in the spring when temperatures are low and become relatively inactive in the summer when temperatures are higher (Biesbrock and Hendrix, 1970a, 1970b; McIntosh, 1964).

V. METHODS OF ISOLATING PYTHIACEOUS FUNGI

Many investigators have struggled with the task of isolating and classifying soil fungi. Spore-forming fungi are readily isolated by most agar techniques. Certain phycomycetous species are difficult to isolate mainly because they are quickly overrun by the ubiquitous and fast-growing spore-forming segments of the fungus population. The difficulty of direct isolation of *Pythium* and *Phytophthora* species is apparent when one considers that in surveys of nursery and cultivated soils, an insignificant number of *Pythium* species were isolated by most agar techniques (Miller *et al.*, 1957). Improved techniques later showed that these two species abound under such situations.

Early workers with *Phytophthora* and *Pythium* species experienced difficulty in isolating these fungi from diseased tissue (Crandall *et al.*, 1945). Apparently, tissues invaded by these fungi are colonized quickly by saprophytes when roots or other plant parts are excised and held for only a few hours before isolation. Such conditions are more favorable for growth of saprophytes than for the survival of Phycomycetes. Tucker (1931) solved the problem by inserting bits of diseased tissue into firm apples. The apple tissue was living material that could be invaded by the parasitic *Phytophthora* species but not by the associated saprophytes. Pure cultures of the parasite could be obtained by transferring bits of the rotted apple to an agar substrate. In the littleleaf investigations, Tucker's apple technique was used successfully to isolate *P. cinamomi* from the feeder roots. Littleleaf research took a giant leap forward when it was discovered that a modification of Tucker's technique effectively isolated *P. cinnamomi* from soil (Campbell, 1949). Soil to be assayed was stuffed into half-inch holes bored two-thirds through a firm apple. The soil was moistened with distilled water, the holes were sealed with cellophane tape, and the apples were incubated at room temperature. Within five to seven days, the fungus produced a firm rot without any outward change in the appearance of the apple except a light surface browning over the invaded area. This same technique can be used on a number of *Phytophthora* species. The authors have successfully isolated *P. cactorum*, *P. drechsleri* Tucker, *P. heveae* Thompson, *P. citricola* Sawada, and *P. parasitica*, in addition to *P. cinnamomi*, from the soil under forest and horticultural plants (Hendrix and Campbell, 1970). Other baiting techniques are in common use. These depend upon the ability of zoospores to infect uninjured fruits and root tips of various plants. Citrus fruits and leaves and avocados have been used (Zentmyer *et al.*, 1962; Grimm and Alexander, 1970). Rooted pineapple crowns are routinely used to trap *Phytophthora* zoospores in soil samples in Hawaii (Anderson, 1951). Lupine radicals are used in Australia for the same purpose (Chee and Newhook, 1965).

The apple technique works equally well with a number of *Pythium* species. The rot produced by these organisms is less characteristic than that associated with *Phytophthora* species because it is softer and browner. *Pythium ultimum* Trow is readily detected in soil by the apple technique (Sleeth, 1953). Other species that can be readily isolated include *P. vexans* DBy, *P. spendens* Braun, *P. helicoides* Drechs., the heterothallic species, and many others (Sutherland *et al.*, 1966; Hendrix and Campbell, 1970).

The main disadvantage of baiting techniques is that quantitative data on the relative abundance of fungus propagules in the soil cannot be obtained. The presence of certain fungi can be detected, but because the method uses a mass of soil from which each bait screens but one or two fungi, the number of successful isolations cannot be related to soil mass.

In the early 1950's selective media were developed by several workers. The newer antibiotics utilized in these media suppressed the growth of most soil saprophytes but did not affect the development of pythiaceous fungi (Schmitthenner, 1970). A modification of a selective medium originating in Australia proved effective for the detection of *Pythium* species and *Phytophthora cinnamomi* propagules in the soil (Kerr, 1963; Hendrix and Kuhlman, 1965). By taking a weighed amount of soil from a composited sample and plating a constant dilution on a predetermined number of agar plates (usually 10), the colony counts could be translated into propagules per gram of dry soil. Another method utilizing gallic acid also proved effective for making quantitative assays of the number of propagules of *Pythium* and *Phytophthora* species in soil (Flowers and Hendrix, 1969). The listing of the chemical constituents of these media is outside the scope of this chapter, and one should consult the original publications for details concerning their formulation.

Extensive work with Kerr's modified *Pythium* medium and the gallic acid medium proved that neither was equally effective in selectively isolating all *Pythium* and *Phytophthora* species. Hence for assays designed to isolate all pythiaceous fungi in the soil it became standard practice to plate each sample on both media and also to assay the samples by the apple technique (Hendrix and Campbell, 1970).

A citrus leaf baiting technique not only proved effective for isolating those pythiaceous fungi recovered by other techniques but demonstrated the presence in the soils of species not recovered by the modified Kerr's medium, the gallic acid medium, or the apple technique (Grimm and Alexander, 1970). This very simple method consists of floating 1/8 to 1/4-inch squares of citrus leaf on water over the soil sample in a waxed container. After a period of several days, zoospores infect the leaf tissue, and the fungi can be isolated by transferring the leaf sections to water agar with or without antibiotics added to reduce bacterial growth.

These methods and others developed earlier, applied to soils from under peaches, pine, turfgrasses, tomatoes (*Lycopersicon esculentum*), corn (*Zea mays*), sugarcane (*Saccharum officinarum*), and other crops, have demonstrated the almost universal association of certain *Pythium* species with specific hosts (Powell *et al.*, 1965; Hendrix and Powell, 1970; Hampton and Buchholtz, 1959; Rands and Dopp, 1938; Lumsden and Haasis, 1964). Each host seemed to encourage specific *Pythium* pathogens and caused a buildup in the soil. This relationship between species and certain crops explains why excessive soil buildup of propagules occurs with the resultant intensification of root diseases year after year.

VI. RELATIVE ABUNDANCE OF *PHYTOPHTHORA* AND *PYTHIUM* SPECIES IN SOIL

Assays of soil samples collected in the Southeast and elsewhere in the United States have given some insight into the relative abundance and specificity of certain species. Approximately 10,000 soil samples from diverse soils and sites and from under a variety of crop and horticultural plants have been assayed for *Pythium* and *Phytophthora* species, and the results expressed in terms of propagules per gram (ppg) of air-dried soil. From inoculation studies and comparisons of fungus populations with host conditions, it is evident that root damage cannot be always correlated with the population. Virulence of fungus species, host susceptibility, and climatic conditions may be more important than the number of propagules in the soil (Schmitthenner, 1970). Counts over 30 ppg have frequently been associated with a noticeable root injury and top symptoms such as growth reduction and chlorosis.

Twenty-one species or species complexes of *Pythium* were recognized (Hendrix and Campbell, 1970). Isolates referable to the *Pythium irregulare-debaryanum* complex were obtained from 90% of the samples and made up 55% of the 10,566 isolates examined. This species complex is universally distributed and is parasitic on many plants. However, its association with any particular host should not automatically suggest that it is a root pathogen. Proof of pathogenicity should be based on inoculation studies.

Species present in 5% to 15% of the samples include those in the *P. dissotocum-perniciosum* complex, *P. spinosum* Saw., *P. splendens*, *P. ultimum*, and the various heterothallic forms. Other species were present in relatively few samples. *Pythium ultimum* is more selective as to host. When present above its rather low normal population level, it should be strongly suspected as a pathogen (Campbell and Sleeth, 1945; Sleeth, 1953). *Pythium ultimum* frequently infects cultivated crops such as tomatoes, beans (*Phaseolus* spp.), and cotton (*Gossypium* spp.) (Averre, 1966; Alicubusan *et al.*, 1965). Wilhelm (1965) proved conclusively that *P. ultimum* in the soil reduced the growth of lettuce (*Lactua sativa*) and other plants even when no evident root damage could be detected. The only possible explanation was obscure injury to the root tips and a gradual dwarfing of the entire root system.

Phytophthora cinnamomi can be found in forest soils throughout the Southeast as far north as Virginia and Kentucky and as far west as eastern Texas (Campbell, 1951 b; Hendrix *et al.*, 1971). *Phytophthora cactorum* and *P. citricola* have been isolated from forest soils in the

southern Appalachian Mountains and *P. parasitica* and *P. drechsleri*
from the Piedmont and Coastal Plain sites. *Phytophthora heveae* has
been found in the soil under several old-growth forest stands in eastern
Tennessee and western North Carolina (Campbell and Gallegly, 1965).
This fungus has not been reported elsewhere in the Western Hemi-
sphere. Its occurrence in the forest stands has not been correlated
with any specific host.

VII. PROOF OF PATHOGENICITY

Any discussion of feeder root diseases would be incomplete without
mentioning some of the frustrations and disappointments of those
attempting to prove pathogenicity by pot inoculation experiments.
In theory such tests should be simple. All one need do is infest sterile
soil with the desired organism and introduce the desired test plant
developed in place from seed or transplanted from seedlings grown
in sterilized soil. After a sufficient period compare growth of tops
and roots with control plants grown under the same conditions in
pathogen-free soil.

The process of soil sterilization is very important because some
methods result in a biological desert unsuitable for both host and
soil pathogens. The soil mixture is important because aeration, water-
holding capacity, and other factors influence both the host and the
pathogen. Temperature fluctuations and extremes may escape general
notice. In greenhouses under poorly controlled conditions, soil temperatures in pots may exceed the limits for the growth and infection
by the pathogen. A frequent cause of failure of inoculation tests is
the form in which the pathogen is introduced into soil. At one time
it was common practice to grow the pathogen on a substrate which
contained excessive amounts of nutrients in addition to the fungus
and its spore stages. In some cases these unused nutrients added with
the inoculum allow saprophytic organisms to grow, and this competition leaves the pathogen in a weakened condition, unable to express
its pathogenic potential. In other cases, the pathogen may utilize the
excess nutrients and quickly increase to very high population levels.
Results of pot inoculation studies conducted without any means of
determining the population level of the pathogen at the beginning
and end of the experiment will be difficult to interpret.

Present procedure is to use the sand-cornmeal method of preparing
fungus inoculum because it reduces the amount of extraneous substrate material and favors the formation of resistant spores rather
than vegetative growth. Sand-cornmeal inoculum also can be readily

assayed to determine propagules per gram of inoculum, and substrate inoculum ratios can be determined to assure uniform inoculum levels in each pot. Soil in the pots should be assayed periodically during the course of the experiment to determine changes in the population level of the pathogen. Without this check, any conclusions about pathogenicity, especially in negative cases, can be misleading.

Pathogenicity tests with organisms capable of infecting the primary roots usually give definitive results. The fungus either kills the plant or it does not. With *Pythium* species causing feeder root necrosis and infecting only the root tips and very juvenile tissue, results of pathogenicity experiments are frequently less than convincing. This is especially true if no effort is made to control water and temperature. Fungi causing feeder root necrosis rarely kill the host. Changes in temperature and water regimes permit some roots to escape infection, and in time the plant becomes pot-bound regardless of the number of feeder roots that have been destroyed. Once pot-bound, the plant may exist indefinitely with few top symptoms if the root system is supplied with sufficient water and nutrients. Under these ideal conditions, a reduced root system can support the plant and mask symptoms that would become apparent if stress conditions occurred. Feeder root necrosis can be detected only before the roots have completely colonized the soil volume in the pots. The amount of damage can be ascertained only by comparison with check plants. Root weight differences may be very slight because the weight of the lost feeder roots represents only a very small fraction of the total weight of the root system (Biesbrock and Hendrix, 1970a, 1970b). Visual comparisons of the carefully washed-out root systems may be the only way to evaluate differences between the checks and the inoculated plants.

VIII. FEEDER ROOT PATHOGENS OTHER THAN PHYCOMYCETES

Many organisms other than those already discussed caused feeder root damage on diverse groups of plants (Benedict and Mountain, 1956; Chupp, 1917; Grogan *et al.*, 1958). Symptoms include stunting, minor or major element deficiency, discoloration, and eventually necrosis. From the partial listing presented in Table 9.1, it is evident that most plants are susceptible in some degree to many pathogenic fungi and nematodes. While distinctive individual symptoms exist, the basic results in terms of plant growth are the same. A detailed description of any one of these diseases would be typical for many of them.

Table 9.1. Organisms causing feeder root diseases of various genera of plants

Crop	Pathogens
Brassica spp.	*Plasmodiophora brassicae, Olpidium brassica, Aphanomyces raphani, Phytophthora negasperma, Phoma lingam, Sclerotinia sclerotiorum, Rhizoctonia solani, Pythium* spp., *Meloidogyne* spp., *Phymatotrichum omnivorum, Pratylenchus* spp., *Fusarium* spp.
Nicotiana spp.	*Meloidogyne* spp., *Phymatotrichum omnivorum, Pythium ultimum* and other species, *Thielaviopsis basicola, Asterocystis radicis, Conopholis americana, Macrophomina phaseoli, Olpidium brassicae, Phytophthora parasitica, Pratylenchus* spp., *Rhizoctonia solani*
Triticum spp.	*Brachycladium spiciferum, Cephalosporium acremonium, Fusarium roseum, F. oxysporum, Gibberella fujikuroi, G. zeae, Helminthosporium* spp., *Meloidogyne* spp., *Naucoria cerealis, Polosporiella verticillata, Pratylenchus* spp., *Pyrenochaeta terrestris, Pythium* spp., *Rhizoctonia solani, Tylenchus* spp., *Wojnowicia graminis*

IX. SURVIVAL OF PLANT PATHOGENS IN SOIL

Some root-infecting fungi are well adapted to saprophytic growth in soil (Garrett, 1956). Sources of nutrients utilized by these fungi include plant debris, both host and nonhost root exudates, and so on (Schroth and Snyder, 1961; Schroth and Hendrix, 1962). These fungi have a survival advantage because of their ability to enter roots before the death of the plant. Weak pathogens are not able to kill root tissue extensively, but have the advantage of being within the tissue in an active physiological state when the root dies from other causes, and are able to grow saprophytically before nonpathogens can become competitive (Hendrix and Nielsen, 1958). Such weak pathogens include many species of *Pythium, Fusarium solani, F. roseum, Curvularia ramosa, Macrophomina phaseoli, Cylindrocarpon* spp., and *Streptomyces scabies. Fusarium solani* germinates, grows, and forms chlamydospores as root tips of nonhost plants, liberating exudates, enter, grow through, and leave the immediate area of existing chlamydospores in soil (Schroth and Hendrix, 1962). *Verticillium albo-atrum* Reinke and Barth. also is capable of this type of saprophytic growth (Emmatty and Green, 1969).

Survival by means of specialized resting bodies is probably the most important single means of survival found in soil fungi. These resting structures, which may be sexual or asexual, always have thickened walls, high stored-food content, and a low rate of metabolic activity. They may survive for 20 or more years in air-dried soil. Chlamydospores, the resting bodies formed by *Pythium* spp., *Fusarium* spp., and *Phytophthora* spp., among others, are single-celled asexual spores, varying in size from 10μ to 80μ, and have thickened walls. Their reserve food is mostly in the form of oil droplets free in the cell and fatty materials deposited in the cell walls. Chlamydospores of *Pythium* and *Phytophthora* may germinate by producing zoospores or mycelium, and those of other fungi germinate by producing mycelium.

Sclerotia, another type of resting structure, are multicelled, and only the walls of the outer cells are thickened, and frequently contain dark pigments. They may vary in size from ten cells to 1.5 cm and in color from tan to black. *Rhizoctonia*, *Verticillium*, *Sclerotinia*, *Sclerotium*, *Helicobasidium*, and *Phymatotrichum* are examples of root-infecting fungi that form sclerotia. Sclerotia of some fungi germinate by producing apothecia and ascospores, and others germinate by producing mycelium. Sclerotia have been reported to survive in field soil for 15 years in the absence of host plants.

Oospores of *Pythium*, *Phytophthora*, and *Aphanomyces* also serve as survival bodies in soil. These sexual spores have thick walls and can survive for 20 years in air-dried soil. They germinate by forming either zoospores or mycelium.

Specialized resting structures remain quiescent until a suitable food source becomes available. Nutrients required for germination of some fungus species are restricted to exudates from roots or seeds of a specified host crop, but for most species the resting structures germinate and grow when sugars and amino acids are exuded from roots or seeds of a wide variety of plants and reach a suitable concentration in the immediate vicinity of the resting bodies. Even plant debris, especially cover crops plowed under as green manure, can trigger the germination of resting structures of fungi. In the absence of host plants, these fungi either lyse or form new resting structures when the food is exhausted (Rovira, 1965).

All stages of plant parasitic nematodes have been reported to survive periods of adverse conditions. The survival stages frequently are also the infectious stages for the first life cycle after a period of dormancy. In general, eggs and second-stage juveniles or sedentary endoparasites survive adverse conditions, whereas most stages of migratory endoparasites and ectoparasites overwinter. Many features of the life cycles of plant parasitic nematodes facilitate survival, which is generally believed to be governed by utilization of stored fats and decreased metabolic activity. In *Heterodera* species the eggs are retained in a cyst and in some cases remain viable up to 8 years in fallow soil. Second-stage juveniles of *Anguina tritici* have survived for 28 years in dried plant material (Van Gundy, 1965; Ellenby, 1969).

X. NONPATHOGENIC HOST INFECTION

Fusarium oxysporum invades and parasitizes the root system of plants without pathogenic production of evident diseases (Hendrix and Nielsen, 1958). When the plant dies, the fungus quickly colonizes the root system prior to invasion by saprophytic organisms. This is a means of long-term survival in the absence of susceptible hosts.

XI. INFLUENCE OF ENVIRONMENT ON FEEDER ROOT NECROSIS

Development of feeder root necrosis is entirely dependent upon environment, even when pathogens and susceptibile hosts are present in proximity to each other. *Pythium vexans* and *P. irregulare* both attack feeder roots of peaches. However, *P. irregulare* was isolated predominantly in the winter and *P. vexans* in the summer (Hendrix *et al.*, 1966). Since this suggested a temperature relationship, the influence of soil temperature and water on root necrosis caused by these species were studied. The amount of root damage by *P. vexans* was increased by periodically saturating the soil, while *P. irregulare* damage was unaffected by soil saturation (Biesbrock and Hendrix, 1970a). *Pythium graminicolum* Subr. was isolated from corn roots in late May and June, but not again until late July. This occurrence was associated with soil moisture and temperature (Hampton and Buchholtz, 1959). Infection of tobacco (*Nicotina tabacum*) by *Phytophthora parasitica* var. *nicotianae* (Breda deHaan) Tucker occurs at a minimum temperature of 20C (Lucas, 1958). Only 32% of plants in flats kept moist were killed, while 71% in flats alternately wet and dried were killed (Wills, 1964). Susceptible tobacco plants were not attacked at pH 4 but were killed at pH 7 (Troutman and LaPrade, 1962).

Root knot larvae developed from infective larvae to egg-laying stage in 17 days at 24.5C but required 57 days at 15.4C. Reproduction, growth and development, and pathogenic capabilities of plant parasitic nematodes are influenced greatly by environmental factors such as soil moisture, pore size, aeration, temperature, osmotic pressure, hydrogen ion concentration, and various parasites and predators (Wallace, 1963).

XII. CONTROL OF FEEDER ROOT NECROSIS

Two general control measures are employed for controlling feeder root necrosis—eradication and biological control. Eradication can be accomplished by treating the soil with fungicides or fumigants or by raising the temperature of soil. Soil fumigation with methyl bromide or methyl bromide-chloropicrin is the most satisfactory method for field control because it kills most pathogens, including nematodes. Injecting the fumigant into the soil and covering the area with a plastic tarp are the most satisfactory means of applying these fumigants (Wilhelm, 1965).

Fungicides, such as captan, when added to soil do not eradicate soil fungi but lower the inoculum level (Powell *et al.*, 1968). These must be combined with a nematicide in order to reduce the population of both groups of organisms.

Use of steam in fields is usually impractical. However, it is used extensively for greenhouse operations. It is necessary to heat soil to 82C for 30 minutes to kill pathogens, weeds, and insects. Thermometers must be placed at the coolest point in the chamber in order to assure uniform sterilization (Baker, 1957).

Methyl bromide is also used extensively to sterilize soil for greenhouse operations. It has the advantages of not requiring a boiler and can be used almost anywhere. The relative merits of these two systems are presented in Table 9.2.

Benches, pots, and utensils in greenhouses also must be sterilized to prevent contamination of treated soil. They can be steamed or treated with methyl bromide. Other methods include washing these articles with a 0.5% solution of sodium hypochlorite.

XIII. SANITATION MEASURES FOLLOWING STERILIZATION

Soil-borne pathogens normally reproduce only to a limited degree in the absence of a host plant because of a phenomenon called *fungistasis*, which probably results from a combination of lack of nutrients and antibiosis. Fungistasis is destroyed when soil is sterilized, regardless of the method used. Thus if pathogens are introduced into sterile soil they grow very rapidly, and the final population is much higher than it was before sterilization (Hendrix and Powell, 1970). Various routes by which reinfestation might occur were investigated by Vaartaja (1967). Basically, anything that moves unsterilized soil will suffice. In greenhouses, one of the most common means is dropping the nozzles of watering hoses on the ground when they are not in use.

Table 9.2. Comparative advantages of steam sterilization and methyl bromide fumigation of soil

Characteristic	Steam (82C–100C for 30 min)	Methyl bromide (4 lb/100 ft^3)
Time required for treatment	About 1 hr	24–48 hr
Time between treatment and planting	About 1–2 hr to cool	24–48 hr
Kills all pathogens, weeds, and insects?	Yes, best treatment; a few weeds survive	Most, but not *Verticillum*; a few survive
When can penetration of material be determined, as a measure of effectiveness?	At once, by measuring soil temperature	Later, by noting reduction of disease or pathogen
Toxic aftereffect to crop?	None with U.C.-type soil mixes	Yes, for carnations and some others
Use near living plants?	Yes	Within 3 ft if adequately ventilated
Destroys organisms in unrotted crop refuse?	Yes	Yes
Can it be used anywhere?	Only if portable boiler is used	Yes
Is its use limited by environment?	Time and cost increase with cold or wet soil, but can be so used	Not recommended below 15C
Ease of application	Easy	Easy
Dangerous to workmen?	No	Yes
Is large capital outlay required?	If boiler unavailable	No
Cost per ft^3 of soil, exclusive of labor	Less than 2¢ including equipment cost	About 2.9¢–3.2¢ excluding equipment cost

Biological control measures include use of resistant varieties, soil amendments, and control of environment. Generally these are not useful for scientific investigations because they interfere with the original objectives. However, when possible, resistant varieties should be used. Use of soil amendments, such as barley straw in the case of Fusarium root rot of beans, or alfalfa meal in the case of Phytophthora root rots, is effective if it does not interfere with the original objectives of the research (Weinke, 1962; Zentmeyer, 1963).

Overwatering and poorly drained soil mixes contribute to root rot development in greenhouses. The best means of avoiding overwatering is to use well-drained soil mixes. The mix used at the University of Georgia is an equal mixture of sand, sandy loam soil, and milled pine bark. Lime and major and minor elements are added to the general mix (Pokorny and Perkins, 1967; Thurman and Pokorny, 1969).

References

Alicbusan, R. V., T. Ichitani, and M. Takahashi. 1965. Ecologic and taxonomic studies on *Pythium* as pathogenic soil fungi. III. Population of *Pythium ultimum* and other microorganisms in rhizosphere. *Bull. Univ. Osaka Prefect.*, ser. B, 16: 59–64.

Anderson, E. J. 1951. A simple method for detecting the presence of *Phytophthora cinnamomi* Rands in soil. *Phytopathology* 41: 187–89.

Averre, C. W. III. 1966. Isolating *Phythium* and *Fusarium* from a limestone soil in subtropical Florida. *Proc. Soil and Crop Soc. Florida.* 26: 279–85.

Baker, K. F. 1957. *The U. C. System for Producing Healthy Container-grown Plants.* Calif. Agr. Exp. Sta. Manual 23.

Benedict. W. G., and W. B. Mountain. 1956. Studies on the etiology of a root rot of winter wheat in southwestern Ontario. *Can. Bot.* 34: 159–74.

Biesbrock, J. A., and F. F. Hendrix, Jr. 1970a. Influence of soil water and temperature on root necrosis of peach caused by *Pythium* spp. *Phytopathology* 60: 880–82.

——. 1970b. Influence of continuous and periodic soil water conditions on root necrosis of holly caused by *Phythium* spp. *Can. Bot.* 48: 1641–45.

Bryan, W. C. 1965. *Testing Shortleaf Pine Seedlings for Resistance to Infection by Phytophthora Cinnamomi.* U.S. Forest Service Research Note SE-50.

Campbell, W. A. 1949. A method of isolating *Phytophthora cinnamomi* directly from soil. *Plant Dis. Reptr.* 33: 134–35.

——. 1951a. Vertical distribution of *Phytophthora cinnamomi* in the soil under littleleaf-diseased shortleaf pine. *Plant Dis. Reptr.* 35: 26–27.

——. 1951b. Fungi associated with the roots of littleleaf-diseased and healthy shortleaf pine. *Phytopathology* 41: 439–46.

——. 1951c. The occurrence of *Phytophthora cinnamomi* in the soil under pine stands in the Southeast. *Phytopathology* 41: 742–46.

Campbell, W. A., and O. L. Copeland, Jr. 1954. *Littleleaf Disease of Shortleaf and Loblolly Pines.* U.S.D.A. Cir. 940.

Campbell, W. A., and M. E. Gallegly. 1965. *Phytophthora heveae* from eastern Tennessee and western North Carolina. *Plant Dis. Reptr.* 49: 233–34.

Campbell, W. A., and J. T. Presley. 1946. *Diseases of Cultivated Guayule and Their Control.* U.S.D.A. Cir. 749.

Campbell, W. A., and B. Sleeth. 1945. A root rot of guayule caused by *Pythium ultimum. Phytopathology* 35: 636–39.

Chandler, W. A., J. H. Owen, and R. L. Livingston. 1962. Sudden decline of peach trees in Georgia. *Plant Dis. Reptr.* 46: 831–34.

Chee, K. H., and F. J. Newhook. 1965. Improved methods for use in studies on *Phytophthora cinnamomi* Rands and other *Phytophthora* species. *New Zea. J. Agr. Res.* 8: 88–95.

Childs, L., and S. M. Zeller. 1929. Observations on Armillaria root rot of orchard trees. *Phytopathology* 19: 869–73.

Chupp, C. 1917. Studies on clubroot of cruciferous plants. *Cornell Univ. Agr. Exp. Sta. Bull.* 387: 419–52.

Copeland, O. L., Jr. 1952. Root mortality of shortleaf and loblolly pine in relation to soils and littleleaf disease. *J. Forest.* 50: 21–25.

Crandall, B. S., G. F. Gravatt, and M. M. Ryan. 1945. Root disease of *Castanea* species and some coniferous and broadleaf nursery stocks, caused by *Phytophthora cinnamomi. Phytopathology* 35: 162–80.

Ellenby, C. 1969. Dormancy and survival in nematodes. In *Dormancy and Survival No. XXIII: Symposia of the Society for Experimental Biology,* ed. H. W. Woolhouse, pp. 83–97. New York: Academic Press.

Emmatty, D. A., and R. J. Green, Jr. 1969. Fungistasis and the behavior of the microsclerotia of *Verticillium albo-atrum* in soil. *Phytopathology* 59: 1590–95.

Flowers, R. A., and J. W. Hendrix. 1969. Gallic acid in a procedure for isolation of *Phytophthora parasitica* var. *nicotianae* and *Pythium* spp. from soil. *Phytopathology* 59: 725–31.

Garrett, S. D. 1956. *Biology of Root-infecting Fungi.* Cambridge: Cambridge University Press. 292 pp.

Grimm, G. R., and A. F. Alexander. 1970. Citrus leaf pieces as traps for soil borne *Phytophthora* spp. *Phytopathology* 60: 1294 (absatr.).

Grogan, R. G., F. W. Wink, W. B. Hewitt, and K. A. Kimble. 1958. The association of *Olpidium* with the big-vein disease of lettuce. *Phytopathology* 48: 292–97.

Hampton, R. O., and W. F. Buchholtz. 1959. Seasonal occurrence of *Phythium gramimicolum* on roots of field-grown corn. *Iowa St. Coll. J. Sci.* 33: 489–95.

Heald, F. D. 1933. *Manual of Plant Diseases.* New York: McGraw-Hill. 953 pp.

Hendrix, F. F., Jr., and W. A. Campbell. 1966. Root rot organisms isolated from ornamental plants in Georgia. *Plant Dis. Reptr.* 50: 393–95.

——. 1968. Pythiaceous fungi isolated from southern forest nursery soils and their pathogenicity to pine seedlings. *Forest. Sci.* 14: 292–97.

——. 1970. Distribution of *Phytophthora* and *Pythium* species in soils in the continental United States. *Can. Bot.* 48: 377–84.

Hendrix, F. F., Jr., W. A. Campbell, and C. Y. Chien. 1971. Some phycomycetes indigenous to soils of old growth forests. *Mycologia* (in press).

Hendrix, F. F., Jr., W. A. Campbell, and J. B. Moncrief. 1970. *Pythium* species associated with golf turfgrasses in the South and Southeast. *Plant Dis. Reptr.* 54: 419–21.

Hendrix, F. F., Jr., and E. G. Kuhlman. 1965. Factors affecting direct recovery of *Phytophthora cinnamomi* from soil. *Phytopathology* 44: 1183–87.

Hendrix, F. F., Jr., and L. W. Nielsen. 1958. Invasion and infection of crops other than the Forma suscept by *Fusarium oxysporium* F. *Batatas* and other formae. *Phytopathology* 48: 224–28.

Hendrix, F. F., Jr., and W. M. Powell. 1968. Nematode and *Pythium* species associated with feeder root necrosis of pecan trees in Georgia. *Plant Dis. Reptr.* 52: 334–35.

——. 1970. Control of root pathogens in peach decline sites. *Phytopathology* 60: 16–19.

Hendrix, F. F., Jr., W. M. Powell, and J. H. Owen. 1966. Relation of root necrosis caused by *Pythium* species to peach tree decline. *Phytopathology* 56: 1229–32.

Hine, R. B. 1961. The role of fungi in the peach replant problem. *Plant Dis. Reptr.* 45: 462–65.

Jackson, L. W. R. 1945. Root defects and fungi associated with the littleleaf disease of southern pines. *Phytopathology* 35: 91–105.

Kerr, A. 1963. The root rot–Fusarium wilt complex of peas. *Aust. J. Biol. Sci.* 16: 55–69.

Lorio, P. L., Jr. 1966. *Phytophthora cinnamomi* and *Pythium* species associated with loblolly pine decline in Louisiana. *Plant Dis. Reptr.* 50: 596–97.

Lucas, G. B. 1958. *Diseases of Tobacco.* New York: Scarecrow Press. 498 pp.

Lumsden, R. D., and F. A. Haasis. 1964. *Pythium Root and Stem Diseases of the Chrysanthemum in North Carolina.* N.C. Agr. Exp. Sta. Tech. Bull. 158.

McIntosh, D. L. 1964. *Phytophthora* spp. in soils of the Okanagan and Similkameen valleys of British Columbia. *Can. J. Bot.* 42: 1411–15.

Marx, D. H. 1969a. The influence of ectotrophic mycorrhizal fungi on the resistance of pine roots to pathogenic infections. I. Antagonism of mycorrhizal fungi to root pathogenic fungi and soil bacteria. *Phytopathology* 59: 153–63.

——. 1969b. The influence of ectotrophic mycorrhizal fungi on the resistance of pine roots to pathogenic infections. II. Production, identification and biological activity of antibiotics produced by *Leucopaxillus cerealis* var. *piceina. Phytopathology* 59: 411–17.

Marx, D. H., and C. B. Davey. 1969a. The influence of ectotrophic mycorrhizal fungi on the resistance of pine roots to pathogenic infections. III. Resistance of aseptically formed mycorrhizae to infection by *Phytophthora cinnamomi. Phytopathology* 59: 549–558.

——. 1969b. The influence of ectotrophic mycorrhizal fungi on the resistance of pine roots to pathogenic infections. IV. Resistance of naturally occurring mycorrhizae to infection by *Phytophthora cinnamomi. Phytopathology* 59: 559–65.

Marx, D. H., and B. Zak. 1965. The effect of pH on mycorrhizal formation of slash pine in aseptic culture. *Forest Sci.* 11: 66–75.

Miller, C. R., W. M. Dowler, D. H. Petersen, and R. P. Ashworth. 1966.

Observations on the mode of infection of *Pythium ultimum* and *Phytophthora cactorum* on young roots of peach. *Phytopathology* 56: 46–49.

Miller, J. H. J. E. Giddens, and A. A. Foster. 1957. A survey of the fungi of forest and cultivated soils of Georgia. *Mycologia* 49: 779–808.

Mircetish, S. M., and H. L. Keil. 1970. *Phytophthora cinnamomi* root rot and stem canker of peach trees. *Phytopathology* 60: 1376–82.

Mulder, D. 1969. The pathogenicity of *Pythium* species to rootlets of apple seedlings. *Neth. J. Plant Pathol.* 75: 178–81.

Oxenham, B. L., and B. L. Winks. 1963. *Phytophthora* root rot of *Pinus* in Queensland. *Queensland J. Agr. Sci.* 20: 355–66.

Podger, F. D., R. F. Doepel, and G. A. Zentmyer. 1965. Association of *Phytophthora cinnamomi* with a disease of *Eucalyptus marginata* forest in western Australia. *Plant Dis. Reptr.* 49: 943–47.

Pokorny, F. A., and H. F. Perkins. 1967. Utilization of milled pine bark for propagating woody ornamental plants. *Forest Prod. J.* 17: 43–48.

Powell, W. M., F. F. Hendrix, Jr., and D. H. Marx. 1968. Chemical control of feeder root necrosis of pecans caused by *Pythium* species and nematodes. *Plant Dis. Reptr.* 52: 577–78.

Powell, W. M., J. H. Owen, and W. A. Campbell. 1965. Association of Phycomycetous fungi with peach decline in Georgia. *Plant Dis. Reptr.* 49: 279.

Rands, R. D., and E. Dopp. 1938. Pythium *Root Rot of Sugarcane.* U.S.D.A. Tech. Bull. 666. 95 pp.

Rhoads, A. S. 1950. Clitocybe *Root Rot of Woody Plants in the Southeastern United States.* U.S.D.A. Cir. 853.

———. 1954. *Clitocybe* root rot found widespread and destructive in Georgia and South Carolian peach orchards. *Plant Dis. Reptr.* 38: 42–46.

Ross, E. W. 1970. Sand pine root rot-pathogen: *Clitocybe tabescens. J. Forest.* 68: 156–58.

Roth, E. R., E. R. Toole, and G. H. Hepting. 1948. Nutritional aspects of the littleleaf disease of pine. *J. Forest.* 46: 578–87.

Rovira, A. D. 1965. Plant root exudates and their influence upon soil microorganisms. In *The Ecology of Soil-borne Organisms*, ed. W. C. Snyder and K. F. Baker, pp. 170–86. Berkeley and Los Angeles: University of California Press.

Ruehle, J. L. 1962. Histopathological studies of pine roots infected with lance and pine cystoid nematodes. *Phytopathology* 52: 68–71.

———. 1968. Pathogenicity of sting nematode on sycamore. *Plant Dis. Reptr.* 52: 523–25.

———. 1969a. Improving seedling growth in longleaf pine plantations with nematicidal soil fumigants. *J. Nematology* 1: 248–53.

———. 1969b. Influence of stubby-root nematode on growth of southern pine seedlings. *Forest Sci.* 15: 130–34.

Ruehle, J. L., and D. H. Marx. 1971. Parasitism of ectomycorrhizal of pine by lance nematode. *Forest Sci.* 17: 31–34.

Schmitthenner, A. F. 1970. Significance of populations of *Pythium* and *Phytophthora* in soil. In *Root Diseases and Soil-borne Pathogens*, ed. T. A. Toussoun, R. V. Bega, and P. E. Nelson, pp. 25–30. Berkeley and Los Angeles: University of California Press.

Schroth, M. N., and F. F. Hendrix, Jr. 1962. Influence of nonsusceptible plants on the survival of *Fusarium solani f. phaseoli* in soil. *Phytopathology* 52: 906–9.

Schroth, M. N., and W. C. Snyder. 1961. Effect of host exudates on chlamydospore germination of the bean root rot fungus, *Fusarium solani f. phaseoli*. *Phytopathology* 51: 389–93.

Scott, F. M. 1963. Root hair zone of soil-grown roots. *Nature* 199: 1009–10.

——. 1965. The anatomy of plant roots. In *Ecology of Soil-borne Plant Pathogens*, ed. W. C. Snyder and K. F. Baker, pp. 145–53. Berkeley and Los Angeles: University of California Press.

Siggers, P. V., and K. D. Doak. 1940. *Littleleaf Disease of Shortleaf Pine*. U.S. Forest Service, South. Forest Exp. Sta. Occas. Paper 95.

Sleeth, B. 1953. Winter Haven decline of citrus. *Plant Dis. Reptr.* 37: 425–26.

Smalley, G. W., and R. L. Scheer. 1963. Black root rot in Florida sandhills. *Plant Dis. Reptr.* 47: 669–71.

Stolzy, L. H., J. Letey, L. J. Klotz, and T. A. Dewolfe. 1965. Soil aeration and root-rotting fungi as factors in decay of citrus feeder roots. *Soil Sci.* 99: 403–6.

Sutherland, C. F., F. J. Newhook, and J. Levy. 1960. The association of *Phytophthora* species with mortality of *Pinus radiata* and other conifers. II. Influence of soil drainage on disease. *New Zeal. Agr. Sci.* 2: 844–58.

Sutherland, J. R., R. E. Adams, and R. P. True. 1966. *Pythium vexans* and other conifer seedbed fungi isolated by the apple technique following treatment to control nematodes. *Plant Dis. Reptr.* 50: 545–47.

Thurman, P. C., and F. A. Pokorny. 1969. The relationship of several amended soils and compaction rates on vegetative growth, root development and cold resistance of "Tifgreen" Bermudagrass. *J. Amer. Soc. Hort. Sci.* 94: 463–65.

Troutman, J. L., and J. L. LaPrade. 1962. *Effect of pH on the Black Shank Disease of Tobacco*. Va. Agr. Exp. Sta. Tech. Bull. 158.

Tucker, C. M. 1931. *Taxonomy of the Genus* Phytophthora *de Bary*. Mo. Agr. Exp. Sta. Res. Bull. 153.

Vaartaja, O. 1967. Reinfestation of sterilized nursery seedbeds by fungi. *Can. J. Microbiol.* 13: 771–76.

Vaartaja, O., and P. J. Salisbury. 1961. Potential pathogenicity of *Pythium* isolates from forest nurseries. *Phytopathology* 51: 505–7.

Van Gundy, S. D. 1965. Factors in survival of nematodes. *Ann. Rev. Phytopathol.* 3: 43–68.

Wallace, H. R. 1963. *The Biology of Plant Parasitic Nematodes*. London: Edward Arnold. 280 pp.

Weinke, K. E. 1962. Influence of nitrogen on the root disease of bean caused by *Fusarium solani f. phaseoli.* Ph.D. Thesis. University of California, Berkeley.

White, R. P. 1937. *Rhododendron Wilt and Root Rot.* N.J. Agr. Exp. Sta. Bull. 615.

Wilhelm, S. 1965. *Pythium ultimum* and the soil fumigation growth response. *Phytopathology* 55: 1016–20.

Wills, W. H. 1965. *Exploratory Investigation of the Ecology of Black Shank Disease of Tobacco.* Va. Agr. Exp. Sta. Tech. Bull. 181. 20 pp.

Zak, B. 1961. *Aeration and Other Soil Factors Affecting Southern Pines as Related to Littleleaf Diseases.* U.S.D.A. Tech. Bull. 1248.

Zentmyer, G. A. 1963. Biological control of *Phytophthora* root rot of avocado with alfalfa meal. *Phytopathology* 53: 1383–87.

——. 1970. Tactic responses of zoospores of *Phytophthora.* In *Root Disease and Soil-borne Pathogens,* ed. T. A. Toussoun, R. V. Bega, and P. E. Nelson,

Zentmyer, G. A., A. O. Paulus, and R. M. Burns. 1962. *Avocado Root Rot.* Calif. Agr. Exp. Sta. Ser. Cir. 511.

Part II

Soil Physical and Chemical

Aspects of the Root Environment

10. Significance of Rooting Pattern to Crop Production and Some Problems of Root Research

Robert W. Pearson

NEARLY 50 years ago, Weaver and co-workers (1922) wrote, "An exact knowledge of the root development of crop plants, of their position, extent, and activity as absorbers of water and solutes at various stages of growth, is of paramount importance to a scientific understanding of plant production. Moreover, a knowledge of modifications produced by variations in the subterranean environment, whether due to natural conditions . . . or to tillage or fertilizers, is of no less importance."

Although a few early researchers attempted to relate various physical parameters of the environment to root growth (Pfeffer, 1893; Beal, 1918; Leitch, 1916; Taubenhaus *et al.*, 1931), progress in this area lagged considerably behind advances in soil chemistry, probably because the first limiting factors of crop production were chemical. However, a surge of interest had occurred by 1940, particularly in mechanical impedance, O_2 supply, and water level effects on root growth. This interest has continued unabated.

The foundation for recent progress in delineation of soil chemical factors governing root growth was laid around 1930 by a number of outstanding researchers, including Pierre *et al.* (1932), McLean and Gilbert (1927), Trenel and Alten (1934), and Sorokin and Sommer (1929). Even so, results of most root growth experiments during that period must be interpreted with caution because the entire root system was usually exposed to the same chemical conditions. Any condition that would affect top growth would indirectly influence root growth, and erroneous conclusions could be easily drawn. Manganese and aluminum toxicities illustrate this point. Funchess (1918), an acute observer, concluded that in acid soils "reduced growth appears to be due chiefly to injury to plant roots from direct action of Mn," but later he (1919)

Contribution from the Soil and Water Conservation Research Division, Agricultural Research Service, U.S. Department of Agriculture, in cooperation with the Alabama Agricultural Experiment Station, Auburn University, Auburn, Alabama 36830.

decided that root damage was not a result of excess Mn. Using split-root experiments, Rios and Pearson (1964) showed that normal root growth can proceed at Mn levels that are highly toxic to tops, whereas top growth may be normal at Al levels that inhibit roots. Yet root growth in the field will no doubt be reduced by high Mn concentration in the soil solution because of damage to the aboveground portion of the plant.

It seems clear, then, that the final root pattern of a given plant, and its rate of achievement, is genetically determined when the environment is favorable but is actually controlled by the physical or chemical environment in most field situations. Mechanical impedance by compacted soil layers, for example, can turn a normally deep-rooted plant into a very shallow-rooted one (Lowry *et al.*, 1970). Thus the field situation very often corresponds to a horizontally layered environment in which most plant roots are growing in the chemical and physical environment of the plow layer while roots below this layer may have to overcome the restraint imposed by unfavorable subsoil conditions.

I. IMPORTANCE OF ROOTING DEPTH IN CROP PRODUCTION

Improvement in root development and function seems to offer considerable promise for raising the yield ceiling imposed by occasional water deficiencies, even in humid regions. The average annual rainfall in the southeastern United States appears to be adequate for maximum crop growth. For example, less than one-half the amount received annually is evapotranspired by a crop of cotton (*Gossypium hirsutum* L.) even under ideal moisture conditions. Under these conditions, cotton will evapotranspire up to 0.7 cm per day, but water movement in unsaturated flow through the soil is far too slow to meet such requirements for a shallow-rooted crop. Thus, in order to meet plant needs and effectively utilize subsoil-stored water, roots must proliferate continuously into unexploited zones between rains. Of course, there is a limit to the effective depth of root proliferation and, therefore, a limit to the duration of rainless periods that crops can tolerate without experiencing serious stress.

The potential rate of root extension in a favorable environment is amazing. For example, a single winter rye (*Secale cereale* L.) plant after 4 months in 0.28 m^3 (1 cubic ft) of fertile, permeable surface soil in the greenhouse produced an average 5.05 km of new roots per day (Richards and Wadleigh, 1952). From the average diameter of the roots and of the root hairs, it was estimated that roots came in contact each day with a cylinder of soil 2 mm in diameter and 5.05 km long. Assuming a sandy loam soil at field capacity, about 1.6 kg of water would have been added to the plant's available water supply each day. This would be more than enough water to meet maximum demand. The problem here is that the same volume of soil was being exploited repeatedly by different roots, but the results do indicate root extension possibilities under unconfined conditions in the field. The extent to which these possibilities are realized will depend upon (a) environmental soil factors opposing the continued exploration of new soil masses by roots, (b) genetic characteristics of the species, and (c) competition from weeds.

Many of our soils, such as those of the Coastal Plains, can retain only about 2.5 cm of available water per 30 cm of profile depth. If a crop's effective rooting depth is restricted to 45 cm, as is frequently the case, the crop would withstand no more than 5 days between rains without suffering some degree of moisture stress, assuming the profile to be at field capacity at the beginning of the period. If effective rooting depth is increased to 90 cm—a reasonable aim—the period between rains that could be tolerated should more than double because subsoil water is not so subject to evaporation as surface soil water is.

The implication from the above illustration is that soil profile recharge may occur often enough during the growing season to provide adequate moisture for deep-rooted crops. Experimental data on frequency and depth of recharge are lacking, but estimates can be made based on longtime rainfall records and a few reasonable assumptions. This has been done at the Wiregrass Substation, Headland, Alabama, where water extraction by cotton was measured on a Norfolk sandy loam with and without root restriction in the subsoil by unfavorable subsoil pH. No water was extracted during the 6-day period at depth below 45 cm in the very acid subsoil (Table 10.1), but an appreciable amount was taken up from the slightly acid subsoil at the 75-cm depth. Also, a much higher percentage of the available water was extracted from the upper subsoil zone (15 to 45 cm) of the slightly acid soil than from that of the very acid soil. Over a 15-year period, there was an average of 4.6 rains of more than 3.8 cm between June 1 and September 15, including wet-weather periods of up to 3 days' duration. Allowing an average 0.4-cm evapotranspiration per day for the 2-day average rain, 3.87 cm would be required to recharge the deep-rooted profile in the absence of runoff (3.07-cm deficit + 0.80-cm evapotranspiration during the rain). Thus, profile recharge could occur between four and five times during June to September. Extrapolating water requirements to a much longer period between rains and assuming all available water extracted to 75-cm depth, 6.14 cm would be required for recharge. Weather records show that, in this case, the profile could be recharged an average of 1.6 times during the growing season. So, even for extreme conditions of profile depletion, rainfall provides the potential for recharge between one and two times

during the summer, which should make an appreciable difference in the
effects of drought periods.

Table 10.1. Water uptake by cotton from a Norfolk sandy loam profile
during a 6-day period in July as affected by acidity of the 15- to 105-
cm zone

Depth (cm)	Available water capacity (cm)	Water uptake during 6 days			
		Very acid subsoil (%)	(cm)	Slightly acid subsoil (%)	(cm)
0–15	0.76	100	0.76	100	0.76
15–30	0.91	30	0.28	75	0.69
30–45	1.27	20	0.25	70	0.86
45–60	1.32	0	0.00	35	0.48
60–75	1.37	0	0.00	15	0.20
75–90	1.37	0	0.00	5	0.08
90–105	1.37	0	0.00	0	0.00
Total	8.37		1.29		3.07

Inherent in a crop's utilization of soil moisture is the relative effectiveness of roots at different soil depths. The observation that plants normally
utilize water nearest the soil surface first and that in the deeper layers
afterward could be interpreted as an indication of different levels of root
effectiveness as well as of different root densities. Recent data of H. M.
Taylor and Betty Klepper (personal communication) show clearly that
absorption of water per unit length of root is the same for deep and
shallow roots. By measuring both root length and water extraction in
the rhizotron, or root observation laboratory, at Auburn, Alabama, they
showed that water uptake per unit length of cotton roots did not decrease
to a depth of at least 150 cm (Table 10.2).

Table 10.2. Extraction of water by cotton roots at different soil depths
during a 3-day period in a drying cycle

Soil depth† (cm)	Water extraction per day (cm³/100 cm of root)
30	1.7
60	1.7
90	1.6
120	1.7
150	1.6

Source: H. M. Taylor and Betty Klepper, personal communication.
† 15-cm layers centered at depths indicated.

In addition to utilizing natural rainfall more effectively, deep rooting might also improve use of suboptimal plant nutrient levels in the soil. Fertilizer is frequently applied in bands rather than being mixed with large volumes of soil. There are good reasons for this, but it raises two questions: (1) Can a very small fraction of the root system in contact with a concentrated fertilizer band absorb at a rate required to meet plant needs during periods of maximum growth? (2) Can the lower level of nutrients in the bulk of the soil restrict root proliferation in that part of the profile? During periods of high evapotranspiration, the surface soil dries quickly. The uptake of nutrients from the dry fertilized zone may then be drastically reduced (Hunter and Kelley, 1946; Lipps and Fox, 1964), and the crop may become much more dependent upon uptake from the subsoil. These questions, of course, relate to the broad problem of fertilizer placement, but the answers are pertinent to crop yield–root pattern relationships.

Finally, the value of deep rooting in the field is supported by crop performance data. In the case of the root restrictions referred to in Table 10.1, seed cotton yields from the deep-rooted treatment ranged up to 1400 kg/ha more than from the shallow-rooted treatment (Fig. 10.1).

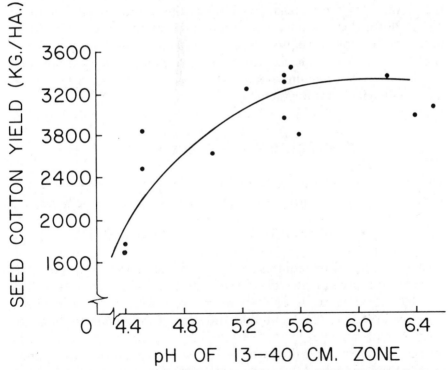

Fig. 10.1. Relationship between seed cotton yield and subsoil pH of a Norfolk sandy loam in a year of about average rainfall during the growing season

As would be expected, yield response varied widely from year to year, depending upon weather conditions. However, the average seed cotton yield increase at this location was 896 kg/ha during a 3-year period (Adams *et al.*, 1967). Using a different approach with corn (*Zea mays* L.), Fehrenbacher and Rust (1956) measured rooting depth in four Indiana soils that had considerably different physical conditions in their subsoils. When compared with longtime average corn yields, deep root development favored higher corn yields (Table 10.3). The chief reason given by

Table 10.3. Relation of effective rooting depth to average corn yield on four Indiana soils

Soil series	Approx. effective root depth (ft)	Available water (inches)	Average yield (bu/acre)
Saybrook	4.5	10.6	77
Ringwood	4.0	9.8	75
Elliott	3.0	7.1	65
Clarence	3.0	6.4	60

Source: Fehrenbacher and Rust, 1956.

the authors for the increased yields was the improved availability of water. Plotting corn yield as a function of the amount of available water in the rooting zone certainly supports this conclusion (Fig. 10.2). These and other results emphasize the need for defining the environmental requirements for optimum root growth in soils, particularly where water may be the yield-limiting factor.

II. SOME PROBLEMS OF ROOT RESEARCH

The various facets of research on plant root growth abound with unanswered questions. The major ones will be examined briefly here.

A. Inability to Define "Optimum" Root System

Because root growth requires the diversion of energy from potential aboveground growth, it seems clear that extravagant root growth is possible. How, then, is an "optimum" root system to be defined? Brouwer (1962) pointed out that different plant organs respond to various environmental factors with different growth rates, resulting in changes in their relative proportions, e.g., shoot:root ratio. Water stress usually results in relatively less shoot than root growth, as illustrated by Figure 10.3 from Troughton (1957). Similarly, N deficiency reduces shoot growth first (Troughton, 1957), but reduced light seems to restrict

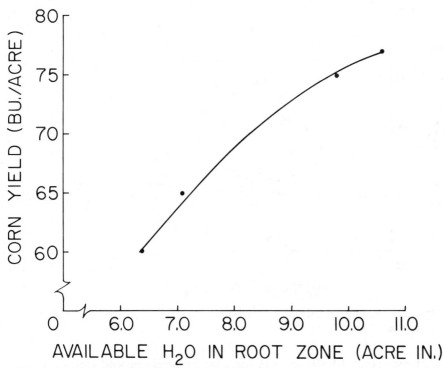

Fig. 10.2. Relationship between long-time corn yields and available water-holding capacity of the effective corn-rooting depth in four Indiana soils. (Reprinted by permission from J. B. Fehrenbacher and R. H. Rust, *Soil Sci.* 82: 369–78, ⓒ 1956, The Williams & Wilkins Co., Baltimore, Md. 21202, U.S.A.)

root growth first (Mitchell, 1954). From this it appears that the definition of a normal, or optimum, shoot:root ratio for a given genotype should include certain elements of the environment.

The question may be asked whether a restricted root system can be harmless, if not actually beneficial, to crop yield under irrigation and intensive fertilization. In furrow irrigation of medium- and coarse-textured soils, the presence of a pan would improve water application efficiency, although it would restrict rooting depth. Experimental attempts to answer this question have been unsuccessful because of inability to ensure adequate water and nutrient supply in widely different rooting volumes without concurrent detrimental decreases in O_2 level.

B. Inability to Characterize Environmental Factors Accurately

Another vexing facet is the inability to define parameters of the root environment except in bulk soil terms. Consider for a moment the two factors with which we are probably best acquainted: pH and mechanical impedance. As indicated earlier, a general relationship exists between

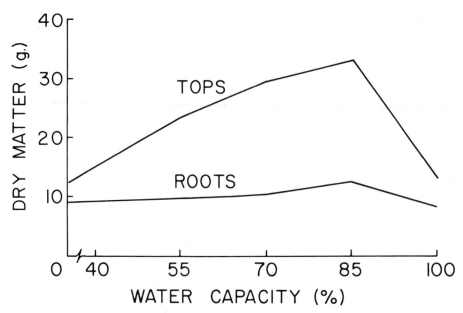

Fig. 10.3. Variation in root and top growth of Italian ryegrass with soil moisture level. (Troughton, 1957)

root growth and bulk soil pH for a given genotype and soil. However, the root itself, through differential absorption of cations and anions, undoubtedly produces important shifts in solution pH at the root-solution interface. The root may actually be growing in a medium of unknown pH. Similarly, absorption of water as the root moves through soil must result in localized increases in soil strength at the point of moisture reduction. Such changes may easily affect root expansion although they probably cannot be discerned by present procedures.

C. Inadequate Techniques for Measuring Root Response

Finally, there is a serious lack of satisfactory indices and techniques by which to evaluate quantitatively soil environmental effects on root behavior. There are several possibilities, all of which have disadvantages. These include (1) root weight, (2) root surface area, (3) root number and/or length, (4) water extraction, (5) tracer methods, and (6) root morphology and root pattern description.

1. Root Weight

Root weight is probably the most commonly used criterion of root growth response to environment. The magnitude of the task of making reasonably satisfactory measurements in the field is a major obstacle in this kind of research.

There are at least two ways to approach the problem of root yields under field conditions. The most common, of course, is to take core samples, wash or pick out the roots, and weigh them. As a rule, the

Fig. 10.4. Appearance of an implanted soil mass after removal from the field for recovery of roots

absolute root yield is of less interest than relative yields of several treatments or conditions. For this reason, a series of cores from different depths at one position near the plant is usually adequate. Variation among observations within a treatment is reduced by using fairly large samples (e.g., 12-cm-diameter cores). A hand-operated posthole digger works well for this purpose down to about 1-m depth and does not mutilate the roots. A clothes-washing machine is used to agitate the samples in water so that roots are washed loose from the soil and swept onto a set of screens. Water flows continuously in at the bottom of the tub and out through an overflow made in the side near the top. Residual soil particles are removed from roots by hand-washing with care.

The "implanted soil mass" technique (Lund *et al.*, 1970) is especially useful for testing direct effects of soil chemical properties on root growth.

It is an adaptation of a method used by Hendrickson and Veihmeyer (1931) to study root extension into dry soil. Holes are made at a predetermined position from individual plants to a depth of about 30 cm using a 10-cm bucket-type soil auger. Cylindrical bags of the same diameter and about 30 cm long, made of a very loosely woven plastic material, are fitted snugly into the holes and filled to about 6 cm of the surface with the treated soil. After the soil is gently firmed to assure good contact with surrounding soil, it is covered over with surrounding field soil and the location is marked with a stake. After about 4 to 6 weeks the bags are removed and the enclosed roots are recovered (Fig. 10.4). This procedure has many applications, including determination of short-term root response by perennial crops to specified soil characteristics. Its chief advantages are that modification of large masses of soil deep in the profile is avoided, results can be obtained in a very short time even with slow-growing plants, and top-root interactions are eliminated since only a small part of the root system is involved.

The interpretation of root weight data depends upon the assumption that root mass is related to root activity in a direct and well-defined way. This assumption is not always valid, for at least two reasons. In the first place, roots grown in soil can never be quantitatively recovered, and by far the poorest recovery is with the fragile and most active roots. Secondly, the morphology of the root system may change with changes in some environmental factors. For example, roots grown at toxic levels of soluble Al do not develop so fine and branched a pattern as roots in the absence of Al (Fig. 10.5). Anaerobiosis and mechanical impedance

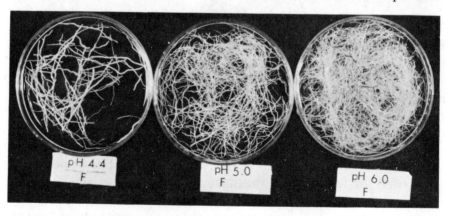

Fig. 10.5. Effect of pH in a Bladen subsoil on cotton root fineness

have similar effects on root character. In less favorable environments, roots are stubbier and thicker and have a lower specific surface. For example, it is possible to have almost no change in root yield as pH increases within a certain range in values, but considerable changes would occur in absorbing surface and root-soil contact. These changes would be reflected in uptake of water and ions. This is illustrated by the results from a split-root stacked-can experiment in the light chamber using a fertile surface soil atop acid subsoils from three different soil series limed to give a range of pH. After the first true leaves emerged, pots were weighed and surface-watered daily for 2 weeks. Within about 48 hours after surface watering was stopped, the surface soil was dry and the plants were forced to obtain water from the subsoils. For the next 6 days, weight loss of the cans was made up daily with water injected into the center of the subsoil mass. Roots in the subsoil were recovered and weighed at the end of the 6-day test period. The results (Fig. 10.6) show little change in water uptake for a given increase in root weight in the range where root growth was severely restricted by low pH. At higher levels of root weight (i.e., higher subsoil pH values), each increment of root weight was associated with the absorption of nearly three times as much water.

2. Root Surface

Some practical method for estimating root surface is badly needed. The most direct method is to measure the diameter of a large number of

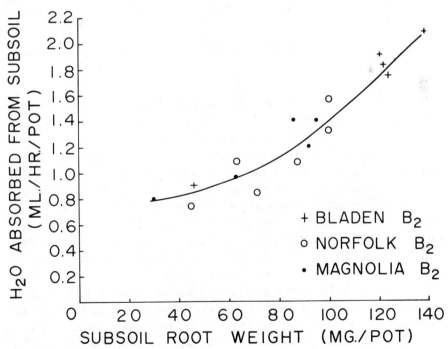

Fig. 10.6. Relationship between subsoil root weight and water absorbed from this zone in a model profile light chamber experiment

individual roots and the total root-length per unit of soil volume. From these data, root surface per unit of soil volume is calculated. Melhuish and Lang (1968) outlined a procedure for root surface in small soil volumes, but the measurements are extremely slow and tedious. There are several approaches to relative root surface estimation. In one (Wilde and Voigt, 1949), roots are dipped in dilute acid and drained, after which the adhering film of acid solution is washed off and titrated. In another (Corley and Watson, 1966), the weight of a relatively viscous solution of $Ca(NO_3)_2$ adhering to the root system is used as a measure of relative surface area. The drawbacks of such procedures include the tendency for fine roots to rope together and behave like coarser strands and the fact that values are only relative among a given series of measurements. Table 10.4, from Corley and Watson (1966), shows the degree of agree-

Table 10.4. Comparison of wheat root dry weight to relative surface as measured by two methods

Relative root yield (%)	Relative root surface	
	Titration (%)	Gravimetric (%)
100	100	100
75	67	50
50	66	68
25	35	30

Source: Corley and Watson, 1966.

ment obtained between these two procedures. The data also indicate that the roots must have varied considerably in fineness, for weight and surface measurements do not agree very well.

Root cation exchange capacity may offer a practical approach to the problem of surface estimation. Data reported by McLean *et al.* (1956) show that root CEC varies with genotype and to some extent with N content as a result of fertilization. Apparently, however, little or no effort has been made to relate root surface to CEC. It seems unlikely that the wide differences in CEC reported by McLean and associates between crop varieties, e.g., 11 and 54 meq/100 gm dry root tissue for corn and soybeans (*Glycine max* L.) respectively, could be a reflection of specific-surface differences since the root fineness of the two species does not appear to differ greatly. However, the possibility certainly exists that CEC is related to specific surface within a genotype, if not among genotypes. Williams (1962) proposed a rapid colorimetric method for estimating anion exchange properties of roots using a negatively charged anion dye. In preliminary experiments we were unable to get reproducible results using this method, apparently because of variations in dye penetration of the root tissue, or because of incomplete replacement from the root surface. The method may be useful for rapid estimation of surface areas if these problems are overcome by careful standardization of exposure time and better control of displacing solution pH.

3. Root Number and/or Length

Root number and/or length as a measure of rooting density at different positions in the soil profile can also be used to characterize root pattern and response to environmental factors. One method involves digging a trench to the depth to which observations are to be made and wide enough for a person to work in. It is helpful to place a metal frame of known dimensions, e.g., a 15- × 15-cm opening, against the trench wall and record the number of visible roots within this area at different horizontal and vertical positions with respect to a plant. Considerable variation among observations is common, and since plants are sacrificed, only one measurement can be made per plant. Also, the trenching method often can be used only at the conclusion of a field experiment. This approach is especially useful for estimating the limit of effective rooting depth where severely unfavorable conditions exist in some horizon of the normal rooting zone. The results from an experiment on subsoil acidity at Headland, Alabama (Table 10.5), provide a good illustration.

Table 10.5. Corn root distribution with depth in Norfolk sandy loam as influenced by acidity

Soil depth (cm)	Strongly acid subsoil		Moderately acid subsoil	
	pH	Root count (% total)	pH	Root count (% total)
0–15	5.5	80	5.9	46
15–30	4.6	10	5.6	21
30–45	4.5	6	5.2	11
45–60	4.4	4	5.2	22

In the strongly acid subsoil effective rooting depth of the corn crop was limited to the surface 30 cm, whereas in the moderately acid subsoil 22% of all roots occurred in the deepest zone examined (45 to 60 cm). There is an important precaution to be observed in using the root count procedure: The area must be free of weeds! Many weeds have greater tolerance for various unfavorable conditions than crop plants, and it is often difficult to identify specific roots.

Newman (1965) suggested a shortcut for estimating root length. A sample of roots is randomly distributed over a known surface area, photographed, and projected onto a translucent grid. Line intersections by roots are counted for an appropriate fraction of the grid length. Reicosky *et al.* (1970) compared this technique with direct measurement and with the use of an inch-counter on projected photographs of root samples and found little difference in accuracy. Time required was widely different, however; the line-intercept and inch-counter procedures required only 1.0 and 1.5 hours per sample, respectively, as compared to 5 hours for direct measurement. Both intercept and inch-counter methods have real possibilities for research use in field experiments.

Root observation boxes with glass fronts have been used in a variety of forms for nearly a century in laboratory studies. The idea was later carried to the field so that roots could be observed over a complete growth cycle and under natural environmental conditions (Rogers, 1934; Pearson and Lund, 1968). The rhizotron (Rogers, 1969; Taylor *et al.*, 1970), is a sophisticated offspring of the glass-walled trench and provides almost unlimited opportunities for the study of roots under field conditions. A common measurement which reflects root length and density is to count roots intersecting horizontal planes at fixed positions throughout the profile at predetermined times; this cumulative root growth can be plotted to show the root pattern and its rate of development (Fig.

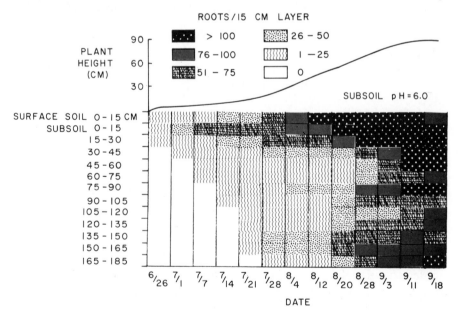

Fig. 10.7. Development of the cotton root pattern as measured by root count in a rhizotron

10.7). Taylor *et al.* (1970) showed that root count, i.e., numbers of roots crossing a horizontal transect, and the root length per cm^2 at the same depth were highly correlated ($r = 0.97$) and that, therefore, the most convenient of the two measurements can be used.

The advantages of the rhizotron include the ability to observe continuously roots of the same plant and the same root growing under field conditions. Also, it enables one to characterize precisely the environments of roots and shoots and to relate the effect of each environmental factor on the roots. On the other hand, several problems are associated with this technique: (1) Only a small part of the total root system is visible, and this may not be representative of roots in the interior. Howard M. Taylor and Betty Klepper (personal communication), however, reported that cotton roots counted at the glass wall gave a reasonably good estimate of root density in the soil mass (Table 10.6). (2) Another question relates to the effect of light on root growth and possible avoidance of the glass-soil interface. We have checked a number of species for light sensitivity and found that corn, tomato (*Lycopersicon esculentum* Mill.), cotton, and soybean roots show no difference in elongation rate due to illumination in glass-front containers or in the rhizotron. In contrast, light appreciably reduced the rate of peanut (*Arachis hypogaea* L.) root extension. When this technique is used, therefore, the effect of light on root growth rate of a particular species must be known in advance. (3) A major problem in using rhizotron compartments with inclined-glass panels is that of stabilizing the soil in the compartment. It is difficult to place soil uniformly in these compartments at a density high enough to avoid slumping with wetting and drying and yet not so high as to restrict root development.

4. Water Extraction Pattern

Determining the pattern of water extraction is probably the soundest way to evaluate root distribution in the field, but it also is beset with

Table 10.6. Cotton root distribution in a rhizotron compartment

Depth (cm)	Root dry weight (mg/250 cm^3 of soil) at		
	Transparent wall	Compartment center	Rear wall
30	24.9	12.4	38.0
60	24.6	26.8	22.0
120	20.8	17.6	39.4

Source: H. M. Taylor and Betty Klepper, personal communication.

problems. The chief advantage is that the characteristic of most concern is directly measured, i.e., the ability of the plant to obtain water from various depths in the soil profile. Thus, no assumptions about root mass, character, rate of extension, and activity are necessary.

Other assumptions must be made, however. It is assumed, for example, that during the period of moisture measurements there is no appreciable transfer of water from one part of the soil profile to another and that there is no water loss except by transpiration. This may be a reasonable assumption for some soils but not for others in which drainage continues at a significant rate after drainage by gravity would normally have ceased. Obviously, if measurements are started before drainage has become insignificant, there will be erroneous indications of water withdrawal from the various horizons (Doss and Taylor, 1970). Water can be lost from soil zones by direct evaporation. However, evaporation does not seem to be a serious source of error after the crop canopy has formed a complete ground cover or after the surface inch or so of soil has dried out. Also, evaporation can be avoided by covering the soil with a plastic sheet following profile recharge.

Another assumption is that the period of water extraction measurement begins with a uniform vertical distribution of available water in the soil. Some degree of compromise is always required because several days are needed for field capacity to be reached in the soil profile. During this time, the surface soil approaches equilibrium first and, because of higher root density, is subjected to a higher evapotranspiration rate than the subsoil. Thus, by the time water extraction measurement can be started, available water level usually varies with depth.

In spite of the pitfalls, however, the water extraction pattern can be used to good advantage as a measure of root distribution, and it is nondestructive. In fact, by using the neutron probe, changes in distribution of roots that occur with time at the same site can be measured. A good example of this procedure, which shows the progressive development of a corn root pattern, is provided by data from Letey and Peters (1949), reproduced in Figure 10.8. Water availability almost uniform with depth was attained at the beginning of the drying cycle by covering the soil immediately after profile recharge. The drying cycle in this case lasted 3 months, a length of time possible only with a soil which has a high water-holding capacity. As the season progressed water was drawn from successively lower soil horizons until on September 10 the entire 5-ft profile was near the wilting point. Effective rooting depth moved from 30 inches on July 1 to 44 inches on July 16 to below 60 inches by September 10.

The results in Table 10.7 show a comparison of cotton root distribution as determined by water uptake and by actual weight of roots recovered.

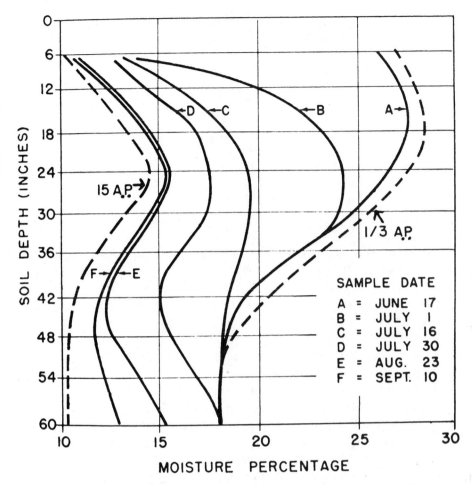

Fig. 10.8. Corn root pattern indicated by water uptake. (Letey and Peters, 1949)

The undue influence of the upper taproot and large laterals on root weight in the surface soil is apparent. Nearly twice as much of the total root system was indicated by weight as by water uptake in this zone.

5. Tracer Methods

Relative activity of roots in ion absorption from different sites in the soil profile has been satisfactorily measured by injecting tracers into the soil at various depths and distances from the plant. Before the availability of radioisotopes, lithium, rubidium, and various dyes were used, but with limited success probably due chiefly to their high mobility in soil. The use of ^{32}P was first proposed by Hall *et al.* (1953) and has since

Table 10.7. Cotton root distribution in a Lucedale fine sandy loam as measured by root yield and by water uptake

Soil depth (cm)	Root yield (% total)	Water uptake (% total)
0–15	55	30
15–30	20	28
30–45	10	22
45–60	7	16
60–75	8	4

Source: B. D. Doss, personal communication.

found rather wide acceptance. This radioisotope technique is simple and quite satisfactory for determining when roots arrive at a given location in the profile or for estimating maximum rooting depth. When it is used to define the root pattern by estimation of relative levels of root activity, it becomes much more involved and at least the following assumptions have to be made:

1. Roots have an equal chance of encountering the tracer in each soil layer or plane injected;
2. Applied material is essentially immobile in the soil but remains equally available at all loci for plant absorption; and,
3. Specific activity of the applied tracer is uniform among the locations injected.

The first assumption can be satisfactorily validated by carefully designing the injection pattern to allow for divergence of roots at different distances from the plant, but this makes the procedure tedious and laborious. The second is acceptable as far as immobility of $^{32}PO_4$ ions is concerned, but in some subsoils precipitation as Fe and Al compounds of low solubility would cause problems of availability to plants. The third assumption always is highly suspect when dealing with $^{32}PO_4$ applied in surface and subsoil layers because of the large difference in the level of soluble native P among the various horizons and the resulting difference in dilution of the tracer by ions from the labile pool at each depth. Furthermore, comparisons of root activity between dates is not possible because of the shift in specific activity of the available P resulting from plant uptake. Although isotopic exchange between added ^{32}P and native soil P in solution is complete within a few hours, plant removal and subsequent replenishment from solid phase soil phosphates causes a continuous decrease in specific activity of soil solution P.

In spite of the acknowledged drawbacks, the radioactive P method frequently is used in the field because it is nondestructive and gives a direct indication of root activity (Burton, 1957). Results reported by Lipps and Fox (1964) illustrate for alfalfa (*Medicago sativa* L.) how a root pattern measured by actual recovered root weight can differ from that indicated by ^{32}P uptake (Fig. 10.9). Far less root activity occurred in the surface soil in late September than the root mass indicated and a great deal more in the zone immediately above a water table. Obviously, soil moisture level influenced P uptake. Such conditions must be taken into account in the interpretation of tracer-root measurements.

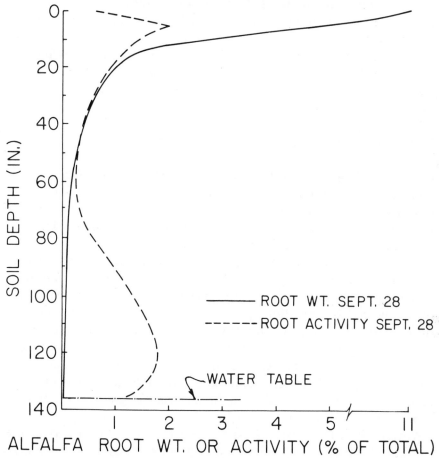

Fig. 10.9. Alfalfa root pattern as measured by root weight and indicated by ^{32}P uptake. (Reprinted by permission from R. C. Lipps and R. L. Fox, *Soil Sci.* 97: 4–12, © 1964, The Williams & Wilkins Co., Baltimore, Md. 21202, U.S.A.)

Some elements are mobile enough in plants to become rather uniformly distributed throughout the root system within a short time after application. Such elements, when injected into the aboveground plant in radioisotopic form, provide a ready means of identifying root locations and determining relative root density at different soil positions. Racz *et al.* (1964) suggested the injection of plants with ^{32}P and subsequent assay of soil-root cores from different locations in the profile as a non-destructive means of evaluating root distribution. They found good agreement in field experiments between relative root distribution among the different soil depths examined and weight of recovered roots ($r^2 = .9$). In separate experiments they showed that uniform distribution of ^{32}P within the root system occurred within 5 days after injection. Russell and Ellis (1968) proposed use of ^{86}Rb instead of ^{32}P because its higher energy radiation would improve assay precision of the soil-root cores. Adequately uniform distribution of ^{86}Rb had taken place in the root system of barley plants within 24 hours after injection into leaf sheaths. As Russell and Ellis pointed out, the procedure has the advantage over root mass measurements of distinguishing between active and dead roots.

6. Qualitative Evaluations

Qualitative evaluations of root systems can be useful in understanding response to environmental factors and in rationalizing quantitative measurements. Observations of root color, fineness, configuration, and general distribution are included under this heading. For example, when roots encounter an acid soil with micromolar levels of Al activity in the soil solution, they become thickened, less branched, and, as Al level increases, discolored (Ligon and Pierre, 1932). Similarly, calcium deficiency typically results first in a translucent appearance of primary root tips, followed by cessation of elongation, after which the root tip turns a brownish color and dies. Roots encountering a high-strength soil zone either become blunted and thickened or turn and grow laterally just above the restricting layer. Examination of thin sections prepared from roots growing under mechanical restriction shows that the cells are shorter and wider, resulting in the thickened appearance of the root (Camp and Lund, 1964).

Various systems of excavating root systems and washing away the soil have been used for nearly a century. In most cases, some system of maintaining the roots in their approximate position is used. The pin-board is one of the most common modifications of this technique and has been used extremely effectively by De Roo (1961) in studying tobacco (*Nicotiana tabacum* L.) root response to tillage. The method is especially suited to the examination of relatively shallow root systems. Samples can be taken from the monolith for root recovery to provide quantitative results in conjunction with the overall picture of root pattern.

III. SUMMARY

Rooting depth of crops is usually much less than the normal growth habit of the genotype would allow if the environment were favorable.

Shallow rooting increases drought hazard to crops.

Profile recharge frequency, especially for soils with relatively low available water-holding capacity, is an important factor in determining potential benefit of deep rooting. This potential is favorable throughout the southeastern United States.

Rationalization of benefits of deep rooting to crop production is supported by evidence from field experiments.

Problems involved in root research include (*a*) inability to define the optimum root system for a given crop, soil, and climatic condition, (*b*) inadequate criteria for evaluating root response to environmental factors, and (*c*) lack of techniques for precise evaluation of root environment. Probably the rhizotron, or its precursor, the glass-walled trench, offers the most immediate promise for definitive research on roots under field conditions.

References

Adams, Fred, R. W. Pearson, and B. D. Doss. 1967. Relative effects of acid subsoils on cotton yields in field experiments and on cotton roots in growth chamber experiments. *Agron. J.* 59: 453–56.

Beal, C. C. 1918. The effect of aeration on roots of *Zea mays* L. *Proc. Ind. Acad. Sci.* 1917: 177–80.

Brouwer, R. 1962. Distribution of dry matter in the plant. *Neth. J. Agr. Sci.* 10: 361–76.

Burton, Glenn W. 1957. Role of tracers in root development investigations. In *Atomic Energy and Agriculture,* ed. C. L. Comar, pp. 71–80. Amer. Assoc. for the Adv. Sci. Publ. no. 49. Baltimore: Horn Shafer Company.

Camp, C. R., and Z. F. Lund. 1964. Effect of soil compaction on cotton roots. *Crops Soils* 17: 13–14.

Corley, H. E., and R. D. Watson. 1966. A new gravimetric method for estimating root-surface areas. *Soil Sci.* 102: 289–91.

De Roo, H. C. 1961. *Deep Tillage and Root Growth.* Conn. Agr. Exp. Sta. Bull. 644.

Doss, Basil D., and Howard M. Taylor. 1970. Evapotranspiration and drainage from the root zone of irrigated Coastal bermudagrass (*Cynodon dactylon* L. Pers.) on Coastal Plain soils. *Trans. Amer. Soc. Agr. Eng.* 13: 426–29.

Fehrenbacher, J. B., and R. H. Rust. 1956. Corn root penetration in soils derived from various textures of Wisconsin-age glacial till. *Soil Sci.* 82: 369–78.

Funchess, M. J. 1918. *The Development of Soluble Manganese in Acid Soils as Influenced by Certain Nitrogenous Fertilizers.* Ala. Agr. Exp. Sta. Bull. 201.
——. 1919. Acid soils and the toxicity of manganese. *Soil Sci.* 8: 69.

Hall, N. S., W. V. Chandler, C. H. M. Von Bavel, P. H. Reid, and J. H. Anderson. 1953. *A Tracer Technique to Measure Growth and Activity of Plant Root Systems.* N. C. Tech. Bull. 101.

Hendrickson, A. H., and F. J. Veihmeyer. 1931. Influence of dry soil on root extension. *Plant Physiol.* 6: 567.

Hunter, A. S., and O. J. Kelley. 1946. The extension of plant roots into dry soil. *Plant Physiol.* 21: 445–51.

Leitch, I. 1916. Some experiments on the influence of temperature on the rate of growth of *Pisum sativum. Ann. Bot.* 30: 25–46.

Letey, J., and D. B. Peters. 1949. Influence of soil moisture levels and seasonal weather on efficiency of water use by corn. *Agron. J.* 49: 362–65.

Ligon, W. S., and W. H. Pierre. 1932. Soluble aluminum studies: II. Minimum concentrations of aluminum found to be toxic to corn, sorghum and barley in culture solutions. *Soil Sci.* 34: 307–21.

Lipps, R. C., and R. L. Fox. 1964. Root activity of subirrigated alfalfa as related to soil moisture, temperature and oxygen supply. *Soil Sci.* 97: 4–12.

Lowry, F. E., H. M. Taylor, and M. G. Huck. 1970. Growth rate and yield of cotton as influenced by depth and bulk density of soil pans. *Soil Sci. Soc. Amer. Proc.* 34: 306–9.

Lund, Zane F., R. W. Pearson, and Gale A. Buchanan. 1970. An implanted soil mass technique to study herbicide effects on root growth. *Weed Sci.* 18: 279–81.

McLean, E. O., D. Adams, and R. E. Franklin, Jr. 1956. Cation exchange capacities of plant roots as related to their nitrogen contents. *Soil Sci. Soc. Amer. Proc.* 20: 345–47.

McLean, F. T., and B. E. Gilbert. 1927. Aluminum tolerance of crop plants. *Soil Sci.* 24: 165–76.

Melhuish, F. M., and A. R. G. Lang. 1968. Quantitative studies of roots in soil. I. Length and diameters of cotton roots in a clay loam soil by analysis of surface-ground blocks of resin-impregnated soil. *Soil Sci.* 106: 16–22.

Mitchell, K. J. 1954. Influence of light and temperature on growth of ryegrass (*Lolium* spp.). III. Pattern and rate of tissue formation. *Physiol. Plant.* 7: 51–65.

Newman, E. I. 1965. A method for estimating the total length of root in a sample. *J. Appl. Ecol.* 2: 139–45.

Pearson, R. W., and Zane F. Lund. 1968. Direct observation of cotton root growth under field conditions. *Agron. J.* 60: 442–43.

Pfeffer, W. 1893. Druck- und Arbeits-leistung durch washsende Pflanzen. *Abhandl. Sachs. Ges. (Akad.) Wiss.* 33:233–474.

Pierre, W. H., G. G. Pohlman, and J. C. McIlwaine. 1932. Soluble aluminum studies: I. The concentration of aluminum in the displaced soil solution of naturally acid soils. *Soil Sci.* 34: 145–60.

Racz, G. J., D. A. Rennie, and W. L. Hutcheon. 1964. The P^{32} injection method for studying the root system of wheat. *Can. J. Soil Sci.* 44: 100–108.

Reicosky, D. C., R. J. Millington, and D. B. Peters. 1970. A comparison of three methods for estimating root length. *Agron. J.* 62: 451–53.

Richards, L. A., and C. H. Wadleigh. 1952. Soil water and plant growth. In *Soil Conditions and Plant Growth,* ed. Byron T. Shaw, pp. 73–251. New York: Academic Press.

Rios, M. A., and R. W. Pearson. 1964. The effect of some chemical environmental factors on cotton behavior. *Soil Sci. Soc. Amer. Proc.* 28: 232–35.

Rogers, W. S. 1934. Root studies IV: A method of observing root growth in the field; illustrated by observations in an irrigated apple orchard in British Columbia. *Rep. East Malling Res. Sta. 1933*, pp. 86–91.

———. 1969. The East Malling root observation laboratories. In *Root Growth*, ed. W. J. Whittington, pp. 361–78. New York: Plenum Press.

Russell, R. Scott, and F. B. Ellis, 1968. Estimation of the distribution of plant roots in soil. *Nature* 217: 582–83.

Sorokin, H., and A. L. Sommer. 1929. Changes in the cells and tissues of root tips induced by the absence of calcium. *Amer. J. Bot.* 16: 23–39.

Taubenhaus, J. J., W. N. Ezekial, and H. E. Rea. 1931. Strangulation of cotton roots. *Plant Physiol.* 6: 161–66.

Taylor, H. M., M. G. Huck, Betty Klepper, and Z. F. Lund. 1970. Measurement of soil-grown roots in a rhizotron. *Soil Sci. Soc. Amer. Proc.* 62: 807–9.

Trenel, M. von, and F. Alten. 1934. Die physiologische bedeutung der mineral-ischen bodenaziditat (worauf beruht die toxische wirkung des aluminium). *Angew. Chem.* 47: 813–20.

Troughton, Arthur. 1957. *The Underground Organs of Herbage Grasses.* Commonwealth Bureau of Pasture and Field Crops Bull. no. 44. Bucks, England: Commonwealth Agr. Bur. Farnham Royal.

Weaver, J. E., F. C. Jean, and J. W. Crist. 1922. *Development and Activities of Roots of Crop Plants.* Carnegie Inst. Wash. Publ. no. 316.

Wilde, S. A., and G. K. Voight. 1949. Absorption-transpiration quotient of nursery stock. *J. Forest.* 47: 643–45.

Williams, D. Emerton. 1962. Anion exchange properties of plant root surfaces. *Science* 138: 153–54.

11. Root Behavior as Affected by Soil Structure and Strength

Howard M. Taylor

AGRICULTURAL workers have long known that soil physical properties affect agricultural usage. For example, Xenophon (about 400 B.C.) recommended spring plowing because the land is more friable at that time. Virgil (70–19 B.C.) described a method for determining soil bulk density and stated that "loose soils provide bounteous vines but dense soils provide reluctant clods and stiff ridges" (Tisdale and Nelson, 1966).

Today's agricultural workers often discuss effects of soil structure on crop growth or tillage energy requirements. However, the term *soil structure* means different soil properties to different people. Some people have considered part or all of the research in soil tilth, consistence, strength, friability, bulk density, aggregation, porosity, fabric, matric, plasma, pedality, and texture to be part of soil structure research. In this discussion, *soil structure* is defined as "the physical constitution of a soil material as expressed by the size, shape and arrangement of the solid particles and voids, including both the primary particles to form compound particles and the compound particles themselves" (Brewer, 1964). *Soil strength* is defined as "the ability or capacity of a particular soil in a particular condition to resist or endure an applied force" (Gill and Vanden Berg, 1967). This article describes effects of both soil structure and soil strength on root behavior.

I. MECHANICS OF ROOT ELONGATION

A. Elongation through Resisting Systems

A particular root increases in length during primary growth when cells of the meristematic region divide, elongate, and push the root tip forward

Contribution from the Soil and Water Conservation Research Division, Agricultural Research Service, U.S. Department of Agriculture, in cooperation with the Department of Agronomy and Soils, Alabama Agricultural Experiment Station, Auburn University, Auburn, Alabama 36830.

through the surrounding material. Turgor pressure in the elongating cells is the driving force and must be sufficient to overcome the cell wall constraints and any additional constraints imposed by the external material (Lockhart, 1965). Thus, cellular turgor pressure, resistance of the cell walls to strain, and resistance of the external medium to deformation are important in evaluating root growth through resisting systems. Soil physical conditions, directly or indirectly, will affect the magnitudes of all three factors.

B. Root Growth Pressure

The pressure available for a root to accomplish work against an external constraint has been termed *root growth pressure* (Gill and Bolt, 1955). Mathematically, root growth pressure is defined as

$$(\Sigma F_t - \Sigma F_{cw})/A,$$

where ΣF_t is the summation of the longitudinal forces in the root that arise as a result of cellular turgor pressure (dynes), ΣF_{cw} is the summation of those longitudinal forces that arise in the cell walls and tend to resist cellular elongation (dynes), and A is the cross-sectional area of the root at the plane where force is determined (cm^2).

It is extremely difficult to measure directly either turgor pressure or cell wall constraint. However, several investigations have determined the longitudinal force that a plant root can exert across a nonencased zone between two sections of encased root. For the first 2 or 3 hours, the force transmitted across this zone was small. Presumably, turgor forces are almost balanced by cell wall forces. After 12 to 18 hours, the force exerted across the nonencased zone reaches a maximum, and the force per unit area of root is about equal to the average osmotic pressure of the cellular contents (Taylor and Ratliff, 1969a). At that time, it appears that the cell wall forces must be small because a force almost equal to the turgor pressure is transmitted to an external sensor (or external media).

1. Magnitudes

When the root was exerting its maximum force, root growth pressures usually ranged from 9 to 13 bars (Fig. 11.1) (Pfeffer, 1893; Stolzy and Barley, 1968; Taylor and Ratliff, 1969a; Eavis *et al.*, 1969).

2. Effects of Environmental Factors

Root growth pressure of cotton (*Gossypium hirsutum* L.) was reduced from 11 bars to 5 bars when air surrounding the roots was reduced in oxygen content from 21 % to 3 % (Eavis *et al.*, 1969). This reduced oxygen supply could affect root growth pressure either by reducing the turgor pressure or by increasing the cell wall constraint. For turgor pressure to decrease, water would have to be transferred longitudinally in the root, or the osmotic concentration in that portion of root would have to be shifted substantially. Huck (1970) showed that a reduced oxygen supply stopped root elongation almost instantaneously. Therefore, it seems somewhat more probable that lack of oxygen decreases root growth pressure by increasing the cell wall constraint. However, definitive measurements have not been used to check this hypothesis.

3. Research Needed

No known investigations have determined variation in growth pressure among the roots on the same plant. In addition, it is not known how

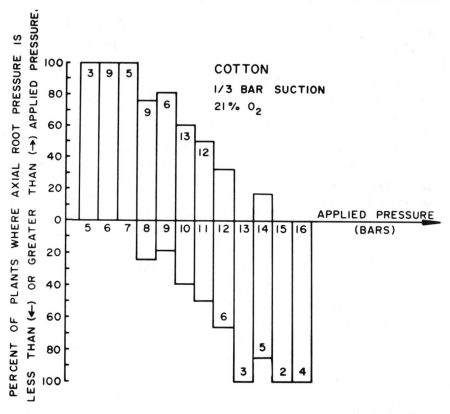

Fig. 11.1. Percentage of cotton (*Gossypium hirsutum* L.) plants with root growth pressure sufficient (*above line*) or insufficient (*below line*) to lift the pressure applied by a dead load. The number of plants used in the test is indicated within each applied pressure interval. (Eavis *et al.*, 1969)

growth pressure of a particular root varies from one time to another, or how growth pressure is affected by hormonal activity, soil chemical environment, or temperature.

II. ROOT ELONGATION THROUGH A SOIL WITH UNIFORM FABRIC

Soil fabric, which is "the physical constitution of a soil material as expressed by the spatial arrangement of the solid particles and voids" (Brewer, 1964), is extremely important in root growth. Soil fabric determines physical behavior and controls water, heat, aeration, and strength relations important in root growth.

A. Soil Strength Effects

If the soil has no continuous pores that are large in relation to the root tip, elongation rates will depend on the magnitude of the external constraint. As an example, Barley (1962) examined the ability of corn (*Zea mays* L.) roots to overcome external constraints by using an apparatus which enabled measurement of length as the roots grew. When cells differentiated and elongated while the apex was compressed, root length increased continuously, but at a rate that declined with increased mechanical stress. Taylor and Ratliff (1969b) showed that the rate of peanut (*Arachis hypogaea* L.) root elongation decreased as soil strength (measured by penetrometer resistance) around the root increased (Fig. 11.2). The elongation rate was 2.7 mm/hr when penetrometer resistance was near zero bars. At a penetrometer resistance of 15 bars, the elongation rate was about 1.5 mm/hr, and it was about 0.8 mm/hr at 30 bars. Soil water potentials between -0.19 and -12.5 bars (water contents between 7.0% and 3.8% by weight) did not affect the root elongation–soil strength relationship. The compaction process used by Taylor and Ratliff left few continuous voids that were larger than the diameter of the peanut root tip; so soil strength controlled elongation rate.

B. Soil Porosity Effects

If enough large vesicles (defined by Brewer, 1964, as voids with walls that consist of smooth, simple curves) or other large pores exist, roots can grow through high-strength material. Aubertin and Kardos (1965a, 1965b) illustrated this point by growing corn in a container

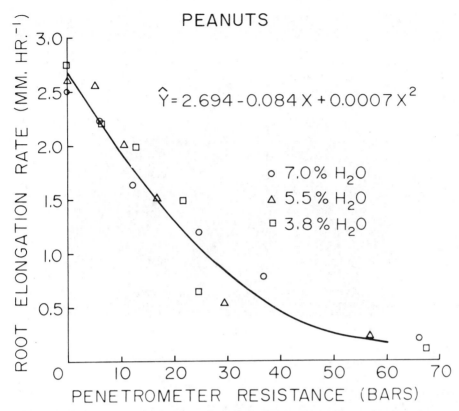

Fig. 11.2. Effect of soil water content and soil strength as measured by penetrometer resistance on peanut (*Arachis hypogaea* L.) root elongation for 40 to 80 hours after radicle emergence. (Reprinted by permission from H. M. Taylor and L. F. Ratliff, *Soil Sci.* 108: 113–19, ©️ 1969, The Williams & Wilkins Co., Baltimore, Md. 21202, U.S.A.)

where a clamping device could alter rigidity of the glass bead matrix. Systems were used whose modal pore sizes ranged from 46μ to 412μ in diameter. In the nonrigid system, roots could grow equally well at 46μ as at 278μ (Fig. 11.3A), but corn roots did not grow into the rigid systems where pore diameters were less than 138μ (Fig. 11.3B). Any reduction in pore diameter below 412μ reduced root growth in the rigid systems.

The diameters of plant roots near the root tips vary greatly. Plants whose roots have small diameters near the tips can penetrate rigid soil volumes that roots with larger tips cannot. On the same plant, tertiary roots often act differently from tap, or seminal, roots. Sometimes their reaction is different because the tertiary roots are smaller; however, the main exploring roots also may have encountered a particular soil volume at a time when its soil water was different from that encountered by the tertiary roots.

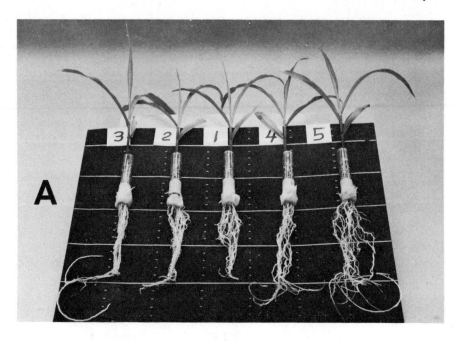

Fig. 11.3.   Corn (*Zea mays* L.) seedlings grown in nonrigid (*A*) and rigid (*B*) glass bead systems with modal pore diameters of: 46μ (3, 8), 87μ (2, 6), 138μ (1, 9), 240μ (4, 10), and 278μ (5, 11). (Data of Aubertin and Kardos, 1965)

C. Systems with Both Soil Strength
and Soil Porosity Effects

In most soils, roots penetrate partly by growing through existing voids and partly by moving aside soil particles (Wiersum, 1957; Aubertin and Kardos, 1965a, 1965b). When a soil is compacted, the modal pore size is reduced, soil strength is increased, and soil aeration is reduced. In a classic early experiment, Veihmeyer and Hendrickson (1948) investigated the effects on root growth of increases in soil bulk density (defined as the mass of oven-dry material per unit volume of soil). They showed that root growth decreased as soil bulk density increased, but their data did not delineate the various factors that might have caused the reduced root growth.

Taylor and Gardner (1963) found that root penetration at a particular soil water potential decreased as bulk density increased (Fig. 11.4). At a specific bulk density, root penetration decreased as soil water potential or water content decreased. They concluded that in their experiment root penetration was reduced as soil strength increased. Taylor *et al.* (1966)

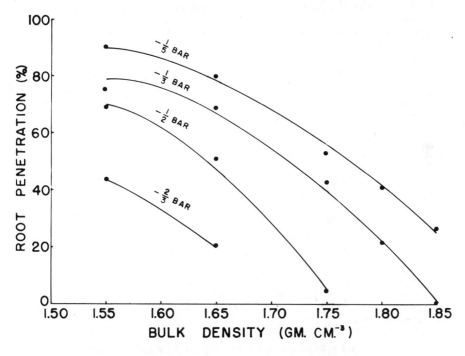

Fig. 11.4. Penetration of cotton (*Gossypium hirsutum* L.) seedling roots through 2.5-cm layers of Amarillo fine sandy loam soil as affected by bulk density and water potential. Each point represents 80 planted seeds. (Reprinted by permission from Taylor and Gardner, *Soil Sci.* 96 : 153–56, ⓒ 1963, The Williams & Wilkins Co., Baltimore, Md. 21202, U.S.A.)

altered penetrometer resistance of four soils by changing water contents
or bulk densities and found that root penetration through the four soils
was inversely related to penetrometer resistance (Fig. 11.5). Independently, Barley (1963) found that soil strength was a controlling factor in
root growth. Barley and Taylor and Gardner (1963) concluded that soil
aeration was not a major factor in their experiments.

Hopkins and Patrick (1970) investigated relations among soil compaction, soil oxygen, and root growth. At low compaction levels, root
growth increased with oxygen concentration. At high compactions, root
growth was only slightly affected by oxygen level, probably because
growth was controlled by soil strength.

Pearson *et al.* (1970) grew cotton seedlings in glass-fronted boxes with
varying soil temperature, pH, and strength levels. Root elongation rate
increased as temperature increased to 32C, then fell sharply as temperature was increased further. The effect of temperature was greatest at low
levels of strength and a pH of 6.2. Similarly, the effect of increased soil
strength was greatest at 32C and a pH of 6.2.

These experiments of Hopkins and Patrick and Pearson *et al.* probably
indicate the general pattern of strength effects on root growth through

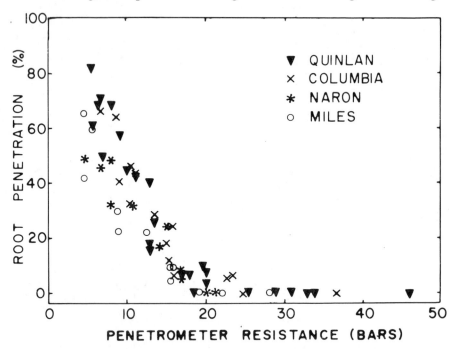

Fig. 11.5.   Relations among root penetration and the penetrometer resistance of four soil
materials. (Reprinted by permission from H. M. Taylor, G. M. Roberson, and J. J. Parker,
Jr., *Soil Sci.* 102: 18–22, ⓒ 1966, The Williams & Wilkins Co., Baltimore, Md. 21202,
U.S.A.)

soil with uniform fabric. When other growth factors are satisfactory, root growth decreases as soil strength increases. However, when other factors limit root growth, soil strength probably exerts little or no effect.

D. Measurement of Soil Strength

Soil strength is affected not only by changes in water content and soil bulk density but also by changes in types and amounts of saturating cations (Gerard, 1965), the number of particle-to-particle contacts (Lotspeich, 1964), the type of clay mineral (Grim, 1962), and the amount and type of organic materials (Jamison, 1954). Consequently, there is no simple way to predict soil strength from measurement of soil texture, organic matter, pH, or other common parameters. The American Society of Agricultural Engineers has published a standard for the design and use of penetrometers in estimating soil strength (American Society of Agricultural Engineers, 1970, ASAE Recommendation R313 Soil Cone Penetrometer, *ASAE Yearbook*, pp. 296–98). In root penetration studies the penetrometer technique is probably as useful as any other technique for assessing soil strength. However, it is emphasized that the various techniques provide different quantitative values when used on the same soil sample. These values, although quantitatively different, often are highly correlated (Taylor and Burnett, 1964).

III. ROOT ELONGATION INTO SOILS WITH WELL-DEVELOPED STRUCTURAL DISCONTINUITIES

Much of the research evaluating effects of soil fabric and soil strength on root elongation has been conducted using unusual methods to achieve uniformity, but nearly all field soils contain some structural development. Where this development has occurred, soil porosities and soil strengths vary from one volume of soil to another. As a result, physical conditions important to root growth will vary from one location to another within the soil profile.

A. Vertical Cracking Pattern Effects

E. Burnett (private communication) investigated the pattern of soil shear planes and plant rooting in Houston Black clay, a Udic Pellustert occurring in the Texas Blacklands. The soil has long continuous planes (sometimes greater than 2 m in length) where one block of soil has moved in relation to its neighbor. These shear planes persist from year to year at the same location. This plane of structural weakness provided a recurring path for root penetration, and roots tended to be concentrated along these shear planes (Fig. 11.6).

V. L. Hauser (private communication) found that vertical shrinkage cracks penetrated at least 5 m in Pullman silty clay loam soil at Bushland, Texas. He found living plant roots, tentatively identified as blueweed (*Helianthus ciliaris* DC), penetrating to a 9-m depth at this site. For most of this depth, the roots tended to follow the shrinkage cracks. Since blueweed is perennial, Hauser could not estimate the time required for root penetration to the 9-m depth.

B. Strong Ped Development Effects

1. Distortion of Rooting Patterns

Many soils contain a three-dimensional network of structural discontinuities. These networks separate soil volumes into peds, defined as "the individual natural aggregates consisting of clusters of primary particles, and separated from adjoining peds by surfaces of weakness which are recognizable as voids or natural surfaces" (Sleeman, 1963).

Edwards *et al.* (1964) studied corn root penetration through Weir silt loam, a Typic Ochraqualf found in Illinois. They found that large corn roots were confined to the larger spaces between peds but that many medium and small roots penetrated about one-half of the discrete peds in the claypan B horizon. Corn roots did not penetrate peds with a bulk density greater than 1.80 g/cm^3.

Fehrenbacher *et al.* (1965) compared the penetration of corn roots with those of alfalfa (*Medicago sativa* L.) through four soils derived from shale. Alfalfa roots penetrated deeper than corn roots. Most of the alfalfa roots followed cracks and cleavage planes in the shale. They cautioned that the deeper alfalfa root penetration could have occurred in the fall when the corn had completed growth.

Growing root hairs can deform clay soils (Champion and Barley, 1969). When pea (*Pisum* sp.) radicles were grown on or in a saturated molded clay, root hairs penetrated the clay mass when the initial voids ratio exceeded 1:1 (bulk density less than 1.3 g/cm^3). The pea radicles penetrated soil materials where root hairs failed to develop. Where ped surfaces are covered with "skins," root hairs must deform the peds to obtain potassium (Soileau *et al.*, 1964) and other nutrients.

Sutton (1969) found that roots of young white spruce (*Picea glauca* Voss) readily penetrated a highly structured Lucas silt loam, but the roots

Fig. 11.6. Plant roots located in a vertical shrinkage crack of Houston Black clay. Note that the roots apparently were unable to readily penetrate the vertical face of the crack. (Photograph courtesy of E. Burnett)

Fig. 11.7. Roots of white spruce (*Picea glauca*) conforming to structural ped surfaces in a Lucas silt loam at Ithaca, N.Y. (Reprinted by permission from R. F. Sutton, *Form and Development of Conifer Root Systems*, ⓒ 1969, Commonwealth Forestry Bureau)

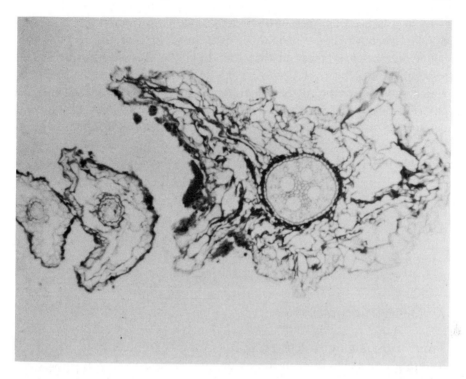

Fig. 11.8. A cross-sectional segment of three sugarcane (*Saccharum officinarum*) roots showing a badly distorted cortex. The stele shows slight distortion. Roots showing this amount of distortion are physiologically active. (Reprinted by permission from A. C. Trouse, Jr., pp. 137–52 in *12th Congr. Int. Sugar Cane Technol. Proc.* © 1965, Elsevier Publishing Company)

occurred only between structural elements. The roots were flattened in cross section, and they zigzagged, conforming with the structural element surfaces (Fig. 11.7). Misshapen roots are common in soils high in clay content (Stephenson and Schuster, 1939; Trouse, 1965). Nevertheless, these deformed roots (Fig. 11.8) are physiologically active (Trouse, 1965).

2. Difference in Growth Conditions between Ped Surfaces and Interiors

A word of caution is necessary here. Currie (1962) has emphasized that ped interiors and ped surfaces have different aeration relations. Therefore, one should not assume that excessive soil strength of the ped interior is the only cause of high concentration of roots at the surface. Also, water, temperature, nutrition, and pH may differ between ped surface and interior. The effects of ped size or shape on total root growth or crop yield are not known.

In structured soils, there is considerable difficulty in actually assessing the soil strength that a root must overcome to penetrate. First, most penetrometers are larger in diameter than the elongating portions of roots. Second, the root tips often have mucigel layers (Leiser, 1968), which may reduce the coefficient of friction between root surface and soil particles below that which occurs between the penetrometer tip and soil particles. Third, the root is easily deformed (Camp and Lund, 1964), but the penetrometer tip is rigid. Fourth, different types of penetrometers used in root penetration studies give different values of soil strength. Thus, measurements of media constraints with penetrometers are, at best, empirical. Measurements of soil strength and other root growth parameters should be made on a scale about equal to the diameter of the root.

C. Horizontal Pan Effects

1. Rooting Patterns on Pans

Most of the highly structured soils are fine textured. However, fabric discontinuities also exist in loams or sandier soils. Some of these horizontal layers, variously called hardpans, plowpans, tillage pans, plow soles, or tillage soles, divert roots and reduce rooting intensity below the pans. Initially, young roots grow downward through soil loosened by tillage. When they encounter a soil pan, part of the roots enter the pan and part are diverted horizontally. Roots that penetrate the pan at least 1 cm exhibit a reduced elongation rate as the soil strength increases. The roots that are diverted laterally may later encounter a vertical crack through which they can penetrate the pan (Taylor and Burnett, 1964). If no crack is encountered, the roots continue to grow horizontally along the pan surface until growth conditions change.

Soil pans sometimes restrict plant rooting to the few centimeters of soil near the surface (Fig. 11.9). As a result, the plants are subjected to extreme drought conditions in semiarid sandy soils. If soils containing pans are chiseled deep enough to disrupt the pans, plants will grow into the chisel slots but not where the soil pan still remains (Fig. 11.10). These soil pans, by reducing the depth of rooting, will reduce the quantity of water available for plant growth (Lowry *et al.*, 1970).

2. Transitory Effects of Pans

Effects of soil pans on root growth often are transitory and depend largely on water content of the soil pan. Taylor *et al.* (1964a) investigated

Fig. 11.9. Root system of a pigweed (*Amaranthus retroflexus*) that grew on soil with a horizontal soil pan of excessive strength. (Reprinted by permission from H. M. Taylor and E. Burnett, *Soil Sci.* 98 : 174–80, ⓒ 1964, The Williams & Wilkins Co., Baltimore, Md. 21202, U.S.A.)

17 root-restricting pans in the Southern Great Plains. They concluded that excessive soil strength caused by drying in the cohesive pan layer was the principal reason for distorted rooting patterns. If pan layers were at water contents near field capacity, most of the roots penetrated the pans. However, few roots penetrated pans that had dried below −1 or −2 bars

Fig. 11.10. On a compacted soil, cotton (*Gossypium hirsutum* L.) plants were established only where the planted row crossed a soil pan fracture created by chiseling to a 30-cm depth. (Taylor and Burnett, 1963)

water potential. Thus, rain or irrigation could change a root-restricting pan to a nonrestricting one. Sometimes, cotton roots penetrated pan layers that later dried sufficiently to girdle plants. If the girdling persisted long enough, the plants died (Mathers and Welch, 1964), probably as a result of reduced transport efficiency for water and nutrients (Taubenhaus *et al.*, 1931). When the pan layer was rewet, the roots again expanded radially (Taylor *et al.*, 1964a).

3. Crop Growth Effects

The root and shoot systems of a plant are dependent on and competitive with each other. Roots absorb water and minerals; leaves provide photosynthates and growth compounds. The proportion of the total supply of water, minerals, photosynthates, and growth compounds used by a particular organ changes with the environment. Therefore, the effect of a given level of soil strength will vary from environment to environment.

Consider a soil pan at a 15-cm depth which is rigid and has no pores larger than the rootcap. If the 15-cm depth above the pan can readily supply the plant's demand for water and nutrients without altering the heat or osmotic balance, yield should not be reduced below that of a nearby soil containing no pan. Similarly, soil pan strength or porosity would show no effect if some other factor, such as high aluminum ion activity, is prohibiting root growth.

4. Crop Yield Effects

Several experiments have shown that yield of cotton can be increased by disrupting high-strength pans. In California's Central Valley, Carter *et al.* (1965) and Carter and Tavernetti (1968) found that seed cotton yield was negatively correlated with penetrometer resistance of a sandy soil. Yields increased when the soil pan immediately below the cotton rows was disrupted in sandy soils (soils whose field capacity water content was below 12% by weight), but they did not increase in finer-textured soils. Presumably, sufficient vertical cracking occurred for adequate rooting through the pans in the fine-textured soils.

Lowry *et al.* (1970) found that cotton yield was reduced as depth to a soil pan decreased, or as strength of the pan increased (Fig. 11.11). Deep plowing or chiseling of compacted soils increased cotton yields in experiments of Grissom *et al.* (1955), Bruce (1960), and Burleson *et al.* (1957).

Yields of corn (Phillips and Kirkham, 1962), four species of grass (Barton *et al.*, 1966), grain sorghum (*Sorghum bicolor*) (Taylor *et al.*, 1964b), and sugar beets (*Beta vulgaris*) (Taylor and Bruce, 1968) were reduced as soil strength increased. It seems probable that high-strength pans will reduce yields of nearly all crops if, and probably only if, the pans substantially increase water stress, mineral deficiencies, or toxicities.

IV. CONCLUDING DISCUSSION

Roots penetrate most soils partly by growing through existing voids and partly by moving soil particles from the path of the root. When no pores larger than the rootcap exist, roots must move substantial quantities of soil from their paths. In this case, increased soil strength will reduce the exploitation of water and nutrients within the soil mass. If large, continuous voids exist, roots can often follow these voids through the soil mass, even in very rigid systems. Size and frequency of the voids will control utilization of water and nutrients in these porous but rigid systems.

The term *soil structure* has been used loosely in contemporary literature on soils. It is suggested that this term be used only when one refers to the arrangement of primary particles into compound natural units and their arrangement within the profile. If force, displacement, or strain is the important consideration, *soil strength* is probably the correct term to be used.

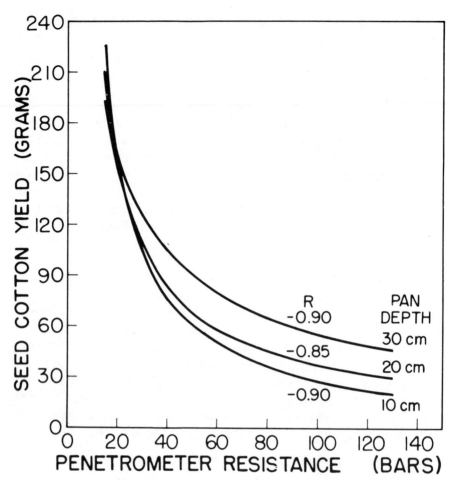

Fig. 11.11.   Relations among soil pan depth, penetrometer resistance of the soil pan 48 hours after wilting, and seed cotton (*Gossypium hirsutum* L.) yield. (Lowry *et al.*, 1970)

Soil structure can be extremely important to root growth in fine-textured soils, but soil strength usually is more important than soil structure in sandy soils. If roots encounter zones of high soil strength, elongation will be reduced. However, there is no direct, simple relationship between root growth and top growth. In many cases, a small proportion of the plant top is harvested and marketed, so there may not be even a simple, direct relationship between plant tops and yield of marketable product. Excessive soil strength usually reduces yield of marketable product by causing plants to undergo additional stress for water or nutrients at critical times. Effects on yield of the various types of structural discontinuities found in fine-textured soils have not been studied extensively.

References

Aubertin, G. M., and L. T. Kardos. 1965a. Root growth through porous media under controlled conditions. I. Effect of pore size and rigidity. *Soil Sci. Soc. Amer. Proc.* 29: 290–93.

———. 1965b. Root growth through porous media under controlled conditions. II. Effects of aeration levels and rigidity. *Soil Sci. Soc. Amer. Proc.* 29: 363–65.

Barley, K. P. 1962. The effects of mechanical stress on the growth of roots. *J. Exp. Bot.* 13: 95–110.

———. 1963. Influence of soil strength on growth of roots. *Soil Sci.* 96: 175–80.

Barton, Howard, W. G. McCully, H. M. Taylor, and J. E. Box, Jr. 1966. Influence of soil compaction on emergence and first-year growth of seeded grasses. *J. Range Manage.* 19: 118–21.

Brewer, R. 1964. Structure and mineral analysis of soils. In *Soil Clay Mineralogy: A Symposium*, ed. C. I. Rich and G. W. Kunze. Chapel Hill, N.C.: University of North Carolina Press.

Bruce, R. R. 1960. Deep tillage of dry soils gives better returns. *Mississippi Farm Res.* 23: 1, 6.

Burleson, C. A., M. E. Bloodworth, and J. W. Biggar. 1957. *Effect of Sub-soiling and Deep Fertilization on the Growth, Root Distribution and Yield of Cotton.* Texas Agr. Exp. Sta. Progress Rep. 1992.

Camp, C. R., and Z. F. Lund. 1964. Effect of soil compaction on cotton roots. *Crops and Soils* 17: 13–14.

Carter, L. M., J. R. Stockton, J. R. Tavernetti, and R. F. Colwick. 1965. Precision tillage for cotton production. *Trans. Amer. Soc. Agr. Eng.* 8: 177–79.

Carter, L. M., and J. R. Tavernetti. 1968. Influence of precision tillage and soil compaction on cotton yields. *Trans. Amer. Soc. Agr. Eng.* 11: 65–67, 73.

Champion, R. A., and K. P. Barley. 1969. Penetration of clay by root hairs. *Soil. Sci.* 108: 402–7.

Currie, J. A. 1962. The importance of aeration in providing the right conditions for plant growth. *J. Sci. Food Agr.* 13: 380–85.

Eavis, B. W., L. F. Ratliff, and H. M. Taylor. 1969. Use of a dead-load technique to determine axial root growth pressure. *Agron. J.* 61: 640–43.

Edwards, W. M., J. B. Fehrenbacher, and J. P. Varva. 1964. The effect of discrete ped density on corn root penetration in a planasol. *Soil Sci. Soc. Amer. Proc.* 28: 560–64.

Fehrenbacher, J. B., B. W. Ray, and W. M. Edwards. 1965. Rooting volume of corn and alfalfa in shale-influenced soils in Northwestern Illinois. *Soil Sci. Soc. Amer. Proc.* 29: 591–94.

Gerard, C. J. 1965. The influence of soil moisture, soil texture, drying conditions and exchangeable cations on soil strength. *Soil Sci. Soc. Amer. Proc.* 29: 641–45.

Gill, W. R., and G. H. Bolt. 1955. Pfeffer's studies of the root growth pressures exerted by plants. *Agron. J.* 47: 166–68.

Gill, W. R., and G. E. Vanden Berg. 1967. *Soil Dynamics in Tillage and Traction*. U.S. Dept. Agr. Handbook no. 316.

Grim, R. E. 1962. *Applied Clay Mineralogy*. New York: McGraw-Hill.

Grissom, P., E. B. Williamson, O. B. Wooten, F. E. Fulgham, and W. A. Raney. 1955. Cotton yields doubled by deep tillage in 1954 tests on hardpan soils in Delta. *Miss. Farm Res.* 18: 1, 2.

Hopkins, R. M., and W. H. Patrick, Jr. 1970. Combined effects of oxygen concentration and soil compaction on root penetration. *Soil Sci.* 108: 408–13.

Huck, M. G. 1970. Variation in taproot elongation rate as influenced by composition of the soil air. *Agron J.* 62: 815–18.

Jamison, V. C. 1954. Effect of some soil conditioners on friability and compactibility of soils. *Soil Sci. Soc. Amer. Proc.* 18: 391–94.

Leiser, A. T. 1968. A mucilaginous root sheath in *Ericaceae. Amer. J. Bot.* 55: 391–98.

Lockhart, J. A. 1965. Cell extension. In *Plant Biochemistry*, ed. J. Bonner and J. E. Varner, pp. 826–49. New York: Academic Press.

Lotspeich, F. B. 1964. Strength and bulk density of compacted mixtures of kaolinite and glass beads. *Soil Sci. Soc. Amer. Proc.* 28: 737–43.

Lowry, F. E., H. M. Taylor, and M. G. Huck. 1970. Growth rate and yield of cotton as influenced by depth and bulk density of soil pans. *Soil Sci. Soc. Amer. Proc.* 34: 306–9.

Mathers, A. C., and N. H. Welch. 1964. Pans in Southern Great Plains soils. II. Effect of duration of radial root restriction on cotton growth and yield. *Agron. J.* 56: 313–15.

Pearson, R. W., L. F. Ratliff, and H. M. Taylor. 1970. Effect of soil temperature, strength and pH on cotton seedling root elongation. *Agron. J.* 62: 243–46.

Pfeffer, W. 1893. Druck-und Arbeit-leistung durch wachsende Pflanzen. *Abhandl. Sachs. Ges. (Akad.) Wiss.* 33: 235–474.

Phillips, R. E., and Don Kirkham. 1962. Soil compaction in the field and corn growth. *Agron. J.* 54: 29–34.

Sleeman, J. R. 1963. Cracks, peds and their surfaces in some soils of the Riverine Plain, N.S.W. *Aust. J. Soil Res.* 1: 91–102.

Soileau, J. M., W. A. Jackson, and R. J. McCracken. 1964. Cutans (clay films) and potassium availability to plants. *Soil Sci.* 15: 117–23.

Stephenson, R. E., and C. E. Schuster. 1939. Physical properties of soils that affect plant nutrition. *Soil Sci.* 44: 22–36.

Stolzy, L. H., and K. P. Barley. 1968. Mechanical resistance encountered by roots entering compact soils. *Soil Sci.* 105: 297–301.

Sutton, R. F. 1969. *Form and Development of Conifer Root Systems.* Commonwealth Forest. Bur. Tech. Commun. no. 7.

Taubenhaus, J. J., W. N. Ezekial, and H. E. Rea. 1931. Strangulation of cotton roots. *Plant Physiol.* 6: 161–66.

Taylor, H. M., and R. R. Bruce. 1968. Effect of soil strength on root growth and crop yield in the Southern United States. *Trans. 9th Int. Congr. Soil Sci.* 1: 803–11.

Taylor, H. M., and E. Burnett. 1963. Some effects of compacted soil pans on plant growth in the Southern Great Plains. *J. Soil Water Conserv.* 18: 235–36.

——. 1964. Influence of soil strength on the root growth habit of plants. *Soil Sci.* 98: 174–80.

Taylor, H. M., and H. R. Gardner. 1963. Penetration of cotton seedling taproots as influenced by bulk density, moisture content and strength of soil. *Soil Sci.* 96: 153–56.

Taylor, H. M., A. C. Mathers, and F. B. Lotspeich. 1964a. Pans in Southern Great Plains soils I. Why root-restricting pans occur. *Agron. J.* 56: 328–32.

Taylor, H. M., L. F. Locke, and J. E. Box, Jr. 1964b. Pans in Southern Great Plains soils III. Their effects on yield of cotton and grain sorghum. *Agron. J.* 56: 542–45.

Taylor, H. M., and L. F. Ratliff. 1969a. Root growth pressures of cotton, peas, and peanuts. *Agron. J.* 61: 398–402.

——. 1969b. Root elongation rates of cotton and peanuts as a function of soil strength and soil water content. *Soil Sci.* 108: 113–19.

Taylor, H. M., G. M. Roberson, and J. J. Parker, Jr. 1966. Soil strength-root penetration relations for medium- to coarse-textured soil materials. *Soil Sci.* 102: 18–22.

Tisdale, S. L., and W. L. Nelson. 1966. *Soil Fertility and Fertilizers*, p. 9. 2nd ed. New York: Macmillan.

Trouse, A. C., Jr. 1965. Effects of soil compression on the development of sugar-cane roots. *12th Congr. Int. Sugar Cane Technol. Proc.*, pp. 137–52.

Veihmeyer, F. J., and A. H. Hendrickson. 1948. Soil density and root penetration. *Soil Sci.* 65: 487–93.

Wiersum, L. K. 1957. The relationship of the size and structural rigidity of pores to their penetration by roots. *Plant Soil* 9: 75–78.

12. Roots and Root Temperatures

Kenneth F. Nielsen

IMPLICIT in a review such as this is a knowledge of the function of roots in plants. What essential roles do the roots play? Unless this is fairly well known it is difficult to be very specific about the influence of root zone temperature on plants.

It is well known that the roots are the major organ absorbing water and nutrients for the plant. It is likewise known that the root is a sink for carbohydrates produced in the tops. There is quite a bit of evidence pointing to the root as a site for assimilating nitrogen and also as a source of growth metabolites. Specific root functions are discussed elsewhere in this volume, but for purposes of this review these aspects of possible root functions are also examined here.

This and an earlier review (Nielsen and Humphries, 1966) have lead me into only slightly familiar fields in thermobiology in search of fundamental principles that could explain the reasons for plant root temperature behavior observed in experimental work. Unfortunately the fragmentation of life sciences has resulted in pertinent information being scattered throughout the literature. Reviews in many areas have been valuable. Thermobiology is still in its infancy, and there is still a healthy lack of unanimity of opinion related to many subjects. This is leading to a careful examination of many evidences and explanations.

We will first examine the practical aspects of the effects that root zone temperature has on plants and the ways that root temperatures can be influenced by field management practices. Then attention will be directed toward more specialized aspects of the physiology of growth responses to root temperatures.

I. OVERVIEW

One of the main climatic factors affecting the distribution of plants in nature is temperature. As Langridge and McWilliam (1967) have pointed out, temperature becomes increasingly more important as a determinant in the geographical distribution of plants the further it deviates from the general biological norm (about 20C). Under tropical conditions it is probably least important, and the supply of plant nutrients and water are dominant. Under temperate and colder climatic conditions, temperature assumes a dominant role in species adaptation over broad geographical areas.

Survival is the basis of selection where temperature is dominant. The diurnal and seasonal temperature variations and means impose stresses upon plants that must be tolerated, or the plants will die. In the evolutionary processes of selection, mechanisms have developed in some plants to give them the ability to withstand cold. Compared to the numbers of species and varieties of plants that thrive in tropical climates, the numbers under temperate and colder conditions are few.

Root temperatures are usually lower than air temperatures during the growing period, and variations in the temperatures of the root zone are less than that of the ambient air to which the tops of plants are subjected. As a result, the roots of plants have a temperature optimum somewhat lower than the tops, have become less adaptive to temperature extremes, and are more sensitive to sudden fluctuations. The roots of most plants would be killed if exposed to the same variations and durations of temperature to which the tops are subjected.

It follows then that root temperatures are the critical temperatures as far as plant survival is concerned. Specific adaptation of roots to temperature conditions have involved certain recognition survival systems that trigger life cycles in the growth of plants. Thus germination commences when certain minimum limits of soil temperature have been exceeded. Likewise the onset of dormancy may be conditioned by temperature of the soil (Ketellapper, 1960). Variation in soil temperatures is evidently necessary to trigger and maintain some of these functions in the plant. Highkin (1958) showed that peas (*Pisum sativum*) grown at constant optimum temperature did not grow so well as those grown under variable temperatures around the optimum.

Temperature of the soil is affected by many factors, of which the more important are:
  1. Air temperature
  2. Intensity, quality, and duration of radiant energy
  2. Precipitation and evaporative potential of air
  4. Color and thermal conductivity of the soil
  5. Surface cover

In addition to their effects upon the soil, these factors may affect plant growth as well; furthermore, they may have an interacting effect on both. For example, radiant energy affects the temperature of the air, and this will affect the temperature of the soil; any management practice that affects the quantity of incident radiation at the soil surface, the insulating effects of the soil surface, or the thermal conductivity of the soil will modify the influence of radiation on the temperature of the root zone. Thus, a well-watered soil with a straw mulch and with a dense crop growing on it will have a lower root zone temperature than a soil in fallow.

Unlike air temperatures there are a number of management factors at the disposal of farmers that can affect the temperature of root zones. There are also some alternatives available to farmers for accommodating the influences of unfavorable soil temperatures.

II. FIELD SOIL TEMPERATURES

Natural soils are not homogeneous with respect to mulch, organic matter content, pore space porosity, or texture. These factors affect the water-holding capacity of the soil and the heat conductivity. Thus the penetration of heat or cold into the soil will vary with different soils. It is well known that sandy soils warm up earlier in the spring than do clayey soils. This is often a direct function of water content, which is in turn affected by organic matter content and pore space.

Summer crops are seeded when temperatures are increasing. Thus the soil near the surface into which the seed is placed is warmer than that below, and germination is quicker in this surface zone. The rate of vertical root growth into the colder horizons is slower than horizontal growth, so that in the early stages of plant growth the roots in the surface horizons will be more branched and plentiful than those in lower horizons

Topography affects soil temperatures considerably and contributes to the variability of seed germination and early growth of plants under field conditions. Sun-facing slopes warm more quickly in the spring and have the advantage of earliness, although the soil may dry out more quickly (Hughes, 1965).

Although there are many uncontrollable factors influencing root zone temperatures, there are also some management techniques that farmers can use to improve the temperatures. It is expected that more attention will be drawn to these techniques at a practical level in the future.

A. Daily and Annual Variations

The root system of a plant growing in soil is subject to a variety of temperatures, some of which are constantly changing. There are differences in the soil temperature profile, and there are usually daily changes in the surface temperature. Table 12.1 shows that when air temperature is falling, the temperature of the soil near the surface may be lower than that at greater depths. When the air temperature is rising, the opposite is true.

There is a considerable lag in temperature changes with depth. The heat load at 100 cm and more is large, as the data in Table 12.2 show. Thus the rooting volume of soil in temperate climates is warmed from both the bottom and the top with the onset of warmer weather following winter.

The diurnal variation is small in the winter (Table 12.1) when snow covers the ground, and it is not manifest at depths greater than about 10 cm. As the weather warms, the variation is greater, and the effects are

Table 12.1. Soil temperature diurnal variations for 1970

Depth (cm)	Temperature (deg C)					
	Jan.		April		Aug.	
	AM	PM	AM	PM	AM	PM
Ottawa, Ontario (longitude 45° 24′N; latitude 75° 43′W; elev. 87 m)†						
5	−2‡		2	5	20	26
10	−2		2	3	20	23
20	−1			2	21	22
50	1		2		20	
100	3		2		17	
150	6		4		15	
300	7		5		14	
Swift Current, Sask. (longitude 50° 16′N; latitude 107° 44′W; elev. 902 m)†						
5	−11	−10§	0	4	17	27
10	−10		0	2	18	23
20	−9		1		19	
50	−6		0		18	
100	−2		0		16	
150	1		0		14	
300	4		2		10	

Source: Canada Department of Transport, Toronto, Ontario, *Monthly Record—Meteorological Observations in Canada*, Jan., April, and Aug. 1970.

Note: Soil temperatures were measured under a 2-inch grass sod with thermistors connected to direct reading indicators. Air temperatures were taken at 1.22 m inside a standard louvered screen.

† Daily air temperature means (minimum and maximum, deg C) were −19 to −10 in January, 1 to 11 in April, and 13 to 26 in August, respectively, at Ottawa, Ontario, and −20 to −12 in January, −2 to 6 in April, and 10 to 28 in August, respectively at Swift Current, Saskatchewan.

‡ Snow cover 5–7 inches in January in Ottawa.

§ Snow cover 1–3 inches in January in Swift Current.

Table 12.2. Average soil temperature profile measurements for 1967–70

Depth (cm)	Temperature (deg C) Jan.	April	Aug.
Ottawa, Ontario			
5	0	4	19
10	0	4	19
20	1	4	19
50	2	4	19
100	3	2	16
150	6	4	15
Swift Current, Sask.			
5	−11	2	17
10	−10	3	18
20	−9	3	19
50	−6	2	18
100	−2	1	16
150	1	0	13

Source: Canada Department of Transport, Toronto, Canada, *Monthly Record—Meteorological Observations in Canada*, Jan., April, and Aug., 1967–70.

The data in Table 12.1 and 12.2 show that definite relationships can exist between surface and subsurface soil temperatures. Wijk (1966) has studied the theory of these relationships. Among other things the time of year is important in the relationship, but it is obvious that information on subsurface temperatures can be obtained through a knowledge of surface temperatures. He has outlined some mathematical relationships. Radiation thermometers are now available that enable the gathering of accurate surface temperatures by remote sensing (Lorenz, 1966; Holmes,

Table 12.3. Diurnal soil temperature variation for Aug. 2, 1965, at Swift Current, Saskatchewan

Depth (cm)	Temperature (deg C) AM	PM
1	22	46
10	22	29
20	23	24
50		22
100		15
150		13

Source: Canada Department of Transport, Toronto, Canada, *Monthly Record—Meteorological Observations in Canada*, Aug. 1965.

1970). Thermal infrared imagery is now very much in use to study various surface phenomena, such as soil conditions, plant diseases, and plant water stresses, that are related to temperature (Idso and Jackson, 1969). evident to at least 20 cm during the summer months. Daily diurnal variations in surface soil temperature can be extreme, as shown in Table 12.3. The measurements were recorded beneath a 2-inch grass sod, which tended to insulate the soil against radiant energy. Bare-ground surface temperature or plant temperatures at ground level could be higher than those shown. Beauchamp and Torrance (1969) have shown that surface temperatures can establish gradients within plants. Both profile variation and diurnal variation are lessened where there is a surface protection such as a mulch or vegetative cover.

B. Effects of Tillage

The purposes of tillage are largely fourfold:
1. To destroy weeds
2. To turn under crop residues
3. To prepare a seedbed
4. To establish surface mulch conditions that will conserve water

Tillage operations naturally influence soil temperatures, at least temporarily. They may reduce the insulating effects of surface crops or mulches. They may hasten drying of the surface soil, thus enabling the penetration of heat or cold to deeper soil layers.

The upper structure of the soil is usually loosened by tillage, and often organic residues are incorporated. This greater porosity makes the tilled zone more responsive to ambient air temperatures because both heat conductivity of the cold from lower depths and volumetric heat capacity are smaller than in the untilled soil. These effects pertain only to soils where the moisture content does not exceed field capacity.

The influence of tillage on the daily and annual variations in soil temperature is not large but shows up more with the daily temperatures than with annual means.

Ridged soil can be warmer than unridged soil, according to Shaw and Bulchele (1957), and any practice that reduces vegetative cover will result in a warmer soil in summer. An extreme example of such a practice is fallowing for a season in order to conserve water. With this practice soil temperatures rise, microbial activity increases, organic matter decomposition is hastened, and many other physical, chemical, and biological reactions that are temperature-dependent are altered. Benefits from summer fallowing that were once attributed to moisture conservation have been traced to a combination of better moisture supply and nitrogen supply resulting from accelerated decomposition of organic matter.

C. Effects of Mulches

Soil temperatures below mulches in temperate climates are usually lower and water contents are usually higher than where mulches are not present. Anderson and Russell (1964) reported a decrease of 0.3C for each 1,000 lb of bright straw spread on the soil surface. Burrows and Larson (1962) obtained a decrease of 0.5C at a depth of 10 cm with every ton of corn (*Zea mays* L.) stover spread on the surface. Willis *et al.* (1957) found similar results and reported reduced yields of corn associated with the lower temperatures. However, Wijk *et al.* (1959) reported no reduction in yields of corn with surface mulch if the soil was warm.

Black (1970) found little effect of surface straw mulch on daily minimum soil temperatures in the spring, but mulched soil was often 10C cooler at maximum temperatures. Working under semiarid conditions in the Northern Great Plains, Black developed a concept he called the "moist-soil degree-day" concept which relates plant growth to moisture availabity at certain temperatures. Surface mulches often cause better water conservation and moisture distribution in the seed zone of soils and result in increased yields of wheat because of the relative importance of water and temperature at the time (Greb *et al.*, 1967).

Under humid conditions surface mulches have been blamed for reduced yields of wheat (McCalla and Army, 1961). Although the supply of plant nutrients, particularly nitrogen, is undoubtedly a factor where mulches are used, lower soil temperature is thought to be a factor as well (Brengle and Whitfield, 1969). These authors reported more heads per plant at 18.3C but more kernels per head at 12.8C, resulting in a higher yield at 12.8C. Thus with an ample supply of water, low soil temperatures may be a limiting factor (Brengle and Whitfield, 1969); whereas when water supply is short, temperature effects are secondary (Black, 1970).

Plastic soil mulches have been used mainly in the production of horticultural crops. Black plastic can effectively control weeds while providing a microwatershed effect as well as being an evaporation barrier. Black films intercept solar radiation and convert it to sensible heat, most of which is then reradiated without having an appreciable influence upon the soil temperature beneath it (Waggoner *et al.*, 1960; Clarkson, 1960). On the other hand, clear plastic mulches allow most of the solar energy to pass through them to the soil, where it is converted to sensible heat and trapped there by the plastic (Paterson *et al.*, 1970). Paterson *et al.* (1970) compared the effects of six mulch treatments on soil temperatures in Texas. Where clear plastic was used as the top layer, temperatures were raised 5C to 7C and caused some injury with sweet-potatoes (*Impomea batatas*). The best mulch was a petroleum mulch which raised the soil temperature from 17C up to 20C. All treatments increased soil temperature.

Ekern (1967) found that paper and plastic mulches raised the average soil temperature in pineapple (*Ananas sativa*) fields by 1.6C during winter months. A marked increase in growth was associated with this increase in soil temperature.

The advantages of earliness obtained with mulches that cause soil warming gradually disappear as the date of seeding is delayed.

D. Effects of Cropping

Crops have an effect on soil temperature similar to that of mulches. They intercept solar radiation and dissipate much of the energy that would otherwise heat the soil. Spring temperatures under sod are always cooler than those under cultivated crops. With annual summer crops it is fortunate that in the early season when soil temperatures are cold the crops do not shade the ground, so that some warming can occur.

The geometry and manner of planting crops has an effect on soil temperature (Larson and Willis, 1957; Shaw and Buchele, 1957). Thus crops such as corn planted in rows in a north-south direction allow more radiation at the soil level than those planted in east-west rows. Solid seeding of crops such as cereals will result in slightly lower soil temperatures than planting the seed in rows. Little information has been obtained about such effects; so it is not known how significant these differences are in terms of yield.

E. Effects of Irrigation

The effect of irrigation water temperature upon soil temperature depends on the temperature of both water and soil and the capacity factor of the soil. Water used for irrigation often comes from low-temperature sources such as snow or glaciers. Its temperature at the time of use is affected by the temperature regimes in the canals and reservoirs through which it passes. As Raney and Mihara (1967) have stated, "Regardless of its initial temperature, water diverted for crop irrigation very rapidly acquires a temperature corresponding to the net balance of energy fluxes in its new environment." Wierenga *et al.* (1971) concluded that the effects of the temperature of irrigation water on soil temperature are small and of short duration. They did report significant decreases in soil temperature by evaporative cooling of irrigation water.

The study of the temperature of water and its effects on growth of crops has naturally advanced further with rice (*Oryza sativa* L.) than with most other crops (Owen, 1971). Low water temperature is considered to be an important factor limiting rice production in Japan (Raney and Mihara,

1967). In spite of much work in the field evaluation of irrigation water temperature, however, the variability in soil, the responses of plants at various stages of growth, and the variety of changing environmental conditions make it difficult to get consistent quantitative data.

Raney and Mihara (1967) pointed out methods of exercising some control over the temperature of irrigation water. These include (*a*) skimming warm water from the surface of reservoirs, (*b*) providing a temporary holding pond for warming the water, and (*c*) using shallow conveyance systems where warming can occur faster than in deep ones. They gave some consideration to the future use of atomic thermal heating. To this point in time, however, not much practical use has been made of these methods.

III. EXERIENCE WITH VARIOUS CROPS

A. General Considerations

Most of the experiments studying the root zone temperature effects on plants have been conducted under controlled environmental conditions. The reasons for this are obvious in that soils and environmental conditions are variable and temperatures are affected by many factors that simultaneously may or may not affect the plants growing in the soil. The emphasis has been to control as many of these variables as possible. Much of the work has been done with potted soil placed in water baths at constant controlled temperatures (e.g., Nielsen *et al.*, 1961) with no replication of soil temperature. Willis *et al.* (1963) described apparatus where soil temperature replication was provided. Cooper *et al.* (1960) found that an insulating layer about an inch thick was needed to keep soil and air temperatures separate, especially when the two temperatures are far apart. Many experiments have been conducted in solution cultures with both intact plants and parts of plants (Biddulph *et al.*, 1958; Humphries, 1967; Kleinendorst and Brouwer, 1970).

In the field there are marked variations in the temperature of surface soils, as can be seen in Tables 12.1 and 12.3. The variation is manifest to a depth of about 20 cm, which encompasses the volume of soil containing most of the plant roots. Under field conditions Mederski and Jones (1963) were able to influence soil temperatures by using heating cables imbedded under rows of corn. Mack (1965) excavated a plot area and laid in pipes to carry both heated and cooled water to condition the soil temperatures for growing barley (*Hordeum vulgare* L.).

All these approaches are needed to elucidate the fundamental and practical effects of root zone temperature on plants. The simplest task is to show what happens under a certain set of environmental conditions.

Much more difficult are the explanations of why the effects occurred and what can be done about them.

This section is devoted mostly to what experience has shown with some crops.

1. Optimum Temperatures

A few words about optimum temperatures are appropriate at this point. From the discussion of rate theory (see Sec. IV below), it is evident that the rate of reactions in plant processes is extended at low temperatures and hastened at higher ones. A common measure of growth rate with temperature is Q_{10}, the temperature coefficient, which is the growth at $T + 10$ compared to that at T in deg C. For nonbiological reactions such as diffusion, Q_{10} is of the order of 1 to 2. With biological reactions the Q_{10} is often in the range 2 to 3 but may be higher with specific enzyme systems. Since root respiration rates increase as temperatures increase, Q_{10} has been found to decrease progressively as root temperatures rise (Bohning *et al.*, 1953; Jensen, 1960).

Given sufficient time, and within certain limits, yields of some crops at low temperatures can be the same as, or better than, those at higher temperatures if other factors are equal. Thus Power *et al.* (1970) found that barley plants grown at low temperatures yielded as well as, or better than, those at higher temperatures. They indicated that yield potential decreased as soil temperature increased from 9C to 22C. Owen (1971) concluded that the same was true with rice.

The optimum soil temperature for plant growth is known to be affected by water and by the supply of nutrients available and their placement in the soil (Brouwer, 1962; Simpson, 1965; Mack and Finn, 1970). It is highest for germination and usually shifts downward with advancing maturity (Army and Miller, 1959; Brouwer, 1962; Radke and Bauer, 1969; Brengle and Whitfield, 1969) and is different in light than in dark (Baker and Jung, 1970; Kleinendorst and Brouwer, 1970). Turnips (*Brassica rapa* L.) had an optimum root temperature of 19C in the autumn, but in the spring it was 27C; the difference in the spring was related to more sunshine (Army and Miller, 1959).

High temperatures increase root branching (Garwood, 1968; Nielsen and Cunningham, 1964), whereas low temperatures encourage new root formation, and these factors in turn affect the uptake of nutrients and water.

Most roots are usually produced at temperatures below the optimum for tops, and Walker (1970) showed that different parts of the corn plant have different optimum root temperatures.

Duration is important in assessing the effects of root temperatures on plants. There are some irreversible effects of low root temperatures (Humphries, 1967; Brouwer and Levi, 1969), but there is also adaptation. Power *et al.* (1970) showed that barley growth was impeded by low root temperatures until about the fourth leaf stage. From then on, rates were similar to those at higher temperatures. It can be seen that reference to optimum temperatures could be misleading depending on what conditions exist and how long they are in effect.

B. Crops

1. Barley

Yields of barley (*Hordeum vulgare*) grown at various soil temperatures have been best at about 18C (Korovin *et al.*, 1961; Williams and Vlamis, 1962; Power *et al.*, 1963, 1964, 1970; Mack, 1965). At earlier stages of maturity, yields of tops have been best at higher root temperatures, but at 27C yields of grain were lowest. Power *et al.* (1970) conclude that the yield potential of barley decreased as root temperature increased, partly because higher temperatures hastened maturity and did not allow for the development of all the plant factors leading to maximum potential. However, the length of the frost-free growing period is a factor which limits the possibility for crops to reach yield potential.

Barley root yields have been maximum at 6C–13C. Dressings of fertilizer phosphorus have alleviated some of the effects of unfavorable root temperatures, and total uptake of nutrients has been positively correlated with root temperatures to the optimum.

2. Oats

Yield of oat (*Avena sativa*) grain has been best at about 15C–20C, depending upon nutrient supply, and lowest above 27C (Nielsen *et al.*, 1960a; Case *et al.*, 1964; Fulton and Findlay, 1966; Brouwer, 1962; Shtrausberg, 1958; Fulton, 1968). In the absence of adequate nutrients the optimum temperature seems invariably to be higher. Most roots have been produced at temperatures about 5 degrees below the optimum for grain (Nielsen *et al.*, 1960a; Puh, 1960; Case *et al.*, 1964). Optimum temperatures for growth of tops shifted downward as the plants aged, and with ample nitrogen, phosphorus, and potassium most roots were produced at 5C (Nielsen *et al.*, 1960a). The sum of cations taken up increased as temperature increased, but potassium and magnesium increased more than calcium. Concentration of N and P increased with temperature.

3. Wheat

Wheat (*Triticum aestivum*) yields have been best at root zone temperatures of about 20C (Stewart and Whitfield, 1965; Varade *et al.*, 1970; Whitfield and Smika, 1971; Warder and Nielsen[1]). The optimum soil temperature was affected by nutrient supply, particularly P, by soil texture, and by state of maturity (Warder and Nielsen[2]). Where ample P was available in the soil, the percentage of P in the plant did not increase with soil temperature, but it did increase when P supply was inadequate. Increasing the soil temperature from 13C to 18C doubled the yield of wheat foliage at 4 weeks with the N, P, and K fertilization but resulted in an increase of less than 40% with the control treatment (Stewart and Whitfield, 1965).

Brengle and Whitfield (1969) reported that wheat grew more slowly at 12.8C than at 18.3C and produced fewer tillers. However at 12.8C there were 50% more kernels per head.

Korovin *et al.* (1963) reported that uptake of P by wheat from a cool fertilized sandy soil was less than from a warm fertilized soil. Synthesis of nucleoprotein was also less. Woolley (1963) reported that the percentage of N in tops and roots decreased as root temperature was raised but the percentage of P increased.

4. Corn

The best root temperature for corn (*Zea mays* L.) appears to be about 25C–30C, depending on other factors such as nutrient supply, soil moisture, and air temperature (Willis *et al.*, 1957; Dormaar and Ketcheson, 1960; Nielsen *et al.*, 1961; Anderson, 1962; Brouwer, 1962; Mederski and Jones, 1963; Allmaras *et al.*, 1964; Beauchamp and Torrance, 1969; Walker, 1969). Corn plants grown in pots at 20C for 35 days wilted when placed into a water bath at 5C and never recovered (Nielsen *et al.*, 1961).

Phosphorus increased yields of corn at low temperatures but had little effect at 26.7C (Allen and Engelstad, 1963). Ketcheson (1966, 1968, 1970) and others (Knoll *et al.*, 1964a, 1964b) concluded that fertilizer P placed with corn seed in a cold soil benefited growth con-

[1] Unpublished Annual Report 1962, Experimental Farm, Swift Current, Saskatchewan, Canada.
[2] *Ibid.*

siderably but even at high rates it will not compensate for the harmful effects of low root temperatures.

The percentage of P and Mg in corn increased as soil temperature increased, but those of Ca and K decreased. However, Jones *et al.* (1963) found that many differences in nutrient composition had disappeared at maturity; so they did not feel that there is a simple relationship between yields and chemical composition. They did not think that low corn yields with low temperatures can be attributed to decreased uptake of nutrients. Having obtained significant varietal responses of corn to soil temperatures, Jones *et al.* (1963) suggested that these responses be used in variety evaluations.

5. Cotton

The optimum root temperature for yield of cotton (*Gossypium hirsutum* L.) seems to be about 28C to 30C (Letey *et al.*, 1961; Pearson *et al.*, 1970). Christiansen (1963) reported that subjecting cotton seedlings to cold temperature regimes caused meristem abortion of the radicle, reductions in growth rate, and death or inactivation of the cortex tissue. Low temperatures caused rapid increases in the percentage of sugar in all parts of the plant (Guinn and Hunter, 1968). These authors postulated a controlling effect of cold roots in that the supply of growth metabolites was severely reduced.

Bloodworth (1960) reported greatly reduced water uptake by cotton plants at root temperatures below 15.5C and above 42C. Arndt (1937) reported similar results.

6. Grasses

Optimum root temperatures for native grasses were higher than for cultivated species (Smoliak and Johnston, 1968). According to Hughes (1965), many temperate grasses yield most at soil temperatures of about 20C–25C, and he reported that soil temperatures frequently have a more marked effect on growth than do air temperatures. Hughes also recommended that more investigations under controlled conditions are needed if field responses are to be understood.

Garwood (1968) reported that there was more root branching of grasses at higher root temperatures but also a concurrent decrease in the new roots formed and in the root diameter.

Orchardgrass (*Dactylis glomerata* L.) yields were higher at air/soil temperatures of 17C/23C than at 23C/23C, 23C/17C, and 17C/17C (Sato and Ito, 1969). Finn and Mack (1964) reported that 20C was the best root temperature for four varities. Application of P raised yields at 10C but not as much as at 20C.

Best yields of *Phalaris tuberosa* at 9 weeks were obtained at a soil temperature of 25C, and it was suggested that the effects of soil temperature mask those of air temperature in vegetative growth (Ketellapper, 1960).

Nielsen *et al.* (1961) reported best yields of brome (*Bromus inermis* Leyss.) foliage at 26.7C with only one cutting. Where two cuttings were made, best total yield was at 19.5C (Nielsen and Warder[3]). Root weights were highest at temperatures 5C–10C lower than the optima for top yields temperatures. A favorable temperature did not fully compensate for the lack of fertilizer P, nor did P offset the effects of unfavorable temperature. As the temperature increased, yield increases were greater than was uptake of nutrients, resulting in a decrease in the percentage of nutrients. Read and Ashford (1968) found best foliage yields at 21C, their highest root temperature.

Best production of ryegrass (*Lolium perenne* L.) foliage was obtained at a soil temperature of about 20C (Nielsen and Cunningham, 1964; Davidson, 1969b; Sato and Ito, 1969; Dijkshoorn and 't Hart, 1957). As soil temperature was raised, the percentages of calcium and magnesium in ryegrass increased markedly, but those of nitrogen, phosphorus, sulfur, and sodium were barely affected (Nielsen and Cunningham, 1964). With increasing soil temperature, amino acids, glutamine, and asparagine decreased (Nowakowski *et al.*, 1965).

Mack and Finn (1970) stated that optimum temperature for timothy (*Phleum pratense*) was obviously related to nutrient and water supply. At low fertility, raising the soil temperature from 10C to 20C increased yields nearly threefold; at high fertility the increase was only by a factor of 1.23. At low root temperature clonal differences were small, but at higher temperatures or with the addition of N and P differences showed up (Mack and Finn, 1970). Smith and Jewiss (1966) state that the chemical composition of timothy herbage was affected more by temperature than by N.

7. Lucerne

The best root temperature for growing lucerne (*Medicago sativa* L.) is about 28C (Nielsen *et al.*, 1960b; Nielsen and Warder[4]; Levesque and Ketcheson, 1963; Heinrichs and Nielsen, 1966). Maximum root yields

[3] Unpublished Annual Report 1961, Experimental Farm, Swift Current, Saskatchewan, Canada.
[4] *Ibid.*

have been obtained at about 20C, except where P was limiting, in which case the optimum temperature is less than 20C.

The percentages of N and P in the tops generally increased with yield to about 28C. All cations accumulated in the roots at 5C (Nielsen *et al.*, 1960b).

There were marked varietal differences in growth of lucerne at different soil temperatures in spite of the fact that all varieties yield best at 27C (Heinrichs and Nielsen, 1966). Nodulation was usually best at 12C. Time to reach flowering was not appreciably affected by soil temperature, and these authors suggested that it was more a function of foliage temperature. This is supported by the data from Smith (1970), who reported that one-year-old vernal lucerne reached flower in 22 days at a day/night temperature of 32C/24C but not for 39 days at 18C/10C.

8. Potatoes

The production of potato (*Solanum tuberosum*) tubers has been best at a soil temperature of about 20C and was usually increased significantly by additions of plant nutrients (Nielsen *et al.*, 1961; Borah and Milthorpe, 1963; Yamaguchi *et al.*, 1964; Epstein, 1966). Yields dropped off at about 30C and were lower at this temperature than at 10C; emergence below 10C was poor (Epstein, 1966). Tuber numbers were highest at 10C and decreased steadily as soil temperature increased. This decrease in numbers of tubers represented a loss in yield potential.

There is a relative preponderance of small tubers at cool temperatures, indicating extended growth rates and less maturity (Yamaguchi *et al.*, 1964). Russeting was reported by these authors and others (Ruf, 1963) to be best at the higher temperatures (i.e., up to about 30C) and practically nonexistent at 10C. Epstein (1966) reported that tubers were oblong in shape when grown at 30C instead of round as at 22C. Specific gravity was highest at 15.5C.

Nutrient uptake, particularly P, generally paralleled soil temperature (Nielsen *et al.*, 1961), and concentration of nutrients in the tubers decreased as soil temperature increased.

9. Rice

The most favorable root zone temperature for rice (*oryza sativa*) seems to be about 25C–30C, according to Owen (1971), who recently reviewed this subject. Place *et al.* (1971) used different day/night regimes and indicated that best growth occurred with the 21C/33C regime. Ehrler and Bernstein (1958) produced more rice straw but less grain from plants grown at root temperature of 18C than at 30C and pointed out that the effect of root temperature varied depending upon whether there was vegetative or reproductive growth. Water temperature at tillering is more important than air temperature because the growing points are submerged. Once tillering has been completed, the yield potential has largely been fixed. At high temperatures tillering rate was increased, but total numbers decreased because of the shorter duration (Owen, 1971). Thus the yield potential in rice tends to decrease as root temperatures increase.

Generally, nutrient uptake by rice is greatest at the root temperature regime giving best growth and grain yield (Owen, 1971).

10. Other Crops

Optimum root temperatures for growth of beans (*Phaseolus vulgaris*) is about 28C (Apple and Butts, 1953; Wallace, 1963; Brouwer, 1964; Mack *et al.*, 1964). Added P has a progressive influence upon growth as root temperatures are lowered but does not fully compensate for the unfavorable temperature.

Coffee (*Coffea arabica*) growth was best where root temperature was 20C at night and 26C during the day (Franco, 1958). Root growth with this temperature regime was double that of constant 28C.

Citrus tops grew best at a root temperature of about 30C and roots at 24C (Wallace, 1958; Labanaskas *et al.*, 1965). Liebig and Chapman (1963) reported best growth of young navel oranges (*Citrus sinensis*) as root temperature increased to 30C from 14C, but flowering decreased at the highest temperature.

Guayule (*Parthenium argentatum*) grew best and yielded the most rubber resin at root temperatures near 28C, according to Benedict (1950). The concentrations of resin in the roots and stems decreased as the temperature was increased.

Peas grown at a root temperature about 21C have better yield than those grown at higher or lower temperatures (Brouwer and van Vliet, 1960; Klacan, 1962; Mack *et al.*, 1964; Adedipe and Ormrod, 1970). Brouwer (1959) related poor growth of peas at low root temperatures to water deficiency within the plant.

Sugar beet (*Beta vulgaris*) germination proceeds best at about 28C, and yields of root are best at 24C (Radke and Bauer, 1969). The percentage of sucrose was highest at root temperature of around 18C (Ito and Takeda, 1963; Radke and Bauer, 1969).

Sugarcane (*Saccharum officinarum*) yields are affected more by root temperature than by air temperature (Burr *et al.*, 1957; Whiteman *et al.*, 1963; Hartt, 1965), and the optimum appears to be about 25C–30C, shifting downward slightly with maturity. At suboptimum root temperatures, Hartt (1965) reported that translocation from the leaves was decreased and congestion occurred which interfered with photosynthesis.

Strawberry (*Fragaria* spp.) plants grew best at root temperatures of 18C–24C (Roberts and Kenworthy, 1956). Letey *et al.* (1962) reported maximum yields of sunflower (*Helianthus annuus*) tops at a root temperature of 23C. Pineapple yields were increased by one-third when a mulch raised the average winter soil temperature by about 2C (Ekern, 1967).

Tobacco (*Nicotiana tabacum* L.) yields were best at 22C, and nicotine content was highest at 30C (Parups and Nielsen, 1960; Parups *et al.*, 1960). Yields were improved with the addition of P when root temperatures were less than 22C.

Best growth of tomatoes (*Lycopersicon esculentum* Mill.) appears to be at a root temperature of about 25C (Shtrausberg, 1958; Rahman and Bierhuizen, 1959; Locascio and Warren, 1960; Davis and Lingle, 1961; Cannell *et al.*, 1963; Kristoffersen, 1963; Martin and Wilcox, 1963), although Canham (1962) reported a lower optimum. Responses to applications of plant nutrients, particularly P, have been greatest at the lowest root temperatures. The effects of unfavorable temperature were not overcome, however, with any nutrient treatment.

IV. RATE OF REACTIONS

An increase of temperature almost invariably increases the rate of a chemical reaction. For a homogeneous process the rate is approximately doubled or trebled for each 10C rise of temperature (Q_{10} of 2 or 3), and even for many heterogeneous reactions the same general relationship holds. It frequently fails, however, for chain reactions.

The application of information obtained with isolated chemical reactions to the complex chemistry of the living plant is difficult; it is obviously misleading to think of plant growth in terms of simple chemical reactions or to apply directly information obtained with isolated enzymes. Nevertheless, all reactions occurring in plant cells follow the basic laws of thermodynamics and rate theory. It is in order, therefore, to discuss briefly the theory of reaction rates at this point.

A. Basic Relationships

In 1889 Arrhenius, building upon the work of van't Hoff, was able to describe his results with the effects of temperature on the rate of hydrolysis of sucrose by the equation

$$V = Ae^{-E/RT}, \qquad [1]$$

where V is the velocity of the reaction, A is a constant, e is the base of the natural logarithm, E is *energy of activation*, R is the gas constant (1.986 cal per deg C per mole), and T is the absolute temperature. Taking the log of this equation gives

$$\log_{10} V = -E/2.303\,RT + \text{constant}. \qquad [2]$$

With most chemical reactions a plot of log V against $1/T$ yields a straight line with a slope of $-E/2.303\,R$ if the temperature range is limited.

Arrhenius suggested that in every system an equilibrium exists between "normal" and "active" molecules and that only the latter are involved in chemical reactions. If heat is absorbed (that is, if the temperature is raised) in the conversion of a normal molecule to an active one, E in equation [2] represents the energy difference between the active and normal states of the molecule. A rise in temperature, therefore, will favor the formation of active molecules, and although the number of collisions between molecules will be affected only slightly, the concentration of active molecules may be doubled or trebled by a 10C rise in temperature. Hence the velocity of the reaction is increased.

B. Activated State

It is believed that when two reactant molecules possessing the necessary energy of activation come together, they first form an activated complex which then decomposes to yield the products of the reaction. In the reaction A + BC → AB + C, as A is brought closer to the diatomic molecule BC, the potential energy of the system increases and reaches a maximum when the possibility is just as great for A to unite with B forming AB + C as it is for B to remain with C. This state is the activated complex or transition state, through which the system A + BC must pass before it can be converted to AB + C or *vice versa*.

Initial state	Activated complex	Final state
A + BC ⟶	A + B + C ⟶	AB + C

Diagramatically the change in potential energy for an exothermic reaction is shown in Figure 12.1. The difference between A + BC and A + B + C is equal to E, the energy of the forward reaction, and that between A + B + C and AB + C is the activation energy E^1 for the reverse reaction. The total energy change is thus equal to ΔE. With endothermic

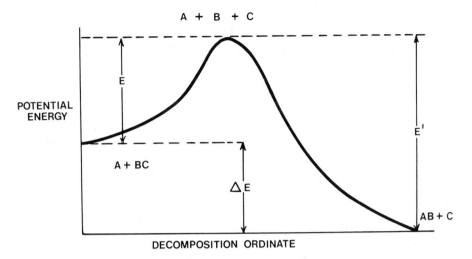

Fig. 12.1. Change of potential energy in a chemical reaction

processes the energy (or heat, H) of activation must be at least equal to the heat absorbed in the reaction. Rates of reactions with large activation energies increase rapidly with increasing temperature, while rates of reaction with low activation energies increase more slowly. (In actual fact it is the free energy of activation, F, which determines the rate of reaction at a given temperature, not the energy of activation, E, but E is a useful approximation.) Catalysis, whether enzymatic or not, has the effect of lowering the activation energy.

Several types of reactions are known not to give a straight line plot of log V against $1/T$ (Farrell and Rose, 1967), but at the present time this nonlinearity is difficult to explain. Nevertheless, this relationship is still extremely useful for comparing the effects of temperature on different physiological processes.

Langridge (1963) reviewed the application of basic kinetics in plant growth and concluded that much of the effect of temperature on plants was manifested through enzyme systems that produced simple, well-known metabolites.

C. Water Viscosity

X-ray studies of ice have shown that each oxygen atom is surrounded by four other oxygen atoms arranged tetrahedrally, the distance between the centers of two adjacent atoms being 2.76Å; presumably a hydrogen atom serves as a link between each pair of oxygen atoms. The fact that the oxygen atom in water has two lone pairs of electrons makes it possible for each such atom to form two hydrogen bonds as well as two covalent bonds. Thus every oxygen atom can take part in the formation of two hydrogen bonds, and so when water freezes the whole ice crystal is virtually one molecule.

A somewhat similar situation exists with liquid water, although the amount of hydrogen bonding decreases as temperature increases and water vapor consists of single molecules. It has only been in the last few years that water has been considered to have a crystalline structure (Glasstone, 1946). It differs from a true crystal, however, in that the hydrogen bonds are being continually broken and remade and that molecules can be interchanged as a result of thermal motions. Although the tendency exists for each oxygen molecule to be tetrahedrally surrounded by four others, it is probable that at any instant each one is joined by hydrogen bonds to only two or three others at ordinary temperature and that at higher temperatures the average number is even less.

It is the presence of a network of hydrogen bonds that accounts for the increasing viscosity of water as temperature decreases. Since water is the medium in which most reactions in plants take place, temperature effects on water viscosity add another dimension to the influence of temperature on the rate of reactions.

V. TOP-ROOT RELATIONSHIPS

A. General Considerations

All the known functions of the root make the roots (and the whole plant) susceptible to temperature. Davidson (1969a) discusses a balanced internal economy of plants under which root activity alters to provide for the tops when the plant is under stress of any kind. In this way the root system of a plant becomes larger relative to tops when the plant is subjected to cold or drought because these factors reduce the efficiency of the root. Brouwer (1964) suggested that root activity (gram of tops per gram of root) is a more meaningful criterion in assessing a functional balance between tops and roots than just weight of roots. Brouwer (1963) suggested that there is a functional equilibrium between shoots and roots which is rhythmical. When the supply of nutrients and water to the tops is limited, carbohydrates accumulate and root growth is increased; this results in the uptake of more water and nutrients and the consequent stimulation of top growth. Davidson (1969b) suggested the presence of a plant mechanism which partitioned carbohydrates between tops and roots. At rapid growth rates of tops there is low diversion to the root.

From a practical point of view, a farmer is concerned about the effects of unfavorable soil temperatures and what he can do about them. Most temperate zone crops are limited by low temperatures during much of their growth period and are bracketed on each end by freezing air temperatures. Tropical crops may be limited by soil temperatures that are too high for certain periods of time. The impact of the unfavorable effects of high root temperatures on crops is less than that of low root temperatures, however. Langridge (1963) reviewed both aspects and found little quantitative data for plants at high temperatures. It is known, however, that higher-than-optimum soil temperatures—

1. accelerate the rate of chemical reactions and hence the rate of requirement for reactants,
2. accelerate the breakdown of metabolites and inactivation of enzymes, and
3. cause rate imbalances where there are branched, coupled, or sequential reactions.

If conditions are favorable to growth, the root system should be structurally and morphologically able to use as much carbohydrate as the tops can supply and in turn supply as much water and nutrients as the tops can use.

B. Anatomic Effects

1. Roots

Root zone temperature affects the morphology and distribution of roots. At optimum temperature cell division is more rapid but of shorter duration than at lower temperatures. At cooler temperatures roots are usually whiter, thicker in diameter, and less branched than at warmer temperatures (Ketellapper, 1960; Brouwer and Hoogland, 1964; Nielsen and Cunningham, 1964; Garwood, 1968), although there are exceptions (Bowen, 1970). At high temperatures roots become filamentous. With cold, root cell maturation is delayed and elongation is favored (Burstom, 1956; Varner *et al.*, 1963). Owen (1971) reported that rice roots became longer at 10C–21C compared with 31C but that there were fewer roots.

The temperature profile in soils as seen in Tables 12.1–12.3 will result in a root distribution pattern that reflects this profile. If soil moisture, root impedance, and aeration were equal, the temperature-dependent root pattern would result in a more branched system near the soil surface, decreasing in numbers with depth but increasing in individual root diameter (e.g., Garwood, 1968). In temperate climates maturation at depths greater than about 20 cm, where diurnal variations are not manifest, will be prolonged, and total root weight will be small compared to that of the part of the system near the surface. Activity, or efficiency, of the lower roots will be less than that of the upper roots.

Epstein (1966) found that at high temperatures (22C and 29C) potato tubers became elongated and that at 29C there were many rhizomes with swollen but undeveloped tips.

2. Shoots

Root temperatures affect leaf growth and anatomy (Humphries and Wheeler, 1963). Humphries (1967) reported that rooted bean leaves accumulated carbohydrates in the mesophyl and palisade cells, resulting in an increase in leaf thickness. Much of the carbohydrate apparently was assimilated in permanent cell structure because mobile sugars and starches accounted for only about half of the increase in dry matter. Phillips and Bukovac (1967) found anatomical changes in the leaves of peas and beans as root temperatures were raised, which resulted in an increased absorption of foliar P and Ca. They also reported a residual effect of the root temperature experienced by plants before the test period. This effect was also found by Brouwer and Levi (1969), who suggested that pretreatment root temperature effects on leaf anatomy were more important than prevailing root temperatures in short-term studies. These authors reported that leaf size increased with root temperature but that this was due to cell elongation since cell numbers remained about the same. Palisade and mesophyll parenchyma cell sizes accounted for the greater leaf thickness at optimum root temperatures.

Herath and Ormrod (1965) reported that an increase in the root temperature of rice from 16C to 32C increased sheath lengths, the size of leaf lamina, and the number and size of stomata. They suggest that structural and functional changes in the tops of plants can be expected if they are subjected to stresses of one kind or another such as water, nutrient, or growth metabolite.

Brouwer and Hoogland (1964) suggested that there was no simple relationship between root growth and shoot growth on the basis of weight. Various physiological activities of tops and roots are modified by changes induced by low root temperatures; thus, they preferred to use root activity as an indication of physiological balance between tops and roots. It is well known that the root:shoot ratio is sensitive to all climatic factors and that it changes with age (Brouwer, 1966).

It can be seen that much is known about the effects of root temperature upon the appearance of plants. Little data could be found, however, on the mechanisms whereby root temperature caused the modifications. Possible explanations include the effects of root temperature on the translocation of carbohydrates for growth or on the changes in endogenous levels of hormones in the tissues.

B. Functions of the Root

1. Water Uptake

The effects of temperature upon the viscosity of water are well known and are discussed in section IV C of this chapter. Since all activity within the plant occurs in a water medium, growth is affected by the temperature of water. As the water temperature decreases, other factors being equal, water viscosity increases and growth activities decrease.

Living protoplasm generally contains about 20% protein and 70%–80% water in a gelatinous condition. Like cellular water, protoplasm may be highly organized and exhibit definite X-ray diffraction patterns, or it may be unorganized and behave more like a sol than a gel (Ling, 1967; Langridge and McWilliam, 1967). Cellular membranes usually become more permeable to water and solutes as temperature increases, although Swanson and Geiger (1967) found no reduction in transport through a 2-cm section of sugar beet petiole cooled to 0C. Cooling plant cells to 0C without freezing causes relatively few changes in cellular organization, and these are reversible as a rule.

At low temperatures the viscosity of protoplasm decreases, although some increases have been reported (Langridge and McWilliam, 1967) with temperature change or with time at a given temperature. Also, the solubility in water of oxygen and carbon dioxide increases, tending to decrease pH and consequently to change enzymatic activity. Kleinendorst and Brouwer (1970) measured osmotic increases in plant cells with decreasing temperatures and noted that these osmotic increases tended to compensate for increased cell impermeability.

A decrease in root zone temperature has resulted in decreased water uptake (Rahman and Bierhuizen, 1959; Nielsen *et al.*, 1961; Cox and Boersma, 1967; Kleinendorst and Brouwer, 1970; Wallace, 1970; Kuiper, 1964), although Power *et al.* (1970) found water use by barley to be the same at low and high soil temperatures. Corn plants whose roots were chilled from 20C to 5C in a matter of 3 hours wilted and never recovered (Nielsen *et al.*, 1961). Resistance to water uptake caused by roots being suddenly chilled could result in a transpiration absorption lag which would impose conditions of water stress in plants. Plants grown at a low root temperature develop a dynamic balance between low absorption and transpiration surface and are able to survive, albeit at low metabolic rates.

Kramer (1956) suggested that the additive effects of temperature on viscosity of water and permeability of protoplasm decrease uptake of water at 5C to one-fourth of that at 25C.

Slatyer (1969) reviewed the effects of internal water stress upon the growth processes in plants. Since low root temperatures appear to impose such stresses in plants, all that is said about plant moisture–growth relationships refers to a secondary influence of root temperatures on plants.

2. Nutrient Uptake

Temperature of the soil will influence the rate of release of nutrients from organic and inorganic forms. It will also affect the uptake of nutrients by plant roots. Finally it will affect the assimilation of the absorbed nutrients into complex organic compounds and subsequent translocation to aboveground portions of the plant.

Prominent among the elements most affected by root temperatures is P. There have been many reports of added P counteracting some of the effects of unfavorably low temperatures (Nielsen and Humphries, 1966). There is ample evidence that while added P may improve the growth of plants in cold soils, it never compensates fully for the effects of the unfavorable temperature. This led Nielsen and Humphries to conclude that it was an oversimplification to attribute the effects of cold soils entirely to a deficiency of a nutrient or of water supply.

Sutton (1969) reviewed the effects of various soil factors that may restrict the supply or availability of P to plants at low temperatures. He concluded that there is good evidence that low soil temperatures can reduce the availability of soil phosphate to plants and that in some soils the quantity may also be reduced. He suggested that the beneficial effects of fertilizer phosphate are primarily due to its being "a source of phosphate that is more available and more able to maintain a higher concentration of phosphate in solution than the native soil phosphate." He cautioned that climatic conditions should be taken into account in the interpretation of soil analysis for purposes of making fertilizer recommendations.

Nielsen and Cunningham (1964) concluded that soil temperature had little effect on the uptake of P from soil rich in this element. The same can probably be said of other nutrients (Parups and Nielsen, 1960; del Valle and Harmon, 1967).

When nutrients are present in the soil in organic combinations, temperature can have a greater influence than with inorganic combinations because of its effect on microorganisms and the decomposition of the organic matter. This may partially explain the different effects of temperature on uptake of phosphorus and zinc (Place *et al.*, 1971) where relatively more phosphorus is organically combined.

According to Bowen (1970) in his work with Monterey pine (*Pinus radiata*), the most important plant factors in phosphate uptake from soil are (1) elongation of roots, (2) production of roots, (3) extent of root hair production, and (4) the extent to which different parts of the root absorb phosphate. In general, uptake of P is at a maximum near or slightly above the optimum root temperature for growth of tops (Nielsen *et al.*, 1960a, 1960b, 1961; Simpson, 1965; Locascio and Warren, 1960; Lingle and Davis, 1959; Power *et al.*, 1964).

Where nutrients are mobile in the soil, root extension is evidently less critical in uptake, and thus low root temperatures are less harmful. Thus additions of nitrogen, sulfur, chlorine, and potassium did not materially improve the growth of ryegrass at low root temperatures (Nielsen and Cunningham, 1964). In a general way the uptake of these nutrients has been found to parallel yield responses.

Wallace (1963) reported that the uptake of potassium, bromine, and rubidinum was temperature dependent only when the concentrations in the soil were small.

Plants may not respond appreciably to changes in root temperature when nutrients are deficient, and it appears that under these conditions root temperature optima are often higher. Stewart and Whitfield (1965) reported that soil temperature had little effect on winter wheat in the absence of N, P, and S. When root temperatures were raised from 12.8C to 18.3C, check yields of 4-week-old winter wheat increased from 1.08 to 1.40 g. With the N, P, and S treatment yields were raised from 4.46 to 8.86 g. In the absence of N and at 10C, forage yields were similar with different species (Hamilton, 1970). When N was added, however, differences began to show.

It is difficult to separate the effects of root temperature on uptake from its effects on assimilation and translocation. Power *et al.* (1970) concluded in their work with barley that uptake of nutrients at 9C was not restricting growth but that translocation of nutrients from roots to tops was impeded. Nielsen *et al.* (1960a) and Read and Ashford (1968) reported a marked accumulation of nutrients in roots at 5C to 10C. Gibson (1969) found that more fixed N remained in the nodule system at 8C than 15C and 22C. Duration of temperature was important. Ionic interactions may also contribute to anomalies in uptake and translocation; the uptake of a disproportionately large amount of divalent cations was linked with the uptake of nitrate-N by ryegrass

(Parks and Fisher, 1958; Nielsen and Cunningham, 1964; Cunningham and Nielsen, 1965).

3. Translocation and Sink Relationships

It is well known that the products of photosynthesis in leaves must be exported or they will result in some congestion in the leaves. Any areas within the plant to which the assimilates can be exported serve as sinks. The establishment of new sinks during plant growth depends on environmental influences (Beevers, 1969).

Low root temperatures will slow down growth processes in the roots and reduce their capacity as sinks for carbohydrates. Davis and Lingle (1961) suggested that the stresses in uptake of water and nutrients by tomatoes could not adequately explain poorer yields at 15C vs. 25C. They theorized that there was retarded movement of assimilates from the top to the root causing a congestion in the shoots which interfered with metabolic activity. Hartt (1965) found that translocation of carbohydrates out of leaves of sugarcane growing at a root temperature of 17C was much less than at a temperature of 22C. She suggested that the accumulation of carbohydrates in the leaves depressed photosynthesis and reduced yields.

Rooted, fully expanded bean leaves increased in thickness and accumulated dry matter when root temperatures were lowered (Humphries, 1967). Humphries stated that net assimilation depended on the size of the carbohydrate sink and that low root temperatures limited both the activity and size of the sink. When the rooted leaves were transferred to warmer root temperatures, lamina carbohydrates were drained, but the mobile starches and sugars accounted for only about half of the dry matter increase. Thus some of the accumulated carbohydrate was used in cell wall synthesis. Guinn and Hunter (1968) reported that low root temperatures caused the carbohydrate content of the tops of cotton plants to increase rapidly but that it was very low in the roots. It seemed remarkable to them that the roots constituted only 12% of the plant weight and yet exerted such a strong influence on the carbohydrate status of the whole plant.

It has been difficult to distinguish the effects of temperature on translocation from its effects on respiration and other physiological processes (Whittle, 1964). The direction of movement of metabolites is related to an intensity factor represented by differences between the source and the sink and inversely to the distance between the two (Biddulph and Cory, 1965). It has been shown recently that as long as there is a gradient, translocation can occur in some plants even though sections of the stem or petiole are cooled to 0C (Swanson and Geiger, 1967; Biddulph, 1969; Weatherly and Watson, 1969). Swanson and Geiger suggested that living phloem cells are essential in translocation only to the extent of maintaining some minimal level of structural integrity.

Wardlaw (1968) in his review suggested that the primary effects of temperature on the distribution of assimilates in plants are associated with growth rather than sugar conduction, although he did not rule out the possibilities of limitation of photosynthetic rate. It seems that the greatest influences of temperature are on sites of metabolic activity and that the translocation system may not be seriously affected by temperatures that are unfavorable to these sites.

4. Assimilation

Uptake of N by plants occurs primarily through the roots. Nitrogen is absorbed as NO_3^- or NH_4^+, but these ionic forms seldom appear in the tops. Most of the N, passing from the root to the shoot is in the organic form (Street, 1966). Kursanov (1958) reported that as much as 50% of labeled photosynthate moved from leaves to roots of *Cucurbita*, where it was rapidly converted to organic acids, amino acids, and amides. Almost half was then returned to the tops.

Unfavorable root temperatures have resulted in accumulation of NO_3-N in plants (Younis *et al.*, 1965; Nowakowski *et al.*, 1965; Watschke *et al.*, 1970). Younis *et al.* (1965) reported that increased temperature caused a reduction in nitrate reductase activity and increased nitrate. Neither moisture nor light greatly altered this basic effect of temperature.

Accumulation of nitrate in ryegrass was associated with low light intensity and high root temperatures (Nowakowski *et al.*, 1965). Williams and Vlamis (1962) reported that the uptake of nitrate by barley was increased tenfold in raising the root temperature from 13C to 25C. In common with all metabolic processes, protein synthesis proceeds at a decreased rate as temperature is decreased. Synthesis of total protein involves numerous compounds and numerous enzyme systems. More seems to be known about the individual enzymes and their reaction to temperature than about how these systems function in the synthesis of total protein (Street, 1966). Street, in an excellent review of the physiology of root growth, stated that the root system as a whole is an active seat of amino acid and protein synthesis. Bollard (1960) had concluded previously that evidence favored the root as a site of N assimilation.

5. Growth Metabolites

Much evidence has accumulated in the last ten years to support the thesis that the production and release of growth metabolites is part of the normal growth process of roots. Osborne (1962) reported that as long as xanthium (cocklebur) leaves were attached to a growing root system, the incorporation of amino acids proceeded normally. Senescence was marked by reductions in protein, DNA, RNA, and chlorophyl, but applications of kinetin to leaves retarded senescence. Oritani (1963) reported that the protein level of shoots was closely correlated with the presence of roots. The removal of tops from roots decreased the ability of shoots to synthesize RNA, and a reduction in protein synthesis followed. He suggested that roots may in some way influence RNA synthesis and thereby control protein level of the leaf with a resultant affect on photosynthetic activity. Korovin *et al.* (1963) reported a reduced synthesis of organic P, particularly nucleoproteins, in potato leaves at 10C compared with higher root temperatures. The same was reported for pumpkin (*Cucurbita pepo*) (Vinokur, 1963), and it was suggested that roots exerted a specific action on the manufacturing processes in leaves.

Itai and Vaadia (1965) suggested that a modification of leaf metabolism in water-stressed sunflower plants may reflect a decreased supply of regulatory root factors for the shoot. Butcher (1963) and Carr *et al.* (1964) presented evidence that roots of tomatoes and some legumes produce gibberellins that are exported to the shoot. Martin and Wilcox (1963) were of the opinion that their data with tomatoes grown at different root temperatures pointed to the existence of a root-produced factor, a caulocaline required for good shoot growth.

Langridge (1963) remarked that it was surprising that thermal inactivation of growth could in many cases be partly overcome by addition of simple well-known metabolites such as glutamic acid, tannic acid, thiomin, biotin, and nicotinic acid.

Street (1966) in his review concluded that the evidence in the literature justifies the conclusion that substances having auxinlike activity are synthesized in the root, probably in the meristem, and are transported to other parts of the plant. He suggested that it is not necessary to postulate that the sole determining effect of roots on the tops is through their function as sinks for carbohydrates produced by the tops. Indeed a wide range of organic compounds is released from growing excised roots, including alkaloids, vitamins, nucleotides, flavones, auxins, amino acids, and organic acids.

Lockard and Grunwald (1970) found evidence that the mechanism controlling stem growth in peas was located in the stem and not in the roots. Brouwer and Kleinendorst (1967) used a number of growth-regulating compounds including gibberellic acid and indoeacetic acid on beans growing at low root temperature. They found no favorable effects but stated that their results do not exclude the possibility that other metabolites would be effective.

Guinn and Hunter (1968) reported marked differences in the carbohydrate status of young cotton seedlings grown at different root temperatures. They suggested that it was reasonable to assume that since low root temperatures slowed down all metabolic activities in roots, the production of growth-regulating compounds such as cytokinins, which influence nucleic acid and protein synthesis and hence the utilization of carbohydrates, would be reduced.

VI. CONCLUSIONS

It would be satisfying if this chapter could end with a comprehensive, point-by-point summary of the effects of root temperatures on plants. Unfortunately this cannot be done. Our knowledge of the precise functions of the root is inadequate; our understanding of the reactions that occur in the roots and how they are influenced by environmental conditions, stage of growth, and so on is incomplete.

However, much good work has been done since Richards *et al.* (1952) made a review of root temperatures and plant growth.

The present review was undertaken with the assumptions that the root (*a*) absorbs water, (*b*) takes up plant nutrients, (*c*) is a sink for carbohydrates made in the tops, (*d*) is a site for assimilating nitrogen and other nutrients, and (*e*) produces growth metabolites.

The information examined is convincing that these are among the essential functions of the root. It is evident that the study of root temperature effects on plants has opened the door to an understanding of what the root does, and it is expected that research along these lines will continue to expand our knowledge of the root. There is no reason to believe that its functions are limited to those listed.

Root zone temperature has a dominant influence on seed germination and early growth of plants. It is of utmost importance in vegetative growth throughout the life of plants and is more critical in survival than is foliage temperature. In triggering reproductive growth above ground, it is less important than foliage temperature, but below ground it is more important.

The main influence of low temperatures is in reducing kinetic activity. This may result only in an extension of the growth period as there is mounting evidence that yield potential is diminished when root temperatures rise.

There are strong interactions between root temperature and other environmental conditions such as water supply and available nutrients; the effects of unfavorable root temperatures can be partially mitigated by favorable supplies of water and nutrients. The stage of growth is important. An evaluation of the reasons for these interactions is leading to a more precise study of the effects of temperature at the molecular and cellular level of plant growth. When root temperature effects are being examined, it is important that other conditions be well defined or controlled if possible.

There is much fruitful work being done by plant breeders in selecting varieties and strains that are more tolerant to unfavorable root temperatures. Root temperatures effects are now being used as a screening tool in plant breeding.

At the field level there are a number of management practices that are effective in influencing root zone temperatures, and it is fully expected that others will be developed.

References

Adedipe, N. O., and D. P. Ormrod. 1970. Air and soil temperature effects on growth response of peas to phosphorus fertilization. *J. Amer. Soc. Hort. Sci.* 95: 111–14.

Allen, S. E., and O. P. Engelstad..1963. Response of corn to added N and P, as affected by soil temperature. *Agron. Abstr.*, p. 34.

Allmaras, R. R., W. C. Burrows, and W. E. Larson. 1964. Early growth of corn as affected by soil temperature. *Soil Sci. Soc. Amer. Proc.* 28: 271–75.

Anderson, D. T., and G. C. Russell. 1964. Effects of various quantities of straw mulch on the growth and yield of spring and winter wheat. *Can. J. Soil Sci.* 44: 109–18.

Anderson, W. B. 1962. Plant growth as affected by aggregate stability, soil moisture and soil temperature. M. Sc. Thesis, Colorado State University, Fort Collins.

Apple, S. B. J., and J. S. Butts. 1953. The effect of soil temperature on growth and phosphorus uptake by pole beans. *Proc. Amer. Soc. Hort. Sci.* 61: 325–32.

Army, T. J., and E. V. Miller. 1959. Effect of lime, soil type and soil temperature on phosphorus deficient soils. *Agron. J.* 51: 376–78.

Arndt, C. H. 1937. Water absorption in the cotton plant as affected by soil and water temperature. *Plant Physiol.* 12: 703–20.

Baker, Barton S., and G. A. Jung. 1970. Effect of environmental conditions on the growth of four perennial grasses III. Nucleic acid concentration as influenced by day-night temperature combinations. *Crop Sci.* 10: 376–78.

Beauchamp, E. G., and J. K. Torrance. 1969. Temperature gradients within young maize plant stalks as influenced by aerial and root zone temperatures. *Plant Soil* 30: 241–51.

Beevers, H. 1969. Metabolic sinks In *Physiological Aspects of Crop Yield*, ed. R. C. Dinauer, pp. 169–84. Madison, Wis.: Amer. Soc. Agron.

Benedict, H. M. 1950. The effect of soil temperature on guayule plants. *Plant Physiol.* 25: 377–80.

Biddulph, O. 1969. Mechanisms of translocation of plant metabolites. In *Physiological Aspects of Crop Yield*, ed. R. C. Dinauer, pp. 189–202. Madison, Wis.: Amer. Soc. Agron.

Biddulph, O., S. Biddulph, R. Cory, and H. Koontz. 1958. Circulation patterns for phosphorus sulfur and calcium in the bean plant. *Plant Physiol.* 33: 293–300.

Biddulph, O., and R. Cory. 1965. Translocation of C^{14} metabolites in the phloem of the bean plant. *Plant Physiol.* 40: 119–29.

Black, A. L. 1970. Soil water and soil temperature influences on dryland winter wheat. *Agron. J.* 62: 797–801.

Bloodworth, Morris E. 1960. Effect of soil temperature on water use by plants. *Trans. 7th Int. Congr. Soil Sci.* 1: 153–63.

Bohning, R. H., W. A. Kendall, and A. J. Linck. 1953. Effect of temperature and sucrose growth on translocation in tomato. *Amer. J. Bot.* 40: 150–53.

Bollard, E. G. 1960. Transport in the xylem. *Ann. Rev. Plant Physiol.* 11: 141–66.

Borah, M. N., and F. L. Milthorpe. 1963. Growth of the potato as affected by temperature. *Indian Plant Physiol.* 5: 53–72.

Bowen, G. D. 1970. Effects of soil temperature on root growth and on phosphate uptake along *Pinus radiata* roots. *Aust. J. Soil Res.* 8: 31–42.

Brengle, K. G., and C. J. Whitfield. 1969. Effect of soil temperature on the growth of spring wheat with and without wheat straw mulch. *Agron. J.* 61: 377–79.

Brouwer, R. 1959. The influence of the root temperature on the growth of peas. *Jaarb. IBS*, pp. 27–36.

——. 1962. Influence of temperature of the root medium on the growth of seedlings of various crop plants. *Jaarb. IBS*, pp. 11–19.

——. 1963. Nutritive influences on the distribution of dry matter in the plant. *Neth. J. Agr. Sci.* 10: 399–408.

——. 1964. Responses of bean plants to root temperatures. I. Root temperatures and growth in the vegetative stage. *Jaarb. IBS*, pp. 11–22.

——. 1966. Root growth of grasses and cereals. In *The Growth of Cereals and Grasses*, ed. F. L. Milthorpe and J. D. Ivins, pp. 1953–1956. London: Butterworths.

Brouwer, R., and Atje Hoogland. 1964. Responses of bean plants to root temperatures. II. Anatomical aspects. *Jaarb. IBS*, pp. 23–31.

Brouwer, R., and A. Kleinendorst. 1967. Responses of bean plants to root temperatures. *Jaarb. IBS*, pp. 11–28.

Brouwer, R., and E. Levi. 1969. Response of bean plants to root temperatures IV. Translocation of Na^{22} applied to the leaves. *Acta Bot. Neerl.* 18: 58–66.

Brouwer, R., and G. van Vliet. 1960. The influence of root temperature on growth and uptake of peas. *Mededel. Inst. Biol. Scheik. Ond. Landbouwg.* 108: 23–36.

Burr, G. O., C. E. Hartt, H. W. Brodie, T. Tanimoto, H. P. Kortschak, D. Takahashi, F. M. Ashton, and R. E. Coleman. 1957. The sugarcane plant. *Ann. Rev. Plant Physiol.* 8: 275–308.

Burrows, W. O., and W. E. Larson. 1962. Effect of amount of mulch on soil temperature and early growth of corn. *Agron. J.* 54: 19–23.

Burstom, H. 1956. Temperature and root cell elongation. *Physiol. Plant.* 9: 682.

Butcher, D. N. 1963. The presence of gibberellins in excised tomato roots. *J. Exp. Bot.* 14: 272–80.

Canham, A. E. 1962. Soil temperature and plant growth. *Advance. Hort. Sci.* 2: 440–51.

Cannell, G. H., F. T. Bingham, J. C. Lingle, and M. J. Garber. 1963. Yield and nutrient composition of tomatoes in relation to soil temperature, moisture and phosphorus levels. *Soil Sci. Soc. Amer. Proc.* 27: 560–65.

Carr, D. J., D. M. Reid, and K. G. M. Skene. 1964. The supply of gibberellins from the root to the shoot. *Planta* 63: 382–92.

Case, V. W., N. C. Brady, and D. J. Lathwell. 1964. The influence of soil temperature and phosphorus fertilizers of different water solubilities on the yield and phosphorus content of oats. *Soil Sci. Soc. Amer. Proc.* 28: 409–12.

Chapman, D. 1967. The effect of heat on membranes and membrane constituents. In *Thermobiology*, ed. A. H. Ross, pp. 123–46. London, New York: Academic Press.

Christiansen, M. N. 1963. Influence of chilling upon seedling development of cotton. *Plant Physiol.* 38: 520–22.

Clarkson, V. A. 1960. Effect of black polyethylene mulch on soil and microclimate temperature and nitrate level. *Agron. J.* 52: 307–9.

Cooper, D. J., K. F. Nielsen, J. W. White, and W. Kalbfleisch. 1960. Apparatus for controlling soil temperatures. *Can. J. Soil Sci.* 40: 105–7.

Cox, L. M., and L. Boersma. 1967. Transpiration as a function of soil temperature and soil water stress. *Plant Physiol.* 42: 550–56.

Cunningham, R. K., and K. F. Nielsen. 1965. Cation-anion relationships in crop nutrition. V. The effects of soil temperature, light intensity and soil water tension. *J. Agr. Sci.* 64: 379–86.

Davidson, R. L. 1969a. Effect of root-leaf temperature differentials on root-shoot ratios in some pasture grasses and clover. *Ann. Bot.* 33: 561–69.

——. 1969b. Effects of edaphic factors on the soluble carbohydrate content of *Lolium perenne* L. and *Trifolium repens* L. *Ann. Bot.* 33: 579–89.

Davis, R. M., and J. C. Lingle. 1961. Basis of shoot response to root temperature in tomato. *Plant Physiol.* 36: 153–62.

del Valle, C. G., and S. A. Harmon. 1967. Collard growth and mineral composition as influenced by soil temperature and two fertility levels. *Proc. Amer. Soc. Hort. Sci.* 91: 347–52.

Dijkshoorn, W., and M. L. 't Hart. 1957. The effect of alteration of temperature upon the cationic composition in perennial ryegrass. *Neth. J. Agr. Sci.* 5: 18–36.

Dormaar, J. F., and J. W. Ketcheson. 1960. The effect of nitrogen form and soil temperature on the growth of phosphorus uptake of corn grown in the greenhouse. *Can. J. Soil Sci.* 40: 177–84.

Ehrler, William, and Leon Bernstein. 1958. Effects of root temperature, mineral nutrition and salinity on the growth and composition of rice. *Bot. Gaz.* 120: 67–74.

Ekern, P. C. 1967. Soil moisture and soil temperature changes with the use of black vapor-barrier mulch and their influence on pineapple growth in Hawaii. *Soil Sci. Soc. Amer. Proc.* 31: 270–75.

Epstein, E. 1966. Effect of soil temperature at different growth stages on growth and development of potato plants. *Agron. J.* 58: 169–71.

Farrell, Judith, and A. H. Rose. 1967. Temperature effects on microorganisms. In *Thermobiology*, ed. A. H. Rose, pp. 147–218. London, New York: Academic Press.

Finn, B. J., and A. R. Mack. 1964. Different response of orchardgrass varieties (*Dactylis glomerata* L.) to nitrogen and phosphorus under controlled soil temperature and moisture conditions. *Soil Sci. Soc. Amer. Proc.* 28: 782–85.

Franco, C. M. 1958. *Influence of Temperature on Growth of Coffee Plant.* I.B.E.C. Res. Inst. Bull. 16. 21. pp.

Fulton, J.M. 1968. Growth and yield of oats as influenced by soil temperature, ambient temperature and soil moisture supply. *Can. J. Soil Sci.* 48: 1–5.

Fulton, J. M., and W. I. Findlay. 1966. Influence of soil moisture and ambient temperature on nutrient percentage of oat tissue. *Can. J. Soil Sci.* 46: 75–81.

Garwood, E. A. 1968. Some effects of soil water conditions and temperature on the roots of grasses and clovers. II. Effects of variation in the soil water content and in soil temperature on root growth. *J. Brit. Grassland Soc.* 23: 117–27.

Gibson, A. H. 1969. Physical environment and symbiotic nitrogen fixation. VI. Nitrogen retention within the nodules of *Trifolium subterraneum* L. VII. Effect of root temperature on nitrogen fixation. *Aust. J. Biol. Sci.* 22: 829–38, 839–46.

Glasstone, Samuel. 1946. *Textbook of physical chemistry.* 2nd ed. Toronto: D. van Nostrand Co.

Greb, B. W., D. E. Smilka, and A. L. Black. 1967. Effect of straw mulch rates on soil water storage during summer fallow in the Great Plains. *Soil Sci. Soc. Amer. Proc.* 31: 556–59.

Guinn, G., and R. E. Hunter. 1968. Root temperature and carbohydrate status of young cotton plants. *Crop Sci.* 8: 67–70.

Hamilton, H. A. 1970. Influence of nitrogen, potassium and root zone temperature on the response of timothy in monoculture and in association with alfalfa and birdsfoot trefoil. *Can. J. Plant Sci.* 50: 401–9.

Hartt, C. E. 1965. The effects of temperature upon translocation of C^{14} in sugarcane. *Plant Physiol.* 40: 74–81.

Heinrichs, D. H., and K. F. Nielsen. 1966. Growth response of alfalfa varieties of diverse genetic origin to different root zone temperatures. *Can. J. Plant Sci.* 46: 291–98.

Herath, W., and Ormrod, D. P. 1965. Some effects of water temperature on the growth and development of rice seedlings. *Agron. J.* 57: 373–76.

Highkin, H. R. 1958. Temperature induced variability in peas. *Amer. J. Bot.* 45: 626–31.

Holmes, R. M. 1970. Meso-scale effects of agriculture and a large prairie lake on the atmospheric boundary layer. *Agron. J.* 62: 546–49.

Hughes, Roy. 1965. Climatic factors in relation to growth and survival of pasture plants. *J. Brit. Grassland Soc.* 20: 263–71.

Humphries, E. C. 1967. The effect of different root temperatures on dry matter and carbohydrate changes in rooted leaves of *Phaseolus* spp. *Ann. Bot.* 31: 59–69.

Humphries, E. C., and A. W. Wheeler. 1963. Physiology of leaf growth. *Ann. Rev. Plant Physiol.* 14: 385–410.

Idso, S. B., and R. D. Jackson. 1969. Comparison of two methods for determining infra-red emittances of bare soils. *J. Appl. Meteorol.* 8: 168–69.

Itai, C., and Y. Vaadia. 1965. Kinetin-like activity in root exudate of water-stressed sunflower plants. *Physiol. Plant.* 18: 941–44.

Ito, K., and T. Takeda. 1963. Studies on the effect of temperature on the sugar production in sugarbeet. I. the effect of temperature on the growth. *Crop Sci. Soc. Jap. Proc.* 31: 272–76.

Jensen, G. 1960. Effects of temperature and shifts in temperature on intact root systems. *Physiol. Plant.* 13: 822–30.

Jones, J., J. Benton, and H. J. Mederski. 1963. Effect of soil temperature on corn development and yield II. Studies with six inbred lines. *Soil Sci. Soc. Amer. Proc.* 27: 189–92.

Ketcheson, J. W. 1966. Influence of controlled soil and air temperature on fertilizer response for corn. *Agron. Abstr.*, p. 76.

——. 1968. Effect of controlled air and soil temperature and starter fertilizer on growth and nutrient composition of corn (*Zea mays* L.) *Soil Sci. Soc. Amer. Proc.* 32: 531–34.

——. 1970. Effects of heating and insulating soil on corn growth. *Can. J. Soil Sci.* 50: 379–84.

Ketellapper, H. J. 1960. The effect of soil temperature on the growth of *Phalaris tuberosa* L. *Physiol. Plant.* 13: 641–47.

Klacan, G. R. 1962. Pod and seed development in canning peas as influenced by mineral nutrition and root temperature. *Diss. Abstr.* 23: 782.

Kleinendorst, A., and R. Brouwer. 1970. The effect of temperature of the root medium and of the growing point of the shoot on growth, water content and sugar content of maize leaves. *Neth. J. Agr. Sci.* 18: 140–48.

Knoll, H. A., N. C. Brady, and D. J. Lathwell. 1964a. Effect of soil temperature and phosphorus fertilization on the growth and phosphorus content of corn. *Agron. J.* 56: 145–47.

Knoll, H. A., D. J. Lathwell, and N. C. Brady. 1964b. Effect of root zone temperature of various stages of the growing period on the growth of corn. *Agron. J.* 56: 143–45.

Korovin, A. I., Z. F. Sycheva, and Z. A. Bystrow. 1961. Effect of soil temperature on utilization of phosphorus by plants. *Dokl. Akad. Nauk.* 137: 458–61 (seen in *Soils Fert.* 24: 4).

——. 1963. Effect of soil temperature on content of forms of phosphorus in plants. *Fiziol. Rast.* 10: 137–41 (seen in *Soils Fert.* 26: 4).

Kramer, P. J. 1956. Physical and physiological aspects of water absorption. In *Encyclopedia Plant Physiology* 3: 124–59. Berlin: Springer-Verlag.

Kristoffersen, T. 1963. Interaction of photoperiod and temperature on growth and development of young tomato plants. *Physiol. Plant.*, suppl. 1: 98.

Kuiper, P. J. C. 1964. Water uptake of higher plants as affected by root temperature. *Mededel. Landbouwh. Wageningen* 64: 1–11.

Kursanov, A. L. 1958. The root system as an organ of metabolism. *Proc. 1st (UNESCO) Int. Conf. Sci. Res.* 4: 494–509.

Labanaskas, C. K., L. H. Stolzy, L. J. Klotz, and T. A. DeWolfe. 1965. Effects of soil temperature and oxygen on the amounts of macro-nutrients and micronutrients in citrus seedlings. *Soil Sci. Soc. Amer. Proc.* 29: 60–64.

Langridge, J. 1963. Biochemical aspects of temperature response. *Ann. Rev. Plant Physiol.* 14: 441–62.

Langridge, J., and J. R. McWilliam. 1967. Heat responses of higher plants. In *Thermobiology*, ed. A. H. Rose, pp. 231–92. London, New York: Academic Press.

Larson, W. E., and W. O. Willis. 1957. Light, soil temperature, soil moisture and alfalfa-red clover distribution between corn rows and various spacings and row directions. *Agron. J.* 49: 422–26.

Letey, J., L. H. Stolzy, G. B. Blank, and O. R. Lunt. 1961. Effect of temperature on oxygen-diffusion rates and subsequent shoot growth and mineral content of two plant species. *Soil Sci.* 92: 314–21.

Letey, J., L. H. Stolzy, N. Valoras, and T. E. Szuszkiewicz. 1962. Influence of oxygen diffusion rate on sunflower growth at various soil and air temperatures. *Agron. J.* 54: 316–19.

Levesque, M., and J. W. Ketcheson. 1963. The influence of variety, soil temperature and phosphorus fertilizers on yield and phosphorus uptake by alfalfa. *Can. J. Plant Sci.* 43: 355–60.

Liebig, G. F., and H. D. Chapman. 1963. The effect of variable root temperatures on behavior of young navel orange trees in a greenhouse. *Proc. Amer. Soc. Hort. Sci.* 82: 204–9.

Ling, G. N. 1967. Effects of temperature on the state of water in the living cell. In *Thermobiology*, ed. A. H. Rose, pp. 5–24. London, New York: Academic Press.

Lingle, J. C., and R. M. Davis. 1959. The influence of soil temperature and phosphorus fertilization on the growth and mineral absorption of tomato seedlings. *Proc. Amer. Soc. Hort. Sci.* 73: 312–22.

Locascio, S. J., and G. F. Warren. 1960. Interaction of soil temperature and phosphorus on growth of tomatoes. *Proc. Amer. Soc. Hort. Sci.* 75: 601–10.

Lockard, R. G., and C. Grunwald. 1970. Grafting and gibberellin effects on the growth of tall and dwarf peas. *Plant Physiol.* 45: 160–62.

Lorenz, D. 1966. The effect of long wave reflectivity of natural surfaces on surface temperature measurements using radiometers. *J. Appl. Meteorol.* 5: 421–30.

McCalla, T. M., and T. J. Army. 1961. Stubble mulch farming. In *Advances in Agronomy*, ed. A. G. Norman, 13: 125–96. New York: Academic Press.

Mack, A. R. 1965. Effect of soil temperature and moisture on yield and nutrient uptake by barley. *Can. J. Soil Sci.* 45: 337–46.

Mack, A. R., and B. J. Finn. 1970. Differential response of timothy clonal lines and cultivars to soil temperature, moisture and fertility. *Can. J. Plant Sci.* 50: 295–305.

Mack, H. J., S. C. Fang, and S. B. Butts. 1964. Effects of soil temperature and phosphorus fertilization on snap beans and peas. *Proc. Amer. Soc. Hort. Sci.* 84: 332–38.

Martin, G. C., and G. E. Wilcox. 1963. Critical soil temperature for tomato plant growth. *Soil Sci. Soc. Amer. Proc.* 27: 565–67.

Mederski, H. J., and J. B. Jones, Jr. 1963. Effect of soil temperature on corn plant development and yield: I. Studies with a corn hybrid. *Soil Sci. Soc. Amer. Proc.* 27: 186–89.

Mitchell, K. J. 1964. Influence of light and temperature on the growth of ryegrass (*Lolium* spp.) III. Pattern and rate of tissue formation. *Physiol. Plant.* 7: 51–58.

Nielsen, K. F., and R. K. Cunningham. 1964. The effects of soil temperature, form and level of N on growth and chemical composition of Italian ryegrass. *Soil Sci. Soc. Amer. Proc.* 28: 213–18.

Nielsen, K. F., R. L. Halstead, A. J. MacLean, S. J. Bourget, and R. M. Holmes. 1961. The influence of soil temperature on the growth and mineral composition of corn, bromegrass and potatoes. *Soil Sci. Soc. Amer. Proc.* 25: 369–72.

Nielsen, K. F., R. L. Halstead, A. J. MacLean, R. M. Holmes, and S. J. Bourget. 1960a. The influence of soil temperature on the growth and mineral composition of oats. *Can. J. Soil Sci.* 40: 255–64.

——. 1960b. Effects of soil temperature on the growth and chemical composition of lucerne. *Proc. 8th Int. Grassland Congr.*, pp. 287–92.

Nielsen, K. F., and E. C. Humphries. 1966. Effects of root temperature on plant growth. *Soils Fert.* 29: 1–7.

Nowakowski, T. Z., R. K. Cunningham, and K. F. Nielsen. 1965. Nitrogen fractions and soluble carbohydrates in Italian ryegrass. I. Effects of soil temperature, form and level of nitrogen. *J. Sci. Food Agr.* 16: 124–34.

Odegbaro, O. A., and O. E. Smith. 1969. Effects of kinetin, salt concentration and temperature on germination and early seedling growth of *Lactuca sativa* L. *J. Amer. Soc. Hort. Sci.* 94: 167–70.

Oritani, T. 1963. The role of root in nitrogen metabolism in crop plants. *Crop Sci. Soc. Jap. Proc.* 31: 277–83.

Osborne, Daphne J. 1962. Effect of kinetin on protein and nucleic acid metabolism in Xanthium leaves during senescence. *Plant Physiol.* 37 : 595–602.

Owen, P. C. 1971. The effects of temperature on the growth and development of rice. *Field Crop Abstr.* 24: 1–8.

Parks, W. L., and W. B. Fisher. 1958. The influence of soil temperature and nitrogen on ryegrass growth and chemical composition. *Soil Sci. Soc. Amer. Proc.* 22: 257–59.

Parups, E. V., and K. F. Nielsen. 1960. The growth of tobacco at certain soil temperature and nutrient levels in the greenhouse. *Can. J. Plant Sci.* 40: 281–87.

Parups, E. V., K. F. Nielsen, and S. J. Bourget. 1960. The growth, nicotine and phosphorus content of tobacco grown at different soil temperature, moisture and phosphorus levels. *Can. J. Plant Sci.* 40: 516–23.

Paterson, D. R., D. E. Speights, and J. E. Larsen. 1970. Some effects of soil moisture and various mulch treatments on the growth and metabolism of sweet potato roots. *J. Amer. Soc. Hort. Sci.* 95: 42–45.

Pearson, R. W., L. F. Ratliff, and H. M. Taylor. 1970. Effect of soil temperature, strength and pH on cotton seedling root elongation. *Agron. J.* 62: 243–46.

Phillips, R. L., and M. J. Bukovac. 1967. Influence of root temperature on absorption of foliar applied radiophosphorus and radiocalcium. *Proc. Amer. Soc. Hort. Sci.* 90: 555–60.

Place, G. A., M. A. Siddique, and B. R. Wells. 1971. Effects of temperature and flooding on rice growing in saline and alkaline soil. *Agron. J.* 63: 62–66.

Power, J. F., D. L. Grunes, G. A. Reichman, and W. O. Willis. 1964. Soil temperature effects upon phosphorus availability. *Agron. J.* 56: 545–48.

——. 1970. Effect of soil temperature on rate of barley development and nutrition. *Agron. J.* 62: 567–71.

Power, J. F., D. L. Grunes, W. O. Willis, and G. A. Reichman. 1963. Soil temperature and phosphorus effects upon barley growth. *Agron. J.* 55: 389–92.

Puh, Y. S. 1960. Lysimeter investigations on the effect of soil temperature and moisture on the movement of cations and their uptake by plants. M.Sc. Thesis. University of New Hampshire.

Radke, J. F., and R. E. Bauer. 1969. Growth of sugarbeets as affected by root temperatures. Part I. Greenhouse studies. *Agron. J.* 61: 860–863.

Rahman, Abd el, and J. F. Bierhuizen. 1959. The effect of temperature and water supply on growth, transpiration and water requirement of tomatoes under controlled conditions. *Mededel. Landbouwh. Wageningen* 59(3): 1–14.

Raney, F. C., and Yoshiaki Mihara. 1967. Water and soil temperature. In *Irrigation of Agriculture Lands*, ed. R. M. Hagan, H. R. Haise, and T. W. Edminster, pp. 1024–36. Agronomy 11. Madison, Wis.: Amer. Soc. Agron.

Read, D. W. L., and R. Ashford. 1968. Effect of varying levels of soil and fertilizer phosphorus and soil temperature on growth and nutrient content of bromegrass and reed canarygrass. *Agron. J.* 60: 680–82.

Richards, S. J., R. M. Hagan, and T. M. McCalls. 1952. Soil temperature and plant growth. In *Agronomy 2, Soil Physical Conditions and Plant Growth*, ed. B. T. Shaw, pp. 303–480. New York: Academic Press.

Roberts, A. N., and A. L. Kenworthy. 1956. Growth and composition of the strawberry plant in relation to root temperature and intensity of nutrition. *Proc. Amer. Soc. Hort. Sci.* 68: 157–68.

Ruf, Robert H., Jr. 1963. Smooth-skin of Russett Burbank potatoes as influenced by soil temperature and moisture. *Amer. Potato J.* 40: 299–302.

Sato, K., and M. Ito. 1969. (Growth responses of orchardgrass and perennial ryegrass to air and soil temperatures.) *Crop Sci. Soc. Jap. Proc.* 38: 313–20 (English summary).

Shaw, W. H., and W. F. Buchele. 1957. The effect of the shape of the soil surface profile on soil temperature and moisture. *Iowa State Coll. J. Sci.* 32: 95–104.

Shtrausberg, D. V. 1958. The utilization of nutrients by plants in the polar region under various temperature conditions. *Fiziol. Rast.* 5: 228–34 (seen in *Soils Fert.* 21:6).

Simpson, K. 1965. The significance of effects of soil moisture and temperature on phosphorus uptake. *Tech. Bull. Min. Agr. and Fish.* 13: 19–29.

Slatyer, R. O. 1969. Physiological significance of internal water relations to crop yield. In *Physiological Aspects of Crop Yield*, ed. R. C. Dinauer, pp. 53–88. Madison, Wis.: Amer. Soc. Agron.

Smith, Dale. 1970. Yield and chemical composition of leaves and stems of alfalfa at intervals up the shoot. *J. Agr. Food Chem.* 18: 652–56.

Smith, D., and O. R. Jewiss. 1966. Effects of temperature and nitrogen supply on the growth of timothy (*Phleum pratense* L.) *Ann. Appl. Biol.* 58: 145–57.

Smoliak, S., and A. Johnston. 1968. Germination and early growth of grasses at four root-zone temperatures. *Can. J. Plant Sci.* 48: 119–27.

Stewart, B. A., and C. J. Whitfield. 1965. Effects of crop residue, soil temperature and sulfur on the growth of winter wheat. *Soil Sci. Soc. Amer. Proc.* 29: 752–55.

Street, H. E. 1966. The physiology of root growth. *Ann. Rev. Plant Physiol.* 17: 315–44.

Sutton, C. D. 1969. Effect of low soil temperature on phosphate nutrition of plants—a review. *J. Sci. Food Agr.* 20: 1–3.

Swanson, C. A., and D. R. Geiger. 1967. Time course of low temperature inhibition of sucrose translocation in sugar beets. *Plant Physiol.* 42: 751–56.

Varade, S. B., L. H. Stolzy, and J. Letey. 1970. Influence of temperature; light intensity and aeration on growth and root porosity of wheat, *Triticum aestivum. Agron. J.* 62: 505–7.

Varner, J. E., L. V. Balce, and R. C. Huang. 1963. Senescence of cotyledons of germinating peas. Influence of axis tissue. *Plant Physiol.* 38: 89–92.

Vinokur, R. L. 1963. Nitrogen metabolism in pumpkin plants depending on the temperature of the root habitat. *Soviet Plant Physiol.* 10: 334–38.

Waggoner, P. E., P. M. Miller, and H. C. De Roo. 1960. *Plastic Mulching: Principles and Benefits.* Conn. Agr. Exp. Sta. Bull. 634.

Walker, John M. 1969. One-degree increments in soil temperatures affect maize seedling behaviour. *Soil Sci. Soc. Amer. Proc.* 33: 729–36.

———. 1970. Effects of alternating versus constant soil temperatures on maize seedling growth. *Soil Sci. Soc. Amer. Proc.* 34: 889–92.

Wallace, A. 1958. Tree physiology studies at U.C.L.A. *U.C.L.A. Special Report,* no. 1: 84–103.

———. 1963. *Solute Uptake by Intact Plants.* Ann Arbor, Mich.: Edwards Bros.

———. 1970. Water use in a glasshouse by *Salsola kali* grown at different soil temperatures and at a limiting soil moisture. *Soil Sci.* 110: 146–49.

Wardlaw, Ian F. 1968. The control and pattern of movement of carbohydrates in plants. *Bot. Rev.* 34: 79–105.

Watschke, T. L., R. E. Schmidt, and R. E. Blaser, 1970. Responses of some Kentucky bluegrasses to high temperature and nitrogen fetility. *Crop Sci.* 10: 372–76.

Weatherly, P. E., and B. T. Watson. 1969. Some low temperature effects on sieve tube translocation in *Salix viminalis. Ann. Bot.* 33: 845–53.

Whiteman, P. C., T. A. Bull, and K. T. Glasziou. 1963. The physiology of sugarcane. VI. Effects of temperature, light and water on set germination and early growth of saccharum spp. *Aust. J. Biol. Sci.* 16: 416–28.

Whitfield, C. J., and D. E. Smika. 1971. Soil temperature and residue effects on growth components and nutrient uptake of four wheat varieties. *Agron. J.* 63: 297–300.

Whittle, C. M. 1964. Translocation and temperature. *Ann. Bot.* 28: 339–44.

Wierenga, P. J., R. M. Hagan, and E. J. Gregory. 1971. Effects of irrigation water temperature on soil temperature. *Agron. J.* 63: 33–36.

Wijk, W. R. van, W. E. Larson, and W. C. Burrows. 1959. Soil temperature and the early growth of corn from mulched and unmulched soils. *Soil Sci. Soc. Amer. Proc.* 23: 428–34.

Wijk, W. R. van. 1966. *Physics of the Plant Environment* 2nd ed. Amsterdam: North-Holland Publishing Co.

Williams, D. E., and J. Vlamis. 1962. Differential cation and anion absorption as affected by climate. *Plant Physiol.* 37: 198–202.

Willis, W. O., W. E. Larson, and D. Kirkham. 1957. Corn growth as affected by soil temperature and mulch. *Agron. J.* 49: 323–27.

Willis, W. O., J. F. Power, G. A. Reichman, and D. L. Grunes. 1963. Constant temperature water baths for plant growth experiments. *Agron. J.* 55: 200.

Woolley, D. G. 1963. Effects of nutrition, osmotic pressure and temperature of the nutrient solution on plant growth and chemical composition. 1. Spring wheat at the 4th to 6th leaf stage. *Can. J. Plant Sci.* 43: 44–50.

Yamaguchi, M., H. Timm, and A. R. Spurr. 1964. Effects of soil temperature on growth and nutrition of potato plants and tuberization, composition and periderm structure of tubers. *Proc. Amer. Soc. Hort. Sci.* 84: 412–23.

Younis, M. A., A. W. Pauli, H. L. Mitchell, and F. C. Stickler. 1965. Temperature and its interaction with light and moisture in nitrogen metabolism of corn (*Zea mays* L.) seedlings. *Crop Sci.* 5: 321–26.

13. Soil Atmosphere

Lewis H. Stolzy

SEVERAL centuries ago Heraclitus (500 B.C.) viewed the universe and concluded that everything is in a state of flux (Russell, 1969). The transient-state flux equation is probably the most meaningful relation for characterizing the various sources, sinks, capacities, and pathways in the environment. This equation is

$$dC/dt = D/CF \cdot d^2C/2x^2, \qquad [1]$$

where the change in concentration with time, dC/dt, is a function of D, the diffusion coefficient; CF, the capacity factor; and d^2C/x^2, the rate of change in concentration gradient with distance.

Oxygen is the one element in the atmosphere essential to most all forms of life. The atmosphere above the soil has an unlimited supply of O_2, which is being constantly supplied to the soil sink in order to support the various forms of organisms and roots living in the system. Carbon dioxide in turn is formed and released to the soil atmosphere by respiring cells. The air around the plant and over the soil surface is both the source and the sink for the O_2 and CO_2 gases used and produced by plants during respiration and photosynthesis.

The importance of the atmosphere above the soil as an infinite reservoir of O_2 and CO_2 for living organisms cannot be overemphasized. We depend upon the O_2 released by plants. Oxygen accounts for 0.09 % of the cosmic atoms and ranks third among the elements in abundance (Gilbert, 1964). The predominant element in the cosmos is hydrogen (86.68 %), and together with helium, it makes up 99.86 % of the total. The elements H, He, and O plus carbon and nitrogen make up 99.99 % of all the atoms. On the other hand, H, O, C, and N make up 99.2 % of the biosphere and are in the same, decreasing order of abundance as in the cosmos. Oxygen is the most abundant atom in the earth's crust (53.77 %), mainly because it reacts with other elements there, such as silicon and iron. Oxygen makes up 90 % of the volume of the earth's crust, but only

0.01 % of all oxygen is molecular oxygen in the atmosphere. A large exchange of O_2 between the hydrosphere and atmosphere takes place due to the exchange of gaseous oxygen in the atmosphere with dissolved oxygen in the ocean. Photosynthesis in the biosphere results in an atmospheric turnover of O_2 about every 5,400 years; other productions of molecular O_2 are negligible in comparison. The most important source of O_2 for land organisms is the atmosphere. The concentration in the atmosphere remains at about 21 % because of the great volume and mobility of the air medium. Therefore, most of the terrestrial environment is provided with a uniform and adequate supply of O_2.

At high altitudes and in some soils, a deficiency of O_2 may exist. As one reaches the higher elevations, although the percentage of O_2 in relation to the other gases remains at about 21%, the atmospheric pressure decreases, the concentration of oxygen (P_{O_2}) in the air decreases, the concentration gradient across membranes is reduced, and O_2 supply to respiring cells becomes limiting. Thus the O_2 percentage and rate of supply in the soil are of great importance to organisms growing in agricultural soils at low elevations.

I. COMPOSITION OF SOIL

A. Soil System

From a physical point of view soil is thought of as a three-phase system: solid, liquid, and gas. Soil variables are bulk density, water content, and gas content, which are often expressed in terms of mass and volume or ratios of the three phases. The solid phase is referred to as the matrix; it controls the form, quantity, and distribution of the other two.

Structural units in many soils determine to a large degree how water moves into and through the soil profile. The soil water in turn influences biological activity and soil gases (Currie, 1965; Greenwood, 1969; Stolzy and Van Gundy, 1968). The pore space in soil can be filled with gases when the soil is very dry, or with water when the soil is saturated or waterlogged. The ideal condition for roots is some ratio of water to air in the soil pores. McLaren and Skujins (1968) pointed out that gases in soil exist in three states: (1) a free state, filling the empty soil pores; (2) dissolved in the water phase; and (3) adsorbed on the solid phase. The quantity of a gas dissolved in the water phase depends on the type of gas, temperature, and salt concentration in the solute phase and the concentration of that gas in the gas phase. Some gases, such as CO_2, NH_3, and H_2S, become ionized in water and are more soluble than the two more abundant gases, O_2 and N_2 (Table 13.1). Collis-George (1959)

Table 13.1. Physical data on certain soil gases

Gas	Diameter (cm × 10^-8)	Mean free path (cm × 10^-6, 750 mm Hg)	Collision frequency (× 10^6/sec, at 20°C)	Average velocity (cm × 10^2/sec, STP)	Density (g/l, STP)	Thermal conductivity (× 10^-5 cal cm^-2 sec^-1 °C^-1 cm^-1, at 0°C)	Solubility (ml gas/ml solution at STP)
N_2	3.15–3.53[†]	8.50[†]	5070[†]	454[‡]	1.251[§]	5.7[‖]	0.016[§]
NH_3	2.97–3.08[†]	5.92[†]	9150[†]	583[†]	0.77126[§]	5.2[‖]	
N_2O					1.9804[§]	3.6[‖]	
CO_2	3.23–3.40[†]	5.56[†]	6120[†]	362[‡]	1.977[§]	3.4[‖]	0.878[§]
CH_4					0.7168[§]	7.3[‖]	0.033[§]
O_2	2.92–2.98[†]	9.05[†]	4430[†]	425[‡]	1.429[§]	5.73	0.031[§]
SO_2					2.9262[§]	2.0[‖]	39.374[§]
H_2S					1.5392[§]	3.0[‖]	

[†] Data from Weast, 1969.
[‡] Data from Dittmer and Grebe, 1958.
[§] Data from Gray, 1963.
[‖] Data from Condon and Odishaw, 1967.
[#] Data from Meites, 1963.

pointed out that equilibrium conditions among water, water vapor, O_2, and CO_2 in pore spaces are greatly affected by diurnal changes in temperature. He described the effects of temperature changes on the equilibria in the soil liquid and gaseous phases with respect to soil organisms.

Oxygen rapidly disappears not only in completely waterlogged soils but also in well-structured and aerated soils under conditions where saturated aggregates are present (Scott and Evans, 1955). By direct measurement, Greenwood and Goodman (1967) showed that O_2 concentrations in water-saturated aggregates can fall from those in air-saturated water to zero over a distance of 0.1 cm if the respiration rates are high. The distribution of water in a soil system has a great effect on the type and concentration of gases, especially O_2. Water release curves of a Yolo loam (Fig. 13.1) with the same texture but different aggregate sizes

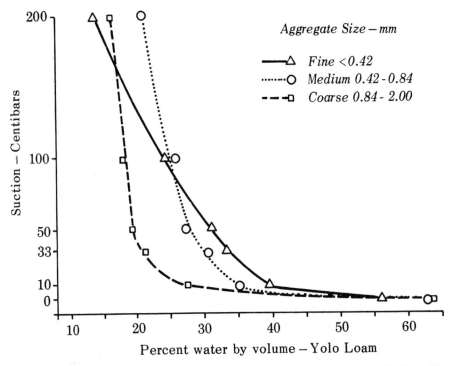

Fig. 13.1.   Water release curves of Krilium-treated Yolo silt loam aggregates in three different size ranges. (Stolzy and Kirkpatrick, data unpublished)

show that aggregation causes the soil to act like a sand, where most of the water is released at a very low suction value, while the fine material (not aggregated) behaves like a finer-textured soil. In a partially drained, well-aggregated soil the model proposed by Greenwood (1969) would very likely occur. That is, when a small amount of water is drained from a saturated soil, the soil consists of water-saturated regions interspersed with a few continuous gas channels to the soil surface. Because diffusivity of gases in water tends to be about 10^{-4} times that in air, the concentration gradients in the liquid phase will be large compared to those in the gas phase. So the O_2 concentration in water-saturated regions will decrease sharply over short distances, leaving zones in which there is no O_2, while gas-filled channels will be similar throughout. The fact that gases extracted from soil at some depths are usually only slightly lower in concentration than those in the air above the soil is evidence for the Greenwood model.

B. Gases in Soil Air

The principal gases of the soil atmosphere at a 6-inch depth are N_2, 79.2%; O_2, 20.6%; CO_2, 0.25%. For atmospheric air these percentages are N_2, 79.0%; O_2, 20.97%; and CO_2, 0.03% (Russell and Appleyard, 1915). These data show that soil air at a 6-inch depth is similar in composition to atmospheric air. However, the composition of soil gases depends on many conditions, and generalization from a few situations should not be used without measured values of the gas. In addition concentration gradients develop in the soil because roots and soil organisms consume O_2 and produce CO_2.

The increased need to recycle more and more water products through the soil has stimulated interest in other metabolically formed trace gases, e.g., methane, nitrous oxide, hydrogen sulfide, and ammonia. Microbial processes in surface soils can be involved in the utilization and formation of such gases. Physical properties of some soil gases are given in Table 13.1. This list does not consider volatile organic components, such as alcohols, acids, esters, aldehydes, and ketones. Cholodny (1953) studied the role of soil gaseous products from certain volatile organic fractions. The soil microflora of the rhizosphere converts these volatile organic components into organics that can be absorbed through the root system of plants. Thus, the volatile and the nonvolatile fractions of organic matter may be recycled through the intermediary action of soil microorganisms. The concentration of these volatile organic compounds in soils is ordinarily low, probably of the order of a few parts per million in total.

Oxygen and CO_2 are the only two soil gases measured and reported in most studies where soil gases are extracted from the soil and analyzed. Nitrogen gas in soil air may be reported, and in most cases it is the residual gas after CO_2 and O_2 have been removed in the analytical process; however, the residual gas includes trace gases. Data on trace gases in soil are scarce because of the lack of methods for sampling the gas in soil, the limited analytical techniques, and a minimum interest by researchers.

Petroleum microbiologists have assayed soil gas as an index of gas seepage from petroleum accumulations in the subsurface (Davis, 1967). Field methods were used to analyze soil gas for ethane, but later soil samples were taken into the laboratory where they were degassed with vacuum. Carbon dioxide and acidic volatile components were trapped in alkali; the other gaseous components were separated with the use of refrigerants. The lightest gas fraction consisted of hydrogen and methane, the next fraction was ethane and propane, and the heaviest fraction contained all other gaseous components. Horvitz (1939) and Rosaire (1939) believed that analysis of soil samples by desorption or extraction of hydrocarbons has many advantages over analyzing the interstitial soil gas obtained from boreholes. Sokolov (1936) and Davis (1967) reported analyses of gases from boreholes with a sensitivity of 10–100 ppb. Some of the analytical methods used by petroleum microbiologists in soil gas analysis include a combustion method, mass spectrometry, and gas chromatography with ether, hydrogen flame, and radiation detectors. Davis (1967) reported studies in which microbially produced ethane, propane, and butane collected over estuary mud sediments amounted to 0.1 %–0.18 % by volume of the total gas phase. According to Harrison and Aiyer (1913, 1916), the gases of swamp rice soils and waterlogged paddy soils are CH_4 with small amounts of CO_2, H_2, and N_2. In most cases only a trace of CH_4 was detected; however, in one waterlogged arable soil, the soil gas was CH_4, 70.7%; CO_2, 27.3%; and N_2, 20% (Skerman and MacRae, 1957). Methane is a typical gas produced by the anaerobic fermentation of organic matter under neutral or slightly alkaline conditions. Partial waterlogging and high organic matter content was used to explain the evolution of CH_4 in such locations.

Restricted aeration in soil horizons provides suitable conditions for denitrification where N_2 or N_2O soil gases are formed. Because this process can cause a major loss of nitrates from soil it has interested researchers. Most of the reported research used laboratory procedures to determine the rates of release of these gases; these studies are similar to soil respiration studies where CO_2 evolution or O_2 uptake from soil samples is measured in the laboratory (Domsch, 1962). These studies give no indication of composition of soil gases in the field.

Burford and Millington (1968) sampled soil gases at depths of 5, 10, 15, 30, 60, and 90 cm in a Urrbrae fine sandy loam. Nitrous oxide was detected at all sampling depths. The levels in the soil air were highest after heavy rainfall. The gas distribution in the profile indicated two main sources in the profile—one at about 10 cm in the A horizon and the other at between 30 and 60 cm in the B horizon. Denitrification occurred in the surface soil only for a few days. Nitrous oxide was detected in the subsoil continuously for three months. The O_2 content of the soil air never fell below 17%, and the maximum CO_2 concentration was 0.5%. There was an observed concentration of N_2 gas of up to 3.5% higher than atmospheric concentrations. The maximum level of N_2O detected was 120 ppm (v/v).

Composition of gas samples taken at depths of 15, 30, and 60 cm in antarctic soil was reported by Cameron and Conrow (1969). In general, the percentages of O_2 and N_2 were only slightly lower in the soil gas than in the atmosphere above the soil. In the Taylor dry valley there was no indication of a buildup of soil CO_2; however, in another area with abundant soil microflora the CO_2 concentration was 2 to 4 times higher than in the atmosphere. Another gas measured was argon, which was about 1% in the air and soil gas. At 60-cm soil depth CH_4 measured 0.21%.

Many studies have been made of the volume of O_2 and CO_2 in soil gases with respect to time, soil depth, soil texture, soil water, climate, and soil management (Wollny, 1886; Russell and Appleyard, 1915; Romell, 1922; Boynton and Reuther, 1939; Furr and Aldrich, 1943). Russell and Appleyard (1915), conducting their study at Rothamsted Experimental Station, England, examined the effect of many of these factors on soil gas composition. The results of Boynton and Reuther (1938) and others indicate that wide variation may be expected in the CO_2 and O_2 content of soil gases within the pores of different soils, depending upon soil conditions. These data showed that the O_2 content in fine-textured soil was normally low during late fall, winter, and early spring (less than 1%), whereas CO_2 content fluctuated over narrower limits and normally reached a maximum (12%) in summer, when the soil temperature and water are favorable for biological activity. The maximum CO_2 content and minimum O_2 content did not occur simultaneously.

J. R. Furr (unpublished data) investigated the O_2 and CO_2 content of soil gases in relation to citrus and avocado root rot (Figs. 13.2 and 13.3) to find out if soil aeration is a factor in the development of a malady called "overirrigation injury." In soils with rapid internal drainage, the O_2 content of the soil gas was rarely less than 15%, and in soils where internal drainage was relatively slow the O_2 content was rarely less than 10%. Soil suction data that I obtained from an orchard with slow internal drainage (Fig. 13.4) show the type of soil condition Furr was describing. The CO_2 content of soil gases usually did not exceed 5% or 6%. Furr took gas samples in several areas where avocados (*Persea americana*) planted on steep terraced hillsides were affected by root rot and decline of the tops. The design of the terraces and the presence of heavy cover

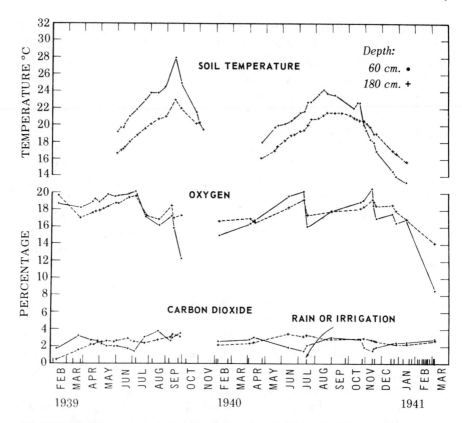

Fig. 13.2.   Soil temperature and volume percentage of O_2 and CO_2 in soil gas samples taken at two depths with time in a dry plot of an orange orchard on Yolo loam soil. (Furr, unpublished data)

from plants caused slow runoff of rainwater. The percentage of O_2 in soil gas samples was as low as 0.1% to 0.2% for 5 days, and for long periods of time during winter months it was around 1% and 2%. The CO_2 content in this same area was as high as 24.2% and was 12% to 16% in April to June.

Studies of gas composition of several soils packed in cylinders under various conditions without plants have been reported (Epstein and Kohnke, 1957; Kristensen and Enoch, 1964; Yamaguchi *et al.*, 1967; and Thomas *et al.*, 1968). Addition of organic materials to the soil and water content has considerable effect on the percentage of O_2 and CO_2 in the soil gases.

A large gas sample may include gases from the soil surface or from very large pore spaces and may not be representative of the sampling site. A classic study by Hack (1956) showed the need for a method of taking only very small soil gas samples. At the same depth in soil beds, he obtained a 0.1-ml gas sample and compared it to a 10-ml gas sample;

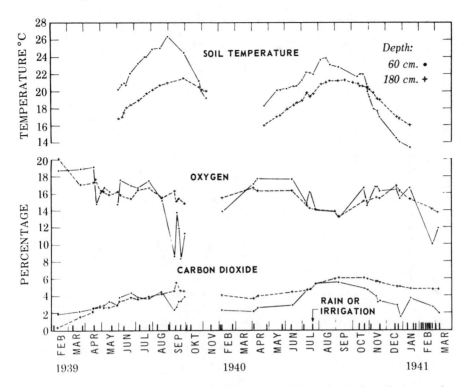

Fig. 13.3. Soil temperature and volume percentage of O_2 and CO_2 in soil gas samples taken at two depths with time in a wet plot of an orange orchard on Yolo loam soil. (Furr, unpublished data)

in most cases the 0.1-ml sample had lower O_2 and higher CO_2. The difference between contents of the large and small samples was greatest when the soil was compacted and wet.

II. MECHANISMS OF SOIL GAS EXCHANGE

The upper surface of the soil is in contact with the atmosphere, and the soil's lower boundary is a water table or some material impervious to gases. Gas exchange within these limits involves both mass-flow and diffusion mechanisms.

A. Mass-Flow

Four factors are considered in mass-flow: (1) soil temperature changes, (2) changes in the volume of pore space due to changes in water content, (3) barometric pressure changes, and (4) wind (Keen, 1931; Romell,

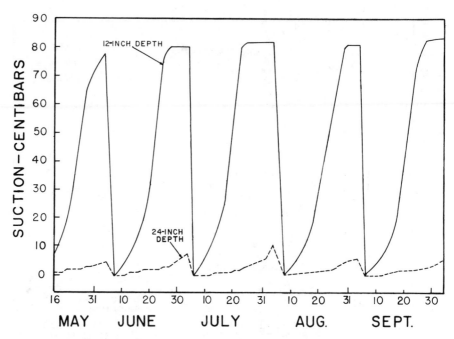

Fig. 13.4. Soil suction value at 12-inch and 24-inch soil depths during irrigation season in an orange orchard

1922, 1935). Russell (1952) concluded that the interchange of gases between the soil and the atmosphere as the result of mass-flow in response to a gradient of the total pressure is of minor significance. Kimball and Lemon (1971) more recently reported the effects of air turbulence upon the exchange of gases between a soil and the atmosphere. They used a specially designed vapor exchange meter to measure the rate of evaporation of liquid heptane from a porous stainless steel plate buried in a porous medium. Natural air turbulence, as indicated by both mean wind speed and root mean square (estimate of standard deviation) air pressure fluctuations, significantly affected the heptane evaporation rates in straw and coarse gravel but had a decreased effect in finer-textured media. These results led them to affirm Romell's (1922) conclusion that diffusion is the main process contributing to soil aeration.

B. Diffusion

Diffusion is a direct consequence of the random thermal motion of molecules. At 25C, oxygen molecules travel at a velocity of about 1,600 miles per hour with a collision frequency of 4,430 per second and a mean free path of 905 Å (Table 13.1). The various soil gases have different molecular sizes and weights and will diffuse at different rates. As a result of this random motion, more gas molecules tend to move from points of high concentration to points of low concentration than vice versa. This net movement of molecular species from points of high to points of low concentration is generally directly proportional to the concentration gradient, the cross-sectional area available for the diffusion, and the time. Because the soil gases tend to contain more CO_2 and less O_2 than the atmosphere, CO_2 diffuses out of the soil and O_2 diffuses into the soil.

For steady-state conditions, equation [1] has the form:

$$Q/At = D(\Delta C/\Delta X), \qquad [2]$$

where Q is the quantity diffusing, A is the area, t is time, and D is the diffusion coefficient of the soil gas. Several researchers have attempted to measure the diffusion coefficient. Keen (1931) considered a soil layer 1 cm thick with a mixture of two gases, O_2 and CO_2, on each side, such that the P_{CO_2} on one side exceeded the P_{CO_2} on the other by 1 mm Hg and similarly for P_{O_2}, so that the total pressure on each side was the same. Carbon dioxide diffused through the soil to the side where the partial pressure was lower, and O_2 diffused in the opposite direction. The magnitude of the effect was measured by the number of cm^3 of one gas passing in one second through 1 cm^2 of the soil layer 1 cm thick. The measured value of the diffusion constant was used for characterizing variations in soil porosity. Buckingham (1904) in his studies used a simplified system. A stream of air was maintained over one surface of a soil, and CO_2 over the opposite surface. The change in composition of the two gases was measured, and this showed that CO_2 and air did not diffuse at the same rate. Buckingham used the term *diffusion constant* to designate the rate of flow of gases through the soil pore space as a result of kinetic movements and showed a definite correlation between gas-filled pore space and the diffusion constant:

$$D = RS^2, \qquad [3]$$

where D is the diffusion constant, S is the gas-filled pore space, R is the diffusion coefficient, which was calculated to be 2.16×10^{-4} cm^2/sec. Porter *et al.* (1960) discussed some of the factors that cause the diffusion coefficient in a porous medium to be less than that in bulk air or water. As a result of the random arrangement of soil particles, the paths available for diffusion are somewhat tortuous. Smith and Brown (1933) tried to measure diffusion in moist, undisturbed soil samples in the laboratory and found that they could not accurately determine the rate of diffusion because of complications resulting from production of CO_2 in the soil. Wesseling (1962) derived theoretical solutions for the steady-state diffusion of CO_2 through soil.

Penman (1940a, 1940b) carried out a series of experiments with a porous material under controlled conditions in the laboratory. He found a linear relationship between D and S. The relationship was $D/D_o = 0.66\,S$, where D is the coefficient of diffusion through the material having a pore space S, and D_o is the diffusion coefficient through free air in the apparatus used, that is, where $S = 1$. When Penman's (1940a and 1940b) curve of diffusion versus porosity was extrapolated to the origin, it had a slope of 0.66 for porosity of 35% to 70% (Wesseling, 1962). Millington (1959) and Marshall (1959) suggested that $D/D_o = S^{4/3}$ and $D/D_o = S^{3/2}$, respectively; while Currie (1960) thought that the relationship had a shape factor involved where $D/D_o = \gamma S^u$. Both γ and u are functions of the material. The lack of homogeneity in the soil causes large deviations from the simple equations used to described homogenous soil. There are various data from Blake and Page (1948), Baver and Farnsworth (1940), and Taylor (1949) in which the diffusion all but stops at porosities of 10% to 15%. In Buckingham's experiments (1904), diffusion became practically zero at air content of 15%. Wyckoff and Botset (1936) in their experiments found that the hydrodynamic permeability was also practically zero if the air-filled pore space was less than 10% by volume. As a first approximation, one can assume that 10% by volume of air-filled pores is the lowest value at which air can be exchanged in soil (Wesseling and Wijk, 1957).

Both Penman (1940a, 1940b) and Hagan (1941) examined gaseous diffusion using carbon disulfide, which has several advantages by virtue of its volatile nature. Blake and Page (1948) used vapors of liquid carbon disulfide on soils in the field. Their data indicated a straight-line relationship between diffusion rates and porosity. They showed that the characteristics of the curve obtained by plotting diffusion rates against porosity may vary among different soils. Diffusion may approach zero in some soils with a porosity above zero. They suggested that this may result from the blocking of some of the air pore space by water films.

Munnecke *et al.* (1969) reported diffusion of methyl bromide in field soil for the control of *Armillaria mellea*. They determined the concentrations of methyl bromide in soil at different depths by withdrawing a small sample through buried stainless steel tubing and measuring the concentration of methyl bromide in the sample with a gas chromatograph. One type of study involved placing four pounds of solid methyl bromide at 100- to 140-cm soil depths and repacking the soil above it (Fig. 13.5). This treatment was applied at two locations in clay loam soils with similar physical properties but different soil water profiles. In the profile where the sudangrass used the water at around the 60-cm depth, a concentration of methyl bromide in parts per million was present at all sample depths. The fact that none was detected near the

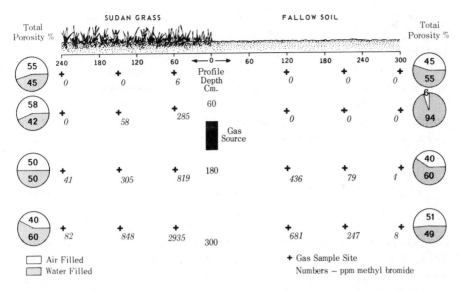

Fig. 13.5. Concentration of methyl bromide from a point source in the profile in parts per million at different locations in field plots with different gas-filled porosites. Kobezen, Munnecke, and Stolzy, unpublished data)

surface at 140 and 240 cm along the soil surface was due to cracks developed in the fine-textured soil. In the fallow soil, with a high water content and low air-filled porosity, no gas was detected at the 30- and 60-cm soil depth. A low soil water content has proved to be very important in fumigation of loams and other finer-textured soils.

III. MEASURING SOIL AERATION

A. Indices of Soil Aeration

Various indices have been used to describe soil aeration: (1) gas porosity, (2) gas composition of open pore space, (3) diffusion in the gas phase, (4) water saturation, (5) air permeability, (6) oxidation-reduction potential, and (7) diffusion through the gas-liquid-solid medium surrounding a cylindrical platinum electrode. Other chemical and biological methods are considered in the articles on soil aeration by Russell (1952), Baver (1956), Domsch (1962), and Black (1968).

B. Methods of Measurement

Letey (1965) in a recent report described in detail many of the methods for measuring soil aeration. Another fine review, by Grable (1966), described the interrelations of soil properties and soil aeration. The oxidation-reduction potential was described by Black (1968) in general and by Ponnamperuma (1965) and Ponnamperuma *et al.* (1966) for flooded soils. Stolzy and Letey (1964a, 1964b), Letey and Stolzy (1964), and McIntyre (1970) have reviewed diffusion of O^2 through the gas-liquid-solid medium surrounding a cylindrical platinum electrode. Some of the authors' ideas and interrelations of the measurements of soil aeration will be considered in the next section.

IV. ROOT RESPONSES TO THE SOIL ATMOSPHERE

Plants vary in their responses to soil physical conditions or changes in conditions that cause anaerobic or microaerobic soil environments. These soil conditions are described as poor or deficient aeration due to a number of related factors in the soil, but most often to excess water in the pore spaces. Kramer (1965, 1969) has described some of the plant symptoms resulting from deficient aeration as wilting, yellowing of leaves, reduction in growth, and death. Other plant responses include changes in mineral uptake (Hammond *et al.*, 1955; Letey *et al.*, 1961; Stolzy *et al.*, 1963), changes in field crop yields (Erickson and Van Doren, 1964; Baver, 1956), ozone damage (Stolzy *et al.*, 1961b, 1964), and wheat seed set (Campbell *et al.*, 1969).

This discussion is mainly concerned with root studies in soils, with some reference to pot culture results. Pearson (1965) concluded that the conditions encountered by growing roots under field conditions are seldom, if ever, optimum for the development of an extensive and effective root system. Consequently, shallow rooting and ineffective utilization of subsoil water are common problems around the world. He pointed out that requirements for several environmental factors have been defined for root growth of a few species. However, little is known of interactions among different soil factors or of the variation in requirements among species and varieties. The effects of soil physical resistance and oxygen supply to roots are two such physical factors. Both of these factors operate together and separately to reduce the size and depth of a root system.

A. Oxygen

1. In Soil Gas

The existence of critical oxygen concentration for root activity in orchards during spring months was established by Boynton and Reuther

(1939) and Compton and Boynton (1944). This is the time of year for trees to form new flushes of root growth. Other work by Boynton *et al.* (1938) on orchard soils gave positive correlation between results from field studies and those conducted under controlled conditions. They recognized four levels of root activity: (1) the subsistence level was 0.1 % to 3.0 % O_2, and below 1 % the root lost weight; (2) O_2 content between 5 % and 10 % was necessary for growth of existing root tips; (3) 12 % O_2 was required for root initiation; (4) below 15 % O_2 there was a progressive decrease in absorption of minerals.

Studies by Furr and Aldrich (1943) of date palms (*Phoenix dactylifera*) in California showed that a continuous O_2 level of 0.4 % to 2.0 % for three weeks did not result in measurable injury. Other studies by Furr (unpublished data) on orange (*Citrus sinensis*) and avocado trees were referred to earlier (Figs. 13.2 and 13.3). In orange orchards a relatively high O_2 concentration and short period of low aeration had no effect on root rot. However, avocado orchards with very low O_2 concentrations had considerable root rot. In controlled studies by Furr, lemon (*Citrus limonia*) roots were not inhibited above 4 % O_2 and were not killed within a two-week period at 1 % O_2.

Root growth of a large number of plant species was part of a study by Cannon (1925). He conducted experiments under controlled conditions in sand and soil cultures for short periods of time. One of his major conclusions was that root growth at different oxygen levels is strongly controlled by temperature. Root growth was inhibited at a 3 % O_2 concentration between 18C and 30C. Normal root growth occurred at 10 % O_2 at 18C, but at 30C root growth was reduced. He attributed this relationship to a decrease in the solubility of O_2 in the soil solution with increasing temperature. The relationship is more likely due to the increase in the respiration demands of the root for O_2. Cannon found differences in the responses of various plant species. Roots of many species maintained a slow growth rate at as low as 0.5 % O_2 for a limited time. All species had some growth at 2 % O_2 if the rate of supply of O_2 was maintained.

Huck (1970) measured the elongation rates of cotton (*Gossypium* spp.) and soybean (*Glycine max*) taproots while the O_2 content of a gas stream passing through the soil surrounding the roots was varied. Root elongation rates were about the same for 10 % and 21 % O_2. Elongation ceased within 2 to 3 minutes after the system was purged of O_2 and then returned to normal shortly after 21 % O_2 was in the soil system. Anaerobiosis for 3 hours killed the primary taproots of cotton, and 5 hours killed the soybean roots. Oxygen levels from 2 % to 5 % resulted in an initial reduction in the rate of taproot extension.

Data on root growth as a function of O_2 concentration over the soil surface are shown in Figure 13.6 (Stolzy *et al.*, 1961a). The extension of

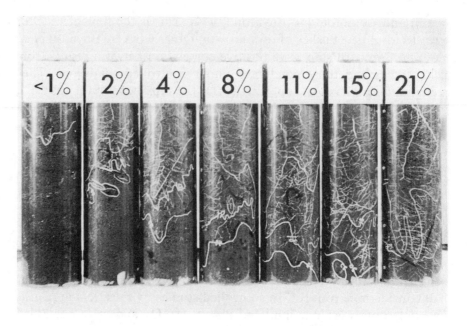

Fig. 13.6. Distribution of snapdragon (*Antirrhinum majus*) roots in columns of a silt loam soil. The numbers in the figure refer to O_2 percentage maintained in the gas at the top of the cylinders. The white markings indicated root penetrations at different times. (Stolzy *et al.*, 1961a)

roots of established plants was very small in an O_2 treatment of less than 1 %. The 2 % O_2 treatment had a small amount of growth in two replications but very little in two other replications.

Russell (1952) indicated that the use of O_2 concentrations in soil gases as a means of characterizing soil aeration for plant growth has not been satisfactory. He thought that the rate at which O_2 can be supplied to the root may be more important. Wiegand and Lemon (1958) attempted to bridge the gap between the O_2 requirement of roots measured in the laboratory and the empiricism that existed in field studies of soil aeration. They sought, through a theoretical approach, to combine field measurement of the O_2 supply in the gaseous and liquid phases of the soil and known quantitative demands by plant roots for O_2 into a single expression derived from Fick's law for radial diffusion. The concentration of O_2 at the root surface was the soil aeration parameter determined. They concluded that at field capacity the concentration of oxygen at certain root surfaces is suboptimal for normal root respiration in a clay soil but optimum in a fine sandy loam. Also, the apparent length of the diffusion path in the liquid phase about plant roots is more often a factor limiting normal root respiration than the gaseous composition in the soil pores.

2. Diffusion through the Gas-Liquid-Solid Media

A method for measuring the rate of O_2 diffusion to a platinum wire electrode inserted into the soil was described by Lemon and Erickson (1952). It was expected that the wire would be in an environment similar to that of a plant root. A high oxygen diffusion rate (ODR) would indicate that a root in the same position could receive O_2 at a high rate through diffusion. Conversely, a low ODR would indicate that the O_2 supply to the root would be received at a relatively low rate. McIntyre's (1970) review was concerned with the fundamental principles underlying the operation of the electrode in soil and how those principles affect application of the method to soil-plant studies. Letey and Stolzy (1964) described the theory and procedure, while Birkle *et al.* (1964), Rickman *et al.* (1968), and Rickman (1968) considered various factors related to the method of measuring ODR.

Past research on O_2 and plant growth has emphasized measuring the minimum demands of root systems upon external soil environments for O_2. More recently, the rapid internal diffusion of O_2 from foliage to the roots has been demonstrated by means of $^{15}O_2$ tracing (Evans and Ebert, 1960; Barber *et al.*, 1962). Recent theoretical studies by Luxmoore *et al.* (1970a, 1970b, 1970c, 1970d) have combined O_2 diffusion in the soil and within the plant root into a predictive model. In maize (*Zea mays*) the supply of O_2 by diffusion from the atmosphere via gas spaces that occur between cells within the plant (plant aeration) accounted for about 30 % of the respired O_2. Luxmoore and Stolzy (1972) predicted that O_2 concentration within the root at 5 cm from the root tip has only a small effect on the O_2 concentration at the root tip, whereas an increase in soil O_2 concentration (0.5 %–15 %) significantly increases the root tip concentration. A related study by Amoore (1961) showed the dependence of mitosis and respiration in pea (*Pisum* sp.) root tips upon oxygen concentrations at the surface root tip. Below 0.005 % O_2 mitosis was completely stopped. Between 0.001 % and 0.02 % O_2 cells initially in mitosis completed division, but no more cells started dividing. Between 0.05 % and 0.2 % O_2 cells initially in interphase entered division but did not finish; while above 0.5 % all cells finished division within four hours. Rate of O_2 uptake was half maximum at about 10 % O_2. He concluded that the respiration of root tips was limited by slow diffusion of O_2 through the tissue. Rate of O_2 uptake by excised barley roots was half maximum at 0.5 % to 0.8 % O_2, while potassium absorption was at half maximum at 2 % (P. C. Jackson, personal communication). Such studies indicate that the tip is the first part of the root to respond to lower O_2 supply and that distinction between rate processes as well as survival should be made.

In two reviews by Stolzy and Letey (1964a, 1964b), it was found that different investigators are in substantial agreement on the relationship between ODR and root growth. The critical ODR value of soils in which roots of many plants will not grow is $0.20\mu g$ cm^{-2} min^{-1}. Oxygen diffusion rates between $0.20\mu g$ and $0.30\mu g$ cm^{-2} min^{-1} retard root growth. However, in a review by Stolzy (1971) on critical ODR for root growth of members of the grass family, it was found that grasses in general had a lower rate of O_2 requirement than other plant species. The order of tolerance to poor aeration for some species are in the following order: rice > corn > bermudagrass > barley > Newport bluegrass. These results were determined independently of the O_2 concentration in gas-filled pore spaces in the soil. The differences in tolerance to lower O_2 conditions could be due partly to plant aeration (Luxmoore *et al.*, 1970c, 1970d).

Letey and Stolzy (1967) studied the limiting distances between a root and the gas phase for adequate O_2 supply. Using a correlation between adequate plant growth and an ODR of $0.20\mu g$ to $0.40\mu g$ cm^{-2} min^{-1}, they obtained a relationship between water film thickness and adequate O_2 for various soil porosities. At a soil porosity of 40% the limiting water film thicknesses were in the range of 0.01 to .025 cm. It is a mistake to compare data obtained by bubbling gases of different O_2 concentrations through solution cultures in order to determine the O_2 level that is limiting for root growth, since bubbling gases with relatively low O_2 content may cause no reduction in plant growth. If, for example, the solution is in equilibrium with 2% O_2, a diffusion layer around the root of 0.0007 cm is necessary to lower the ODR to $0.20\mu g$ cm^{-2} min^{-1}.

Cannon (1925) attributed the necessity for higher soil O_2 concentrations to maintain normal root growth at higher temperatures to the decreased solubility of O_2 in the soil solution. Solubility, however, is only one of the factors determining the rate of O_2 supply to the root surface. Under natural conditions O_2 diffuses from the external atmosphere through the gas-filled pore spaces, dissolves in the water films surrounding the root, and diffuses through the water to the root surface.

Although the solubility of O_2 decreases with increasing temperature, the diffusion coefficient through both gas and liquid increases. Diffusion of O_2 through water increases in the range of 3% to 4% per deg C (Carlson, 1911; Millington, 1955; Lammann and Jessen, 1929), and the solubility decreases approximately 1.6% per deg C, indicating a net increase in the rate of O_2 supply with increased temperature. The respiratory consumption of O_2 greatly increases with a rise in temperature, and a Q_{10} of 2 is characteristic; i.e., the respiration rate doubles with every 10C rise in temperature. For a given O_2 concentration in soil pore space, the O_2 concentration at the root surface will be lower at hgiher temperatures since the increase in oxygen consumption exceeds the increase in rate of supply (Luxmoore and Stolzy, 1972). To maintain maximum respiration, the O_2 concentration in a soil pore must increase more than proportionately to the increase in respiration rate with increased temperature.

B. Carbon Dioxide

According to Woolley (1965), there are many contradictory reports concerning the levels of CO_2 that limit root growth. Generally, excess CO_2 is not believed to limit respiration so much as O_2 deficiency does. Some studies show that CO_2 concentrations higher than 10% severely limit metabolism, while others show that CO_2 stimulates root growth (Woolley, 1965). Others show that concentrations of 20% do not harm cotton and soybeans (Kramar, 1969). Labanauskas *et al.* (1971) showed that 12% CO_2 in the root zone for 60 days significantly reduced root weight and increased top growth. Intermittent CO_2 in the root zone for the same period had no effect on plants. It appears that the usual CO_2 concentration in the soil gases is not high enough to cause injury, but O_2 is often low enough to be inhibitory (Kramer, 1969).

C. Flooded and Waterlogged Soil

When water fills soil pore spaces, it not only displaces the soil gases but also obstructs gaseous diffusion. Oxygen in wet soils is limited by the small amount of O_2 dissolved in water and by the extremely slow rate of diffusion through such soils. Just as O_2 cannot enter the soil, other gases are prevented from escaping, until they build up pressures and bubble out to the surface. Carbon dioxide is one of these gases that are of considerable importance to both plant growth and soil chemistry. According to Ponnamperuma *et al.* (1966), bacteria can produce as much as 10 to 40 tons of CO_2 per hectare in 28 days. In recently flooded soil, more than 50% of all gases dissolved in the soil water was CO_2 (Russell and Appleyard, 1915). Most plant roots are unable to tolerate such soil conditions. After the O_2 has been consumed, organisms utilize soil components such as nitrate, manganic oxides, ferric oxide, sulfate, and certain organic products. The stepwise reduction of various oxidized materials in a waterlogged soil, beginning with CO_2, has been suggested by Ponnamperuma *et al.* (1966). Oxidation-reduction potentials have the advantage of providing a measure of the intensity of reduction in soils containing no dissolved O_2 and can be useful in characterizing aeration conditions where ODR values have no meaning, such as water-logged soils (Black, 1968). However, Black (1968) indicated there is no evidence that the oxidation-reduction potential of soil affects plants directly. It seems that plants respond to the differences in chemical environment with which the potentials are associated. In a review of crop responses at excessively high soil water levels, van't Woudt and Hagan (1957) indicated a very limited number of cases where a positive relationship was established between crop production and oxidation-reduction potentials.

In a pot experiment with tobacco (*Nicotiana*), Kramer and Jackson (1954) studied the extent of plant damage in flooded soils and found that artificial aeration reduced plant injury but did not eliminate it. Part of the plant damage was due to chemical factors or toxic gases produced in these soils. Van't Woudt and Hagan (1957) pointed out several studies of flooding effects on crops but showed very little data on roots. One can assume that under the conditions reported, roots were the first to be adversely affected.

Williamson (1970) reported on flooding effects in soil and comparisons related to lack of O_2 on tobacco roots. Both treatments for 24 hours prevented further root growth and after 48 hours prevented root death. Stolzy *et al.* (1965, 1967) studied the effects of saturated soil and soil aeration on root decay of oranges and avocados. In citrus seedlings *Phytophthora parasitica* and *Phytophthora citrophthora* root rot increased with increased duration of flooding, and the duration was more important than the frequency of saturation for root decay. However, little or no decay of roots occurred unless the soil was saturated. Poor soil aeration prevented growth and regeneration of roots. In the avocados more root decay occurred in saturated soils, and oxygen supply to roots was the most significant factor associated with this decay. Plants growing in soils with an ODR of $0.17\mu g \ cm^{-2} \ min^{-1}$ and less had 44% to 100% of their root systems in a state of decay.

Crop plants growing in soil with water tables at some depth in the profile or in soils saturated for a short period of time due to rainfall or irrigation will be affected in many ways, but in most cases the plants will continue to survive. In waterlogged soils, certain crop plants such as rice produce better than in a well-drained soil, but most of the cultured plants will not survive under these conditions.

Under prolonged flooded conditions, in pot studies, corn and sunflower plants showed less severe damage than tomato and barley plants (Yu *et al.*, 1969). Of two varieties of wheat, Inia and Pato (dwarf high-yield strains developed in Mexico), the Pato variety tolerated the flooded soil better than the Inia. The corn, sunflower, and Pato wheat plants showed higher root porosities after being in flooded soils than the other plants; this was suggested as the reason for their toleration of flooding. A field study of irrigated Inia wheat near Ciudad Obregon, Sonora, Mexico, involved 4-day flooding treatments at 15, 30, and 45 days after planting (Luxmoore *et al.*, 1971). Neither the root porosities nor the yields of these flooded plots showed any significant differences. The differences between these pot and field studies may be attributed to the portion of water-filled pores. In pot studies flooding fills most of the pore space, whereas flooding in the field may saturate only a few centimeters of surface soil because of surface sealing and swelling phenomena.

References

Amoore, J. E. 1961. Dependence of mitosis and respiration in roots upon oxygen tension. *Proc. Roy. Soc.* (London), ser. B, 154: 109–29.

Barber, D. A., M. Ebert, and N. T. S. Evans. 1962. The movement of 15O$_2$ through barley and rice plants. *J. Exp. Bot.* 13: 397–403.

Baver, L. D. 1956. *Soil Physics.* 3rd ed. New York: John Wiley & Sons.

Baver, L. D., and R. B. Farnsworth. 1940. Soil structure effects on the growth of sugarbeets. *Soil Sci. Soc. Amer. Proc.* 5: 45–48.

Birkle, D. E., J. Letey, L. H. Stolzy, and T. E. Szuszkiewicz. 1964. Measurement of oxygen diffusion rates with the platinum micro-electrode. II. Factors influencing the measurement. *Hilgardia* 35: 555–66.

Black, C. A. 1968. *Soil-Plant Relationships*, pp. 153–97. New York. John Wiley & Sons.

Blake, G. R., and J. B. Page. 1948. Direct measurement of gaseous diffusion in soils. *Soil Sci. Soc. Amer. Proc.* 13: 37–42.

Boynton, D., J. I. DeVilliers, and W. Reuther, 1938. Are there different critical oxygen levels for the different phases of root activity? *Science* 88: 569–70.

Boynton, D., and W. Reuther. 1938. A way of sampling soil gases in dense subsoils and some of its advantages and limitations. *Soil Sci. Soc. Amer. Proc.* 3: 37–42.

———. 1939. Seasonal variation of oxygen and carbon dioxide in three different orchard soils during 1938 and its possible significance. *Proc. Amer. Soc. Hort. Sci.* 36: 1–6.

Buckingham, E. 1904. Contribution to our knowledge of the aeration of soils. *U.S. Dept. Agr., Bureau of Soils, Bull.* 25: 1–52.

Burford, J. R., and R. J. Millington. 1968. Nitrous oxide in the atmosphere of a red-brown earth. *Trans. 9th Int. Congr. Soil Sci.*, 2: 505–12.

Cameron, R. E., and H. P. Conrow. 1969. Antarctic dry valley soil microbial incubation and gas composition. *Antarctic J. U.S.* 5: 28–33.

Campbell, C. A., D. S. McBean, and D. G. Green. 1969. Influence of moisture stress, relative humidity, and oxygen diffusion rate on seed set and yield of wheat. *Can. J. Plant Sci.* 49: 29–37.

Cannon, W. A. 1925. Physiological features of roots with especial reference to the relation of roots to the aeration of soil. *Carnegie Inst., Wash., Pub.* 368: 1–168.

Carlson, T. 1911. The diffusion of oxygen in water. *J. Amer. Chem. Soc.* 33: 1027–32.

Cholodny, N. 1953. Soil gases and their biological significance. *Priroda* 42: 37–47.

Collis-George, N. 1959. The physical environment of soil animals. *Ecology* 40: 550–57.

Compton, O. C., and D. Boynton. 1944. Normal seasonal changes of oxygen and carbon dioxide percentages in gas from the larger pores of three orchard subsoils. *Soil Sci.* 57: 107–17.

Condon; E. U., and Hugh Odishaw (ed.). 1967. *Handbook of Physics.* New York: McGraw-Hill.

Currie, J. A. 1960. Gaseous diffusion in porous media. I. A non-steady state method. *Brit. J. Appl. Phys.* 11: 314–17.

——. 1965. Diffusion within soil microstructure, a structural parameter for soils. *J. Soil Sci.* 16: 279–89.

Davis, J. B. 1967. *Petroleum Microbiology*, pp. 93–143. London: Elsevier Publishing.

Dittmer, D. S., and R. M. Grebe (ed.). 1958. *Handbook of Respiration.* Philadelphia: W. B. Saunders Co.

Domsch, K. H. 1962. Bodenatmung, sammelbericht uber methoden und ergbnisse. *Zentralbe. Bakteriol. Parasitenk.* 116: 33–78.

Epstein, E., and H. Kohnke. 1957. Soil aeration as affected by organic matter application. *Soil Sci. Soc. Amer. Proc.* 21: 585–88.

Erickson, A. E., and D. M. Van Doren. 1964. The relation of plant growth and yield to soil oxygen availability. *Trans. 7th Int. Congr. Soil Sci.* 3: 428–34.

Evans, N. T. S., and M. Ebert. 1960. Radioactive oxygen in the study of gas transport down the root of *Vicia faba. J. Exp. Bot.* 11: 246–57.

Furr, J. R., and W. W. Aldrich, 1943. Oxygen and carbon dioxide changes in the soil atmosphere of an irrigated date garden on calcareous very fine sandy loam soil. *Proc. Amer. Soc. Hort. Sci.* 42: 46–52.

Gilbert, D. L. 1964. Atmosphere and evolution. In *Oxygen in the Animal Organism*, ed. F. Dickens and E. Neil, pp. 641–53. New York: Macmillan.

Grable, A. R. 1966. Soil aeration and plant growth. In *Advances in Agronomy*, ed. A. G. Norman, 18: 57–106. New York: Academic Press.

Gray, D. E. (coord. ed.). 1963. *American Institute of Physics Handbook.* New York: McGraw-Hill.

Greenwood, D. J. 1969. Effect of oxygen distribution in the soil on plant growth. In *Root Growth.* ed. W. J. Whittington, pp. 202–23. New York: Plenum Press.

Greenwood, D. J., and D. Goodman. 1967. Direct measurements of the distribution of oxygen in soil aggregates and in columns of fine soil crumbs. *J. Soil Sci.* 18: 182–86.

Hack, H. R. B. 1956. An application of a method of gas microanalysis to the study of soil air. *Soil Sci.* 82: 217–31.

Hagan, R. M. 1941. Movement of carbon disulfide vapor in solids. *Hilgardia* 14: 83–118.

Hammond, L. C., W. H. Allaway, and W. E. Loomis. 1955. Effects of oxygen and carbon dioxide levels upon absorption of potassium by plants. *Plant Physiol.* 30: 155–61.

Harrison, W. H., and P. A. S. Aiyer. 1913. The gases of swamp rice soils. I. Their composition and relation to the crop. *Mem. Dept. Agr. India*, Chem. ser. 3: 3–65.

——. 1916. The gases of swamp rice soils. IV. The source of gaseous soil nitrogen. *Mem. Dept. Agr. India*, Chem. ser. 5: 1–31.

Horvitz, L. 1939. On geochemical prospecting. 1. *Geophysics* 4: 210–28.

Huck, M. G. 1970. Variation in taproot elongation rate as influenced by composition of the soil air. *Agron. J.* 62: 815–18.

Keen, B. A. 1931. The physical properties of the soil. London: Longman, Green.

Kimball, B. A., and E. R. Lemon. 1971. Air turbulence effects upon soil gas exchange. *Soil Sci. Soc. Amer. Proc.* 35: 16–20.

Kramer, P. J. 1965. Effects of deficient aeration on the roots of plants. In *Intersociety Conference on Irrigation and Drainage*, pp. 13–14, 23. Proc. Amer. Soc. Agr. Eng.

——. 1969. *Plant and Soil Water Relationships*, pp. 104–49. New York: McGraw-Hill.

Kramer, P. J., and W. T. Jackson. 1954. Causes of injury to flooded tobacco plants. *Plant Physiol.* 29: 241–45.

Kristensen, K. J., and H. Enoch. 1964. Soil air composition and oxygen diffusion rate in soil columns at different heights above a water table. *Trans. 8th Int. Congr. Soil Sci.* 2: 159–70.

Labanauskas, C. K., L. H. Stolzy, L. J. Klotz, and T. A. DeWolfe. 1971. Soil carbon dioxide and mineral accumulation in citrus seedlings (*Citrus sinensis* var. *bessie*). *Plant Soils* (in press).

Lammann, G., and V. Jessen. 1929. Diffusion coefficients of gases in water and their temperature dependence. *Z. Anorg. Allg. Chem.* 179: 125–44.

Lemon, E. R., and A. E. Erickson. 1952. The measurement of oxygen diffusion in the soil with a platinum microelectrode. *Soil Sci. Soc. Amer. Proc.* 16: 160–63.

Letey, J. 1965. Measuring aeration. In *Intersociety Conference on Irrigation and Drainage*, pp. 6–10. Proc. Amer. Soc. Agr. Eng.

Letey, J., O. R. Lunt, L. H. Stolzy, and T. E. Szuszkiewicz. 1961. Plant growth, water use and nutritional response to rhizosphere differentials of oxygen concentration. *Soil Sci. Soc. Amer. Proc.* 25: 183–86.

Letey, J., and L. H. Stolzy. 1964. Measurement of oxygen diffusion rates with the platinum microelectrode. I. Theory and equipment. *Hilgardia* 35: 545–54.

——. 1967. Limiting distances between root and gas phase for adequate oxygen supply. *Soil Sci.* 103: 404–9.

Luxmoore, R. J., R. E. Sojka, and L. H. Stolzy. 1971. Root porosity and growth responses of wheat to aeration and light intensity. *Soil Sci.* (in press).

Luxmoore, R. J., and L. H. Stolzy. 1972. Oxygen concentration and temperature effects on oxygen relations predicted for maize roots. *Soil Sci. Soc. Amer. Proc.* (in press).

Luxmoore, R. J., L. H. Stolzy, and J. Letey. 1970a. Oxygen diffusion in the soil-plant system. I. A model. *Agron. J.* 62: 317–21.

——. 1970b. Oxygen diffusion in the soil-plant system. II. Respiration rate, permeability, and porosity of consecutive excised segments of maize and rice roots. *Agron. J.* 62: 322–25.

——. 1970c. Oxygen diffusion in the soil plant system. III. Oxygen concentration profiles, respiration rates, and the significance of plant aeration predicted for maize roots. *Agron. J.* 62: 325–29.

——. 1970d. Oxygen diffusion in the soil-plant system. IV. Oxygen concentration profiles, respiration rates and radial oxygen losses predicted for rice roots. *Agron. J.* 62: 329–32.

McIntyre, D. S. 1970. The platinum microelectrode method for soil aeration measurement. In *Advances in Agronomy*, ed. N. C. Brady, 22: 235–83. New York: Academic Press.

McLaren, A. D., and J. Skujins. 1968. The physical environment of micro-organisms in soil. In *The Ecology of Soil Bacteria*, ed. T. R. G. Gray and D. Parkinson, pp. 3–24. Toronto: University of Toronto Press.

Marshall, T. J. 1959. The diffusion of gases through porous media. *J. Soil Sci.* 10: 79–82.

Meites, L. (ed.). 1963. *Handbook of Analytical Chemistry*. New York: McGraw-Hill.

Millington, R. J. 1955. Diffusion constant and diffusion coefficient. *Science* 122: 1090.

——. 1959. Gas diffusion in porous media. *Science* 130: 100–102.

Munnecke, D. E., M. J. Kolbezen, and L. H. Stolzy. 1969. Factors affecting field fumigation of citrus soils for control of *Armillaria mellea*. *Proc. 1st Int. Citrus Symposium* 3: 1273–83.

Pearson, R. W. 1965. Soil environment and root development. In *Plant Environment and Efficient Water Use*, ed. W. H. Pierre, D. Kirkham, J. Pesek, and R. Shaw, chap. 6, pp. 95–126. Madison, Wis: Amer. Soc. Agron. and Soil Sci. Soc. Amer.

Penman, H. L. 1940a. Gas vapor movements in the soil. I. The diffusion of vapors through porous solids. *J. Agr. Sci.* 30: 437–62.

——. 1940b. Gas vapor movements in the soil. II. The diffusion of carbon dioxide through porous solids. *J. Agr. Sci.* 30: 570–81.

Ponnamperuma, F. N. 1965. Dynamic aspects of flooded soils and the nutrition of the rice plant. In *Symposium on the Mineral Nutrition of the Rice Plant*, pp. 295–328. Baltimore: Johns Hopkins Press.

Ponnamperuma, F. N., E. Martinez, and T. Loy. 1966. Influence of redox potential and partial pressure of carbon dioxide on pH values and the suspension effect of flooded soils. *Soil Sci.* 101: 421–31.

Porter, L. K., W. D. Kemper, R. D. Jackson, and B. A. Stewart. 1960. Chloride diffusion in soils as influenced by moisture content. *Soil Sci. Soc. Amer. Proc.* 24: 460–63.

Rickman, R. W. 1968. Effects of salts on oxygen diffusion rate measurements in unsaturated soils. *Soil Sci. Soc. Amer. Proc.* 32: 618–22.

Rickman, R. W., J. Letey, G. M. Aubertin, and L. H. Stolzy. 1968. Platinum microelectrode poisoning factors. *Soil Sci. Soc. Amer. Proc.* 32: 204–8.

Romell, L. G. 1922. Luftvaxlingen i marken som ekologisk faktor. *Medd. fran. Statens Skoysforsoksanstalz* 19: 125–27, 334, 335, 348.

——. 1935. Mechanism of soil aeration. *Ann. Agron.*, n.s. 5: 373–84.

Rosaire, E. E. 1939. *The Handbook of Geochemical Processing.* Houston: Gulf Publ. Co. 61 pp.

Russell, B. 1969. *History of Western Philosophy.* London: George Allen & Unwin.

Russell, E. J., and A. Appleyard, 1915. The atmosphere of the soil, its composition and the causes of variation. *J. Agr. Sci.* 7: 1–48.

Russell, M. B. 1952. Soil aeration and plant growth. In *Soil Physical Conditions and Plant Growth*, ed. B. T. Shaw, chap. 4, pp. 253–301. New York: Academic Press.

Scott, A. D., and D. D. Evans. 1955. Dissolved oxygen in saturated soil. *Soil Sci. Soc. Amer. Proc.* 19: 7–12.

Skerman, V. B. D., and I. C. McRae. 1957. The influence of oxygen availability on the degree of nitrate reduction by *Pseudomonas denitrificans. Can. J. Microbiol.* 3: 215–30.

Smith, F. B., and P. E. Brown. 1933. Diffusion of carbon dioxide through soils. *Soil Sci.* 35: 413–23.

Sokolov, V. A. 1936. *Gas Surveying.* Moscow: Gostoptekhizdat. 269 pp.

Stolzy, L. H. 1971. Soil aeration and gas exchange in relation to grasses. In *The Biology Utilization of Grasses*, ed. V. B. Youngner and C. M. McKell, vol. 1. New York: Academic Press (in press).

Stolzy, L. H., and J. Letey. 1964a. Characterizing soil oxygen conditions with a platinum microelectrode. In *Advances in Agronomy*, ed. A. G. Norman, 16: 249–79. New York: Academic Press.

——. 1964b. Measurement of oxygen diffusion rates with the platinum microelectrode. III. Correlation of plant response to soil oxygen diffusion rates. *Hilgardia* 35: 567–76.

Stolzy, L. H., J. Letey, L. J. Klotz, and C. K. Labanauskas. 1965. Water and aeration as factors in root decay of *Citrus sinensis. Phytopathology* 55: 270–75.

Stolzy, L. H., J. Letey, T. E. Szuszkiewicz, and O. R. Lunt. 1961a. Root growth and diffusion rates as functions of oxygen concentration. *Soil Sci. Soc. Amer. Proc.* 25: 463–67.

Stolzy, L. H., O. C. Taylor, W. M. Dugger, Jr., and J. D. Mersereau. 1964. Physiological changes in and ozone susceptibility of the tomato plant after short periods of inadequate oxygen diffusion to the roots. *Soil Sci. Soc. Amer. Proc.* 28: 305–8.

Stolzy, L. H., O. C. Taylor, J. Letey, and T. Szuszkiewicz. 1961b. Influence of soil-oxygen diffusion rates on susceptibility of tomato plants to air-borne oxidants. *Soil Sci.* 91: 151–55.

Stolzy, L. H., and S. D. Van Gundy. 1968. The soil as an environment for microflora and microfauna. *Phytopathology* 58: 889–99.

Stolzy, L. H., S. D. Van Gundy, C. K. Labanuskas, and T. E. Szuszkiewicz. 1963. Response of *Tylenchulus semipenetrans* infected citrus seedlings to soil aeration and temperature. *Soil Sci.* 96: 292–98.

Stolzy, L. H., G. A. Zentmyer, L. H. Klotz and C. K. Labanauskas. 1967. Oxygen diffusion, water and *Phytophthora cinnamomi* in root decay and nutrition of avocados. *Proc. Amer. Soc. Hort. Sci.* 90: 67–76.

Taylor, S. A. 1949. Oxygen diffusion in porous media as affected by compaction and moisture content as a measure of soil aeration. *Soil Sci. Soc. Amer. Proc.* 14: 55–61.

Thomas, R. E., W. A. Schwartz, and T. W. Bendixem. 1968. Pore gas composition under sewage spreading. *Soil Sci. Soc. Amer. Proc.* 32: 419–23.

van't Woudt, B. D., and R. M. Hagan. 1957. Land drainage in relation to soils and crops. III. Crop responses at excessively high soil moisture levels. In *Drainage of Agricultural Lands*, ed. J. N. Luthin, chap. 5, pp. 514–78. Madison Wis: Amer. Soc. Agron.

Weast, R. C. (ed.). 1969. *Handbook of Chemistry and Physics*. Cleveland: Chem. Publishing Co.

Wesseling, J. 1962. Some solutions of the steady state diffusion of carbon dioxide through soils. *Neth. J. Agr. Sci.* 10: 109–17.

Wesseling, J., and W. R. van Wijk. 1957. Land drainage in relation to soils and crops. I. Soil physical conditions in relation to drain depth. In *Drainage of Agricultural Lands*, ed. J. N. Luthin, chap. 5, pp. 461–504. Madison, Wis.: Amer. Soc. Agron.

Wiegand, C. L., and E. R. Lemon. 1958. A field study of some plant-soil relations in aeration. *Soil Sci. Soc. Amer. Proc.* 22: 216–21.

Williamson, R. E. 1970. Effect of soil gas composition and flooding on growth of *Nicotiana tabacum* L. *Agron. J.* 60: 365–68.

Woolley, J. T. 1965. Drainage requirements of plants. In *Intersociety Conference on Irrigation and Drainage*, pp. 2–5. Proc. Amer. Soc. Agr. Eng.

Wollny, E. 1886. Untersuchungen uber den Einfluss der physikalischen Eigenschaften des bodens auf dessen Gehalt an freier Kohlensaure. *Forsch. gebeite Agr.-Phys.* 9: 165–94.

Wyckoff, R. D., and H. G. Botset. 1936. The flow of gas-liquid mixtures through unconsolidated sands. *Physics* 7: 325–45.

Yamaguchi, M., W. J. Flocker, and F. D. Howard. 1967. Soil atmosphere as influenced by temperature and moisture. *Soil Sci. Soc. Amer. Proc.* 31: 164–67.

Yu, P. T., L. H. Stolzy, and J. Letey. 1969. Survival of plants under prolonged flooded conditions. *Agron. J.* 61: 844–47.

14.   Root and Soil Water Relations

E. I. Newman

THIS chapter discusses the water relations of both soil and root. It thus contrasts with the treatment in this volume of such topics as mineral nutrients, where processes in the root are dealt with separately from those in the soil. This unified treatment of water is in keeping with the view that water movement through soil, plant, and atmosphere can be viewed as a continuum. This approach is particularly appropriate when only root and soil are considered. Soil, cell walls and xylem, and probably also cell membranes are each porous systems, though the pores vary markedly in size and shape, as well as in the material forming their walls. It is thus very illuminating to examine both the similarities and the differences between the behavior of water in these four systems.

This is a selective review. Some topics receive detailed discussion, while others of equal importance get only a brief mention or none at all. The topics have been chosen for two main reasons: first, because they interest me ; second, because I think they have received inadequate attention in the recent literature, or because I think current views are open to question. This chapter is largely concerned with water movement through soil to roots and through roots up to the base of the stem. Even within this field the treatment is selective. Those requiring a more balanced review of the water relations of the root should see the relevant chapters of Slatyer (1967) or Kramer (1969). There is also much relevant information in the symposia edited by Fogg (1965). Hagan *et al.* (1967), and Kozlowski (1968), and in shorter reviews by Gardner (1965a), Philip (1966), and Weatherley (1970).

I shall use the water potential terminology, which is explained by Slatyer (1967), Kramer (1969), and others. Water moves from a higher to a lower potential. The principal components of water potential are matric potential (capillary and surface forces), osmotic potential, and pressure potential (hydrostatic pressure). The water potential of plants and soils is usually below zero. Potentials will usually be measured in bars, more rarely in centimeters (water) or dyne/cm^2. 1 bar $= 1.00 \times 10^6$ dyne/cm$^2 = 1.00 \times 10^5$ Newtons/m$^2 = 0.987$ atm $= 1,020$ cm water $=$ 75.0 cm Hg $= 14.50$ lb/in^2.

I. SOME PROPERTIES OF WATER

A. Structure

Water is in many respects an anomalous substance, and this can be related to its structure. In the water molecule the average H-O bond length is 0.10 nm, and the average distance between the centers of the two H atoms is 0.15 nm. These distances vary slightly between the solid, liquid, and gaseous states, and with temperature. In ice the molecules form a regular tetrahedral lattice, linked by hydrogen bonds, with oxygen atoms of neighboring molecules 0.28 nm apart. The distance between the center of neighboring molecules in the liquid is also about 0.28 nm, increasing slightly with temperature. The properties of liquid water are much influenced by the dipoles formed within the molecule and by the strong tendency to form hydrogen bonds between molecules. When ice melts its lattice structure is not completely lost. Although this fact is generally accepted, there is controversy over what type of structure occurs in the liquid. Frank (1970) has reviewed the three main theories. These are: (1) that there are regions of liquid and regions of solid inter-mingled; (2) that there is a complete ice framework, but voids in it are occupied by free individual molecules; (3) that all molecules remain in the framework, but the intermolecular bonds become bent and the framework more irregular. Frank considers that the evidence currently available favors theory (2). Theory (1) seems the least likely, since X-ray studies have failed to show the variation in density that would be expected.

Hydrogen bonds can also form between water and other substances, for example, clays and proteins. This leads to a layer of water very tightly bound to the surface of the substance. According to Bernal (1965), the water within 1–2 nm of a hydrophilic surface is so tightly bound and strongly structured that it excludes ions, and up to 10 nm it is sufficiently bound to restrict flow. There is certainly evidence that water in thin films differs from bulk water. For example, Oster and Low (1964) found that as wet clay lost water (and the water films became thinner), the specific heat of the water in it rose above the bulk value (1.00), reaching 1.05–1.08 when the water content was 10%. At the water content at which the specific heat first rose significantly above 1.00, the water film thickness was estimated to be 1.2–2.0 nm.

B. Water in Capillaries and Films

One of the anomalous features of water is its high surface tension. It is 72.75 dyne/cm at 20C and varies little with temperature between 0C and 40C. The pressure (P) necessary to remove a liquid from a capillary tube of radius r is given by

$$P = 2\gamma/r, \qquad [1]$$

where γ is the surface tension (dyne/cm), P is in dyne/cm^2, and r is in centimeters. Putting $\gamma = 72.75$ gives

$$P = 1.45/r, \qquad [2]$$

where P is now in bars and r in microns. Although the pores in soil and cell walls are irregular and varied in size and shape, this formula can be applied to them in a very approximate way. Since soils almost always contain pores more than 15μ in radius, one would expect them to contain some air-filled pores at -0.1-bar matric potential, which is about field capacity, a fact which is of course well known as well as important to the plant. In contrast, cell walls probably contain few pores larger than 10 or 20 nm in radius (Gaff *et al.*, 1964); so no air would be expected to be drawn in until the matric potential reached about -100 bars, by which time the plant would almost certainly be dead. Figure 14.1 shows the moisture characteristics of two soils and of a root cell wall preparation. The different shapes of the curves are attributable partly to their different pore sizes and partly to the much greater rigidity of soil. Water loss from soil involves mainly evacuation of pores, with relatively little shrinkage, and the marked water loss between -0.1 and -1 bar indicates many pores between about 1.5μ and 15μ radius in these soils. In contrast, water loss from cell walls involves contraction of the framework.

In experiments with capillaries, the tube is considered as either full or empty; surface films are ignored. However, in soil films may play an important part in water flow (and nutrient movement), especially in drier soils. The thickness of a water film depends very much on the properties of the wall, and no useful generalization can be made. The determination of film thicknesses in soil is not easy, but Kemper (1960) has estimated that at -0.3-bar matric potential the films are less than 100 nm thick, and at -1 bar less than 30 nm thick.

C. Flow

1. Liquids in General

A useful review for nonphysicists on some aspects of flow is given by Scott Blair (1969). Scheidegger (1957, 1966) is specifically concerned with porous systems, but is mathematically more advanced.

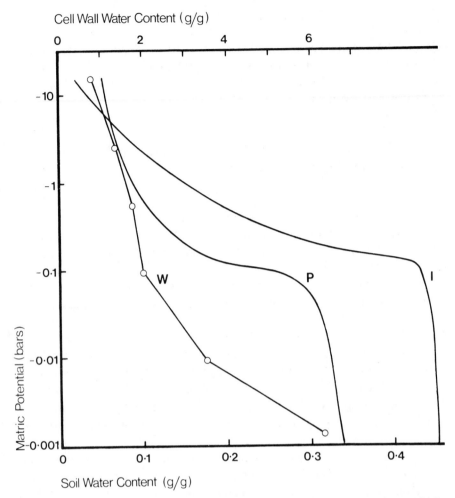

Fig. 14.1. Moisture characteristics of two soils and one cell wall preparation. *I*: Indio loam; *P*: Pachappa sandy loam. (Data from Gardner, 1960) *W*: cell wall material from *Vicia faba* root. (Data from Teoh *et al.*, 1967) Note the different water content scales for soil and cell wall.

The principle most fundamental to flow through porous systems is Darcy's law, which states that the rate of flow of a liquid through a porous medium is proportional to the pressure difference applied (line *A* in Fig. 14.2). This can be written

$$F = k \cdot \Delta P, \qquad [3]$$

where F is the rate of flow, ΔP the difference in hydrostatic pressure between the two ends of the system, and k is a constant, the conductivity. In this paper the dimensions of F, k, and ΔP will vary according to the system being studied. For a whole root system F would be in units of volume of water per plant per time; for a membrane or for soil it would

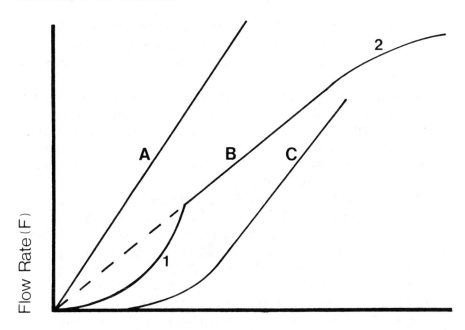

Fig. 14.2. Diagrammatic representation of three possible relationships between the rate of flow through a resistance and the difference in hydrostatic pressure between the two ends. (Data for *B* and *C* from Scott Blair, 1969)

be volume per unit area per time. For a whole root or a membrane ΔP would be a simple difference of pressure between the two sides; but in soil a pressure gradient, pressure difference per unit length, might be involved. The units of k will vary accordingly. In equation [3] k is affected by the liquid being used as well as the properties of the porous medium. An alternative way of stating the law is

$$F = (k' \cdot \Delta P)/\eta, \qquad\qquad [4]$$

where η is the viscosity of the liquid, and k' is dependent on the properties of the porous medium but is independent of what liquid is flowing (provided all the pores are filled). This equation makes it clear that Darcy's law is closely related to Newton's law of flow, which is (put somewhat crudely) that the viscosity of a liquid is unaffected by the pressure gradient. Liquids that obey this law are called Newtonian liquids.

If the "porous medium" is a single capillary tube, k' of equation [4] can be further defined by the Poiseuille-Hagen equation as

$$F = (\Delta P \cdot \pi r^4)/8L\eta, \qquad\qquad [5]$$

where r is the tube radius, and L its length (units: F, cm^3/sec; ΔP, dyne/cm^2; r and L, cm; η, g cm^{-1} sec^{-1}). It can be shown (Scott Blair,

1969) that any Newtonian liquid should obey this law provided that (1) the layer of molecules next to the tube wall is always stationary and (2) the diameter of the tube is very much larger than the diameter of the liquid molecules.

The relationship of flow rate to pore size when the pores are not straight uniform capillaries has been the subject of theoretical analyses (Scheidegger, 1957), but the mathematics becomes complicated. Since we have inadequate quantitative information on the sizes and shapes of the pores in soils, cell walls, or membranes, I doubt if anything useful can be gained from this approach at the moment.

Not all liquids obey these laws. Some give flow rates in different tube sizes that are not proportional to r^4. The rate in small tubes may be either higher or lower than expected, depending on the liquid involved. More surprising, perhaps, some liquids give a flow rate which is not proportional to ΔP. Lines B and C in Figure 14.2 show diagrammatically two sorts of behavior that are found. In line B the rate of flow increases faster than expected in region 1, at higher ΔP it follows expectation, and at very high ΔP it falls below expectation (region 2). This latter deviation is due to turbulence. All liquids become turbulent when the velocity is sufficiently high; so lines A and C must also show this curvature eventually, somewhere off the graph to the top right. Line C shows another phenomenon, a threshold value of ΔP below which no movement at all occurs. This is characteristic of solids but is nevertheless shown by some substances that would normally be called liquids.

2. Water Flow

The viscosity of water at 20C is 1.00×10^{-2} g cm^{-1} sec^{-1}. It decreases quite markedly with rising temperature: the Q_{10} of 1/viscosity is 1.37 from 0C–10C, and 1.22 from 30C–40C.

I consider now the question, How far does water deviate from the basic laws outlined above? In capillaries larger than about 1μ diameter it obeys the laws closely (except for turbulence effects), but in smaller tubes it does not always do so.

a. Relation to Tube Diameter. Rates of water flow through small pores sometimes agree closely with Poiseuille-Hagen predictions. For example, Beck and Schultz (1970) found close agreement for water flow through pores about 10 nm in diameter in mica film. In contrast, Derjaguin (1965) reported that flow in glass microcapillaries 40 nm in diameter is several times faster than predicted by the Poiseuille-Hagen equation, perhaps because the molecules next to the wall are not, as assumed, stationary, but slide. Derjaguin and Krylov, as reported by Henniker (1952), found that water flow through a ceramic disc was seven times faster after the pore walls had been coated with oleic acid, a clear demonstration of the importance of surface effects in very small pores. Henniker himself found significant, though much smaller, increases in flow rate through a porcelain filter candle, of pore diameter 300 nm, when it was treated with an amine. This, like oleic acid, would make the surface hydrophobic. He concluded that the increase was not due to slippage but rather that previously the water near the wall had had abnormally high viscosity, due to bonding between the water and the wall, and this was reduced or eliminated by making the walls nonwettable.

The relationship of flow to pore radius thus appears to be rather variable. The results evidently depend on the nature of the wall material, but if flow is not proportional to ΔP in small pores, this would also lead to variable results.

b. Deviations from Darcy's Law: High Velocity. The high velocity deviations from Darcy's law (point 2 in Fig. 14.2) are due to turbulence. The minimum velocity at which turbulence occurs is defined by the Reynolds number, *Re:*

$$Re = F\rho d/\eta, \qquad [6]$$

where F is the flow rate in $cm^3\ cm^{-2}\ sec^{-1}$, d the pore diameter in cm, and ρ the density of the liquid. For water, with $\rho = 1\ g/cm^3$ and $\eta = 10^{-2}\ g\ cm^{-1}sec^{-1}$,

$$Re = 100Fd. \qquad [7]$$

In straight capillary tubes the critical Reynolds number is about 2,000, but in irregular pores the flow may decline below the expected rate at much lower velocities, when the Reynolds number is anything from 0.1 to 75 (Scheidegger, 1966). When the pores are irregular, d is some sort of average diameter which is not easy to define precisely. However, it seems unlikely that turbulence will be at all important in water flow through either soil or roots. Table 14.1 shows the minimum flow rates needed to give turbulence at the lowest reported Reynolds number, 0.1. In soil at field capacity the largest water-filled pores are about 10μ in diameter. The flow rate required to give turbulence, $1\ cm^3\ cm^{-2}\ sec^{-1}$, is extremely improbable in soil at field capacity. Turbulence could occur when water is moving rapidly through large pores or cracks, say 1 mm across, as it may after rain. In cell walls, where the pores are less than 0.1μ across, turbulence is clearly out of the question. In xylem vessels, the critical Reynolds number is likely to be 10^3-10^4 times larger than assumed in Table 14.1. The diameter of vessels is often $10\mu-100\mu$; so even at a very

Table 14.1.   Water flow rates necessary to give turbulence when the
critical Reynolds number is 0.1; calculated from equation [7]

Pore diameter		Flow rate
cm	μ	$(cm^3\ cm^{-2}sec^{-1})$
1	10^4	10^{-3}
10^{-1}	10^3	10^{-2}
10^{-2}	10^2	10^{-1}
10^{-3}	10	1
10^{-4}	1	10
10^{-5}	10^{-1}	100

high flow rate of 1 $cm^3\ cm^{-2}\ sec^{-1}$, turbulence would not be expected,
except perhaps at cross walls, if they occur. Thus high velocity departures
from Darcy's law are unlikely to be common in soil or root.

c. Deviations from Darcy's Law: Low Velocity. Some botanists have
gained the impression from the writings of Hylmö (1955, 1958) that water
flow phenomena similar to line C or region 1 of line B in Figure 14.2
are well known and well understood. Hylmö stated that Erbe (1933),
studying flow through micropore filters, had shown that flow in any
given pore did not commence until a certain threshold pressure was
reached and that the threshold pressure could be related to the pore
diameter by a simple formula. However, in Erbe's experiments there
were two different liquids, immiscible with each other, on the two sides
of the filter; absence of flow in small pores was due to surface tension
between the two liquids. The formula Erbe used was in fact equation [1],
which shows the pressure difference needed to draw the liquid of lower
surface tension into the pore. This work has no bearing on our situation,
where water is present on both sides and no other liquid is involved.

Wiegand and Swanson (1966) measured the flow rate through 13
membranes of steel, ceramic material, cellulose, or glass, with various
pore sizes. The pressure difference across the membrane was varied from
0.1 to 0.8 bar. Ten of the membranes gave close approximations to
Darcy's law; their mean pore diameters ranged from 0.45μ to 15μ. With
the three remaining membranes, however, flow rate increased proportionately faster than ΔP, as in region 1 of line B in Figure 14.2. One of
them gave no measurable flow at the lowest pressure, suggesting a threshold gradient as in curve C. The pore diameters of these three membranes
were not known, but were probably below 0.45μ.

Because of the small size of the pores and the relatively slow rates of
flow, experiments such as this involve technical problems. Jackson (1967)

performed a similar experiment, using a ceramic filter with pore diameter 0.1μ, and concluded that the departure from Darcy's law which he found was an artifact caused by an osmotic effect produced by dissolved silicates on one side of the plate. However, his largest pressure difference was only 0.025 bar, much smaller than those of Wiegand and Swanson.

Direct experimental evidence of low velocity non-Darcy behavior by water is thus scanty. The main evidence is indirect: if water near the wall is highly structured, as has been suggested above, it would be expected to give a flow relationship like lines *B* or *C* of Figure 14.2. But it would clearly be better if there was more extensive and rigorous direct evidence on flow rates through fine pores.

d. Effect of Ions. Dissolved ions tend to break down the structure of water. We might expect, therefore, that ions would increase flow rates in narrow tubes and reduce deviations from Darcy's law. Henniker (1952) found that the flow rate through a porcelain filter candle, pore diameter 0.3μ, was 16% faster for 1 mM KCl than for pure water. Saturated flow through sandstones and clays shows progressively less deviation from Darcy's law as the electrolyte concentration is increased (Swartzendruber, 1962). Although this may be due to the effect of the ions on the structure of the water, other explanations are possible, particularly with clays (see section X).

Normally in both soils and roots the pore walls carry permanent negative charges, and this can affect water flow. It results in an excess of cations over anions in the water within 10 or 15 nm of the wall, and this is associated with the phenomenon of streaming potential. If an ionized solution is moved by pressure through a pore whose walls are charged either negatively or positively, an electrical potential which retards water flow is set up between the two ends of the pore. According to calculations by Kemper (1960), in soil pores 30 nm in diameter streaming potential could retard flow by as much as 50%.

II. THE CATENARY HYPOTHESIS

Perhaps the most influential paper on plant water relations in the last thirty years was that by Honert (1948) in which he set out the catenary hypothesis. He pointed out first that water movement between any two points depends on the difference in water potential and on the resistance to flow:

$$F = (\psi_1 - \psi_2)/R, \qquad\qquad [8]$$

where F is the rate of water flow between the two points, ψ_1 and ψ_2 are the water potentials at the two points, and R is the resistance to flow. This is a statement of Darcy's law. The soil, plant, and atmosphere can be conceived as resistances in series. If the rate of transpiration is steady, F is the same throughout the system, and so

$$F = (\psi_s - \psi_r)/R_s = (\psi_r - \psi_e)/R_p = (\psi_e - \psi_a)/R_a, \quad [9]$$

where ψ_s, ψ_r, ψ_e, and ψ_a are the water potentials of the soil at some distance from the root, of the root surface, of the evaporating surfaces in the leaves, and of bulk air, respectively; and R_s, R_p, and R_a are the resistances of the soil, plant, and vapor pathways, respectively. (Honert did not in fact include the soil in his discussion, but the same principle can be applied to it.) The plant can be further divided, for example, into root, stem, and leaf.

Honert expected only very limited accuracy from equation [9]. He was mainly concerned to show that R_a is always much larger than R_s and R_p, so that the soil and plant resistances cannot influence the transpiration rate directly, a conclusion now generally accepted. Various later authors (including myself) have assumed much greater quantitative accuracy for the equation, but here caution is necessary. The criticism most commonly leveled is that vapor movement is not proportional to water potential but to water vapor concentration, making the right-hand part of the equation incorrect; this does not concern us here. Another criticism is that equation [9] is only true if there are no temperature gradients. Again, this is most serious in the vapor pathway. The temperature gradients across an individual root, and in its immediate vicinity, are unlikely to have any appreciable influence on water movement. There are two equally serious sources of error that do concern us. First, the driving forces are not always water potential and are not the same in different parts of the soil-root pathway. In the soil and the xylem, the driving forces are differences of matric potential and pressure potential. There may be differences in solute concentration, but these will exert negligible driving forces on the water. In contrast, when water is crossing root membranes, osmosis may be important. Second, equations [8] and [9] assume that soils and plants obey Darcy's law. I shall initially assume this to be so, but later I shall examine cases of departure from Darcy's law and discuss their significance.

Equation [9] also assumes that active water pumps, using metabolic energy, do not contribute significantly to water movement within the plant, an assumption which will be made throughout this paper. The balance of evidence supports this (e.g., Slatyer, 1967), but suggestions of active water movement still appear in the literature quite frequently, and the question is by no means closed.

III. PATHWAY OF WATER MOVEMENT
ACROSS THE ROOT

It is generally agreed that during its movement from the outside of the root to the xylem, water at some point crosses living material. The strongest evidence for this is indirect: first, treatments that affect metabolic rate often affect permeability (section IV B); second, the root acts as a selective and semipermeable membrane. Neither of these statements proves that *all* the water must pass through living material, only that much of it does. The reflection coefficient of roots may be less than 1; according to data of Mees and Weatherley (1957b), for tomato (*Lycopersicon esculentum*) it was on average 0.76. (The reflection coefficient, σ, is a measure of a membrane's efficiency as a semipermeable membrane. When $\sigma = 0.76$, a difference of 1 bar between the osmotic potentials on the two sides is required to balance a difference in pressure potential of 0.76 bar in the opposite direction, so that no flow occurs.) Mees and Weatherley took this as evidence that there were pathways through the root by which some of the water avoided crossing any semipermeable membrane. This is quite possible; but it could equally be that the membranes themselves had a reflection coefficient of 0.76.

It is generally assumed that the free space of the root is blocked at the endodermis by the Casparian strips. The direct evidence for this is not entirely satisfactory. The chemical nature of the Casparian strip is still uncertain. When Scott and Priestley (1928) dipped roots of broad beans (*Vicia faba*) into dyes, most of the dye penetrated the cell walls only as far as the endodermis, but it did sometimes penetrate into the stele at certain points. Nevertheless, it is hard to doubt that the free space is usually blocked at the endodermis; if it were not, many of the water uptake and salt uptake phenomena of the root would be very difficult to explain.

There are basically three possible pathways across the root, as shown in Figure 14.3. In the early years of this century, pathway 1 (the "vacuolar pathway") was favored. Later (Scott and Priestley, 1928) pathway 3 was put forward. In this one (the "free-space pathway"), water is assumed to move in the free space of the walls and in the intercellular spaces if they are present and not air-filled, except at the endodermis, where it has to cross the plasmalemma and pass through the cytoplasm. This theory is the one now most widely accepted, but I suggest that pathway 2 deserves serious consideration.

A. Free-Space Pathway

The evidence put forward in support of the free-space pathway is as follows (see also Weatherley, 1970). (1) As mentioned above, when a root

Endodermis   Cortex

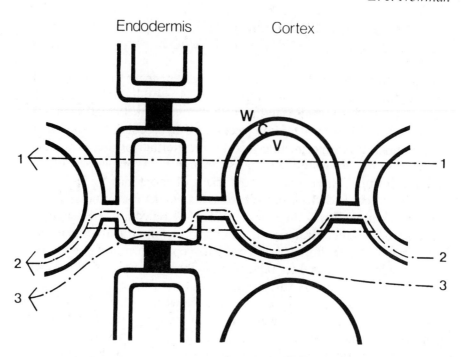

Fig. 14.3. Diagrammatic transverse section of part of a root, showing possible pathways for water movement. *C*: cytoplasm; *V*: vacuole; *W*: cell wall

is dipped in dyes, later anatomical investigation shows that the dye is in the cell walls as far in as the endodermis. This actually proves nothing about water flow: the dye could have moved in by diffusion. (2) In the analogous situation in the leaf, where water passes from the xylem across parenchyma cells to the sites of evaporation, there is evidence that movement is mainly in the walls (Weatherley, 1963). However, this situation differs in an important respect: in the leaf the water does not have to cross any membrane to reach the evaporation surfaces, whereas in the root it must cross at least two plasmalemmas. This evidence therefore seems to me irrelevant to the root.

(3) Estimates of the conductivities of pathways 1 and 3 have been made by Russell and Woolley (1961) and Tyree (1969). These depend on the relative volumes of cell wall and vacuole, which determine the area they provide for water movement, and on permeability estimates for cell wall material and membranes. Russell and Woolley estimated that in the cortex the free-space pathway would have a conductivity 20 times that of the vacuolar pathway (per unit area of cortex), so that most water would flow in the free space; but according to Tyree the vacuolar pathway would have a conductivity about twice that of the free space. The discrepancy between these two conclusions is largely due to the values used in the calculations, which are as follows, those of Russell and

Woolley first in each case: cell wall, 4%, 2.5% of total cross-sectional area; cell diameter, 25μ, 100μ (this determines how many membranes have to be crossed by the vacuolar pathway); conductivity of cell wall material, 2.8×10^{-7}, $1.4 \times 10^{-7}\,\mathrm{cm^2\,sec^{-1}\,bar^{-1}}$; conductivity of membrane (plasmalemma, cytoplasm, and tonoplast combined), 6×10^{-7}, $10 \times 10^{-7}\,\mathrm{cm\,sec^{-1}\,bar^{-1}}$. Each of these differences is within the range of likely variation between species, though cortical cells as large as 100μ are probably rare. It seems possible that both conclusions are correct for different species. The least satisfactory data are those on wall conductivity. Russell and Woolley used data for artificial cellulose membranes; Tyree, from internodal cells of the alga *Nitella flexilis*. Kamiya *et al.* (1962) obtained the lower value of $0.5 \times 10^{-7}\,\mathrm{cm^2\,sec^{-1}}$ $\mathrm{bar^{-1}}$ for the same species by a similar technique. It would be desirable to have data for wall material from angiosperm roots. The conductivity of the cell wall is likely to be influenced by matric potential. Figure 14.1 shows that broad bean wall material declined in water content by a factor of 3 between -0.001 and -0.1 bar matric potential, and by a further factor of 2 to -10 bars. This involved shrinkage of the wall, and if the Poiseuille-Hagen equation is obeyed, conductivity would fall by a factor of 9 between -0.001 and -0.1 bar, and a further factor of 4 to -10 bars. Hence conductivity measurements (such as those quoted) made on wall material in saturated conditions would not apply to normal conditions in the soil. There is evidence from another source against the vacuolar pathway. House and Findlay (1966a) found the conductivity of a root of corn (*Zea mays*) for movement of water from the outside to the xylem to be $6 \times 10^{-7}\,\mathrm{cm\,sec^{-1}\,bar^{-1}}$, i.e., about the same as for entry to a single cell. The data of Brouwer (1953a) indicate a similar conductivity for broad bean roots at low transpiration rates. This would fit very well with the hypothesis that entry to the root involves crossing no tonoplasts and only two plasmalemmas, if two plasmalemmas provide about the same resistance as one plasmalemma plus one tonoplast. In contrast, the vacuolar pathway, in corn roots of this size, would involve crossing about 12 cells, which would be expected to give a conductivity about 1/24th that for entry to an individual cell.

On balance the evidence suggests that the vacuolar pathway is not often important, but its operation in some cases cannot be excluded. It may be important in root hairs (see section IX). Further data on conductivities of cell walls and membranes in higher plants are needed to help settle this question.

B. Symplasm Pathway

Russell and Woolley (1961) explicitly stated that they could not decide on the relative importance of pathways 2 and 3 because they did not know the permeability of the plasmalemma or cytoplasm. At that time there was doubt as to whether the plasmalemma was a selective membrane. We now know that it is as selective as the tonoplast; so it seems more reasonable to assume that they have similar resistance to water; but we still have no definite evidence on this. Russell and Woolley assumed that water could only pass from the cytoplasm in one cell to that of the next by crossing the plasmalemmas. However, there is an alternative possibility, the "symplasm pathway," in which water passes from cell to cell mainly through the plasmodesmata. Thus it would need to cross only two plasmalemmas, once to enter the symplasm and once to leave it (inside the stele). Compared with the free-space pathway, the area of plasmalemma would be greatly enhanced; the whole of the cortex could be envisaged as providing an extension of the outer membrane of the endodermis, and the parenchyma of the stele as an extension of the inner endodermal membrane. Thus the effective conductivity of the endodermis would be greatly increased.

The conductivity of a single plasmodesma can be estimated by the Poiseuille-Hagen equation (equation [5]); hence, if one knows the frequency of plasmodesmata, one can estimate the conductivity per unit area of wall for flow from one cell to the next through the plasmodesmata. Table 14.2 shows three alternative calculations. The first column uses

Table 14.2. Estimates of the conductivity, per unit area of wall, of the plasmodesmatal pathway from one cell to the next

Frequency (number/mm^2)	5×10^6	0.67×10^6	0.67×10^6
Length (μ)	0.35	0.5	0.5
Radius (nm)	15	10	5
Conductivity (cm sec^{-1} bar^{-1})	1.4×10^{-3}	2.6×10^{-5}	1.6×10^{-6}

Note: Assumes liquid viscosity of 2×10^{-2} g cm^{-1} sec^{-1} and the stated plasmodesma dimensions and frequencies.

the length, radius, and frequency of plasmodesmata in the tangential wall of the endodermis of young barley (*Hordeum vulgare*) roots, as found by Helder and Boerma (1969); the frequency in their material was estimated by Clarkson *et al.* (1971). The conductivity is estimated as 1.4×10^{-3} cm sec^{-1} bar^{-1}; so if water had to pass through ten consecutive cells, a normal number for cortex, endodermis, and pericycle, the overall conductivity would be 1.4×10^{-4} cm sec^{-1} bar^{-1}. This is about 200 times larger than the measured conductivity of a corn root, 6×10^{-7} cm sec^{-1} bar^{-1} (House and Findlay, 1966a), which suggests that plasmodesmata could provide a pathway of low resistance for movement from cell to cell. However, there are several points we need to consider concerning both the resistance of individual plasmodesmata and their frequency and distribution. (1) In electron micrographs each plasmodesma usually appears to have a central column, though this is sometimes interpreted as an artifact. According to a formula of Tyree (1970), a column of radius 5 nm would reduce the conductivity in a tube of radius 15 nm by a factor of 8. (2) Poiseuille's law may not be obeyed in tubes of this size. The living membrane is probably partly or entirely hydrophilic (see section V), and flow might be considerably retarded by interaction of water with the walls (see section I C). (3) Data on the frequency of plasmodesmata in roots are limited. Tyree (1970) quotes $1.5 \times 10^6/\text{mm}^2$ for cortical cells of onion (*Allium cepa*) roots. In barley Helder and Boerma (1969) found plasmodesmata to be frequent on the inner and outer tangential walls of the endodermis and in the radial walls of the pericycle, but much less frequent in the cortex. In contrast, in corn Anderson and House (1967) found frequent plasmodesmata between cortical cells but none between stelar parenchyma cells. (4) The radius is very critical, because the conductivity is proportional to r^4. Tyree (1970) reports values larger than I have assumed, but most of the examples he cites are in other parts of the plant. Clarkson *et al.* (1971) examined old barley roots, in which the inner wall of the endodermal cells was heavily thickened. There were pits in this wall, in the bottom of which numerous plasmodesmata passed through to the pericycle. Their frequency and dimensions are shown in Table 14.2, column 2. Compared with the data for younger roots of the same species, in column 1, the plasmodesmata are fewer and narrower. Nevertheless, the conductivity of this layer is still high compared with that of a whole root. (5) Plasmodesmata may not be open tubes. The endoplasmic reticulum sometimes appears to lie against the ends and block them. Another theory (Robards, 1968) is that the plasmodesma is completely filled at its two ends by a "desmotubule." This Robards has most clearly shown in willow stems, but there was some suggestion of it in the electron micrographs of barley roots of Clarkson *et al.* (1971); so they performed alternative calculations assuming the desmotubule to be a hollow tube and all flow to occur through it. The result is shown in column 3 of Table 14.2. If a row of ten cells had plasmodesmata like this, the conductivity would be less than that of a whole root. (6) If the plasmodesma is an open tube, the protoplasm may provide considerable resistance to flow. I have assumed a viscosity of 2×10^{-2} g cm^{-1} sec^{-1}, twice that of water. Tyree (1970), after a discussion of protoplasmic viscosity, takes values 25 and 100 times higher. The viscosity of protoplasm is not itself relevant: the protoplasm does not flow en masse; the water flows through it. No data on the resistance to flow through protoplasm are available, but it could be much higher than I have supposed.

On the basis of these observations and calculations several alternative theories are possible. (1) The symplasm pathway may play little part in water movement, because the plasmodesmata have a high resistance. (2) The free-space and the symplasm pathways may both have low resistance, and the main resistance may be in the plasmalemma. Water would then enter the symplasm all over the plasmalemmas of the epidermis, cortex, and outer endodermal wall and leave it again by the inner endodermal wall and stelar parenchyma. (3) Only some of these cells may be connected by frequent plasmodesmata; so the symplasm pathway may be restricted. (4) The resistances of the free-space and symplasm pathways may be of the same order of magnitude, and their relative contributions to water movement may vary. There are several ways in which the conductivity of the symplasm pathway might vary in relation to cell metabolic rates: the conductivity of the cytoplasm might change; the membrane lining the plasmodesma might become more (or less) hydrophilic; the endoplasmic reticulum or desmotubule, if either of them blocks the plasmodesma, may not always do so, but may contract or move. (5) If the protoplasm in the plasmodesmata provides a high resistance to water flow, it might accelerate flow by protoplasmic streaming through the plasmodesmata. Presumably protoplasm would stream in opposite directions in different plasmodesmata, but the amount of water carried in the two directions would differ if the water potential in the two cells differed. There is no evidence on whether streaming does occur in plasmodesmata, or whether, if it does, it would accelerate water movement to a significant extent.

Thus calculations leave us with a whole spectrum of alternative theories.

C. Experimental Evidence

1. Mass-Flow

Ginsburg and Ginzburg (1970) have published some interesting results obtained with "sleeves" from corn roots. These were prepared by removing the steles from root segments about 3 cm long; the break occurred at the endodermis. Separate solutions were placed inside and outside the sleeve, and rates of osmotic flow across the sleeve were determined. In several respects, the sleeves behaved like intact roots: the reflection coefficient was near 1; the conductivity was similar to that of intact corn roots; both KCN and dinitrophenol caused a marked reduction in conductivity. It is not clear how the sleeves remained semipermeable, and precise interpretation of the results is difficult; but it seems highly probable that the metabolism of the endodermis was disrupted. If in the intact root all water crosses the endodermal plasma-lemmas, and they provide the main resistance to flow across the root, these results with sleeves would be difficult to explain. Thus, they provide some support for the symplasm theory.

House and Findlay (1966b) and Anderson and House (1967) produced evidence, from experiments too complex to describe here, that the volume inside the semipermeable barrier in corn roots is approximately equal to that of the xylem vessels and much less than the total volume of the stele. From this they concluded that the semipermeable membrane is not at the endodermis. However, whatever the pathway of water across the roots, the effective inner volume is not the whole volume within the endodermis, but the free space within the endodermis. If this comprises the xylem lumina plus the cell walls of all the stelar cells, its volume will be little greater than that of the xylem alone.

Anderson and House (1967) observed on electron micrographs of corn roots that the xylem vessels could retain cytoplasm and membranes for some distance behind the root tip. The disruption of cytoplasm began at about 2 cm from the tip but was not completed in all vessels until 10 or 11 cm back. They suggested that the membranes of the vessels themselves are the semipermeable membranes of the root, which was in agreement with their experimental evidence just mentioned. The possibility that the vessels retain a semipermeable membrane for some time is an interesting one; but this cannot be the only semipermeable and selective membrane in the root. For one thing, many roots continue to act as osmometers and to take up ions actively far beyond 10 cm from the tip. Second, the sleeves prepared from corn roots by Ginsburg and Ginzburg (1970) had a semipermeable membrane of very high reflectivity, although all the xylem had been removed.

2. Diffusion of Labeled Water

The rate at which ^3H_2O, or other labeled water, diffuses through a tissue is not necessarily proportional to its permeability to mass-flow water movement. Nevertheless, this technique may give some qualitative indication of differences in mass-flow permeability. If detached roots are placed in ^3H_2O solution, and samples removed at intervals and assayed for ^3H content, the rate of ^3H entry can be assessed by the half-time, i.e., the time taken for the ^3H concentration inside to reach half that in the external solution. This is mainly a measure of rate of entry into vacuoles, since these occupy much of the volume of the root. The half-time for several species is about 0.5–1 minute (Ordin and Kramer, 1956; Ordin and Gairon, 1961; Raney and Vaadia, 1965a; Jarvis and House, 1967). Raney and Vaadia found that the half-time for sunflower (*Helianthus annuus*) roots was about half a minute if the roots were detached or attached to a shoot in the dark ; but if the shoot was in light, the half-time was about 30 minutes. This indicates a very large increase in resistance to entry into vacuoles. This could be caused in some way by the high transpiration rate of the plants in the light; but the possibility should not be excluded that it is related to some other process in the shoot that is stimulated by light.

The ^3H_2O entry into the xylem is more difficult to study. Woolley (1965) immersed detached roots in ^3H_2O, then separated the stele from the rest, and determined the ^3H concentration in both parts. The half-time for the stele was about 2 minutes, compared to about 25 seconds for the whole root. Woolley stated that the rate of entry into the stele was as expected if there was no special barrier at the endodermis, the delay being due simply to its position. Ginsburg and Ginzburg (1970) measured ^3H_2O diffusion rates across their cortical sleeves and found the diffusion coefficient to be 1.2×10^{-6} cm^2/sec. Values reported for movement into whole intact roots of the same species are $1.2-5 \times 10^{-6}$ (Ordin and Gairon, 1961; Woolley, 1965). This would suggest a barrier to movement across the cortical sleeves no greater than that for entry into single cells. In contrast to these rates, those of ^3H_2O appearance in xylem sap and in stems are very slow. The half-time for xylem exudate is nearly 2 hours in onion roots (Hodges and Vaadia, 1964) and 1 hour in corn (Anderson and House, 1967). For stem bases, when the root system is immersed in ^3H_2O, the half-time is 0.75 hour for dwarf bean (*Phaseolus vulgaris*) (Biddulph *et al.*, 1961) and 1 hour for sunflower (Raney and Vaadia, 1965b). In these cases there will be delays due to time for movement along the xylem and to equilibrium with tissues above the water line, but these alone could not account for the very long half-times. All these results taken together could be explained by a strong barrier to ^3H_2O diffusion lying inside the endodermis, so that most of the stelar volume is outside it, though the xylem is inside it. If movement is mainly by the symplasm, this barrier could be the plasmalemmas of the stelar parenchyma. However, more work on ^3H_2O movement into the xylem is needed before definite conclusions can be drawn.

D. Conclusion

Knowledge of the pathway of water movement across the root is fundamental to the understanding of many other problems in water uptake, including most of those discussed in subsequent sections. At present the question is still very much open. There is some evidence supporting the symplasm pathway, but the free-space pathway certainly cannot be discounted, and even the vacuolar pathway is a possibility under certain circumstances. One point raised here, which will be touched on again later, is that the widespread assumption that the main resistance to flow is in membranes may not be correct.

IV. ROOT CONDUCTIVITY

A. Methods of Measurement

Usually these methods have provided only the conductivity per root system, so that the value obtained is dependent on the size of the root system, among other things. However, there are no great technical problems in determining the surface area, and hence calculating an average conductivity per unit area.

1. Transpiration Rate

The rate of transpiration, or water uptake, by intact plants has sometimes been used to estimate changes in root conductivity. This is based on the assumption that if a certain treatment applied to the roots increases the transpiration rate by a factor of 2 (for example), then it must necessarily have increased the root conductivity by a factor of 2. This assumption is not justified and, indeed, is probably rarely correct. Experimental evidence for this can be obtained by altering the effective amount of root by root pruning or some other technique. Almost always the relationship between amount of root on the plant and transpiration is not linear (Bialoglowski, 1936; Parker, 1949; Totsuka and Monsi, 1960; Christersson and Pettersson, 1968). Transpiration rate is controlled in the vapor phase, and the root can affect it only indirectly, through changes of leaf water status and hence stomatal aperture; so a simple relationship between root conductivity and transpiration rate would not be expected. This topic has been discussed in more detail by Weatherley (1970).

2. Applied Suction

In this method the shoot is cut off, and water is drawn through the root by a constant suction applied to the cut stem or by a constant pressure applied to the root system and its bathing medium. The rate of water uptake is assumed to be proportional to the root conductivity. This method is often satisfactory, but significant error can arise if no allowance is made for the contribution of osmosis to water movement. In some cases quite a different conclusion is reached if the osmotic and pressure components of the driving force are taken into account, instead of the pressure alone; for example, see Figure 14.11b. It is therefore desirable to collect the xylem exudate and determine its osmotic potential.

3. Osmotic Uptake

As a modification of the last method, a detopped root system may be allowed to take up water by osmosis alone, without applied suction. The rate of exudation and the osmotic potential of the xylem exudate are determined. One possible source of error is that the concentration of the exudate may be appreciably different from that in the xylem in some parts of the root system (see section VIII B). This is less likely if the root has been in the same conditions for some hours.

4. Changes of External Solution

Arisz *et al.* (1951) devised a technique which overcomes this last problem. They determined the rate of exudation from a root system, then quickly changed the external solution to a different osmotic potential, and measured the new rate of uptake or efflux over periods of 30 seconds. An example of their results is shown in Figure 14.4. Immediately after the change of solution the rate was much reduced, or even reversed, but there was a subsequent recovery, which they attributed to osmotic adjustment in the xylem: If inward salt secretion continues at the same rate as before, while water uptake is reduced or reversed, the xylem sap will become more concentrated. This adjustment took some minutes, and they provided evidence that for their first flow rate measurement immediately after the change of external solution, the xylem osmotic potential, ω_x, was similar to that before the change. Just before the change,

$$F_1 = k({}_1\omega_e - \omega_x), \qquad [10]$$

and immediately after the change,

$$F_2 = k({}_2\omega_e - \omega_x), \qquad [11]$$

where F is the water uptake rate, k the root conductivity, and ω_e the osmotic potential of the external solution; subscripts 1 and 2 refer to before and after the change of solution, respectively. Combining equations [10] and [11], and rearranging,

$$k = (F_1 - F_2)/({}_1\omega_e - {}_2\omega_e). \qquad [12]$$

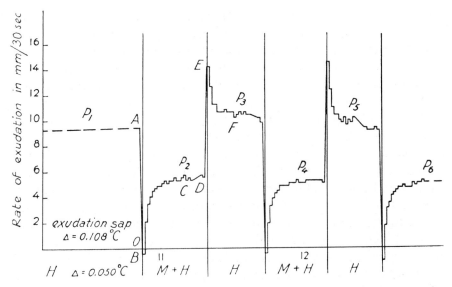

Fig. 14.4. Rate of exudation from a detopped tomato plant, measured during successive 30-second intervals. Horizontal axis: time of day (hr). External solution: *H*, Hoagland; *M + H*, mannitol + Hoagland. Δ = freezing-point depression, as a measure of osmotic potential. (Arisz *et al.* 1951)

Brewig (1937) developed a method for determining root conductivity in intact plants. The water potential of the sylem was determined by finding the osmotic potential in the external solution necessary to give zero uptake. This was done on a small portion of one root, so that the xylem water potential would not be altered. Rate of water uptake from water or nutrient solution was also determined, and the root conductivity could thus be calculated.

B. Some Factors Affecting Root Conductivity

Root conductivity is influenced by many factors, and I shall not discuss them all in detail. The effect of external osmotic potential is considered in section VIII B and the effect of pressure differences in section X B. The present section contains a few remarks relating particularly to the site of the main resistance in the root.

If a root is killed, its conductivity to water increases markedly; but if its metabolism is inhibited without doing permanent damage, the conductivity is usually reduced. These facts have been known for so long that it is easy to forget how surprising they are; one would surely expect killing to have an effect in the same direction as metabolic inhibition, only more extreme. No satisfactory explanation of this phenomenon has ever been offered.

One question raised in section III is whether the main resistance to water movement across the root is provided by membranes or by something else, e.g., plasmodesmatal protoplasm. One possible way of answering this would be to find a chemical which had a specific effect on membrane structure. Such a substance would be a very useful research tool, provided it could be shown not to affect other parts of the cell. Chibnall (1923) found very rapid loss of turgor in leaves treated with ether or chloroform, and Currier (1951) found similar effects with benzene. These substances, which dissolve fats, may well be acting directly on the membranes. But their effect is very drastic, they cause irreversible damage and death, and they probably affect protein structure and hence enzyme activity. Their effect on permeability is similar to that of very hot water (for example), and they do not tell us anything about the significance of membranes. Kuiper (1964b) found that 1mM decenylsuccinic acid increased the permeability of bean roots, and he stated that it acted by incorporation of the molecules into the lipid layer of the cytoplasmic membrane. However, Kramer and I showed that this increase in permeability was associated with death of the roots and that at 0.1 mM the permeability was decreased (Newman and Kramer, 1966). From this and other evidence we concluded that this substance was acting as a metabolic inhibitor, rather than directly on the membrane. Subsequently, Kuiper (1967) compared the effects of several groups of surface-active compounds on growth and permeability. In all cases he found a close relationship between increased root permeability and reduced growth, indicating that irreversible damage was involved.

Various metabolic inhibitors, at sublethal concentrations, are known to reduce root permeability. Almost invariably it takes some time, often 1–3 hours, for the maximum effect to be reached (Brouwer, 1954a; Jackson and Weatherley, 1962; Lopushinsky, 1964). This is probably not due to delay in penetration, since temperature change shows a similar delayed effect (Jackson and Weatherley, 1962; Kuiper, 1964a). These observations suggest that these substances, and temperature, are not acting directly on the membrane but are causing adjustments in metabolic rate, which then affect the permeability. Kuiper (1964a) discussed temperature effects in some detail. He concluded that temperature was acting directly on the structure of membranes, and he suggested that there is a critical temperature above which the membranes have permanently water-filled pores, whereas below this temperature there are only temporary pores and temperature effects are consequently much more pronounced. His conclusions were based almost entirely on temperature effects on transpiration. Transpiration rate is not a measure of root conductivity (see section IV A1 above). Brouwer (1965) has compared the effect of temperature on transpiration and root conductivity in the same species and has shown that they are not the same. Indeed, the effect on transpiration depended on the conditions of the aerial environment.

In conclusion, the permeability of the root is strongly influenced by metabolic processes. It is therefore influenced by metabolic inhibitors and by temperature. So far no substances have been found that are known to alter membrane permeability directly without affecting metabolism; so this approach provides no information on whether membranes are a major resistance to flow across the root.

V. MEMBRANE STRUCTURE

There are questions about the structure of cell membranes whose answers would contribute greatly to our understanding of water movement through the root. It would be very useful to know whether the plasmalemma and tonoplast have permanently water-filled pores, and if so how big they are, how abundant, and whether their size and abundance is affected by root metabolism or external environmental factors. No such pores have so far been detected by electron microscopy. This is probably not due to lack of resolving power: The limits of resolution of the best electron microscopes are now near the diameter of a single water molecule. Pores may nevertheless exist but be undetected, either because they are not straight and perpendicular to the membrane face or because they are lost during preparation of the membrane for examination. The structure of membranes is still a matter for debate. Staehelin and Probine (1970) show diagrams of 15 different membrane models. Part of this variation may result from genuine differences between different membranes, but part is probably due to differences of technique or interpretation. Some models show the membrane as a "sandwich," a central lipid layer bounded on each side by a layer of protein. However, much of the electron microscope study during the last ten years has shown the membrane protein as being not smoothly spread but in particles about 5–18 nm in diameter. Such particles are seen on both sides of the plasmalemma and the tonoplast in root cells (Northcote and Lewis, 1968). Some models show these particles stuck on the surface of the lipid layer; in others they are partly embedded; in others totally embedded. Which of these models is correct will affect how hydrophilic the membrane surface is; this will have an influence on rates of water flow within the cytoplasm, since no point in the cytoplasm is likely to be more than a fraction of a micron from a membrane. If the protein particles are partly or totally embedded they may form positions at which water can more easily cross the membrane. There have been a few reports that the frequency of the particles may increase with faster metabolic activity (Staehelin and Probine, 1970); this would be a possible means by which metabolism could alter membrane permeability. However, this is a highly speculative suggestion; the particles may themselves be involved in metabolic activity, perhaps in cell wall synthesis, and may have no relation to water movement.

VI. RATES OF WATER UPTAKE

Discussions of water regimes in the immediate environs of the root require a knowledge of rates of water uptake by the individual root. Uptake per unit length is of special interest. There are two ways to obtain such data. (1) Uptake by a short portion of a single root can be measured, using a micropotometer. (2) Uptake by a whole plant can be measured and then divided by the length of root on the plant. The latter method can also be applied to an area of vegetation in the field. In the first case, we need to know whether the root portion studied is representative of the root system as a whole. In the second case, we need to know whether an average rate of uptake is meaningful or whether some parts of the root system are contributing very little to water uptake. Thus, before actual figures are considered, it is relevant to discuss how rates of uptake may vary within a single root system.

A. Variation within a Single Root System

The following are possible causes of variation:
1. Root diameter. This is often related to morphological status, i.e., main axis, primary lateral, secondary lateral, and so on.
2. Distance from the stem base, which may or may not involve differences of depth.
3. Root age or stage of development.
4. Conditions in particular regions of the soil, including soil water conditions.

Of these the last could clearly be important, but it will get no further discussion at the moment.

1. Root Diameter

There appears to be virtually no data on uptake rates by roots of different diameters. The only data I know are by Kramer and Bullock (1966), for suberized roots of loblolly pine (*Pinus taeda*), where uptake per unit surface area increased markedly with increasing diameter. Potometer measurements on unsuberized roots have always been made on the fat main roots. There are obvious practical reasons for this, but it is unfortunate, because after the early seedling stage the main roots form only a tiny part of the total surface area or length of the root system. It is commonly assumed that uptake rates will be proportional to root surface area, other things being equal. If the main resistance is at the endodermis we should expect this, provided the endodermis area has a constant relationship to the root surface area in roots of different sizes. But if water enters the symplasm all over the cortical plasmalemma (see section III), then uptake rate might be more nearly proportional to root volume than to surface area. This merits investigation.

2. Distance from Stem Base

Direct evidence on uptake in relation to depth or distance from the stem base is also scanty. There are many reported cases when soil moisture content was initially uniform down a profile and then decreased most rapidly near the surface, but I know of no case where this has been clearly shown to be due to difference in uptake per unit amount of root rather than to differences in root density or to evaporation. Movement of water against gravity involves a pressure potential gradient of 0.1 bar/m. In a transpiring plant the potential difference between the outside of the root and the xylem is likely to be at least 1 bar; so one would expect uptake at 1-m depth to be reduced less than 10 % by the gravity effect. Only with very deep roots would an appreciable effect be expected.

Emerson (1954) measured the rates of flow through single detached main roots of three grass species under a pressure gradient of 1 cm/cm. Representative rates were 5 mm^3/hr for *Phleum pratense*, 2 mm^3/hr for *Lolium perenne* and 0.2 mm^3/hr for *Dactylis glomerata*. The rates agreed well with predictions by the Poiseuille-Hagen equation. He found that under grass plots there were 5.1 *Phleum* main roots and 9.3 *Lolium* main roots per cm^2. At these densities the pressure gradient along the xylem needed to supply water fast enough to balance a transpiration rate of 1 mm/hr would be 4 cm/cm for *Phleum* and 5 cm/cm for *Lolium*. These root densities were for roots extending at least 6 cm below the surface. At greater depths the densities of main roots would probably be lower and the gradients required correspondingly greater. The bulk conductivities provided by the roots in the top 6 cm would be 0.6 and 0.5 cm/day for *Phleum* and *Lolium* respectively, similar to the conductivity of soil at −0.1-bar matric potential (see Fig. 14.7). Wind (1955) pointed out that in very wet soil it would be easier for water to flow upward through the soil than through the roots. This does not prove, however, that all water would then enter the roots near the soil surface; if the main resistance is in crossing from the root surface to the xylem, water will tend to enter all over the root system.

Slavikova (1963) measured the water potential of fine roots of an ash tree (*Fraxinus excelsior*) at various horizontal distances, up to 4 m, from the trunk. When the soil was moist and its moisture content uniform throughout this distance, the root water potential varied with distance by about 10 cm/cm.

These measurements suggest that gradients up to 10 cm/cm (1 bar/m) can occur. More data are needed, especially on the density of main roots at various depths, to indicate whether larger gradients sometimes occur.

The results of Taylor and Klepper, quoted by Pearson (Chapter 10), indicate that when cotton is taking up water from a uniformly moist soil, the rate of uptake per centimeter of root does not vary appreciably with depth, at least down to 2 m. This contrasts with the results of Davis (1940). He grew a corn plant at one end of a long, fairly shallow box of soil. After the roots had become evenly distributed throughout, the soil was evenly wetted, and uptake at 10-cm intervals along the box determined by tensiometer readings. Uptake was initially fastest near the base of the stem, and the zone of rapid uptake gradually extended along the box. Soil near the plant base was depleted to the permanent wilting point while at the other end of the box, about a meter away, it was still at about field capacity. Main roots of corn have several very large xylem vessels, about 100μ in diameter (Essau, 1965, p. 713). If flow along these roots follows expectation from the Poiseuille-Hagen equation it is very difficult to explain Davis's results; the potential difference along the roots should be small compared with that across the root. One possibility is that there are cross walls in the xylem strongly impeding flow. Another is that tension in the xylem results in increased resistance, perhaps due to bubble formation in the larger vessels. There is some evidence that this happens in stems and petioles. Gibbs (1935, 1939) found that the water content of trunks of birch (*Betula alba*) and poplar (*Populus tremuloides*) fell markedly during the summer and rose again in the autumn, suggesting that some vessels became gas-filled. In contrast, three conifer species showed little variation, perhaps because they have no vessels. Haines (1935) observed bubbles in the xylem of various tree and shrub species when they were under conditions of moisture stress. One could object that the shock of removing the bark to make the observations might have promoted bubble formation. Milburn and Johnson (1966) developed an ingenious method of detecting bubble formation in undisturbed petioles. They attached an amplifier apparatus to the plant, and under conditions of water stress "clicks" could be heard, which they interpreted as due to formation of individual bubbles in the xylem. This interpretation was supported by other evidence on varying petiole conductivity (Milburn, 1966).

Thus the evidence, though scanty, suggests that if roots extend about a meter or more from the stem base, the resistance to flow along the xylem can be large enough to cause appreciable decrease in water uptake rate with increasing distance.

3. Root Age

Conclusions about the variation of uptake with root age first came from anatomical investigations (Scott and Priestley, 1928). First, it was concluded that uptake would be slow just behind the tip, because the xylem is not yet differentiated. This is no doubt true, but the region involved is usually less than 1 mm, and at most only a few millimeters long. Second, it was concluded that after a region of maximum uptake, the rate would decrease again further from the tip, due to suberization. There are three common processes that are all loosely referred to as suberization, and they do not necessarily affect water uptake in the same way.

1. Suberin may be deposited on the inside of the cell wall of the endodermis. On this may be superimposed a cellulose layer, and the walls may later become lignified. In some species certain endodermal cells (the passage cells) remain unsuberized, but this is not always so. The cortex normally remains alive. It has generally been assumed that the suberized endodermal cells are completely impermeable. However, Clarkson *et al.* (1971) have shown by electron microscopy that the thickened endodermal wall of barley roots is penetrated by numerous plasmodesmata, which may form a pathway for water flow of relatively low resistance (see section III). This interesting discovery necessitates a complete rethinking about the significance of endodermal suberization.

2. A periderm may form in the pericycle. The endodermis and cortex are then sloughed off. Again such roots have generally been assumed to be impermeable, but entry could occur through lenticels and other breaks.

3. Suberin may be deposited in the walls of the exodermis, a layer just beneath the epidermis. Often only alternate cells are suberized, and this would probably affect root permeability very little. Even where all are suberized there can be pits, which, as with the endodermis, leaves the possibility of water movement via plasmodesmata.

Thus the effects of suberization on water uptake are not as obvious as has been supposed. Measurements of uptake in different regions of the root have been accumulating for more than thirty years. The only simple generalization that can be drawn from them is that no simple generalization is possible. Brouwer (1953a, 1953b, 1954b) has performed some of the technically most satisfactory experiments, using broad beans. In these, measurements of uptake were made along a single root, while the rest of the root system was in a separate vessel. The plant was transpiring normally. Figure 14.5 shows the effect of transferring the

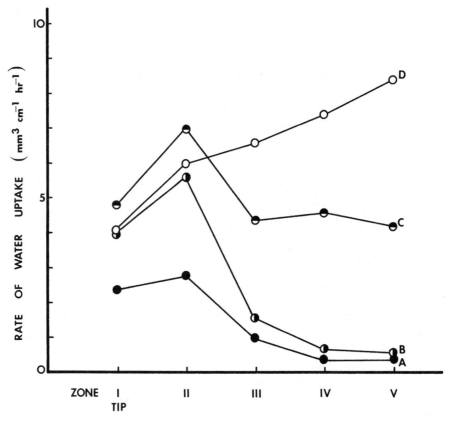

Fig. 14.5. Water uptake rate (mm^3 cm^{-1} hr^{-1}) in five zones, each 2.5 cm long, of a *Vicia faba* root. *A*: all roots in water; *B*, *C*, and *D*: 20 min, 1 hr, 3 hr, respectively, after all other roots were transferred to salt solution. (Reprinted by permission from R. Brouwer, *Proc. Kon. Neer. Acad. Wet.* (C) 56: 130, copyright 1953, Royal Netherlands Acadamy of Arts and Sciences)

bulk of the root system (but not the test root) from water to a salt solution of osmotic potential of -2 bars. This stimulated uptake by the test root. Before the change of treatment (line *A*), uptake was much faster near the tip than farther back, and this pattern was still maintained

20 minutes after the change (*B*). But with time the distribution changed, until after 3 hours (*D*) uptake was fastest farthest from the tip. Brouwer obtained similar results by altering the aerial environment. Similar results were previously obtained by Sierp and Brewig (1935) and Brewig (1936b), also using broad beans.

Figure 14.6 shows another example calculated from data of Meiri and Anderson (1970b). This is based on the rate of exudation by detached

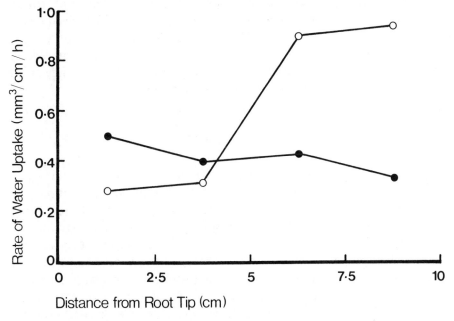

Fig. 14.6. Water uptake by different regions of detached main root of corn, calculated from exudation by different lengths of root. External solution: *open circles*, 1 mM KCl; *black circles*, 25 mM KCl. (Data from Meiri and Anderson, 1970b)

corn roots of various lengths. It shows that the distribution of uptake is influenced by the external salt concentration. This was not related to differences in salt uptake but to differences in root permeability. Anderson and House (1967) reported that in similar corn roots suberization and thickening of the endodermis was completed by 8 cm from the tip. Evidently this did not reduce permeability significantly.

These two examples are sufficient to show that no generalization can be made about variation of uptake within the tip 10 cm or so. These roots of course had no periderm. Kramer and Bullock (1966) are the only workers who have compared uptake, under the same conditions, by roots with and without periderm. They studied loblolly pine and yellow poplar (*Liriodendron tulipifera*), measuring uptake rates of small portions of root under an applied suction of 0.4 bar. The results were very variable, those for yellow poplar so much so that no meaningful average rate could be calculated. This suggests that uptake by roots with periderm may depend on lenticels or other breaks in the periderm. In the pine, uptake by roots grown in soil was 4–97 $mm^3 cm^{-2} hr^{-1}$ for unsuberized roots and 3–190 for suberized. So there was no indication that suberization (i.e., formation of a periderm) affected root permeability significantly. In suberized roots uptake per unit area increased with increasing diameter, and rates above 40 $mm^3 cm^{-2} hr^{-1}$ were only found in roots more than 3 mm in diameter, larger than any unsuberized roots.

In conclusion, there is no justification for the generalization that uptake occurs mainly in younger parts of the root system. It seems likely that uptake may vary markedly with root diameter and distance from the stem base, but evidence on this is inadequate.

B. Quantitative Data

Table 14.3 summarizes data from three sources on rates of uptake per unit amount of root. Part (*a*) gives data from micropotometer studies on single attached roots. Results are included only if they meet the following requirements: (1) the plants were transpiring; (2) roots other than the test root were left intact and were in similar conditions to the test roots. This second requirement in fact eliminates most of the published data. Part (*b*) is based on data from these experiments where plants were grown in containers of soil. *Dactylis* and wheat (*Triticum*) were in growth chambers; leek plants were outdoors under a polyethylene-covered frame. The transpiration rates, averaged over the 24 hr, have been divided by the total length of root to calculate the average uptake rate per unit length. Here and in section VII, I have recalculated the results of Perrier *et al.* for *Dactylis* from their primary data; they made several mistakes (see Newman, 1969a).

Since data from these two methods are so scanty, a more general type of estimate is made in part (*c*). The values of L_A, the root length under 1 cm^2 of ground surface, are taken from the summary table of Newman (1969a), where the original sources are listed. To these I have added data for *Stylosanthes humilis* (Townsville lucerne) from Torssell *et al.* (1968) and *Pseudotsuga taxifolia* (Douglas fir) from Reynolds (1970). Uptake rates are calculated for two transpiration rates; the higher of these, 1 mm/hr, represents the highest rate likely to be sustained during the midday part of a sunny summer day in temperate regions (Monteith, 1965). Higher values could occur in special circumstances, e.g., in irrigated crops in arid regions.

Comparing the three parts of the table, the rate by *Pinus echinata* in part (*a*) falls within the range for woody species in part (*c*), and the *Citrus*

Table 14.3. Rates of water uptake per unit amount of root, determined by three methods

a. Uptake by short length of single root, attached to transpiring plant

		Root diameter (mm)	Rate of uptake	
Species	Root condition		per unit area (mm^3 cm^{-2} hr^{-1})	per unit length (mm^3 cm^{-1} hr^{-1})
Pinus echinata	suberized	3–5	2	2
Citrus sp.	suberized	?	4	
Vicia faba	unsuberized	[1]†	[20]	6

b. Calculated from transpiration rate/root length

	Rate of uptake (mm^3 cm^{-1} hr^{-1})	
Species	Soil at about field capacity	SMP‡ = −15 bars
Dactylis glomerata (orchard grass)		
subsp. *judaica*	0.03	0.0025
subsp. *lusitanica*	0.05	0.005
Triticum aestivum (wheat)	0.07	0.012
Allium porrum (leek)	0.8	

c. Root lengths under unit area of ground surface (L_A) and corresponding rates of uptake at two transpiration rates

			Rate of uptake (mm^3 cm^{-1} hr^{-1})	
Plant group	Number of species examined	L_A (cm/cm^2)	Transpiration = 1 mm/hr	Transpiration = 0.1 mm/hr
Gramineae	15	100–4000	1–0.025	0.1–0.0025
Other herbaceous species	8	52–670	2–0.15	0.2–0.015
Woody species	5	[5–]17–110	[20–]6–0.9	[2–]0.6–0.09

Sources: Pinus, Kramer, 1946; *Citrus,* Hayward *et al.,* 1942; *Vicia,* Brouwer, 1953a; *Dactylis,* Perrier *et al.,* 1961; *Triticum,* Andrews and Newman, 1969; *Allium,* Brewster and Tinker, 1970; part (c), see text.
† Figures in brackets are approximate.
‡ SMP = soil matric potential.

would probably have had a similar rate per centimeter. Plants of *Vicia* have a rate higher than the range for herbaceous dicotyledons. This was probably due, at least in part, to the test root being a large main axis, whose diameter might well be 5 times the average for a whole mature plant. The transpiration rate may also have been higher than would occur in the field. In part (*b*) all three species fall within the corresponding ranges in part (*c*). *Dactylis* and wheat are both toward the low end, but their transpiration rates at field capacity were between 0.1 and 0.2 mm/hr, not very high rates.

It thus seems reasonable to take the rates in part (*c*) of the table as indications of the likely values in the field, remembering always that higher rates could occur on fat main roots, and also if uptake was restricted to part of the root system by dry soil elsewhere or for some other reason.

VII. MATRIC POTENTIAL GRADIENTS IN THE RHIZOSPHERE

A. Estimates

Water will move through the soil to the root surface only if there is a gradient of matric potential. There has been argument for many years over how large the gradients of matric potential are in the immediate vicinity of a root (rhizosphere). The distances involved are very small, a few millimeters at most, and direct measurements of water content or matric potential differences at this scale are very difficult. So far no such measurements have been published, though there are people actively working on this. Considerable weight therefore has to be placed on estimates, which depend particularly on knowledge of two things, soil conductivity and rate of water uptake per unit length of root.

Soil conductivity declines as the soil dries, a fact which has been known for many years. The concept of field capacity, introduced many years ago, involves soil conductivity: When the conductivity becomes too low for appreciable water movement by gravity, field capacity has been reached. This idea was extended by some people to the further, unjustified, conclusion that in soils appreciably drier than field capacity, water movement is negligible under any driving force. This appeared to be supported by early determinations of soil conductivity, which could only cover soil moisture contents from saturation down to about field capacity. When such data are plotted on arithmetic scales they look like Figure 14.7b, where conductivity appears to be effectively zero at −0.2 bar. This led to the idea that a root might take up all the available water in its root hair zone, so that the soil matric potential there would reach −15 bars; yet the soil just outside would remain at field capacity, and no water would move into the root hair zone because it was effectively impermeable. Thus the plant could only get more water by root extension. This idea received

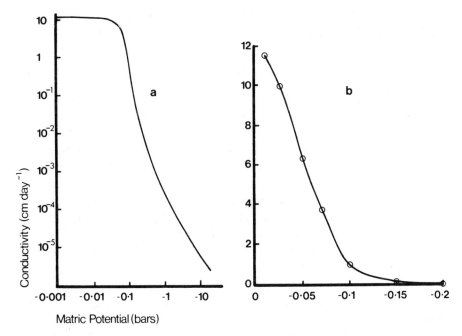

Fig. 14.7. Relationship between conductivity and matric potential for Pachappa sandy loam. Part *a* plotted on log-log scale; *b* part of the same, on arithmetic scale. (Data from Gardner, 1960)

support from a calculation by Kramer and Coile (1940) that the rate of root extension of a rye (*Secale cereale*) plant would lead to interception of enough water to balance transpiration losses. We should note that rye has a profuse root system, especially when, as in this case, it is a winter rye plant which has been grown in the warm conditions of a greenhouse. If their calculations had been applied to almost any nongraminaceous species, a very different conclusion would have been reached. In any case, as we shall see, the idea that water cannot move to roots is mistaken.

A turning point came in 1956, when two methods of measuring soil conductivity were proposed that could be used over the whole range of matric potential down to −15 bars or below. These were the pressure plate outflow method of Gardner (1956) and the method of inflow into dry soil of Bruce and Klute (1956). With subsequent improvements, these still form the basis of current methods. For a review of these methods, see Klute (1965) or Swartzendruber (1969). Figure 14.7a shows data of Gardner (1960) for Pachappa sandy loam, obtained by the pressure plate outflow method. It is plotted on a log/log scale and gives a rather different impression from Figure 14.7b, although the data are the same.

Using such data, Gardner (1960) estimated the gradients of soil matric potential in the rhizosphere. His steady-state formula applies, strictly speaking, where all water reaching the root comes from beyond a distance

b (measured from the root axis). The root is considered to be a cylinder of uniform radius, r, without root hairs. Then,

$$\tau_b - \tau_r = (q/2\pi k) \cdot \ln(b/r), \qquad [13]$$

where τ_b and τ_r are the matric potentials at distance b and at the root surface respectively, q is the rate of uptake (cm^3 water per cm root per day), and k is the soil conductivity (cm/day); τ_b, τ_r, b, and r are in centimeters. Obviously not all the water does come from beyond a given distance; the soil near the root is losing moisture. However, because of the cylindrical geometry of the root and of the soil zone from which it draws water, the steepest potential gradients lie near the root, whereas most of the water comes from some distance away; so the approximations in this model are not too critical. Cowan (1965) used an alternative set of assumptions, that the root withdraws water from a cylinder of soil around it, the water content of this soil declines at a steady rate, and no water moves into it from outside. If the radius of the cylinder is approximately half the mean distance between neighboring roots, this seems a more realistic model than the steady-state one; but the resulting mathematics is rather more complicated, and the actual results obtained from the two models are usually similar (Passioura and Cowan, 1968).

Figure 14.8 shows estimates by Cowan (1965) of the matric potentials in the vicinity of a root 2 mm in diameter which is taking up water at 0.16 cm^3 cm^{-1} day^{-1} (about 7 mm^3 cm^{-1} hr^{-1}). Gardner (1960) made fairly similar predictions. According to these calculations, when the soil is no drier than -2 or -3 bars, quite small gradients occur; but as the soil dries further, the gradients required to maintain the same rate of flow become large. This does not support the idea that when the bulk soil is at field capacity the root surface can be at -15 bars, but it does suggest that the plant could experience soil matric potentials markedly different from those in the bulk soil. However, the gradients are proportional to the rate of uptake (as shown by equation [13]). If the uptake rate per centimeter of root is very low, then only a small gradient of matric potential is required to move the water at the required rate, even in quite dry soil. The crucial question, therefore, is whether the rates assumed by Gardner and Cowan, 4 and 7 mm^3 cm^{-1} hr^{-1}, are common in field conditions. In Table 14.3, these rates are at the very top end of the range shown in part (c), likely to occur only at very high transpiration rates, and then only in plants with very sparse root systems. Furthermore, by the time the soil matric potential has fallen to several bars below zero, the plants would probably not be transpiring at the potential rate, because of stomatal closure. The transpiration rate is thus much more likely to be about 0.1 mm/hr or less, as in the right-hand column of Table 14.3c. The rates there are below those used by Gardner and Cowan, by factors ranging

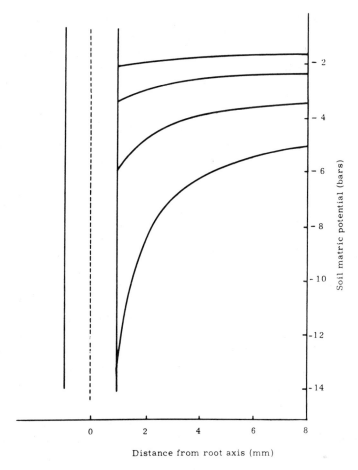

Fig. 14.8. Predictions by Cowan (1965) of matric potentials in the vicinity of a root taking up water at 0.16 cm³ cm⁻¹ day⁻¹

from about 2 to 2,000, which indicates that the gradients predicted by them are 2–2,000 times too steep.

In the experiments with *Dactylis* and wheat (Table 14.3b), equation [13] can be applied to the data for various soil matric potentials. At − 15 bars the predicted difference in matric potential between the root surface and the point midway between neighboring roots was: *Dactylis glomerata judaica*, 0.25×10^{-3} bar; *D. glomerata lusitanica*, 0.5×10^{-3} bar; wheat, 20×10^{-3} bar. In moister soil the values were even smaller.

Another approach is to compare the resistance to water flow in the soil and in the plant. (Plant resistance here refers only to the pathway from the root surface to the evaporation sites in the leaf; the subsequent vapor pathway is not included.) If the soil resistance (R_s) is small compared with that of the plant (R_p), then it will have little effect on leaf water deficits or transpiration rate. If it is assumed that both R_s and R_p are unaffected by the rate of water flow (often an unjustified assumption; see section X), then this avoids the difficulty of predicting transpiration rate at different soil matric potentials. I have made estimates (Newman, 1969a), based on equation [13], using k values for Pachappa sandy loam (Fig. 14.7); b/r is taken as 250, and plant resistance is assumed constant at 5×10^3 days. Table 14.4 shows the prediction of how low the soil matric potential

Table 14.4. Estimates, for vegetation with various values of L_A (length of root under unit area of ground surface), of the soil matric potential at which the soil (rhizosphere) resistance equals one-fourth of the plant resistance

L_A (cm/cm^2)	1	10	100	1000
Critical soil matric potential (bars)	-0.7	-2	-10	< -40

Note: For details of calculation, see Newman, 1969a.

would need to fall for soil resistance to be as high as one-fourth of the plant resistance. Comparing this with the data on L_A in Table 14.3c suggests that in many species the soil resistance will not become appreciable, compared with the plant resistance, until near or below the permanent wilting point. However, some species, especially some woody plants, have sparse enough root systems for soil resistance to become appreciable when the soil matric potential is a few bars below zero, though never at about field capacity.

Thus all these calculations lead to the conclusion that most plants have such a great length of root, and the uptake per centimeter is therefore so slow, that the gradients required to drive the water across the rhizosphere are generally very small even in quite dry soil; or to put it another way, the soil resistance generally remains small compared with that of the plant. We must remember, however, the possibility that uptake rates may be higher than we have assumed. First, uptake may vary within the root system, for reasons discussed in the previous section. Second, uptake may be restricted to certain soil layers by drying of soil elsewhere. In both cases there would be a tendency to self-correction; if the fast uptake led to big potential gradients and a marked fall in water potential at the root surface, this would probably reduce uptake there and raise it elsewhere in the root system.

I have ignored the problems raised by temporal fluctuations in rate of uptake. The rates of uptake used in the calculations are intended to apply to the period of fastest uptake during the day; if this rate is maintained for too short a time for equilibrium to be achieved, then the potential differences will be smaller than predicted. If there is a change in the rate of uptake by the root, the time taken for this to influence the rate of movement through the soil a distance L away is given by

$$t = L^2/D_w \qquad\qquad [14]$$

where t is the time and D_w the diffusivity[1] of water (Philip, 1966). In the main absorbing zone, root density is rarely less than 1 cm/cm^3; so the distance to the midpoint between neighboring roots will rarely be more than 5 mm. For Pachappa sandy loam, $D_w = 7$ cm^2/day at -15 bars matric potential (Gardner, 1960). For this value of D_w, and $L = 0.5$ cm, equation [14] gives $t = 50$ minutes. This suggests that even in dry soil the lag will be small compared with the time involved in the diurnal cycle of uptake on a clear or overcast day, though not compared with short-term fluctuations on a partly cloudy day. An interesting discussion of lags in adjustment within the plant is given by Weatherley (1970).

B. Experimental Evidence

Over the last twenty years, various sorts of experimental evidence have been put forward that are claimed to show appreciable matric potential gradients in the rhizosphere. All have used the plant in some manner to indicate that the matric potential at its root surface is not the same as that in the bulk soil. I have reviewed this work (Newman, 1969b) and have shown that none of it is acceptable as evidence, except for one paper by Sykes and Loomis (1967), which suggests that soil conductivity can affect the permanent wilting point, a conclusion not in conflict with the calculations above. I shall not review this evidence again here. Inevitably it was indirect or involved certain assumptions, and often close examination of the results produced internal evidence that the authors' interpretations were incorrect. In other cases there was a crucial misunderstanding of the importance of scale. This I can illustrate by the following example. Imagine a single plant growing in a large container of soil. Its roots ramify within a cylindrical volume of soil 5 cm in diameter and 20 cm deep. Outside this there is untapped soil. If the plant is transpiring rapidly, a potential difference of several bars could build up between the rooting zone and a point in the soil 10 cm away from it. There is no doubt about this; calculations predict it, and it can be measured (Macklon and Weatherley, 1965). The mistake that has been made is to assume that one can scale this down, making the cylinder a single root 0.5 mm in diameter, and with the same potential difference now occurring across 1 mm in the rhizosphere. This is correct *only* if the rate of uptake per unit length remains the same. In other words, the plant would have only a single root 20 cm long through which all its water uptake occurred. I have called any movement of water which is not to an individual root *pararhizal*, the case

[1] Meaning of diffusivity (D_w): if the driving force for water movement through soil is taken as a gradient of water content, $d\theta/dz$, then the rate of flow, $F = D_w(d\theta/dz)$.

cited above, with its cylinder of exploited soil, being an example. It is important to realize the differences between rhizosphere and pararhizal pathways (though there is no sharp dividing line between them). With pararhizal movement, the rates of flow per cm^2 are likely to be much faster, and the distances longer. Also the flow is often parallel (e.g., movement up from a water table), rather than converging toward the root. All these features mean that much bigger differences in potential can occur in a pararhizal pathway than will ever occur in the rhizosphere. All the conclusions of this section apply only to the individual root. There is no doubt that the soil resistance can be significant when water moves over long distances to the rooting zone. Equally, root interception can be important if the roots grow as an advancing front into unexploited soil, rather than (as I have assumed) among existing roots.

Recently three papers have reported measurements of water potentials of both roots and soil (Kaufmann, 1968; Vartanian and Vieira-da-Silva, 1968; De Roo, 1969). All three reported that the difference in potential between root and soil, $\psi_s - \psi_r$, could be large. This could be taken as evidence of a large difference of water potential across the rhizosphere. Table 14.5 shows some of Kaufmann's data. Measurement of root water

Table 14.5. Root water potential of loblolly pine (*Pinus taeda*) at various soil water potentials

Water potential (bars)	
Soil	Root
−1.0	−4.0
−4.0	−5.6
−8.0	−8.5
−12.0	−11.5
−16.0	−15.0

Source: Data from Kaufmann, 1968.

potential presents technical problems, and in my opinion all of these investigations are open to objection on technical grounds. Kaufmann, working with pines, measured the potentials of roots that were not in the soil but hanging below it in "essentially saturated" air. Even a small change in the relative humidity of the air could appreciably alter the root water potential. Vartanian and Vieira-da-Silva measured the water potential of the tip of the main root of *Sinapis alba* and related it to the potential of the soil immediately around it. However, the main bulk of the root system was in drier soil, and this might well affect the water potential of the root measured (Slavikova, 1967). De Roo measured root water potential of tobacco (*Nicotiana* sp.) by inserting into a pressure bomb the entire soil plus root system from a small pot or a main root with its surrounding soil. In my view he was measuring the water potential, not of the roots, but of some sort of average of roots and soil. How far this average would be weighted toward the soil or the roots is difficult to predict. Even ignoring technical criticism, the results do not suggest large rhizosphere gradients. One must remember that there will be differences in water potential across the root and that the measured potential will be some sort of average of these. The results suggest that the measured difference $(\psi_s - \psi_r)$ was mainly due to these gradients in the root. First, all three showed large $(\psi_s - \psi_r)$ when the soil was at about field capacity. Large rhizosphere gradients are very unlikely under these conditions. Furthermore, as the soil dried, $(\psi_s - \psi_r)$ decreased in most cases (Table 14.5). It follows from equation [9] that $\psi_s - \psi_r = RF$, where R is the combined resistance of the relevant parts of the soil and plant pathways. If R is approximately constant, as it might well be if the resistance is mainly in the root, $(\psi_s - \psi_r)$ would decline as transpiration declined. But if R was mainly in the soil, the conductivity would decline so sharply as the soil dried that a rise in $(\psi_s - \psi_r)$ would be expected. Therefore these results suggest that the main resistance was in the root.

In Kaufmann's results (Table 14.5), $(\psi_s - \psi_r)$ became negative in dry soils. This reversal was even more marked in De Roo's experiments. It is uncertain whether this is due to experimental error or is genuine.

C. Significance for the Plant

Two matters meriting brief discussion arise from the conclusion that rhizosphere gradients are generally small. First, if this is true, it follows that most plants have a far higher root density than they need for water uptake. In a crude way, comparison of species supports this. A lawn growing near a stand of pine may have 100 times the length of root, per unit area of ground, but the pine survives summer droughts all right. Of course, drought resistance involves many characteristics of the plant; so such comparisons between different species are of limited value. Derera *et al.* (1969) compared 15 varieties of wheat in Australia but could find no consistent correlation between root density and drought tolerance. The most direct evidence comes from the experiments performed by Miss Andrews and myself in which we used a root-pruning technique to alter the root densities of wheat plants growing in pots of soil (Andrews and Newman, 1968). Some pots were then allowed to dry out, while others were kept at field capacity. In growth response, one might have expected the pruned plants to suffer more from drought, but in fact the reverse was true; the drying treatment slowed down the growth of the unpruned plants more than the pruned. This was only statistically significant in one of the two experiments; but certainly the pruning did

not increase drought susceptibility. The transpiration rate was significantly decreased by root pruning when the soil was at field capacity; but
as the soil dried, this difference between pruned and unpruned became
less, and by −15-bar soil matric potential it was virtually nil. If soil
resistance was appreciable, one would have expected an increasing difference in transpiration between pruned and unpruned as the soil dried
(Cowan, 1965). Thus these experiments are in agreement with the prediction that rhizosphere gradients are very small. Of course, wheat is a
profusely rooted plant; all species would not necessarily give the same
results.

A second matter worth discussion is why only a slight fall in soil matric
potential below field capacity, e.g., allowing it to reach −1 bar occasionally, can cause a significant decrease in plant growth (Sands and
Rutter, 1959; Jarvis, 1963). If the leaves and shoot tips are at −4 bars
water potential (for example), it seems odd that a fall to, say, −5 bars
should have a marked influence on their metabolism. In the past, some
people have suggested that the effect is due to the large decrease in soil
conductivity. What they are suggesting, though they do not always seem
to have realized it, is that the plant water potential falls much faster than
the soil water potential as the soil dries. This is in fact rarely true (see
section X); more often it falls in parallel, or even less rapidly. I can suggest
three possible alternative explanations. (1) Stomata often close sharply
over a rather narrow range of leaf water potential (Gardner and Ehlig,
1963). Hence a small change in leaf water potential may cause a large
change in CO_2 uptake by the leaf. (2) The water potential of the root
affects its kinin production (Itai and Vaadia, 1965), and kinins affect
various metabolic processes. A small drop in root water potential might
in this way affect plant growth, though this has not been clearly demonstrated, as far as I know. (3) A fall in soil matric potential from −0.1 to
only −0.5 bar would be accompanied by a marked decrease in the diffusion coefficient of ions in the soil and could thus reduce nutrient supply
to the roots.

VIII. OSMOTIC EFFECTS

Osmotic potentials in the soil and the contribution of osmosis to water
uptake are often ignored. The reasons given for this are generally that
(1) the osmotic component of soil water potential is usually small compared with the matric component; (2) osmosis has little effect on water
uptake compared with transpiration pull; (3) plant osmotic potential
adjusts to any changes in soil osmotic potential. A fourth reason, not
usually stated explicitly, is that the catenary hypothesis does not work so

well if osmosis is allowed to contribute to water movement across the root. Each of these reasons has some force, but none is completely true. This section is mainly concerned with discussing points (1) and (2).

A. Osmotic Potentials in the Rhizosphere

In saline soils the osmotic potential can be very significant. The special problems of saline soils will not be discussed here. In most nonsaline soils, the bulk osmotic potential is above (less negative than) -1 bar at field capacity. Greenhouse staff are sometimes so enthusiastic about adding fertilizers that an osmotic potential well below -1 bar can develop. In one experiment I found -3 bars at field capacity for this reason. The plants grew excellently.

There is one situation very common on farms where local extreme osmotic potentials occur, namely in the vicinity of bands or particles of phosphorus fertilizer. An extremely concentrated solution develops, which, because of the low diffusion coefficient of phosphate, is very slow to spread out. Blanchar and Caldwell (1966) simulated a fertilizer band in the laboratory, using $Ca(H_2PO_4)_2 \cdot H_2O$, the principal P compound in superphosphate fertilizer. They found that after two weeks the soil solution 0–1 cm from the band contained 71 g/1 P, which assuming equivalent amounts of cations would give an osmotic potential of about -80 bars. Lindsay and Stephenson (1959) obtained similar results; even after six weeks the soil solution 2 cm from the band contained 28 g/1 P, indicating about a -30-bar osmotic potential.[2] It is hardly surprising that root tips growing into the vicinity of the band were killed (Blanchar and Caldwell, 1966; Duncan and Ohlrogge, 1958). Apart from the extreme osmotic potential, the pH is often below 3, and the soluble aluminum content can be very high. What is more surprising is that roots can ever penetrate fertilizer bands. Duncan and Ohlrogge (1958) observed that although the tips of the main roots of corn were killed by the band, lateral branch roots could grow into it unharmed.

When water moves through the soil to the root surface, it brings with it ions and other solutes. If these arrive at the root faster than they are taken up, they will tend to accumulate in the rhizosphere (see Chapter 18). The question arises whether this could cause significant osmotic potentials. As the concentration builds up, diffusion away from the root will increase, until a steady state is reached when diffusion away just balances arrival by mass-flow. If we take the extreme case where none of the ion is taken

[2] Note added in proof: according to W. L. Lindsay (personal communication), normal field practice would involve less concentrated application of fertilizer than was made in these experiments, and the concentration in solution would therefore decrease more rapidly.

up by the root, the steady-state concentration can be calculated by this formula, given by Gardner (1965b):

$$C_r/C_b = (b/r)^{q/2\pi D}, \qquad\qquad [15]$$

where C_r is the concentration at the root surface, C_b the concentration at a distance b from the root axis, r the root radius, q the rate of water uptake (cm^3 cm^{-1} sec^{-1}), and D the diffusion coefficient of the ion in soil (cm^2/sec). This assumes that all ions come from beyond distance b. The assumption of no uptake is never strictly true, but it can be approximately true for such ions as Na^+, Cl^-, and SO_4^{2-}. Table 14.6 gives values of

Table 14.6. Ratio of ion concentration at root surface to ion concentration in soil at distance equal to five times the root radius

Rate of water uptake		Diffusion coefficient (D) of ion (cm^2/sec)	
mm^3 cm^{-1} hr^{-1}	cm^3 cm^{-1} day^{-1}	10^{-6}	10^{-7}
42	1	19	8×10^{12}
4.2	0.1	1.3	19
0.42	0.01	1.03	1.3
0.042	0.001	1.003	1.03

Note: Calculated from equation [15]. Assumes steady state and no ion uptake by the root.

C_r/C_b, when $b/r = 5$, for various values of q. Because of the low values of D (much lower than those for water), a steady state will normally take some days to attain, and q should therefore be average uptake rates over 24 hours. The diffusion coefficient for many anions, including Cl^- and NO_3^-, is of the order of 10^{-6} cm^2/sec, but for most cations it is nearer 10^{-7} (Fried and Broeshart, 1967). Because C_r/C_b is exponentially related to q, it increases very suddenly as q rises beyond a certain level. If an uptake rate of 1 cm^3 cm^{-1} day^{-1} could be sustained, enormous concentrations of cations are predicted. Of course, the steady value of 8×10^{12} could never be attained; this means that accumulation would continue virtually indefinitely. However, I very much doubt whether rates as high as 1 cm^3 cm^{-1} day^{-1} are ever attained, and even 0.1 would be extremely high as a 24-hr average. From the evolutionary point of view, one may speculate that natural selection has ensured that plants always have enough roots to avoid such accumulations in the rhizosphere. It seems possible that C_r/C_b may be as high as 10 under certain circumstances, e.g., some woody plants with sparse root systems, or young seedlings, or plants whose roots have been damaged by parasites or transplanting. It is possible that in such circumstances the restricted root system leads to water stress not so much through low matric potential, as is normally supposed, but mainly through low osmotic potential.[3]

It has also been suggested that concentration may occur within the root, at the endodermis, by the same process. If the symplasm pathway operates (see section III), the effect would be smaller; but to estimate the maximum possible effect, let us assume that all the water passes through the free space to the endodermis. In formula [15] b is now the root radius, r the radius of the endodermal cylinder, and D the diffusion coefficient in the cortex as a whole, which is approximately D in water (about 10^{-5}) multiplied by the proportion of the cortex volume occupied by free space, say 0.05. Often b/r is in the range 2–3, but it may be as high as 7, as in *Ranunculus acris*. Table 14.7 shows the ratio of the concentration at the

Table 14.7. Ratio of ion concentration at the endodermis (radius r) to ion concentration at the surface of the root (radius b)

Rate of water uptake		$b/r = 2$	$b/r = 7$
mm^3 cm^{-1} hr^{-1}	cm^3 cm^{-1} day^{-1}		
4.2	0.1	1.29	2.05
0.42	0.01	1.02	1.07
0.042	0.001	1.002	1.007

Note: Calculated from equation [15]. Assumes steady state, that all water crosses cortex by free space, $D = 5 \times 10^{-7}$ cm^2/sec, and no uptake of ion into root symplasm.

endodermis to that at the outside of the root for $D = 5 \times 10^{-7}$ cm^2/sec. When $b/r = 2$, even the very high uptake rate of 0.1 cm^3 cm^{-1} day^{-1} gives only about a 30% increase in concentration at the endodermis, though with $b/r = 7$, the increase is about 100%. In general, it seems unlikely that concentration gradients within the root will be important. The main uncertainty in these calculations is that D might be much lower than I have assumed, due to interaction of the ions with the wall material. Bernstein and Nieman (1960) provide some experimental evidence on solute accumulation. They placed the root systems of intact plants of four species in NaCl or mannitol, then determined the amount that had entered the free space by allowing it to diffuse out into water. Surface contamination was corrected for. They found that a tenfold increase in

[3] Since writing this review I have read the paper by Passioura and Frere (1967, *Aust. J. Soil Res.* 5: 149–59), in which they predict ion concentrations at the root surface (C_r) by a much more sophisticated numerical method, using a computer. They support the conclusion that appreciable accumulation near the root will only occur at extremely high rates of water uptake. They give estimates of the diurnal fluctuation of C_r associated with diurnal fluctuation of water uptake, which make it clear that the higher values of C_r/C_b in Table 14.6 would in reality not be steady states but 24-hr averages over wide diurnal fluctuations. They also point out that as soil dries, C_r could rise markedly because of the sharp fall in D.

transpiration rate caused no significant increase in the amount of solute in the free space, except for one experiment with corn where the increase was about 30%. This supports the conclusion that accumulation in the free space due to mass-flow is generally small.

It is interesting to compare this radial movement of water and ions with the longitudinal movement in the xylem. House and Findlay (1966b) found that in corn main roots of 1-mm diameter, the volume of the xylem vessels was about 0.28 mm^3/cm root. If the root was taking up water at the moderate rate of 0.01 cm^3 cm^{-1} day^{-1} along a 10-cm apical portion, water movement in the xylem 10 cm from the tip would be 360 cm/day. Even if there were very marked ionic concentration differences in the xylem sap, longitudinal diffusion rates would be hundreds or thousands of times slower than mass-flow movement. Thus outside the endodermis mass-flow would be too slow (at this uptake rate) to have any significant effect on ion concentration; yet inside the endodermis ion concentration would be under the control of mass-flow and would be very much affected by the rate of water uptake (as, indeed, measurements show; see Table 14.10). So an increase in transpiration rate is likely to reduce the ion concentrations inside the endodermis without increasing appreciably the concentrations outside the endodermis, thus altering the osmotic driving force for water entry.

B. Osmosis and Water Uptake

1. Osmotic Driving Force

It is commonly stated that osmosis contributes little to water uptake compared with the transpiration pull. Probably this is often true, but not always. Brouwer (1965) gives estimates of the xylem sap concentration in intact transpiring plants, according to which there can be a positive osmotic potential difference between the medium and the xylem sap of 1 bar or more. (A positive potential difference tends to drive water inwards.) This could make a significant contribution to water uptake.

As the transpiration rate increases, the xylem sap usually becomes more dilute. This can be seen in experiments with detopped root systems, where transpiration pull is simulated by suction from above or pressure from below. An example is given in Table 14.10. At low applied pressure there was a large positive osmotic potential difference in tomatoes, but this had declined to zero at 1.5-bar applied pressure. A more usual situation is for the xylem sap to become more dilute than the external medium. For example, in a similar experiment, also with tomatoes, Lopushinsky (1964) found that at 0.33-bar applied pressure, the concentration of the xylem exudate was almost exactly the same as that of the external solution; but by the time the applied pressure had reached 2 bars, the xylem exudate concentration was only 8% of the external concentration.

Under these circumstances osmosis could significantly retard water uptake. The maximum effect is limited by the osmotic potential of the external medium, and this is one reason why an external osmotic potential of -1 bar, say, could be important.

It is quite common in the field for one part of the root system to be in soil at field capacity while another part is in drier soil (e.g., near the surface). In these circumstances one would often expect water entering the root system from moist soil to pass out again from another part of the root system into the dry soil. Sometimes this does happen (Breazeale, 1930; Hunter and Kelley, 1946; Thorup, 1969). In these cases the soil that gained water was extremely dry, well below the permanent wilting point, but even so the gain was slow. I suggest that local portions of the root system can partially adjust their xylem water potential osmotically when the soil around them becomes dry. Figure 14.4 shows that when the osmotic potential around the root system is lowered suddenly, water may at first be drawn out of the roots, but the rate of loss soon falls, and usually the plant resumes uptake after a time. Presumably this is due to osmotic adjustment. If water is withdrawn from the xylem but most of the solutes remain behind, the xylem sap will become more and more concentrated until the osmotic gradient is again inwards and a new steady state can be set up. I have performed experiments with dwarf bean (*Phaseolus vulgaris*) root systems in which downward suction was applied by attaching a suction pump to the vessel containing the roots. The results resembled qualitatively those in Figure 14.4. To take one example, a plant four weeks old exuded at 0.8 mm^3/min with no applied suction. Suction of 0.6 bar upwards produced 3.9 mm^3/min. Therefore one might at first sight expect 0.6 bar downwards to result in considerable loss of water to the medium. In fact there was water loss at first, but after about 15 minutes it had stopped and a steady uptake of 0.15 mm^3/min was finally reached. This was not a "high-salt" plant; it had been grown in tap water with no added nutrients and transferred the day before the experiment to 0.01-strength Hoagland solution. High-salt plants give similar results. If such adjustment occurs locally in a root system, it would prevent or reduce water loss to drier regions of soil. However, data of Slavikova (1967) suggest that this does not occur in *Fraxinus excelsior*. She grew young trees with divided root systems and found that when the two halves were in soils of very different moisture contents, the root water potentials were closely similar.

2. Effect on Permeability

Solutes outside the root could affect water entry not only by their osmotic potential but also by altering the permeability of the root.

Arisz *et al.* (1951) determined the conductivity of tomato roots by the rates of exudation before and after changing the external solution (see section IV A). They found that both mannitol and Hoagland solution reduced the conductivity. Five minutes in mannitol had no effect, but half an hour had a marked effect.

O'Leary (1969) grew kidney beans (*Phaseolus vulgaris*) in nutrient solutions with and without NaCl and then determined the root conductivity by the rate of water flow through the root system under a controlled pressure. NaCl solution caused a marked reduction in conductivity, even when its osmotic potential was only -2 bars.

In contrast to these reductions, the permeability of individual cells of roots is increased by lowering their osmotic potential. This was shown by Myers (1951) with beet (*Beta vulgaris*) roots using sucrose as an osmoticum. He thought the increased permeability was related to plasmolysis. However, Ordin and Gairon (1961) found that ^3H_2O entry into broad bean root segments was accelerated by pretreatment with mannitol, even when the cells were not plasmolyzed by the treatment. The rate of ^3H_2O entry into corn roots was unaffected by mannitol.

Other methods of investigating this question run into technical difficulties. Brouwer (1954a) placed sucrose solution outside one 2.5-cm section of a broad bean root for 1 hr, and on returning it to water found its uptake rate was for a time higher than before the treatment. This he interpreted as due to increased conductivity. However, there is no evidence of increased water entry into the xylem; at least some of the increased uptake was presumably into cortical cells whose water potential had been lowered by the sucrose treatment.

Another method that has been used is to determine the rate of uptake (F) and the osmotic potential of the xylem exudate (ω_x) when detached roots are placed in various solutions. Then the conductivity is given by $k = F/(\omega_e - \omega_x)$, where ω_e is the external osmotic potential. Sometimes the exuding sap is found to have a higher (less negative) osmotic potential than the external solution; yet water uptake continues (Klepper, 1967). The formula would then make k negative, which is meaningless. The explanation could be either active water uptake or that the exuded sap is more dilute than that elsewhere in the xylem due to reabsorption of salts by other root cells. The latter explanation is generally favored, though the former has not been conclusively disproved. Even when $(\omega_e - \omega_x)$ is positive, the possibility of salt reabsorption makes the method liable to large errors. Whole plant conductivities have also been determined by measuring leaf water potential (ψ_l) and using the formula $k = F/(\omega_e - \psi_l)$. Again, errors could arise from salt reabsorption from the xylem, and reversals of potential gradient have been reported, even when water uptake and transpiration were going on. In some cases the measurement of leaf water potential may be at fault, but this is not always the explanation (Barrs, 1966).

It is worth mentioning some results obtained by these techniques, if only to show how variable the results are. Klepper (1967) grew corn seedlings in Hoagland solution with various amounts of NaCl added. Using the root exudation technique, she found that increasing NaCl concentration tended to lower the conductivity. The data of Meiri and Anderson (1970b), also for corn, show that 25 mM KCl lowered the conductivity in the tip 5 cm of root but raised it in the 5–10 cm region.

Table 14.8 summarizes the results of some experiments on whole plants.

Table 14.8. Summary of the apparent effects of external osmotic potential on whole-plant conductivity

Species	Osmoticum	Lowest osmotic potential (bars)	Effect on conductivity
Tomato	mannitol	− 10.7	Decrease
	sucrose ⎫ NaCl ⎬ KNO$_3$ ⎭	− 10.7	Increase
Castor	polyethylene glycol	− 8	Increase
Pepper	polyethylene glycol	− 7.5	Little change
Cotton	NaCl	− 8.5	Little change

Sources: Data on tomato (*Lycopersicon esculentum*), Slatyer, 1961; on castor (*Ricinus communis*), Macklon and Weatherley, 1965; on pepper (*Capsicum frutescens*), Janes, 1970; on cotton (*Gossypium hirsutum*), Boyer, 1965.

Note: Conductivity determined from: (transpiration rate)/(solution osmotic potential − leaf water potential).

Slatyer (1961) found different effects with different osmotica, but the same osmoticum can have different effects in different experiments. It is not clear how far the differences are genuine permeability effects and how far they are due to the error mentioned above. It is doubtful whether any solute has been found that acts solely through its osmotic effect and not through its chemical nature. Polyethylene glycol is one of the most favored as a "neutral" osmoticum, but it can be toxic (Leshem, 1966). A further source of confusion is that deionized water may reduce root permeability as compared with water containing some ions (Drew, 1967). Authors often do not state whether the water they used was deionized or not.

In conclusion, the experiments of Arisz *et al.* (1951) and O'Leary (1969) provide good evidence that there are three osmotica that can cause reduction in the permeability of the water pathway across the root. In contrast, the permeability of individual root cells can be increased by similar treatment. In view of the variable results obtained by other methods, it should not be assumed that the same effects necessarily occur with other solutes or other species.

IX. THE ROOT-SOIL INTERFACE

In the previous two sections, the discussion of matric and osmotic potentials in the rhizosphere involved several assumptions, three of which I shall now consider. (1) The soil was taken as a homogeneous medium, having a certain water content, a certain conductivity, and other bulk properties. This macroscopic approach is perfectly valid if one is considering soil masses of centimeter dimensions or larger. But the larger soil pores are as large as root hairs in diameter and not much smaller than fine roots. Thus we should consider whether this approach could lead to errors when applied to the immediate vicinity of a root. (2) The root was considered as a cylinder of uniform diameter. In particular this ignores root hairs, and I shall now discuss whether they are significant in water uptake. (3) It was assumed that contact between soil and root was good enough to prevent any additional resistance arising at the interface itself. There are reasons for thinking this may not always be so.

A. Water Flow in Individual Pores

If water arrives at the root mainly at a few points, instead of evenly all over, this could be important. Although the free space of the cortex is permeable compared with membranes, it is not when compared with soil. The values from Russell and Woolley (1961) that were quoted in section III A put the conductivity of cell wall material at 2.4×10^{-5} cm/day, and the cortex as a whole at 1×10^{-6} cm/day. According to the alternative calculations of Tyree (1969), the free-space conductivity would be lower, but there would also be flow across the vacuoles, and the overall conductivity of the cortex again comes to about 1×10^{-6} cm/day. This is lower than the conductivity of soils at -15-bar matric potential, and about 10^4 times lower than soils at field capacity.

The physics of flow through porous media has been considered on a microscopic scale from the theoretical point of view, but the mathematics rapidly becomes complex (see Scheidegger, 1957). At present we have too little quantitative information about the shapes and sizes of soil pores to be able to get very far. However, a very crude model is sufficient to illustrate a few points about microscopic flow.

Imagine the soil to be made up of numerous straight capillary tubes, random in direction and arrangement. Each capillary is of uniform diameter, but there is a range of diameters among the different tubes. We can estimate the number of tubes in a certain size range, and the flow along them, from data for a real soil. Table 14.9 shows data for Pachappa sandy loam at selected matric potentials, taken from Figures 14.1 and 14.7. This shows that as the matric potential falls from -0.2 to -0.3 bar, the water content declines by 15% (of its value at -0.2 bar), but the conductivity declines by 70%. This implies that the pores which were water-filled at -0.2 bar but empty at -0.3 bar carried 70% of the flow. By equation [2] their radius, in our capillary model, would be 4.8μ—7.3μ. Knowing the total volume of pores and their average radius, we can calculate their frequency: about $120/mm^2$ would cross a random plane, and this frequency would be expected at the root surface.

Table 14.9. Selected data for Pachappa sandy loam

Matric potential (bars)	Conductivity (cm/day)	Water content (% by weight)	Pore radius† (μ)
-0.2	27×10^{-3}	15.1	7.3
-0.3	8×10^{-3}	12.9	4.8
difference	19×10^{-3}	2.2	
-10	6.6×10^{-6}	9.1	0.15
-15	3.5×10^{-6}	8.2	0.10
difference	3.1×10^{-6}	0.9	

Source: Data from Gardner, 1960.
† Radius of largest water-filled pores, calculated by equation [2].

Hence, they would be about 100μ apart on average. Thus, we may envisage the water arriving at the root unevenly: 70% of it at these points 100μ apart, only the remaining 30% elsewhere. Will this affect water uptake? Even in the finest roots the distance from the root surface to the xylem is about 100μ. Delivery of the water at only a few points might increase this distance, if some of the water moved laterally; but since no point on the root surface would be much more than 50μ from a point of delivery, the increase in distance will be less than twofold. If the resistance to movement across the cortex is small compared with that involved in crossing membranes, the effect is likely to be negligible.

Data are also given in Table 14.9 for -10 and -15 bars. A similar calculation leads to a prediction of 10^5 pores of this size per mm^2. So these pores are about 1,000 times more frequent (in this soil) than the larger ones and would be only about 3μ apart on the root surface.

Solutes would also arrive more rapidly by mass-flow through the largest pores. But again diffusion is rapid in the cortex (section VIII), and significant differences in concentration are unlikely.

This model is crude. But in practice differences in flow rate between different pores are likely to be smaller, because of cross-connections and irregular diameter. So the conclusion that they do not affect the plant is almost certainly correct.

B. Root Hairs

It was once assumed that because root hairs increase the surface area of the root, they must necessarily increase its water uptake. This would be true only if the epidermis were a major source of resistance to water flow, which it is not. Another suggestion is that root hairs influence water uptake by improving contact between root surface and soil. Rosene (1943) demonstrated that root hairs can take up water. In her experiments only the single hair under investigation was in contact with liquid water; the rest of the root surface was in moist air. Therefore her results give no indication of the rates of uptake by root hairs in soil.

It has sometimes been stated that root hairs are short-lived. This is certainly not always true. Abundant root hairs have been reported on wheat roots 10 weeks old, on sugarcane (*Saccharum*) roots several months old, and on roots more than a year old on several Compositae (Weaver, 1925; Evans, 1938; Whitaker, 1923). The view that soil moisture stress kills root hairs seems to be widespread, but in my opinion needs confirmation.

Dittmer (1949) measured the root hairs of 37 species from a range of families. He found that the length varied from 80μ to $1,500\mu$, but the range of diameters was only $5\mu–17\mu$. This comparative uniformity in diameter may be of functional significance, because this is just about the size of the largest water-filled pores at field capacity. Some root hair densities reported for Gramineae, per millimeter root length, are: wheat, 50; rye, 19–53; *Poa pratensis*, 47–88 (480 on main roots); *Lolium multiflorum*, 98 (Lewis and Quirk, 1967; Dittmer, 1937, 1938; Drew and Nye, 1969). These indicate a maximum density per unit of root surface area of about $50–100/mm^2$. Kozlowski and Scholtes (quoted by Kramer [1969]) found only 5.2 hairs/mm^2 on *Robinia pseudoacacia*, and 2.2/mm^2 on loblolly pine. The density of $50–100/mm^2$ on Gramineae is similar to the number of pores of the relevant size estimated above. The question thus arises whether the root hairs "seek out" these pores. There is no definite evidence on this. Root hairs grow at the tip, one would expect them to follow the path of least resistance, and root hairs washed out of soil often are irregular rather than straight. On the other hand, they can make holes in soil, if its density is not very high (Champion and Barley, 1969); and they can grow straight across air gaps, so there is no evidence that they avoid larger air-filled pores. Thus it is not clear whether or not root hairs tend to occupy pores that are likely to be water-filled.

It has always been assumed that water movement along a root hair is easy, but this is not obvious. The possible pathways for movement are the same as those across the cortex (see section III), namely the free-space (wall), the cytoplasmic (symplasmic), and vacuolar pathways. The data of Rosene (1943) provide some evidence against the free-space pathway. She inserted individual root hairs of radish (*Raphanus sativus*) into micro-capillaries and measured the rate of water uptake when different proportions of the hair were immersed. If the resistance to water uptake by the hair was mainly in longitudinal movement, we should expect uptake to be inversely proportional to the distance from the immersed region to the base. In fact, increasing the proportion of the hair immersed had only a slight effect on uptake. This could be explained in either of two ways: (1) the main resistance is in entering the hair, and the tip is the most permeable part; or (2) the main resistance is in leaving the root hair cell at the base. Often the surface area of the basal (epidermal) portion of the cell is less than that of the hair itself, which would fit the second of these explanations.

We can make predictions of the permeability of the free-space and vacuolar pathways in the hair in a similar way to the calculations for the cortex in section III. I use the values for wall and membrane conductivity proposed by Russell and Woolley (1961), $2.8 \times 10^{-7} \text{ cm}^2 \text{ sec}^{-1} \text{ bar}^{-1}$ and $6 \times 10^{-7} \text{ cm sec}^{-1} \text{ bar}^{-1}$ respectively. Let us consider a hair 0.5 mm long, 10μ in diameter, with walls 0.5μ thick. (The thickness of the wall of a root hair is difficult to measure with a light microscope, but it is certainly not more than 1μ. An electron micrograph by Belford and Preston [1961] of a *Sinapis alba* root hair shows a wall 0.2μ thick.) Suppose also that all the water enters by the tip 0.1 mm of the hair and that the basal (epidermal) portion of the cell has a surface area of $4,000\mu^2$. Then the predicted rate of flow by the wall (free-space) pathway is $3 \times 10^{-6} \text{ mm}^3/\text{hr}$ per bar difference between the two ends of the hair. By the vacuolar pathway it would be $3 \times 10^{-5} \text{ mm}^3 \text{ hr}^{-1} \text{ bar}^{-1}$, about ten times as fast. Thus in the root hair the vacuolar pathway is predicted to have a much higher conductivity than the free-space pathway, in contrast to the conclusion for the cortex. The difference is mainly due to the length of the cell. The symplasm pathway may give even faster flow if the conductivity of the cytoplasm is high and plasmodesmata are frequent in the basal portion; I do not know if this is so.

Now consider a root 0.2 mm in diameter bearing 100 hairs/mm, each of them 0.5 mm long, each capable of the flow rate predicted above, $3 \times 10^{-5} \text{ mm}^3 \text{ hr}^{-1} \text{ bar}^{-1}$. We can compare the conductivity of the soil in the 0.5-mm layer round the root with that of the 100 hairs that penetrate it. If the soil was Pachappa sandy loam at -0.2-bar matric potential, its conductivity would be more than 10^5 times higher than the root hair pathway, and even at -15 bars it would be 20 times higher. This is of course much influenced by the small area for flow provided by the root hairs compared with the soil lying between them. So either root hairs play little part in water uptake, or my calculations are wrong. The latter could well be true if the symplasm pathway is involved in the hair. But it seems a distinct possibility that even in dry soil it is easier for the water to move through the soil than along the root hairs.

C. Soil-Root Gap

I argued above that if water was delivered at points 100μ apart on the root surface, this would be almost as effective as an even distribution all over the surface. But if the points of delivery were several millimeters or centimeters apart, this would not necessarily be so. The experimental evidence of Meiri and Anderson (1970b) indicates that little water movement along the cortex is to be expected over a distance of several centimeters. They compared the rate of exudation by a 5-cm long corn root tip with that of a 10-cm tip whose proximal 5 cm were coated with petrolatum. There was on average little difference between their exudation rates, from which Meiri and Anderson concluded that the cortical resistance was large enough to prevent much water from flowing longitudinally along the cortex between the petrolatum and endodermis and then crossing the endodermis in the coated portion. Therefore it is likely that if there are sizable portions of the root not in contact with the soil, this could seriously restrict uptake.

One possible cause of failure of contact is the minor irregularities of the soil. If a soil-root gap is only a few microns wide it may be filled by mucilage, which is secreted by roots of at least some species (Jenny and Grossenbacher, 1963). Gaps a few microns wide would be water-filled anyway at field capacity, but mucilage could be important in drier soils. Root hairs may be better able to follow irregular soil surfaces than roots (see above). But there is evidence that much larger gaps can occur. Sheikh and Rutter (1969) made a careful study of the roots of *Erica tetralix* (a subshrub) and the grass *Molinia caerulea* growing in wet heath in southern England. They prepared thin sections of soil in which they measured the diameters of roots and of the pores in which they lay. For both species the average ratio of pore diameter to root diameter was within the range $2.4-2.8$; so the roots fitted very loosely into their pores in most cases. This might be related to impeded drainage at this site, which made aeration a critical factor. Similar data from other soils would be very desirable.

Huck *et al.* (1970) have reported marked diurnal changes in diameter of cotton roots. They made time-lapse films of individual roots growing in soil against an observation window. In the extreme case a root's diameter decreased by 40% during the day but recovered during the following night. If such behavior is common, it clearly has far-reaching implications for both water and nutrient uptake. Huck *et al.* noted that maximum root contraction lagged behind maximum transpiration (estimated by radiation measurements). This suggests that contraction is not directly related to the pressure potential difference between the outside and the xylem but rather to the water status of the root cells.

In order to find out whether xylem pressure potential could itself cause an immediate root contraction, I have performed some experiments on dwarf beans. Root systems in water were detopped, and 0.8-bar suction was applied to the cut stump. Direct observation through a microscope detected no change in diameter when the suction was stopped. A change of 3% would have been detectable. I therefore used an indirect method. An 0.8-bar suction was applied downwards (i.e., to the vessel enclosing the roots), and the instantaneous change in the level in a tube attached to the cut stump was measured when the suction was stopped. The volume of the root system was then determined by water displacement. The experiment was performed on four bean plants one month old. The volume change averaged 0.22% (range between plants 0.20%–0.24%), indicating a diameter change of about 0.1%, or about 0.3μ for a bean root of average diameter. Larger suctions than this can occur in intact plants, but it seems likely that the diurnal variations caused directly by changes in xylem tension will cause gaps so small that they will usually be filled by water or mucilage. Thus, root shrinkage is probably caused mainly by a fall in water potential of root cells, particularly in the cortex, since this constitutes much of the volume of the root. If the endodermis provides the main resistance to water flow, the cortex, being outside it, should have a similar water potential to the surrounding soil; hence diurnal shrinkage would not be expected, except in the rare cases where a large rhizosphere gradient develops. In contrast, if the symplasm pathway operates, the cortex water potential may lie between that of the soil and the xylem. In either case, if a small vapor gap does develop, due to slight shrinkage of either root or soil, this would impede water supply to the cortex, and the rate of root shrinkage would increase. It would be very desirable to have further information on how widespread diurnal root shrinkage is.

One question as yet unanswered is whether, if the root is in contact with soil at the start and has root hairs, the contraction will cause a gap between root and soil, or whether the root hairs will bind the soil near the root so that the break comes in the soil farther away and probably more irregularly. It seems likely that the root will maintain contact with the soil at least on one side for much of its length. The significance of root hairs in water relations may lie here as much as in their actual conducting ability.

Secondarily-thickened roots have no hairs and presumably no mucilage. Furthermore, their water uptake may be restricted to localized regions of their surface, such as lenticels. In these roots, therefore, contact problems may be particularly severe. Young trees that have been transplanted often suffer from water stress. This is commonly attributed to their small amount of root, but I think poor soil-root contact is a more likely cause.

Cowan and Milthorpe (1968) have discussed the possible contribution of water vapor movement across a gap in supplying the plant's water requirement. If the roots are taking up water at rate q (mm^3 cm^{-1} hr^{-1}), then the difference in water potential, $\Delta\psi$ (bars), between soil and root surface needed to move the water at this rate across a vapor gap w cm wide is given by their formula,

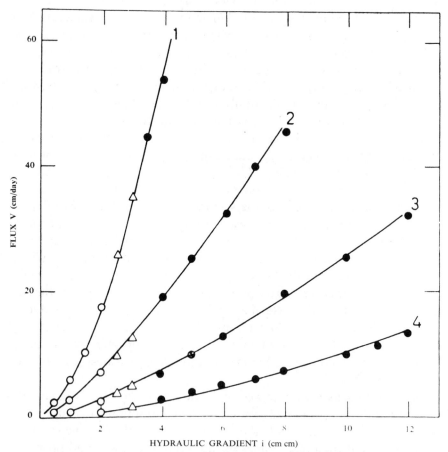

Fig. 14.9. Relationship between flow rate (V) and matric potential gradient (i) for Ksalon clay at four matric potentials (1–4). (Hadas, 1964)

$$\Delta\psi = 45q \cdot \ln(1 + w/r), \qquad [16]$$

where r is the root radius. If w/r is 0.5 and q the moderate rate of 0.1 mm^3 cm^{-1} hr^{-1}, then $\Delta\psi$ would be 1.8 bars. With a profusely rooted plant, q, and hence $\Delta\psi$, could be much lower. If, for the sake of comparison, we take w, the gap width, as 0.5 mm, and if there are 100 root hairs/mm^2 emerging from the root surface and crossing the gap, and their conductivity is 3×10^{-5} mm^3 hr^{-1} bar^{-1} as predicted earlier, $\Delta\psi$ of 1.8 bars would cause a water flow along them to the root of 0.3 mm^3 cm $root^{-1}$ hr^{-1}, three times the predicted rate by vapor movement. Thus either root hairs or vapor movement could provide a pathway for water movement across a gap to the root; but in both cases the potential difference across the rhizosphere would be far larger than is expected in soil when good contact is maintained. It is hard to predict the relative contributions of vapor, root hairs, and normal soil-epidermis flow in a situation where only occasional portions of the root surface contact the soil. We need in particular to know more about resistances to longitudinal flow in the cortex.

X.   DEPARTURE FROM DARCY'S LAW

Much of what has been written so far in this review is based on the assumption that Darcy's law applies to soils and roots; i.e., that the rate of water flow through them is proportional to the driving force (see section I C). There is evidence, however, that both soils and root systems do sometimes depart from Darcy's law.

A. Soil

Such soil departures from Darcy's law have been reviewed in detail by Swartzendruber (1966, 1968, 1969); the reader should consult these papers for further details and a full list of references.

1. Examples

Figure 14.9 shows the rate of flow of water through a clay soil under various gradients of matric potential; four different average matric potentials were compared. The experiment involved steady-state flow. If Darcy's law were obeyed, the graph would show four straight lines through the origin. Their upward curvature indicates that in fact flow rate increases more rapidly than expected as the potential gradient is increased. Similar departures from Darcy's law have been reported for various other soils, both saturated and unsaturated. Evidence from steady-state experiments is confined to soils at about field capacity or wetter, but evidence on drier soils comes from study of water flow into columns of dry soils. Figure 14.10 shows results from such an experiment.

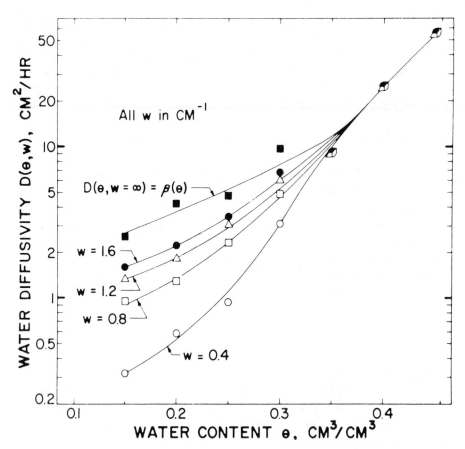

Fig. 14.10. Relationship between diffusivity and water content for various values of water-content gradient, w (units, cm^3 cm^{-3} cm^{-1}), for Salkum silty clay loam. (Data of Rawlins and Gardner as analyzed by Swartzendruber, 1968. For definition of *diffusivity*, see section VII A, n. 2)

When the soil was moist, Darcy's law was obeyed, but in drier soil the diffusivity (and conductivity) became increasingly influenced by the water-content gradient. After surveying these and other results, Swartzendruber (1968) concluded that changes in gradient are often associated with 2- to 4-fold changes in conductivity, and sometimes up to 5- to 15-fold.

In Figure 14.9 the relationship of flow rate to potential gradient is similar to line $B1$ in Figure 14.2, a curve passing through the origin. However, a relationship of type C in Figure 14.2, indicating a threshold gradient required to initiate flow, has been reported for saturated flow through clay (Miller and Low, 1963). The threshold gradient depended on temperature: it was 70 cm/cm at 10C, but less than 13 cm/cm at 15C.

In some soils Darcy's law is obeyed. Swartzendruber (1968) has suggested an association between presence of clay particles in the soil and non-Darcy behavior. However, this is not a clear-cut distinction; the loam and silt loams used by Richards and Weeks (1953) and Vachaud (1967) presumably contained clay, but Darcy's law was obeyed. The reason why some soils obey Darcy's law and others do not remains unclear.

2. Possible Causes

Possible causes of this non-Darcy behavior have been discussed in some detail by Swartzendruber (1966, 1969) and will be covered only very briefly here. The deviations may be experimental artifacts. This seems unlikely for most of the steady-state experiments but is more difficult to exclude for the transient-state inflow experiments. Another possibility in the inflow experiments is that the relationship between water content and matric potential changes during the experiment, but again this could hardly explain the steady-state results. If the non-Darcy behavior is accepted as real, then there are three basic explanations possible. (1) There may be some other driving force operating that has not been taken into account. The one most frequently suggested is streaming potential. Although this probably can be quite large in soil under certain circumstances (see section I), it is not clear that it would cause deviations from Darcy's law. (2) The properties or arrangement of the soil particles may alter. Reorientation of fine particles in relation to flow rate is possible but has not been demonstrated. It would have to be reversible to explain the observed non-Darcy behavior. (3) The viscosity of water may change. As section I showed, there is some evidence, but by no means conclusive, that water flow through fine rigid pores does depart from Darcy's law. This would explain why non-Darcy behavior is most pronounced in drier soils, where the water-filled pores and films are thinner; also why Miller and Low (1963) found a greater departure from Darcy at lower temperatures, where water structure is more stable. Ions in the soil solution tend to reduce departure from Darcy's law (Swartzendruber, 1969), and this could be due to their tendency to break down water structure; however, it could equally well be an effect on clay particle structure. Evidence against this explanation is the finding by Nielsen *et al.* (1962) that the flow of oil through a silt loam deviated from Darcy in a similar way to water.

3. Significance for the Plant

In pararhizal flow, if water moves distances of several centimeters or more to the roots, soil conductivity will have a significant and probably controlling effect on the rate of water movement. Under these conditions departures from Darcy's law are likely to be very important; the increased conductivity at high gradients will enhance the flow at times of peak demand. In contrast, I have argued that the resistance to water flow across the rhizosphere is generally too small to be of any significance to the plant; this would suggest that here non-Darcy behavior matters little. However, the estimates of gradients in the rhizosphere made in section VII themselves involved the assumption that Darcy's law is valid; so it might be asked how they will be affected if it is invalid. Basically, the requirement is that the soil conductivity values used in the calculation should have been obtained at potential gradients similar to those occurring in the rhizosphere. This is so, very approximately; the values came from pressure plate outflow determinations, and this is one reason for preferring this method to the inflow method, where the gradients are much steeper. Although the conclusion that rhizosphere gradients are usually small is probably valid, the quantitative accuracy of individual estimates of rhizosphere gradients is doubtful. In any case, determination of soil conductivity is liable to large errors (Jackson *et al.*, 1963). This whole discussion of non-Darcy behavior in the rhizosphere is speculative, because there is no definite evidence on whether or not Darcy's law is obeyed under the relevant conditions, namely, fairly dry soil (when rhizosphere gradients become most interesting) and fairly small gradients, generally less than 1 bar/mm. Data from steady-state methods are confined to moist soils. The data of Figure 14.10 cover quite dry soil; but the gradients involved were very large, the smallest being 0.4 cm^{-1}, or 40% water content over a distance of 1 cm. A new method of testing Darcy's law is needed for the conditions relevant to the rhizosphere.

B. Plant

In this section the symbols ΔP, $\Delta \omega$, and $\Delta \psi$ will be used to denote the difference between the external solution and the xylem sap in pressure potential (hydrostatic pressure), osmotic potential, and total water potential, respectively.

1. Examples

Figure 14.11 shows examples of the relationship between the rate of water uptake and the driving force, determined by various methods. Part (*a*) shows the rate of movement through detopped root systems of dwarf beans, with various suctions applied to the cut stump. The roots were in water or in KNO_3 solution. Part (*b*) shows similar results for

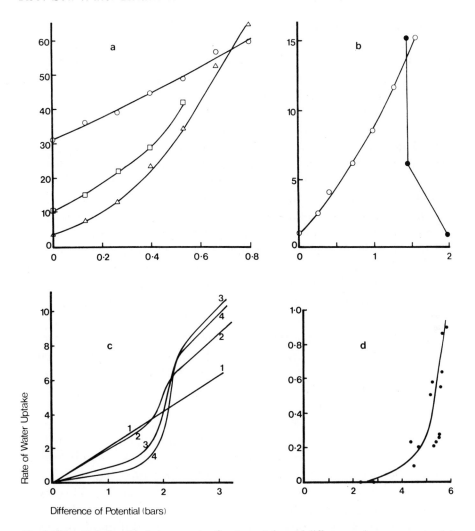

Fig. 14.11. Relationship between rate of water uptake and difference of pressure potential or water potential across the root. Part *a*, external solution: *circles*, water; *squares*, 0.01 M KNO_3; *triangles*, 0.05 M KNO_3. Units of water uptake rate: *a*: mm^3/min; *b* arbitrary; *c*: mm^3/cm root/hr; *d*: cm^3/dm^2 leaf/h. (Data for *a* from Kuiper, 1963; *b*, Mees and Weatherley, 1957a; *c*, Brouwer as analyzed by Hylmö, 1958; *d*, Tinklin and Weatherley, 1966)

detopped tomatoes, but in this case pressure was applied to the vessel containing the roots, instead of suction from above. The open circles show the relationship between water flow and applied pressure (ΔP). The exudate was collected and its osmotic potential determined, allowing an estimate of the total water potential difference across the root, $\Delta \psi$ (black circles). Part (*c*) shows the uptake by 2.5-cm sections of a single broad bean root; the sections are numbered consecutively, 1 being the tip. Here the horizontal axis is $\Delta \psi$; xylem water potential was deter-

mined by countersuction (see section IV A). The technique and results
are closely related to those of Figure 14.5. Part (d) shows the transpiration rate of whole *Ricinus communis* plants, plotted against the difference
between the osmotic potential of the nutrient solution around the roots
(which was constant) and the water potential of the leaves. The authors
thought the line became exactly vertical for much of its length, but their
actual data do not support this. From additional experiments they concluded that the main resistance to water flow lay in the roots. In this
experiment the leaf water potential was markedly different from the solution osmotic potential at zero transpiration. This the authors attributed
to two sources of error in leaf water potential determination, one of
which they thought would not operate when transpiration was rapid.

In most cases there was a marked departure from a linear relationship,
the rate of flow increasing faster than expected as the driving force increased. The exception is the tip section, no. 1, of the broad bean root
in part (c). In part (a) the line for the root system in water shows only
slight curvature. In all four experiments there is a tendency for the line
to become straight toward the top right.

In parts (c) and (d) the horizontal axis represents a difference in overall
water potential, $\Delta\psi$, but in parts (a) and (b) it represents pressure potential (hydrostatic pressure), ΔP. In these latter cases a curved line could
arise even with constant root conductivity. Figure 14.12 illustrates this
for a hypothetical detopped plant to whose stump suction is applied.
Lines A, B, and C could all, in theory, occur with constant conductivity.
Line B would occur if osmosis made no contribution to water flow. Line
A is the expected relationship if osmosis made a constant contribution.
For this to occur, the xylem sap concentration would have to remain
constant as the rate of water uptake increased due to suction. This rarely
happens; usually the sap becomes more dilute. Suppose instead that the
rate of ion uptake was constant, so that the concentration of the xylem
sap was inversely related to water uptake. Then curve C is the expected
relationship. This approaches line D asymptotically but would only
reach it when the xylem sap became pure water. Line D cuts the hori-
zontal axis at $\sigma\omega_e$, where σ is the reflection coefficient of the root and
ω_e the external osmotic potential. In practice ion uptake is usually faster
when water uptake is faster, and the curve would lie somewhere between
lines A and C. Nevertheless, if with real data we extrapolate the top part
of the curve back and find that it cuts the axis to the right of ω_e, this is
evidence that the curvature is not due to simple osmotic adjustment. By
this criterion the results of Figure 14.11a could be due to osmotic adjustment; and the fact that the plants in water, whose xylem sap may have
been very dilute at all flow rates, gave a nearly linear relationship might
seem to support this. However, the crossing of the lines at the top right
could not be explained by simple osmotic adjustment. In part (b) the

xylem exudate was collected, $\Delta\psi$ was calculated, and it is represented by black circles; further details are given in Table 14.10. According to these calculations, the conductivity of the root rose about 30-fold over this range of ΔP. This assumes a reflection coefficient of 1 in calculating k; the data of Mees and Weatherley (1957b) suggest an average value of 0.76, which alters the effective $\Delta\psi$ and hence the calculated value of k, but k would still change more than 20-fold, as shown in the bottom line of the table.

It is sometimes assumed that a straight portion of any line represents constant conductivity. This is not always so. For example, the upper part of Figure 14.11d is approximately a straight line, but it clearly represents rising conductivity. When interpreting this graph, one must remember the effects that can be produced by active movement of ions into or out of the xylem, so that the water flow is not necessarily proportional to the potential difference between the leaves and the medium (see section VIII B). But in this case the departure from expectation is so large that a variation in root conductivity seems certain.

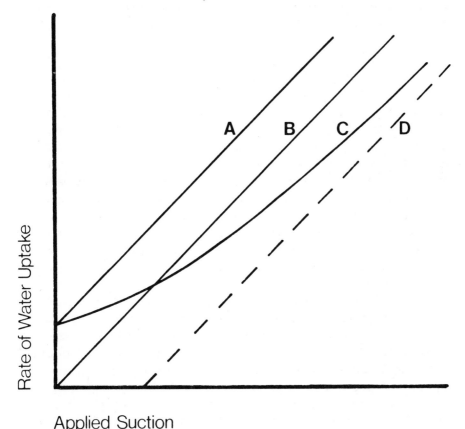

Fig. 14.12. Hypothetical relationships between rate of water uptake and applied suction for a detopped root system of constant conductivity

Table 14.10. Further details on the experiment with tomato featured in Figure 14.11b

	Potentials (bars)		
Applied pressure, ΔP	0	0.67	1.48
Osmotic potential of xylem exudate, ω_x	-2.87	-1.60	-0.77
Osmotic potential of external medium, ω_e	-0.89	-0.89	-0.89
$\Delta\omega = \omega_e - \omega_x$	1.98	0.71	-0.12
Water potential difference, $\Delta\psi = \Delta P + \Delta\omega$	1.98	1.38	1.36
Water uptake rate, F (arbitrary units)	0.75	6.05	15.2
Root conductivity, $k = F/\Delta\psi$	0.38	4.4	11.2
$\Delta\psi' = \Delta P + \sigma\Delta\omega$, where $\sigma = 0.76$	1.50	1.21	1.39
$k' = F/\Delta\psi'$	0.50	5.0	10.9

Source: Data from Mees and Weatherley, 1957a.

Since these qualitatively similar departures from Darcy's law have been demonstrated by several different procedures, in both intact plants and detached root systems, there is little danger that they are experimental artifacts.

2. Some Features of Non-Darcy Behavior in Roots

1. Departures from Darcy's law have been reported for the following species: tomatoes (*Lycopericon esculentum*), cotton (*Gossypium*), sunflowers (*Helianthus annuus*), *Vicia faba, Phaseolus vulgaris, P. multiflorus, Coleus hybridus, Ricinus communis* (Köhnlein, 1930; Brewig, 1937; Brouwer, 1953a; Mees and Weatherley, 1957a; Kuiper, 1963; Lopushinsky, 1964; Tinklin and Weatherley, 1966; Weatherley, 1970). The departure from expectation is always in the same direction, a greater increase in flow than expected as the driving force is increased. I know of no case where a species definitely obeyed Darcy's law, but this possibility should not be excluded. It is worth noting that the list of species above contains no monocotyledons.

2. There is evidence that the non-Darcy behavior is only shown by plants that are alive and metabolizing actively. Stoker (reported by Weatherley, 1970) found that when root systems of cotton were killed, an experiment similar to that of Figure 14.11d gave a linear relationship. Lopushinsky (1964) found a linear relationship between water uptake and applied pressure for a tomato root system treated with 1 mM sodium azide. On the other hand Brouwer (1954a) found that boiled, non-aerated water did not prevent the conductivity increase in broad bean roots; and Tinklin and Weatherley (1966) found very similar results to Figure 14.11d with plants in temporarily waterlogged soil.

3. One would expect, from point (2) above, that flow in xylem and cell walls would obey Darcy's law, and the available evidence supports this. Tyree and Fensom (1970) found that flow in the xylem of *Heracleum* petiole obeys Darcy's law. Peel (1965) found a decline below expected flow rates in stems of three tree species at high pressure gradients. However, the departure from expectation was small, and the gradients of pressure were much larger than would normally occur in living plants. Kamiya *et al.* (1962) found that flow in the cell wall of *Nitella flexilis* obeys Darcy's law.

4. Sometimes the change in conductivity takes some time to reach its maximum. This is shown clearly in Figure 14.5, where the uptake rates only reached a steady state after about 3 hours. Brewig (1936b) obtained similar results when the transpiration rate was suddenly altered. Jackson and Weatherley (1962) found that exudation took 2 hours to reach a maximum rate after pressure was applied to detopped *Ricinus*. Analysis of the xylem exudate showed that osmotic adjustment was not the explanation. On the other hand, Lopushinsky (1964) found marked non-Darcy behavior when pressure was applied to tomato root systems for only 15 minutes; so the effect apparently commences fairly quickly, though it may take much longer to reach its maximum.

5. Brouwer (1954a) found that a change in root conductivity induced by increasing $\Delta\psi$ was not accompanied by a change in root respiration rate.

6. The conductivity is probably not influenced by $\Delta\psi$ but rather by ΔP, or some other related factor. This was first suggested by an experiment of Brewig (1936a) with *Vicia*. He was measuring uptake from micropotometers, and he placed in one of the potometers a sucrose solution strong enough to cause withdrawal of water from the root. The transpiration rate could be reduced by placing a bell jar over the plant. Some of his results are:

	Transpiration rate	
	high	low
Uptake (mm^3/hr) from water-filled potometer	39	7
Loss (mm^3/hr) to sucrose-solution potometer	20	4

When the transpiration rate was reduced, and presumably the pressure potential in the xylem rose (became less negative), he expected water to be withdrawn more rapidly by the sucrose, but the opposite occurred, indicating a marked decline in root conductivity. The $\Delta\psi$ was presumably greater at the low transpiration rate; so the conductivity must have been influenced by something else.

The results of Mees and Weatherley (1957a) given in Figure 14.11b and Table 14.10 also support this view. In this particular tomato plant (though not in others used by the same authors), the increase in ΔP was almost exactly balanced by a fall in $\Delta\omega$, due to the xylem sap becoming more dilute. Thus the increase in water uptake appears to occur without any increase (or even with a slight decrease) in the overall driving force $\Delta\psi$. One explanation of this would be a low reflection coefficient, but later results (Mees and Weatherley, 1957b) indicate an average reflection coefficient (σ) of 0.76, which would still leave the effective $\Delta\psi$ (i.e., ΔP + $\sigma\Delta\omega$) virtually constant while uptake increased (see Table 14.10, next-to-bottom line). It is possible that the xylem exudate does not represent the concentration elsewhere in the xylem. But even allowing for this, it seems very likely that the increase in water flow with increasing ΔP is largely due to increasing conductivity, which itself can hardly be due to change in $\Delta\psi$. This would also help to explain the very steep portions of the curves in Figure 14.11c and d. Here the uptake rate rose dramatically while $\Delta\psi$ changed only very slightly; but ΔP may have been changing much more than $\Delta\psi$, partly balanced by $\Delta\omega$ changing in the opposite direction. Hence it seems much more likely that the conductivity is related to ΔP or $\Delta\omega$ than to $\Delta\psi$.

3. Significance for the Plant

The main significance for the plant is obvious. In times of high transpiration rate the increased conductivity will mean much smaller increases in water deficit of the shoot than would otherwise have occurred. Another point deserves closer scrutiny, the relation between the water potentials of the leaf and the soil as the soil dries and transpiration declines. The equation

$$F = (\psi_s - \psi_l)/(R_s + R_p) \qquad [17]$$

can be derived from equation [9]; where ψ_s and ψ_l are the water potentials of soil and leaf, and R_s and R_p the resistances to water flow in soil and plant, respectively. If R_s is commonly much smaller than R_p (see section VII), and R_p is constant, then as the transpiration rate (and hence F) declines, $(\psi_s - \psi_l)$ should decrease in proportion. Sometimes $(\psi_s - \psi_l)$ does decrease markedly, e.g., in the data of Yang and de Jong (1968) for wheat. But more often it declines only a little, stays about constant, or even increases somewhat (Slatyer and Gardner, 1965; Gavande and Taylor, 1967; Kaufmann, 1968). In the past this has been ascribed to increase of R_s; this was most explicitly argued by Gardner and Ehlig (1962), but it can be shown that in their experiment the results do not fit quantitatively with the assumption that the increasing resistance was in the soil (Newman, 1969b). The alternative is that R_p increases. This might be caused by the fall of plant water potential. As far as I know, the only definite evidence that water potential affects plant resistance is that of Kramer (1950), where the water stress imposed was severe; but there is some evidence also for bubble formation in the xylem under stress (see section VI A). Increasing plant resistance could also be a non-Darcy phenomenon; nearly constant $\Delta\psi$ while water uptake rate declines is a feature of Figure 14.11, parts (b) (black points) and (d). This would assume that as transpiration declines, although ($\psi_s - \psi_l$), and also $\Delta\psi$ across the root, may be constant, ΔP and $\Delta\omega$ are changing in opposite directions, with the result that root resistance changes.

4. Possible Causes

Two causes of this non-Darcy behavior that have been suggested can be quickly dismissed. Brouwer (1954a) suggested that the root water potential was the operative factor, perhaps acting through cell turgor. If this were true, increasing pressure from below should have the opposite effect on root conductivity to suction from above; in fact the effect is very similar. Later Brouwer (1965) came to favor more the theory of Hylmö (see below). Meiri and Anderson (1970a) suggested that increasing suction increases conductivity by removing bubbles from intercellular spaces in the cortex. This would be important only if the free space were the major resistance to water flow across the root, which it is not.

The most widely quoted theory is that non-Newtonian behavior of water in small pores is the cause. As I showed in section I, there is plenty of evidence that many properties of water are anomalous near solid surfaces, which makes anomalous flow properties quite likely; but the direct evidence that water disobeys Darcy's law in small rigid pores is scanty. The strongest protagonist of this theory is Hylmö (1955, 1958), who analyzed Brouwer's data for broad beans in the manner of Figure 14.11c. He interpreted each curve (except no. 1) as two straight lines pointing to the origin, separated by a sudden "jump" at about 2 bars. The original individual points, not shown on my graph, were widely scattered. The lower part of each line could equally well have been a smooth upward curve with no straight section; this was Brouwer's own interpretation. Hylmö likened these results to those of Erbe (1933) for micropore filters. He said that between 0 and about 1.5 bars flow was only through large pores, and there was no movement in smaller pores, but at about 2 bars there was a sudden increase in conductivity because smaller pores began to conduct also. As pointed out in section I, Erbe was concerned with flow when there are two immiscible liquids involved, and his work has no bearing on the situation in the root. Hylmö's main piece of evidence is therefore invalid. There are two positive reasons for thinking that

Hylmö was wrong. First, his theory would not explain the long time taken for conductivity to change. The relationship of Figure 14.11c was only attained 2 or 3 hours after $\Delta\psi$ was altered. Changes in the viscosity of water should be almost instantaneous. Second, one would expect flow anomalies to be responsive to the total driving force, which would be $\Delta\psi$ if the pores are in membranes. But in fact, as we have seen, the change in conductivity is related to ΔP or $\Delta\omega$. Therefore non-Newtonian behavior of water cannot be the only cause. However, we must consider the possibility that there are two separate causes: (1) non-Newtonian behavior of water, related to $\Delta\psi$ and responding instantaneously, and (2) another cause, responding more slowly and related to ΔP or $\Delta\omega$. The best way of trying to detect an instantaneous change in conductivity is by the quick solution-change method of Arisz et al. (1951) (see section IV A). Their equation [12] is not suitable, since it assumes that k remains the same. Instead, let us take the conductivities before and after the solution change to be k_1 and k_2. Then equations [10] and [11] become

$$F_1 = k_1({}_1\omega_e - \omega_x) \tag{18}$$

and

$$F_2 = k_2({}_2\omega_e - \omega_x), \tag{19}$$

where F_1 and F_2 are the flow rates immediately before and immediately after the solution change, respectively, and ${}_1\omega_e$ and ${}_2\omega_e$ the osmotic potentials of the two external solutions. Combining equations [18] and [19], we get

$$F_1/k_1 - F_2/k_2 = {}_1\omega_e - {}_2\omega_e. \tag{20}$$

In an experiment with tomatoes, Arisz et al. transferred the plant to a solution of mannitol, recorded the new rate of flow, then within 5 minutes returned it to water. After flow had returned to its original rate, the plant was transferred to another mannitol solution of different osmotic potential from the first, and this procedure was repeated several times. They performed a similar experiment starting with roots in Hoagland solution instead of water. From the results one can determine by interpolation the value of ${}_2\omega_e$ which gives $F_2 = 0$. At that point $F_1/k_1 = {}_1\omega_e - {}_2\omega_e$; so k_1 can be determined. Inserting this in equation [20], one can then calculate k_2 for other values of ${}_2\omega_e$. Figure 14.13 shows the values of k plotted against $\Delta\omega$, the difference between internal and external osmotic potential. Sometimes $\Delta\omega$ and F were negative (i.e., flow downwards), sometimes positive; this has been ignored in plotting the points, but negative values are denoted by black symbols. In both cases k does appear to increase with increasing $\Delta\omega$. In neither case is the correlation statistically significant, and the change in k is small for the plant in Hoagland; so no definite conclusion can be drawn. But it seems possi-

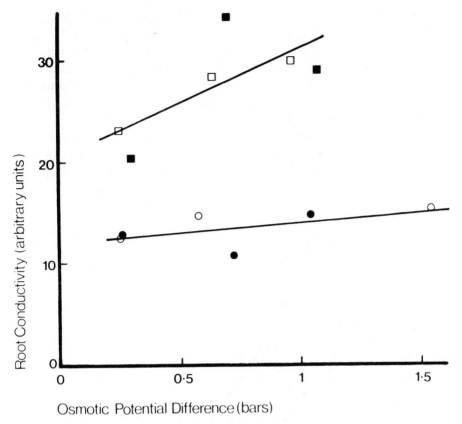

Fig. 14.13.   Root conductivity of tomatoes in relation to Δω, the osmotic potential difference between external medium and xylem. External solution water (*squares*) or Hoagland
(*circles*), plus various amounts of mannitol. *Open symbols*: water uptake occurred; *black
symbols*: water lost to solution. (Calculated from data of Arisz *et al.*, 1951, Fig. 4)

ble that there is an instantaneous change of k in response to change of
$\Delta\psi$ rather than ΔP.

To find the other cause of conductivity change is more difficult. The
fact that it is slow suggests a relation to metabolic rate. Brouwer (1954a)
could detect no change in respiration rate accompanying conductivity
change, but this does not prove that no change in any metabolic process
occurs. The change is apparently related to ΔP; $\Delta\omega$ usually changes
inversely to ΔP; so they are difficult to separate, but the counterosmosis
experiment of Brewig (1936a) (see above) favors ΔP. It is surprising that
ΔP should affect the living material of the plant. The plasmalemmas of
the root normally have a pressure difference across them of several bars,
due to cell turgor, and it is difficult to understand how a change of a
fraction of a bar could have an appreciable affect. It is also surprising
that the conductivity change should often be greatest over a small range
of ΔP, conductivity reaching a new fairly steady value at higher ΔP, as

it does in Figure 14.11c and in some results of Mees and Weatherley (1957a) and Lopushinsky (1964). If water flows from cell to cell in the symplasm, gradients of pressure will be set up there (e.g., in the plasmodesmata) where probably none occur in the absence of water uptake. But the same gradients would be set up whether the difference between the xylem and the outside was in pressure, osmotic potential, or a combination of the two; so the specific effects of ΔP would be unexplained.

An alternative explanation is that conductivity is not influenced by ΔP directly but by the osmotic potential in the xylem, which almost always varies inversely to ΔP. Figure 14.14 shows the relationship be-

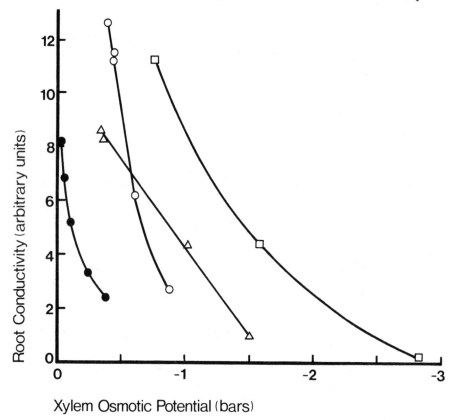

Xylem Osmotic Potential (bars)

Fig. 14.14. Relationship between root conductivity and osmotic potential of xylem exudate, in four experiments with tomatoes, in which water was forced through detopped root systems by applied pressure. (*Black circles*: data of Lopushinsky, 1964; *open symbols*: data of Mees and Weatherley, 1957a; *open squares*: same experiment as Fig. 14.11b and Table 14.10)

tween the osmotic potential of the xylem exudate and the root conductivity in four experiments with tomatoes. The fact that the points lie on smooth curves fits with the theory, though it does not, of course, prove a causal relationship. In the experiments indicated by open circles and triangles, the root conductivity became approximately constant at high ΔP; this corresponds with virtually constant xylem sap concentration also, thus neatly explaining the failure of increasing ΔP to have any further effect. It is known that osmotic potential can influence root conductivity, and the technically most satisfactory experiments reviewed in section VIII showed the effect in the same direction as in Figure 14.14. However, there is no definite evidence that the osmotic potential in the xylem, rather than outside the root, is the critical factor. In addition, there is indirect evidence that I have been too hasty in dismissing a direct effect of ΔP. Brouwer (1953c) and Jackson and Weatherley (1962) raised ΔP and lowered $\Delta \omega$ simultaneously in such a way that the water uptake rate was little altered. They found that salt uptake into the xylem was sometimes significantly increased. This shows that the root can respond to increases in ΔP when the xylem osmotic potential does not decrease; alternatively the plants may have been responding to $\Delta \omega$ or to external osmotic potential.

Although this non-Darcy behavior of roots has been known for more than 50 years, we are still a long way from fully understanding it. Like many of the problems I have discussed, its solution would be easier if we knew definitely the pathway of water movement across the root and the site of the main resistance to flow. That seems to me the most fundamental of all the unanswered questions about the water relations of the root.

References

Anderson, W. P., and C. R. House. 1967. A correlation between structure and function in the root of *Zea mays*. *J. Exp. Bot.* 18: 544–55.

Andrews, R. E., and E. I. Newman. 1968. The influence of root pruning on the growth and transpiration of wheat under different soil moisture conditions. *New Phytol.* 67: 617–30.

——. 1969. Resistance to water flow in soil and plant. III. Evidence from experiments with wheat. *New Phytol.* 68: 1051–58.

Arisz, W. H., R. J. Helder, and R. van Nie. 1951. Analysis of the exudation process in tomato plants. *J. Exp. Bot.* 2: 257–97.

Barrs, H. D. 1966. Root pressure and leaf water potential. *Science* 152: 1266–68.

Beck, R. E., and J. S. Schultz. 1970. Hindered diffusion in microporous membranes with known pore geometry. *Science* 170: 1302–5.

Belford, D. S., and R. D. Preston. 1961. The structure and growth of root hairs. *J. Exp. Bot.* 12: 157–68.

Bernal, J. D. 1965. The structure of water and its biological implications. In *The State and Movement of Water in Living Organisms*, ed. G. E. Fogg, pp. 17–32. Symp. Soc. Exp. Biol. 19. Cambridge.

Bernstein, L., and R. H. Nieman. 1960. Apparent free space of plant roots. *Plant Physiol.* 35: 589–98.

Bialoglowski, J. 1936. Effect of extent and temperature of roots on transpiration of rooted lemon cuttings. *Proc. Amer. Soc. Hort. Sci.* 34: 96–102.

Biddulph, O., F. S. Nakayama, and R. Cory. 1961. Transpiration stream and ascension of calcium. *Plant Physiol.* 36: 429–36.

Blanchar, W. R., and A. C. Caldwell. 1966. Phosphate-ammonium-moisture relationships in soils. I. Ion concentrations in static fertilizer zones and effects on plants. *Soil Sci. Soc. Amer. Proc.* 30: 39–43.

Boyer, J. S. 1965. Effects of osmotic water stress on metabolic rates of cotton plants with open stomata. *Plant Physiol.* 40: 229–34.

Breazeale, J. F. 1930. Maintenance of moisture-equilibrium and nutrition of plants at and below the wilting percentage. *Tech. Bull. Univ. Ariz. Agr. Exp. Sta.* 29: 137–77.

Brewig, A. 1936a. Beobachtungen über den Einfluss der Spross-Saugung auf die Stoffdurchlässigkeit der Wurzel. *Ber. Deut. Bot. Ges.* 54 (supplement): 80–85.

——. 1936b. Regulationserscheinungen bei der Wasseraufnahme und die Wasserleitgeschwindigkeit in *Vicia faba*-Wurzeln. *Jahresber. Wiss. Bot.* 82: 803–28.

——. 1937. Permeabilitatsanderungen der Wurzelgewebe, die vom Spross beeinflusst werden. *Z. Bot.* 31: 481–540.

Brewster, J. L., and P. B. Tinker. 1970. Nutrient cation flows in soil around plant roots. *Soil Sci. Soc. Amer. Proc.* 34: 421–26.

Brouwer, R. 1953a. Water absorption by the roots of *Vicia faba* at various transpiration strengths. I. Analysis of the uptake and the factors determining it. *Proc. Kongr. Ned. Akad. Wetensch.* (C) 56: 106–15.

——. 1953b. Water absorption by the roots of *Vicia faba* at various transpiration strengths. II. Causal relation between suction tension, resistance and uptake. *Proc. Kon. Ned. Akad. Wet.* (C) 56: 129–36.

——. 1953c. Transpiration and anion uptake. *Proc. Kon. Ned. Akad. Wet.* (C) 56: 639–49.

——. 1954a. Water absorption by the roots of *Vicia faba* at various transpiration strengths. III. Changes in water conductivity artificially obtained. *Proc. Kon. Ned. Akad. Wet.* (C) 57: 68–80.

——. 1954b. The regulating influence of transpiration and suction tension on the water and salt uptake by the roots of intact *Vicia faba* plants. *Acta Bot. Neerl.* 3: 264–312.

——. 1965. Water movement across the root. In *The State and Movement of Water in Living Organisms*, ed. G. E. Fogg, pp. 131–49. Symp. Soc. Exp. Biol. 19. Cambridge.

Bruce, R. R., and A. Klute. 1956. The measurement of soil moisture diffusivity. *Soil Sci. Soc. Amer. Proc.* 20: 458–62.

Champion, R. A., and K. P. Barley. 1969. Penetration of clay by root hairs. *Soil Sci.* 108: 402–7.

Chibnall, A. C. 1923. A new method for the separate extraction of vacuole and protoplasmic material from leaf cells. *J. Biol. Chem.* 55: 333–42.

Christersson, L., and S. Pettersson. 1968. Water and sulphate uptake at root reduction in *Ricinus communis. Physiol. Plant.* 21: 414–22.

Clarkson, D. T., A. W. Robards, and J. Sanderson. 1971. The tertiary endodermis in barley roots: fine structure in relation to radial transport of ions and water. *Planta* 96: 292–305.

Cowan, I. R. 1965. Transport of water in the soil-plant-atmosphere system. *J. Appl. Ecol.* 2: 221–39.

Cowan, I. R., and F. L. Milthorpe. 1968. Plant factors influencing the water status of plant tissues. In *Water Deficits and Plant Growth*, vol. 1, ed. T. T. Kozlowski, pp. 137–93. New York: Academic Press.

Currier, H. B. 1951. Herbicidal properties of benzene and certain methyl derivatives. *Hilgardia* 20: 383–406.

Davis, C. H. 1940. Absorption of soil moisture by maize roots. *Bot. Gaz.* 101: 791–805.

Derera, N. F., D. R. Marshall, and L. N. Balaam. 1969. Genetic variability in root development in relation to drought tolerance in spring wheats. *Exp. Agr.* 5: 327–37.

Derjaguin, B. V. 1965. Recent research into the properties of water in thin films and in microcapillaries. In *The State and Movement of Water in Living Organisms*, ed. G. E. Fogg, pp. 55–60. Symp. Soc. Exp. Biol. 19. Cambridge.

De Roo, H. C. 1969. Water stress gradients in plants and soil-root systems. *Agron. J.* 61: 511–15.

Dittmer, H. J. 1937. A quantitative study of the roots and root hairs of a winter rye plant (*Secale cereale*). *Amer. J. Bot.* 24: 417–20.

———. 1938. A quantitative study of the subterranean members of three field grasses. *Amer. J. Bot.* 25: 654–57.

———. 1949. Root hair variations in plant species. *Amer. J. Bot.* 36: 152–55.

Drew, D. H. 1967. Mineral nutrition and the water relations of plants. II. Some relationships between mineral nutrition, root function and transpiration rate. *Plant Soil* 26: 469–80.

Drew, M. C., and P. H. Nye. 1969. The supply of nutrient ions by diffusion to plant roots in soil. II. The effect of root hairs on the uptake of potassium by roots of rye grass (*Lolium multiflorum*). *Plant Soil* 31: 407–24.

Duncan, W. G., and A. J. Ohlrogge. 1958. Principles of nutrient uptake from fertilizer bands. II. Root development in the band. *Agron. J.* 50: 605–8.

Erbe, F. 1933. Die Bestimmung der Porenverteilung nach ihrer Grösse in Filtern and Ultrafiltern. *Kolloid-Z.* 63: 277–85.

Esau, K. 1965. *Plant Anatomy*. New York: Wiley.

Evans, H. 1938. Studies on the absorbing surface of sugar-cane root systems. I. Method of study with some preliminary results. *Ann. Bot.*, n.s. 2: 159–82.

Fogg, G. E. (ed.). 1965. *The State and Movement of Water in Living Organisms*. Symp. Soc. Exp. Biol. 19. Cambridge.

Frank, H. S. 1970. The structure of ordinary water. *Science* 169: 635–41.

Fried, M., and H. Broeshart. 1967. *The Soil-Plant System in Relation to Inorganic Nutrition*. New York: Academic Press.

Gaff, D. F., T. C. Chambers, and K. Markus. 1964. Studies of extra-fascicular movement of water in the leaf. *Aust. J. Biol. Sci.* 17: 581–86.

Gardner, W. R. 1956. Calculation of capillary conductivity from pressure plate outflow data. *Soil Sci. Soc. Amer. Proc.* 20: 317–20.

———. 1960. Dynamic aspects of water availability to plants. *Soil Sci.* 89: 63–73.

———. 1965a. Dynamic aspects of soil-water availability to plants. *Ann. Rev. Plant Physiol.* 16: 323–42.

———. 1965b. Movement of nitrogen in soil. In *Soil Nitrogen*, ed. W. V. Bartholomew and F. E. Clark, pp. 550–72. Madison, Wis.: Amer. Soc. Agron.

Gardner, W. R., and C. F. Ehlig. 1962. Impedance to water movement in soil and plant. *Science* 138: 522–23.

———. 1963. The influence of soil water on transpiration by plants. *J. Geophys. Res.* 68: 5719–24.

Gavande, S. A., and S. A. Taylor. 1967. Influence of soil water potential and atmospheric evaporative demand on transpiration and the energy status of water in plants. *Agron. J.* 59: 4–7.

Gibbs, R. D. 1935. Studies of wood. II. On the water content of certain Canadian trees and on changes in the water-gas system during seasoning and flotation. *Can. J. Res.* 12: 727–60.

——. 1939. Studies in tree physiology. I. General introduction. Water content of certain Canadian trees. *Can. J. Res.* 17: 460–82.

Ginsburg, H., and B. Z. Ginzburg. 1970. Radial water and solute flows in roots of *Zea mays*. I. Water flow. *J. Exp. Bot.* 21: 580–92.

Hadas, A. 1964. Deviations from Darcy's Law for the flow of water in unsaturated soils. *Isr. J. Agr. Res.* 14: 159–68.

Hagan, R. H., H. R. Haise, and T. C. Edminster (eds.). 1967. *Irrigation of Agricultural Lands*. Madison, Wis.: Amer. Soc. Agron.

Haines, F. M. 1935. Observations on the occurrence of air in conducting tracts. *Ann. Bot.* 49: 367–79.

Hayward, H. E., W. M. Blair, and P. E. Skaling. 1942. Device for measuring entry of water into roots. *Bot. Gaz.* 104: 152–60.

Helder, R. J., and J. Boerma. 1969. An electron-microscopical study of the plasmodesmata in the roots of young barley seedlings. *Acta Bot. Neerl.* 18: 99–107.

Henniker, J. C. 1952. Retardation of flow in narrow capillaries. *J. Colloid Sci.* 7: 443–46.

Hodges, T. K., and Y. Vaadia. 1964. Uptake and transport of radiochloride and tritiated water by various zones of onion roots of different chloride status. *Plant Physiol.* 39: 104–8.

Honert, T. H. van den. 1948. Water transport in plants as a catenary process. *Disc. Faraday Soc.* 3: 146–53.

House, C. R., and N. Findlay. 1966a. Water transport in isolated maize roots. *J. Exp. Bot.* 17: 344–54.

——. 1966b. Analysis of transient changes in fluid exudation from isolated maize roots. *J. Exp. Bot.* 17: 627–40.

Huck, M. G., B. Klepper, and H. M. Taylor. 1970. Diurnal variations in root diameter. *Plant Physiol.* 45: 529–30.

Hunter, A. S., and O. J. Kelley. 1946. The extension of plant roots into dry soil. *Plant Physiol.* 21: 445–51.

Hylmö, B. 1955. Passive components in the ion absorption of the plant. I. The zonal ion and water absorption in Brouwer's experiments. *Physiol. Plant.* 8: 433–49.

——. 1958. Passive components in the ion absorption of the plant. II. The zonal water flow, ion passage, and pore size in roots of *Vicia faba*. *Physiol. Plant.* 11: 382–400.

Itai, C., and Y. Vaadia. 1965. Kinetin-like activity in root exudate of water-stressed sunflower plants. *Physiol. Plant.* 18: 941–44.

Jackson, J. E., and P. E. Weatherley. 1962. The effect of hydrostatic pressure gradients on the movement of potassium across the cortex. *J. Exp. Bot.* 13: 128–43.

Jackson, R. D. 1967. Osmotic effects on water flow through a ceramic filter. *Soil Sci. Soc. Amer. Proc.* 31: 713–15.

Jackson, R. D., C. H. M. van Bavel, and R. J. Reginato. 1963. Examination of the pressure-plate outflow method for measuring capillary conductivity.

Soil Sci. 96: 249–56.

Janes, B. E. 1970. Effect of carbon dioxide, osmotic potential of nutrient solution, and light intensity on transpiration and resistance to flow of water in pepper plants. *Plant Physiol.* 45: 95–103.

Jarvis, M. S. 1963. A comparison between the water relations of species with contrasting types of geographical distribution in the British Isles. In *The Water Relations of Plants*, ed. A. J. Rutter and F. H. Whitehead, pp. 289–312. London: Blackwell.

Jarvis, P., and C. R. House. 1967. The radial exchange of labelled water in maize roots. *J. Exp. Bot.* 18: 695–706.

Jenny, H., and K. Grossenbacher. 1963. Root-soil boundary zones as seen in the electron microscope. *Soil Sci. Soc. Amer. Proc.* 27: 273–77.

Kamiya, N., M. Tazawa, and T. Takata. 1962. Water permeability of the cell wall in *Nitella. Plant Cell Physiol.* 3: 285–92.

Kaufmann, M. R. 1968. Water relations of pine seedlings in relation to root and shoot growth. *Plant Physiol.* 43: 281–88.

Kemper, W. D. 1960. Water and ion movement in thin films as influenced by the electrostatic charge and diffuse layer of cations associated with clay mineral surfaces. *Soil Sci. Soc. Amer. Proc.* 24: 10–16.

Klepper, B. 1967. Effects of osmotic pressure on exudation from corn roots. *Aust. J. Biol. Sci.* 20: 723–35.

Klute, A. 1965. Laboratory measurement of hydraulic conductivity of unsaturated soil. Water diffusivity. In *Methods of Soil Analysis*, vol. 1, ed. C. A. Black, pp. 253–72. Madison, Wis.: Amer. Soc. Agron.

Köhnlein, E. 1930. Untersuchungen über die Höhe des Wurzelwiderstandes und die Bedeutung activer Wurzeltätigkeit für die Wasserversorgung der Pflanzen. *Planta* 10: 381–423.

Kozlowski, T. T. (ed.). 1968. *Water Deficits and Plant Growth*. New York: Academic Press.

Kramer, P. J. 1946. Absorption of water through suberized roots of trees. *Plant Physiol.* 21: 37–41.

——. 1950. Effects of wilting on the subsequent intake of water by plants. *Amer. J. Bot.* 37: 280–84.

——. 1969. *Plant and Soil Water Relationships: A Modern Synthesis*. New York: McGraw-Hill.

Kramer, P. J., and H. C. Bullock. 1966. Seasonal variations in the proportions of suberized and unsuberized roots of trees in relation to the absorption of water. *Amer. J. Bot.* 53: 200–204.

Kramer, P. J., and T. S. Coile. 1940. An estimation of the volume of water made available by root extension. *Plant Physiol.* 15: 743–47.

Kuiper, P. J. C. 1963. Some considerations on water transport across living cell membranes. *Bull. Conn. Agr. Exp. Sta.* 664: 59–67.

——. 1964a. Water uptake of higher plants as affected by root temperature. *Mededel. Landbouwh. Wageningen* 64(4): 1–11.

——. 1964b. Water transport across root cell membranes: effect of alkenylsuccinic acids. *Science* 143: 690–91.

——. 1967. Surface-active chemicals as regulators of plant growth, membrane permeability and resistance to freezing. *Mededel. Landbouwh. Wageningen* 67(3): 1–23.

Leshem, B. 1966. Toxic effects of carbowaxes (polyethylene glycols) on *Pinus halepensis* Mill. seedlings. *Plant Soil* 24: 322–23.

Lewis, D. G., and J. P. Quirk. 1967. Phosphate diffusion in soil and uptake by plants. III. P^{31} movement and uptake by plants as indicated by P^{32} autoradiography. *Plant Soil* 26: 445–53.

Lindsay, W. A., and H. F. Stephenson. 1959. Nature of the reactions of monocalcium phosphate monohydrate in soils. I. The solution that reacts with the soil. *Soil Sci. Soc. Amer. Proc.* 23: 12–18.

Lopushinsky, W. 1964. Effect of water movement on ion movement into the xylem of tomato roots. *Plant Physiol.* 39: 494–501.

Macklon, A. E. S., and P. E. Weatherley. 1965. Controlled environment studies of the nature and origins of water deficits in plants. *New Phytol.* 64: 414–27.

Mees, G. C., and P. E. Weatherley. 1957a. The mechanism of water absorption by roots. I. Preliminary studies on the effects of hydrostatic pressure gradients. *Proc. Roy. Soc.* (London), ser. B, 147: 367–80.

——. 1957b. The mechanism of water absorption by roots. II. The role of hydrostatic pressure gradients across the cortex. *Proc. Roy. Soc.* (London), ser. B, 147: 381–91.

Meiri, A., and W. P. Anderson. 1970a. Observations on the effects of pressure differences between the bathing media and exudates of excised maize roots. *J. Exp. Bot.* 21: 899–907.

——. 1970b. Observations on the exchange of salt between the xylem and neighbouring cells in *Zea mays* primary roots. *J. Exp. Bot.* 21: 908–14.

Milburn, J. A. 1966. The conduction of sap. I. Water conduction and cavitation in water stressed leaves. *Planta* 69: 34–42.

Milburn, J. A., and R. P. C. Johnson. 1966. The conduction of sap. II. Detection of vibrations produced by sap cavitation in *Ricinus* xylem. *Planta* 69: 43–52.

Miller, R. J., and P. F. Low. 1963. Threshold gradient for water flow in clay systems. *Soil Sci. Soc. Amer. Proc.* 27: 605–9.

Monteith, J. L. 1965. Evaporation and environment. In *The State and Movement of Water in Living Organisms*, ed. G. E. Fogg, pp. 205–34. Symp. Soc. Exp. Biol. 19. Cambridge.

Myers, G. M. P. 1951. The water permeability of unplasmolysed tissues. *J. Exp. Bot.* 2: 129–44.

Newman, E. I. 1969a. Resistance to water flow in soil and plant. I. Soil resistance in relation to amounts of root: theoretical estimates. *J. Appl. Ecol.* 6: 1–12.

——. 1969b. Resistance to water flow in soil and plant. II. A review of experimental evidence on the rhizosphere resistance. *J. Appl. Ecol.* 6: 261–72.

Newman, E. I., and P. J. Kramer. 1966. Effects of decenylsuccinic acid on the permeability and growth of bean roots. *Plant Physiol.* 41: 606–9.

Nielsen, D. R., J. W. Biggar, and J. M. Davidson. 1962. Experimental consideration of diffusion analysis in unsaturated flow problems. *Soil Sci. Soc. Amer. Proc.* 26: 107–11.

Northcote, D. H., and D. R. Lewis. 1968. Freeze-etched surfaces of membranes and organelles in the cells of pea root tips. *J. Cell Sci.* 3: 199–206.

O'Leary, J. W. 1969. The effect of salinity on permeability of roots to water. *Isr. J. Bot.* 18: 1–9.

Ordin, L., and S. Gairon. 1961. Diffusion of tritiated water into roots as influenced by water status of tissue. *Plant Physiol.* 36: 331–35.

Ordin, L., and P. J. Kramer. 1956. Permeability of *Vicia faba* root segments to water as measured by diffusion of deuterium hydroxide. *Plant Physiol.* 31: 468–71.

Oster, J. D., and P. F. Low. 1964. Heat capacities of clay and clay-water mixtures. *Soil Sci. Soc. Amer. Proc.* 28: 605–9.

Parker, J. 1949. Effects of variations in root-leaf ratio on transpiration rate. *Plant Physiol.* 24: 739–43.

Passioura, J. B., and I. R. Cowan. 1968. On solving the non-linear diffusion equation for the radial flow of water to roots. *Agr. Meteorol.* 5: 129–34.

Peel, A. J. 1965. On the conductivity of xylem in trees. *Ann. Bot.*, n.s. 29: 119–30.

Perrier, E. R., C. M. McKell, and J. M. Davidson. 1961. Plant-soil-water relations of two subspecies of orchard grass. *Soil Sci.* 92: 413–20.

Philip, J. R. 1966. Plant water relations: some physical aspects. *Ann. Rev. Plant Physiol.* 17: 245–68.

Raney, F., and Y. Vaadia. 1965a. Movement of tritiated water in the root system of *Helianthus annuus* in the presence and absence of transpiration. *Plant Physiol.* 40: 378–82.

——. 1965b. Movement and distribution of THO in tissue water and vapor transpired by shoots of *Helianthus* and *Nicotiana*. *Plant Physiol.* 40: 383–88.

Reynolds, E. R. C. 1970. Root distribution and the cause of its spatial variability in *Pseudotsuga taxifolia* (Poir.) Britt. *Plant Soil* 32: 501–17.

Richards, S. J., and L. V. Weeks. 1953. Capillary conductivity values from moisture yield and tension measurements in soil columns. *Soil Sci. Soc. Amer. Proc.* 17: 206–9.

Robards, A. W. 1968. A new interpretation of plasmodesmatal ultrastructure. *Planta* 82: 200–210.

Rosene, H. F. 1943. Quantitative measurement of the velocity of water absorption in individual root hairs by a microtechnique. *Plant Physiol.* 18: 588–607.

Russell, M. B., and J. T. Woolley. 1961. Transport processes in the soil-plant system. In *Growth in Living Systems*, ed. M. X. Zarrow, pp. 695–721. New York: Basic Books.

Sands, K., and A. J. Rutter. 1959. Studies in the growth of young plants of *Pinus sylvestris* L. II. The relation of growth to soil moisture tension. *Ann. Bot.*, n.s. 23: 269–84.

Scheidegger, A. E. 1957. The *Physics of Flow through Porous Media*. London: Macmillan.

——. 1966. Flow through porous media. In *Applied Mechanics Surveys*, ed. H. N. Abramson, H. Liebowitz, J. M. Crowley, and S. Juhasz, pp. 893–900. Washington, D.C.: Spartan Books.

Scott, L. I., and J. H. Priestley. 1928. The root as an absorbing organ. I. A reconsideration of the entry of water and salts in the absorbing region. *New Phytol.* 27: 125–40.

Scott Blair, G. W. 1969. *Elementary Rheology*. New York: Academic Press.

Sheikh, K. H., and A. J. Rutter. 1969. The responses of *Molinia caerulea* and *Erica tetralix* to soil aeration and related factors. I. Root distribution in relation to soil porosity. *J. Ecol.* 57: 713–26.

Sierp, H., and A. Brewig. 1935. Quantitative Untersuchungen über die Wasserabsorptionszone der Wurzeln. *Jahresber. Wiss. Bot.* 82: 99–122.

Slatyer, R. O. 1961. Effects of several osmotic substrates on the water relationships of tomato. *Aust. J. Biol. Sci.* 14: 519–40.

——. 1967. *Plant-Water Relationships*. New York: Academic Press.

Slatyer, R. O., and W. R. Gardner. 1965. Overall aspects of water movement in plants and soils. In *The State and Movement of Water in Living Organisms*, ed. G. E. Fogg, pp. 113–29. Symp. Soc. Exp. Biol. 19. Cambridge.

Slavikova, J. 1963. Horizontaler Gradient der Saugkraft eines Wurzelastes und sein Zusammenhang mit dem Wassertransport in der Wurzel. *Acta Hort. Bot. Pragensis* 1963: 73–79.

——. 1967. Compensation of root suction force within a single root system. *Biol. Plant.* 9: 20–27.

Staehelin, L. A., and M. C. Probine. 1970. Structural aspects of cell membranes. *Adv. Bot. Res.* 3: 1–52.

Swartzendruber, D. 1962. Non-Darcy flow behavior in liquid-saturated porous media. *J. Geophys. Res.* 67: 5205–13.

——. 1966. Soil-water behavior as described by transport coefficients and functions. In *Advances in Agronomy*, ed. A. G. Norman, 18: 327–370. New York: Academic Press.

——. 1968. The applicability of Darcy's Law. *Soil Sci. Soc. Amer. Proc.* 32: 11–18.

——. 1969. The flow of water in unsaturated soils. In *Flow through Porous Media*, ed. R. J. M. De Wiest, pp. 215–92. New York: Academic Press.

Sykes, D. J., and W. E. Loomis. 1967. Plant and soil factors in permanent wilting percentages and field capacity storage. *Soil Sci.* 104: 163–73.

Teoh, T. S., L. A. G. Aylmore, and J. P. Quirk. 1967. Retention of water by plant cell walls and implications for drought resistance. *Aust. J. Biol. Sci.* 20: 41–50.

Thorup, R. M. 1969. Root development and phosphorus uptake by tomato plants under controlled soil moisture conditions. *Agron. J.* 61: 808–11.

Tinklin, R., and P. E. Weatherley. 1966. On the relationship between transpiration rate and leaf water potential. *New Phytol.* 65: 509–17.

Torssell, B. W. R., J. E. Begg, C. W. Rose, and G. F. Byrne. 1968. Stand morphology of Townsville lucerne (*Stylosanthes humilis*). Seasonal growth and root development. *Aust. J. Exp. Agr. Anim. Husb.* 8: 533–43.

Totsuka, T., and M. Monsi. 1960. Effect of water economy on plant growth. 2. An analysis of water economy of water-cultured tobacco plant. *Bot. Mag. Tokyo* 73: 14–21.

Tyree, M. T. 1969. The thermodynamics of short-distance translocation in plants. *J. Exp. Bot.* 20: 341–49.

——. 1970. The symplast concept. A general theory of symplastic transport according to the thermodynamics of irreversible processes. *J. Theor. Biol.* 26: 181–214.

Tyree, M. T., and D. S. Fensom. 1970. Some experimental and theoretical observations concerning mass flow in the vascular bundles of *Heracleum*. *J. Exp. Bot.* 21: 304–24.

Vachaud, G. 1967. Determination of the hydraulic conductivity of unsaturated soils from an analysis of transient flow data. *Water Resources Res.* 3: 697–705.

Vartanian, N., and J. B. Vieira-da-Silva. 1968. Evolution du potentiel hydrique dans la plante en relation avec le potentiel du sol et l'humidité atmosphérique. *C. R. Acad. Sci.* (Paris), ser. D, 266: 2341–44.

Weatherley, P. E. 1963. The pathway of water movement across the root cortex and leaf mesophyll of transpiring plants. In *Water Relations of Plants*, ed. A. J. Rutter and F. H. Whitehead, pp. 85–100. London: Blackwell.

——. 1970. Some aspects of water relations. *Adv. Bot. Res.* 3: 171–206.

Weaver, J. E. 1925. Investigations on the root habits of plants. *Amer. J. Bot.* 12: 502–9.

Whitaker, E. S. 1923. Root hairs and secondary thickening in the Compositae. *Bot. Gaz.* 76: 30–59.

Wiegand, C. L., and W. A. Swanson. 1966. Water transmission by various ceramic, cellulose, glass and steel membranes. *Soil Sci. Soc. Amer. Proc.* 30: 124–26.

Wind, G. P. 1955. Flow of water through plant roots. *Neth. J. Agr. Sci.* 3: 259–64.

Woolley, J. T. 1965. Radial exchange of labelled water in intact maize roots. *Plant Physiol.* 40: 711–17.

Yang, S. J., and E. de Jong. 1968. Measurement of internal water stress in wheat plants. *Can. J. Plant Sci.* 48: 89–95.

15.   Soil Solution

Fred Adams

THE Soil Science Society of America (1965) defines *soil solution* as "the aqueous liquid phase of the soil and its solutes consisting of ions dissociated from the surfaces of the soil particles and of other soluble materials." A simpler definition would be: "The soil solution is the aqueous component of a soil at field-moisture contents." Chemically, it is the soil water and its dissolved electrolytes plus small quantities of dissolved gases and water-soluble compounds. Thus, the soil solution is conveniently viewed as a dilute solution of electrolytes at equilibrium with definable solid phase and gas phase components of the soil.

The concept of soil solution and its role in plant-soil interrelations dates back more than a century. Attempts to validate this concept through experimentation, however, have suffered numerous failures because of inability to define the chemical composition of soil solutions. Early failures are readily ascribable to the fact that the experiments predated the modern theory of dissolved electrolytes, initiated by the work of Debye and Hückel in 1923, which provided the means for treating electrolytic solutions with the methods of thermodynamics.

If the soil solution is conceived as being the water held by a soil at field capacity or less, it may also be viewed as the medium in which most soil chemical reactions occur. It bathes absorbing plant roots and is the medium from which roots obtain inorganic nutrients. Thus the soil solution provides the chemical environment of plant roots, and defining plant-soil interrelations in quantitative terms requires a complete and accurate knowledge of soil solution chemistry.

Since presently it is not feasible to measure quantitatively the chemical components of a soil solution *in situ*, the major task of measuring soil solution composition was recognized early as being the removal of the solution from the soil without altering the solution's composition.

I. OBTAINING THE SOIL SOLUTION

A. Methods

Many efforts have been made to obtain unaltered soil solutions. Richards (1941) grouped them into five categories: (1) suction, (2) displacement, (3) compaction, (4) centrifugation, and (5) molecular adsorption. The latter three methods suffered from serious deficiencies and were discarded soon after being proposed. Displacement, on the other hand, first used in France by Schloesing in 1866 (Parker, 1921), has since been used successfully by numerous investigators to obtain unaltered soil solutions. It must be considered as the reference method by which other methods are compared and tested.

1. Suction

The suction principle has survived primarily in the form of the pressure membrane or pressure plate method (Richards, 1941; Reitemeier and Richards, 1944). Although solutions obtained by this method are similar in many respects to soil solutions obtained by the displacement method, there are important differences: (1) the pressure membrane or porous plate effectively removes phosphate from the solution; (2) the membrane may add contaminants to the solution (washing the membrane usually prevents this); (3) excess water frequently must be added to the soil-membrane or soil-plate interface so that water films will be contiguous from membrane to soil particles (this dilution effect cannot be corrected quantitatively); and (4) 15 to 18 atm of gas pressure within the chamber will alter CO_2 solubility and, hence, solution pH. The most successful application of the pressure membrane method has been not in collecting soil solution composition data but in defining moisture retention curves for soils at moisture contents between field capacity and permanent wilting percentage.

The suction principle also applies to porous ceramic cups embedded in moist soil to obtain soil solutions. Wagner (1962) described a procedure for sampling soil water in lysimeters via ceramic cups. Reeve and Doering (1965) adapted the same technique for withdrawing soil solution samples from various soil depths in a study of sodic soil reclamation. The procedure involves evacuating an embedded ceramic cup by suction and allowing soil water to move into the evacuated cup. Then, normal air pressure is used to force the water in the ceramic cup up a tube to a collection bottle. Solutions obtained in this manner probably have been altered from the true soil solution by the sieving action of the ceramic clay and by altered gas pressures.

Kemper (1959) and Shimshi (1966) also used porous ceramic to obtain a measure of the electrolytic content of soil solutions. Kemper connected platinum wires to units of ceramic so that they served as conductivity cells when embedded in moist soil and connected to a resistance bridge. Shimshi embedded dry ceramic pieces in moist soil, removed them after an equilibration time, weighed them for moisture content, washed them with water, and analyzed the wash water. Data were not presented to verify that this method yielded the true solution.

One of the more imaginative methods for obtaining soil solutions was that of Lauritzen (1934). He grew beans in soil to pod development stage, cut off the tops, and attached rubber tubing to the stems. He then applied pressure to the soil and collected the stem exudate. He had to conclude, however, "The exact nature of the solution forced from the cut stem . . . is not known. There can be no doubt that it is not un-altered soil solution."

2. Displacement

Credit for general acceptance of the displacement procedure goes to Parker (1921). He described his soil solution displacement method as follows: "The method consists of packing the moist soil in a cylinder provided with an outlet at the bottom. The displacing liquid is then poured on top of the soil column and as it penetrates the soil it displaces some of the soil solution which forms a zone of saturation below the displacing liquid. This zone increases in depth as it is continually forced downward by the pressure of the liquid above. When the saturated zone reaches the bottom of the soil column the clear soil solution, free of the displacing liquid, drops from the soil as gravitational water."

Although Parker accurately described what occurs during displacement, the mechanics of soil solution displacement is not so obvious. However, one must conclude that the collected soil solution is displaced by other soil solution farther up the soil column. The zone saturated with soil solution that forms immediately below the displacing liquid is moved downward by the hydrostatic pressure from the displacing solution. The displacing liquid should not mix extensively with the soil solution except in a narrow zone where the two liquids meet. Because the soil solution in one zone of a soil column displaces that below it, the first portion of soil solution to fall from the column has moved the shortest distance through the soil. The last portion collected, on the other hand, may have moved several inches and may have been moving downward for several hours. If the collected soil solution is displaced from the soil by other soil solution, then the collected solution should be unaltered and should be the true soil solution.

B. Veracity of Displaced Soil Solutions

Providing experimental proof that the displaced soil solution was, in fact, the true soil solution was a challenge from the beginning. Schloesing's early effort (Parker, 1921) included (1) coloring the displacing water with carmine and observing its descent through the soil column and (2) collecting the displaced solutions in successive small increments and analyzing them for nitrate content.

Validation by analyzing the displaced soil solution in successive increments is based on the reasonable premise that soil solution composition in a well-mixed soil is the same throughout. In its downward movement through the soil, then, the soil solution only encounters and mixes with other soil solution of identical composition. Thus, each successive increment of displaced solution will have the same composition as the previous one until the displacing solution appears in the effluent.

A method of measuring total-salt constancy was introduced by a Russian scientist, Ischerekov (Parker, 1921), in 1907. He displaced the soil solution with ethanol, collected successive increments, dried each, and weighed the residue as "total salts."

Parker (1921) presented convincing evidence that the displaced solution was the true soil solution. He made three separate tests for constancy of composition of successive increments of soil solution: (1) weight of total salt after evaporation of solution at 105C, (2) weight of total salt after ignition to constant weight, and (3) freezing-point depression. Parker also tested for the presence of the displacing solution in the effluent. He used the iodoform reaction when ethanol was the displacing solution and tested with $AgNO_3$ when dilute NaCl was used. As the data in Table 15.1 show, there was excellent agreement among the three tests.

Burd and Martin (1923) further tested the thesis that a properly displaced solution is indeed the true soil solution. One series of their tests utilized mixtures of pure silica sand and solutions of known composition. In one experiment, a sand solution mixture containing 10% moisture was placed in a glass tube without packing and covered with distilled water. It gave six successive 13-ml portions of displaced solution that were identical to the original solution (this was 78% of the total moisture present in the mixture). In a second experiment, a sand solution mixture at 4% moisture was covered with a NaCl solution. No Cl^- appeared in the displaced solution until more than 50% of the original solution had been displaced. An even more efficient displacement of the original solution would have resulted if the moist sand had been compacted in the column before adding the displacing solution.

Table 15.1. Freezing-point depression and total salts in successive 10-ml increments of displaced soil solution

Solution increment no.	Freezing-point depression† (°C)	Total salts in solution (ppm)	
		Dried at 105C	After ignition
1 and 2	0.0235	—	—
3 and 4	0.0245	352	168
5 and 6	0.0230	360	176
7 and 8	0.0235	324	172
9 and 10	0.0870	344	180

Source: Reprinted by permission from F. W. Parker, *Soil Sci.* 12: 209–32, © 1921, The Williams & Wilkins Co., Baltimore.
† Average values.

A second series of tests by Burd and Martin utilized displaced soil solution as the displacing solution. Moist soil samples were packed in six glass columns. Water was added on top of the soil in four columns, and the displaced solution was collected in 10-ml portions. These increments were then composited and used as the displacing liquid on the other two columns. Soil solution constancy was tested by measuring the specific resistance of each successive 10-ml increment of displaced solution and by analyzing occasional increments for calcium, magnesium, and chlorine. The results (Table 15.2) showed that (1) 25 consecutive 10-ml portions of equal specific resistance were collected from soil columns displaced with water and (2) 31 successive 10-ml portions of unaltered soil solution were collected from columns displaced with

Table 15.2. Constancy of composition of successive 10-ml increments of soil solution when displaced by water and by soil solution

Solution increment no.†	Water displacement				Soil solution displacement			
	Sp. res. (ohm)	Ca	Mg (ppm)	Cl	Sp. res. (ohm)	Ca	Mg (ppm)	Cl
2	190	720	226	290	192	750	247	295
11	192	—	—	—	191	—	—	—
21	192	730	223	290	192	700	228	295
24	191	—	—	—	—	—	—	—
26	201	—	—	—	192	—	—	—
31	873	—	—	—	192	—	—	—

Source: Burd and Martin, 1923.
† Total volume of soil water was 310 ml per soil column.

soil solution. They concluded that "it hardly seems necessary to point out that the concordance of analytical results with specific resistances makes it quite certain that the displaced solutions have not been diluted."

The third series of tests by Burd and Martin, which were designed to validate displaced solutions as true soil solutions, tested the premise that successive increments of displaced soil solutions should contain equal quantities of all inorganic constituents. They displaced solutions from replicate columns of several soils and collected the effluent in 5-ml increments. Table 15.3 shows examples of their results for two soils. Considering the experimental errors involved with the small solution volumes available for analysis and the analytical methods used at that time, there is little doubt that the successive solutions were of uniform concentrations. Burd and Martin (1924) later observed that there is "little doubt that such displaced solutions are substantially identical with the true soil solution."

Pierre (1925) provided still another critical test for the veracity of displaced soil solutions. In addition to measuring the total-salt weight of each soil solution portion, he measured the pH of each solution increment. He found no change in solution pH or total salts until after the solution had become contaminated with the displacing solution (Table 15.4). Parenthetically, the change in pH with increased time (beyond 7.5 hours) required for displacement was undoubtedly caused by changes occurring within the soil column, such as denitrification or increased CO_2 pressure in the soil atmosphere, rather than by contamination with the displacing solution.

The latest confirmation of the veracity of the displaced soil solution has been provided by the work of Adams and co-workers (Howard and Adams, 1965; Adams and Lund, 1966; Adams, 1966; Bennett and Adams, 1970a, 1970b). Their results are summarized in Table 15.5. Initial growth rates of primary roots of cotton (*Gossypium* spp.) seedlings were measured in nutrient solutions and in soils with the separate variables of calcium, aluminum, and ammonia. Soil solutions were

Table 15.3. Analyses of successive 5-ml increments of displaced soil solution

Solution increment no.	Soil no. 8						Soil no. 9					
	Sp. res. (ohm)	Ca	Mg	K	NO₃	Cl	Sp. res. (ohm)	Ca	Mg	K	NO₃	Cl
			(ppm)						(ppm)			
2	206	640	216	152	2430	360	292	460	148	85	1042	640
7	206	640	192	142	2332	320	289	480	120	85	1116	640
12	206	640	206	158	2480	320	289	480	124	85	1042	620

Source: Burd and Martin, 1923.

Table 15.4. Total-salt content and pH of successive 10-ml increments of a displaced soil solution

Solution increment no.	Time of displacement (hr)	Total salts dried at 105C (mg)	Solution pH
1	$4\frac{1}{2}$	6.8	6.65
2	$5\frac{1}{2}$	6.2	6.65
3	$6\frac{1}{2}$	6.2	6.65
4†	$7\frac{1}{2}$	6.5	6.65
6†	29	6.0	6.75
10†	29	13.8	6.00

Source: Reprinted by permission from W. H. Pierre, *Soil Sci.* 20: 285–305, © 1925, The Williams & Wilkins Co., Baltimore.
† Displacing solution was detected in the displaced solution.

Table 15.5. "Critical" levels of Ca, Al, and NH$_3$ in both culture solutions and in soil solutions *in situ* for growth of primary cotton roots

Plant reaction	Critical level
Ca deficiency	$a_{Ca^{++}}/\Sigma a_{cation\ i} \cong 0.15$
Al toxicity	$a_{Al^{3+}} \cong 2\mu M$
NH$_3$ toxicity	$a_{NH_4OH} \cong 0.2\ mM$

Sources: Data for Ca: Howard and Adams, 1965; Adams, 1966; Bennett and Adams, 1970b. Data for Al: Adams and Lund, 1966. NH$_3$: Bennett and Adams, 1970a.

displaced and analyzed. The results showed that the "critical," deficient level of Ca was the same in nutrient solutions and in soil solutions *in situ* (see Table 15.5). Similarly, "critical" toxicity concentrations for Al and NH$_3$ were identical in nutrient solutions and in soil solutions *in situ*.

The uniqueness of these experiments is that they provided biological proof as additional and convincing evidence that a properly displaced soil solution is the true soil solution. Furthermore, the experiments showed that the chemical composition of the soil solution was indeed the root's chemical environment in the soil.

C. Soil Solution Displacement Procedure

There is clear evidence that a properly displaced soil solution is the unaltered, true soil solution. However, obtaining that solution has proved to be a challenge to the novice. The operator's skill is the key to proper packing of the soil column. It was a chief concern of the pioneers in the field, typified by Parker's observation: "After a little experience one can readily determine the proper degree of packing for any soil at a given moisture content." The technique is easily mastered by a skilled technician if the following principles are properly applied to the procedure.

1. The Displacing Solution

The basic technique of identifying contaminants in displaced soil solutions has not changed appreciably since Schloesing in 1886 (Russell, 1950) poured water, colored with carmine dye, on soil in an inverted bell jar. The purpose of the dye was to identify the displacing solution in the effluent. Dyes proved inadequate because they are extensively adsorbed by soil particles and fail to identify accurately when the effluent is first contaminated.

Parker (1921) found ethanol to be an efficient displacing solution and used the iodoform reaction for identifying it in the effluent. When sodium hypoiodite (NaOI) is added to ethanol and the solution is warmed, a precipitate of iodoform (CHI_3) is immediately formed. This test is neither convenient nor highly sensitive. In fact, Pierre (1925) concluded that displacement with ethanol was unsatisfactory because the iodoform reaction was not sensitive enough to determine when the effluent contained small amounts of ethanol.

Use of a displacing solution that differed greatly in electrical conductivity from the soil solution enabled Burd and Martin (1923) to measure specific resistance of the effluent and thus to determine when the displacing solution had become a part of the displaced solution. This method is convenient, rapid, and sufficiently sensitive to detect significant changes in the ionic composition of the effluent.

Pierre (1925) used a 0.5% KCNS solution as the displacing solution because CNS^- can be identified readily in very dilute concentrations. When thiocyanate is added to an acid ferric solution, an intensely colored, blood-red complex of ferric thiocyanate of uncertain composition, possibly $FeCN^{2+}$, is formed. Pierre noted that this characteristic red color was obtained when one drop of $FeCl_3$ solution was added to 5 ml of 0.005% KCNS. Thus, the presence of one drop of a 0.5% KCNS solution (the displacing solution) in 5 ml of displaced solution can be detected by the addition of one drop of a $FeCl_3$ solution. Pierre further showed that detectable amounts of CNS^- appeared in the displaced solution before changes in soil solution composition were measurable. The ferric thiocyanate complex is probably the most sensitive measure reported for detecting the presence of the displacing solution in the effluent.

The most common liquids used for displacement have been water and ethanol. Ischerekov (Parker, 1921) used ethanol in 1907 to obtain what he considered to be the true soil solution. Parker (1921) compared several liquids and found that liquids immiscible with water were unsatisfactory because they passed through the soil without displacing more than a trace of soil solution. Of the solutions Parker found effective, ethanol yielded the most soil solution, acetone yielded the least, and methanol and water yielded intermediate amounts.

In the recent experiments of Adams and co-workers (Howard and Adams, 1965; Adams, 1966; Adams and Lund, 1966; Khasawneh and Adams, 1967; Bennett and Adams, 1970a, 1970b), a saturated $CaSO_4$ solution containing 0.4% KCNS was used as the displacing solution. Evans and Kamprath (1970) used ethanol as the displacing liquid (private communication from C. E. Evans).

2. Column Size

Parker (1921) experimented with columns that varied from 5 to 11 cm in diameter and 30 to 60 cm in height. He found that the time required for complete displacement increased with increasing column height and that the percentage of soil solution displaced was inversely related to the diameter of the column.

Burd and Martin (1923) and Pierre (1925) used cylinders that held about 1 kg of soil. Glass columns, 60 × 5 cm, were found highly satisfactory by Howard and Adams (1965) for obtaining an effective displacement of ample soil solution for analysis.

A large column was used by Conrad *et al.* (1930) to obtain the large volume of soil solution needed for their analyses. They needed a minimum of 555 ml of soil solution per sample. This amount was obtained by displacement from 8 kg of moist soil in an iron tube that measured 15 cm in diameter and 46 cm long.

3. Column Packing

The manner in which the soil is packed in the tube or cylinder is the most critical step of the entire procedure. Packing moist soil to an ideal state of compaction is truly an art. Too little packing results in the early appearance of the displacing liquid in the displaced effluent. This is prevented by compacting the soil. An obvious result of soil compaction in the tube is to slow down the rate of displacement. Severe compaction results in imperviousness; less compaction may displace solution so slowly as to afford ample time for biological changes (denitrification, for example) to occur during displacement.

The general procedure is to pack moist soil uniformly in a brass or glass cylinder. The degree of packing is determined by the soil texture and the moisture content. Sandy soils and peats can be packed relatively firmly without danger of puddling, which renders the soil impervious. Parker (1921), Burd and Martin (1923), and Pierre (1925) all found that finer-textured soils had to be packed at below-optimum moisture contents and with care to prevent puddling and imperviousness.

The condition of the soil at optimum moisture for plant growth obviously deserves primary consideration in soil-plant interrelations. This corresponds to the imprecise but useful concept of field capacity. The methods of Parker (1921) and Burd and Martin (1923) successfully displaced solution from sandy loam soils at field capacity, but they could be used on finer-textured soils only at moisture contents where such soils did not puddle during packing. This is less than the field capacity moisture content on many fine-textured soils.

Howard and Adams (1965) devised a method that successfully displaces solutions from fine-textured soils at field capacity, or about 1/3-bar moisture percentage. The method consists of wetting the soil to 1/3-bar moisture, or to as near that as practicable without soil puddling. The moist soil is then sieved through a 5- or 10-mm screen to give a highly granular structure and placed loosely in glass tubes. For soils that must be placed in the tubes at higher moisture tensions, additional water may be added in 10- to 20-ml increments at 1- to 2-hour intervals after the soil is placed in the column and while the soil is still very loose and open; the moist soil is then allowed to equilibrate overnight. This procedure works well for sandy soils and soils with enough clay to impart a granular structure to the moist soil. However, it has not worked satisfactorily for some soils with high silt but low clay contents because the displaced solution tends to contain silt-size particles.

Although all soils must be packed firmly, the degree of compaction is different for each soil and each moisture content. Whereas Parker (1921) and Burd and Martin (1923) used wooden mallets to tamp increments of moist soil into tubes, Howard and Adams (1965) used an entirely different application of force. They attached a rubber cushion to the bottom of the glass tube filled with moist soil and repeatedly raised the tube a few inches above the desk top and then slammed the assembly against the desk top until proper compaction of soil was attained. This action forces the soil aggregates closer together while maintaining the granular soil structure and its associated permeability.

4. Time

The time required for displacing a soil solution varies widely. In general, more time is required (1) at lower soil moisture contents, (2) with finer-textured soils, (3) at higher degrees of compaction, and (4) with taller soil columns. Parker (1921) obtained complete displacement in less than 12 hours generally. Pierre (1925) achieved displacement in less than 8 hours. However, soil solution displacement may extend over a period of 2 or 3 days if there is excessive soil compaction (Burd and Martin, 1923; Reitemeier, 1945).

Soil solution displacement should be completed within a practicable minimum time period, such as 8 to 10 hours, so that microbiological activity does not change soil solution composition appreciably. Nitrification can be expected upon wetting a dry soil; dentrification can occur in the packed tubes when displacement time is unduly prolonged. Evidence that microbiological activity may be significant during displacement is an increased level of CO_2 in the displaced soil air. Pierre (1925) indirectly observed this when he reported that soil solution pH changed upon exposure to the atmosphere (he then devised a means of measuring the pH without first exposing the displaced solution to the atmosphere).

Because the factors that increase the amount of soil solution that can be displaced also increase the time required for displacement, a compromise must be made between high displacement values and time required for displacement. The experience of this laboratory has been that 25% to 75% of the total soil solution can be displaced within an 8-hour period.

II. INTERPRETATIONS OF SOIL SOLUTION DATA

The greatest advantage of a displaced soil solution is that it gives a correct measure of the electrolytic content of soil solution at equilibrium with the other soil phases. Its disadvantages are (1) the time required, (2) the necessity for an experienced operator, and (3) the need for large soil samples. It appears unfortunate that soil solution displacement was essentially abandoned after the mid-1930's. The reason is obscure, but it likely resulted from procedural difficulties encountered by novices, especially with fine-textured soils.

The original concept of soil solution merits retention and revitalization as a basic principle of soil chemistry and plant nutrition. Considerable information regarding soil chemical reactions and soil-plant interrelations may be gained by its use. Soil solutions should not be confused with soil extracts; they are not synonymous. It is possible that better information about soil solution compositions and reactions would improve the understanding of the voluminous information obtained by water extractions and other methods.

Because of numerous difficulties encountered when attempts are made to separate the true soil solution from its harboring solid phase, it is only natural that attempts be made to define soil solution composition in terms of water extract composition. It is also immediately apparent that immense difficulties will be encountered in this effort because soil solution electrolytes are supplied from (1) free, soluble salts, (2) absorbed salts, (3) solid phase compounds of slight solubility, and (4) exchangeable ions. The addition of water affects the equilibrium concentrations of soil solution ions differently because dilution results in ion exchange, precipitation, and dissolution reactions. The net effect of all these reactions is hardly predictable.

A. Soil Solution Approximations

1. Water Extracts

A much-used approximation of the soil solution is the saturated-paste extract, made popular by the U.S. Salinity Laboratory at Riverside, California. It proved invaluable in qualitative assessments of soil salinity intensity and management practices, but it was not intended to be an accurate facsimile of the true soil solution. With the passage of time, however, original limitations became blurred, and the assumption has been frequently made that soil solution ionic concentrations are inversely proportional to ionic concentrations of saturated pastes. The inherent error in that assumption is exemplified by Reitemeier's (1945) review of the literature on the effect of dilution on ionic concentrations in soil solutions. He noted the preponderance of experimental data that show the following general ionic behavior:

Nonsaline soils:

1. Total potassium in solution increases with dilution;
2. Total Ca and Mg in solution frequently increases with dilution while the ratio of Ca:Mg in solution changes;
3. Amount of phosphate usually increases proportionally to dilution, suggesting a saturated solution of phosphate;
4. Quantity of nitrate may slightly increase or slightly decrease.

Alkali, calcareous, and gypsiferous soils:

In virtually all cases, dilution results in increased amounts of Ca, Mg, Na, K, CO_3-HCO_3, SO_4, PO_4, and SiO_3.

Thus, the ionic concentrations in soil solutions and soil extracts are not inversely proportional to the amount of water present.

Burd and Martin (1923, 1931) compared ionic concentrations of displaced soil solutions with those of water extracts for several California soils. Results of two such comparisons are given in Table 15.6. There were decreases in the amounts of Ca and Mg with the first dilution in both soils; however, further dilution increased the amounts of both. The amounts of K and PO_4 progressively increased with dilution. Changes in NO_3^- and Cl^- were relatively small but irregular.

Burgess (1922) used a direct pressure method to obtain soil solutions in which up to 16,000 pounds of pressure per square inch was applied to force out the soil solution. He compared the ionic contents of such soil solutions with 1:5 water extracts for eight soils, two of which are reported in Table 15.7. He found the quantities of Ca and Mg in solution increased with one soil but remained almost constant with the other. Large increases occurred in the amounts of solution K, PO_4, and SO_4. There was a small decrease in solution NO_3^- with dilution in both soils.

Two significant chemical reactions occur in the soil as water is added: (1) the amounts of sparingly soluble salts in solution are increased because of the solubility product principle; (2) cation exchange reactions between solution cations occur because the relative activities of the ions are altered by dilution. Exhaustive information about a soil is required if ionic concentrations of a water extract are to be corrected accurately to soil solution concentrations. Only limited advances have been made in this area since Burgess (1922) observed, "While it is doubtless a fact that those substances present in the true soil solution form a certain

Table 15.6. Comparison of ionic concentrations of displaced soil solutions and water extracts, corrected to equivalent soil solution volumes

	Solution composition (ppm)					
Solution	Ca	Mg	K	NO_3	Cl	PO_4
Soil no. 8 at 14.8% moisture						
Displaced	640	204	150	2460	335	8.5
1:1 H_2O extract	592	184	322	2160	350	69.0
1:5 H_2O extract	627	178	408	2105	340	115.0
Soil no. 1C at 22.5% moisture						
Displaced	660	600	80	2726	—	5.6
1:1 H_2O extract	500	345	135	2639	—	1.0
1:5 H_2O extract	525	373	200	3145	—	24.0

Source: Burd and Martin, 1923.

Table 15.7. Comparison of ionic concentrations of pressure-extracted soil solutions and water extracts, corrected to water-free soil basis

Solution	Solution composition (ppm)†					
	Ca	Mg	K	NO_3	PO_4	SO_4
Soil no. 7						
Soil solution	52	12	8	74	0.4	33
1:5 H_2O extract	83	20	27	66	9.0	65
Soil no. 11						
Soil solution	80	27	20	135	1.0	52
1:5 H_2O extract	78	29	62	123	50.0	79

Source: Reprinted by permission from P. S. Burgess, *Soil Sci.* 14: 191–216, © 1922, The Williams & Wilkins Co., Baltimore.
† Corrected to water-free soil basis.

definite fraction of those dissolved in a 1:5 pure water extract of that same soil, we are hardly able to calculate just how great a part of the solutes in such an extract formerly belonged to the true soil solution and just how much has been subsequently dissolved by the greatly diluted solution formed by the addition of so great an excess of pure water."

A modification of the water extract procedure was the introduction of 0.01 M $CaCl_2$ extracts for phosphate solubility studies (Aslying, 1954) and so-called lime potentials, i.e., pH − 1/2 pCa, for acid soils (Schofield, 1952). The use of 0.01 M $CaCl_2$ assumed that this solution closely approximates the ionic composition of most soil solutions and therefore causes only minor alterations in true soil solutions.

2. Ratio Law

By restricting a soil-water system to cation exchange reactions, Schofield (1947) proposed that dilution causes changes in solution cationic concentrations that follow a ratio law. The ratio law proposed that the concentrations (later revised to activities) of solution cations upon dilution remain at constant ratios relative to one another, provided their concentrations are raised to the reciprocal power of their valences. For example, the ion-pair K-Ca in solution would behave according to

$$a_K/(a_{Ca})^{1/2} = k, \quad\quad\quad [1]$$

where a is molar activity and k is a constant.

The ratio law must be cautiously applied because it assumes that (1) adsorbed cations remain constant and (2) dissolution or precipitation reactions are absent. These conditions are not met in soils except under very restricted conditions.

It is natural that extensions of the ratio law would be made for K-Ca systems that are less restrictive than Schofield's original system (Beckett and Craig, 1964; Khasawneh and Adams, 1967; Moss, 1969). Although this approach has practical implications in soil fertility evaluations, its quantitative application remains highly restricted and cannot be applied indiscriminately to estimate soil solution composition.

B. True Soil Solution Data

The ionic composition of soil solutions within the field moisture range is of great agronomic significance. It was this obvious significance that resulted in numerous interpretive evaluations of soil solution compositions following the successful displacement procedures by Parker (1921) and Burd and Martin (1923).

The most comprehensive collection of soil solution data was compiled over a 15-year period at the University of California (Berkeley) beginning in 1915. Bulk samples of 14 soils were placed in large galvanized iron containers, and each soil was subjected to three treatments: (1) cropped, (2) fallowed, and (3) stored air-dry. Crop yields and soil solution compositions were measured periodically. The results, parts of which were published from time to time, were summarized in 1931 by Burd and Martin. Their data effectively demonstrated that (1) the ionic concentrations of soil solutions fluctuate seasonally, (2) continuous cropping decreases soil solution ionic concentrations, (3) the qualitative composition of a soil solution is constantly changing, (4) fluctuations in soil solution phosphate concentration are greatly affected by pH changes, and (5) soil solution composition of air-dried soil changes with time. Burd and Martin (1931) further concluded from these experiments that "it can hardly be controverted that high concentrations of dissolved constituents (except Na) usually connote a high degree of fertility. That the converse of this proposition does not necessarily hold and that soils of low concentration in the liquid phase are often very fertile . . . is confirmed . . . in culture solution experiments." Thus began the effort to relate plant growth in soil solutions *in situ* with plant growth in culture solutions of controlled composition.

An extensive field experiment with orchard trees was started in 1922 at the University of California (Davis) with the objective of defining soil fertility in terms of soil solution composition (Proebsting, 1929, 1930, 1933). The magnitude of the experiment can be judged from the fact that five management treatments were applied to each of the following: almond (*Prunus amygdalus* Batsch.), apple (*Malus pumila*

Mill.), apricot (*Prunus* ssp.), cherry (*Prunus* ssp.), peach (*Prunus persica*), pear (*Pyrus* ssp.), plum (*Prunus* ssp.), and prune (*Prunus* ssp.). Soil samples for displacement were taken weekly from May to September at depths of 0–90 cm and 90–180 cm. Large volumes of soil (8 kg per sample) were required because a minimum of 550 ml of soil solution was required for a complete soil solution analysis (Conrad *et al.*, 1930). In later years, soil solutions were obtained weekly from soil depths of 0–60, 60–120, 120–180, and 180–240 cm (Proebsting, 1933). In spite of the considerable work required for this experiment, the author failed to establish any quantitative relationship between soil solution composition and crop yields.

1. Phosphorus

Because many soil solutions contain very low concentrations of phosphate, early researchers were intrigued by the observation that plants were able to obtain adequate phosphate from such soils. Some examples of soil solution phosphorus concentrations from 20 soils studied by Pierre and Parker (1927) are listed in Table 15.8. These data showed no consistent relationship between soil solution P and 1:5 water extract P nor between inorganic P and organic P. However, the authors did show that (1) plants failed to absorb organic P from either solution but absorbed all the inorganic P and (2) corn (*Zea mays* L.) seedlings made no growth on soil extracts containing only organic P but made good growth on extracts containing inorganic P. Fudge (1928) subsequently reported soil solution inorganic and organic phosphate concentrations similar to those in Table 15.8.

In 1930 Tidmore made a significant effort to define deficient levels of phosphate in soil solutions for good plant growth. He grew corn, sorghum (*Sorghum vulgare* Pers.), and tomatoes (*Lycopersicon esculentum*) in two soils and in several culture solutions. Then he displaced and analyzed the soil solutions for inorganic and organic phosphate concentrations. He found that plants did not respond to low P concentrations equally in the two media; they grew better in soils at low solution P concentrations than in culture solutions. Thus, he reasoned that "plants in these soils . . . have access to a greater phosphate concentration than is found in the displaced solution. . . . There is no apparent reason for assuming that plants will grow at a lower concentration of phosphate in the soil than in culture solutions. . . . This would mean that the displaced solution is not the true soil solution."

Tidmore was misled in his interpretation because (1) he probably failed to maintain a constant low P concentration in culture solution,

Table 15.8. Comparison of inorganic and organic phosphate concentrations in displaced soil solutions and 1 : 5 water extracts of several soils

Soil no.	Soil solution (ppm)[†]		1 : 5 water extract (ppm)[‡]	
	Inorg. PO_4	Org. PO_4	Inorg. PO_4	Org. PO_4
1	0.03	0.23	0.06	0.10
2	0.14	0.54	0.38	0.26
3	0.02	0.28	<0.02	0.13
5	<0.02	0.42	0.05	0.05
8	0.05	0.15	<0.02	0.06
10	0.32	0.60	1.72	0.57
13	<0.02	0.51	0.35	0.45

Source: Reprinted by permission from W. H. Pierre and F. W. Parker, *Soil Sci.* 24: 119–28, ⓒ 1927, The Williams & Wilkins Co., Baltimore.
† Soil solution displaced by method of Burd and Martin (1923).
‡ These are actual concentrations; no correction to constant moisture contents were made.

such as occurs in a soil, and (2) he failed to recognize that solution ions behave according to their activities and not according to their concentrations.

2. Aluminum

Soil solution Al was suspected early as a probable cause of the infertility of strongly acid soils. McGeorge (1924) attempted to relate yield of sugarcane (*Szecharum* spp.) in Hawaii with soil solution concentrations of iron, aluminum, and manganese. His results were inconclusive.

Magistad (1925) followed this with a comprehensive study of $Al(OH)_3$ solubility and soil solution Al concentrations. He found high amounts of Al in soil solutions at both low and high pH values. A general, qualitative relationship was evident between plant growth and soil solution Al concentration. Magistad was probably the first to suggest that poor plant growth in high-pH soils, as well as in low-pH soils, was caused by toxic concentrations of Al.

Pierre (1931) and Pierre *et al.* (1932) generally substantiated the findings of Magistad on naturally acid soils. They found a general relationship between soil solution Al and plant growth, but many inconsistencies were evident among soils of equal acidity or equal soil solution Al concentrations.

Although soil Al, especially exchangeable Al, has received much attention since the late 1940's, it was not until 1966 that soil solution Al was reevaluated. Adams and Lund (1966) measured growth rate of cotton primary roots in culture solutions containing Al and in acid soils of varying Al contents. They found that if culture solution Al and soil solution Al concentrations were corrected to molar activities, there was no difference in root growth rate in the two media at equal solution Al activities. The "critical" value of solution Al for cotton roots was an Al^{3+} activity of about $2\mu M$ (for convenience, all solution Al was incorrectly assumed to be present as Al^{3+}). This critical value for soil solution Al was subsequently substantiated by Adams *et al.* (1967) and Richburg and Adams (1970). The real significance of these experiments was that they further substantiated the thesis that the displaced solution is the true soil solution and that the soil solution composition is the chemical environment of roots.

Evans and Kamprath (1970) also found a general, albeit qualitative, relationship between plant growth in acid soils and displaced soil solution Al concentrations. They did not, however, correct Al concentrations to Al activities to compensate for differences in ionic strength of the different soil solutions.

3. Fertilizers

It was popular during the 1930's and 1940's to conduct field experiments comparing different fertilizers. In an effort to understand the results of such experiments, White and Ross (1936) determined the effect of various mixtures of fertilizer salts on total electrolyte content of displaced soil solutions, as measured by freezing-point depressions and specific resistances. They found that (1) total salts were greater in a sandy soil than in a clay soil at equal fertilizer rates, (2) the decreasing order of soil solution salts at equivalent fertilizer rates was $NaNO_3 > (NH_4)_2SO_4 > KCl > K_2SO_4 > NH_4H_2PO_4 > CA(H_2PO_4)_2 \cdot 2H_2O$, and (3) doubling the fertilizer rate did not necessarily double the soil solution salt concentration. They concluded, "The influence of a mixed fertilizer on the effective concentration of a soil solution depends on its composition and not on its total content of the plant-food elements . . . . The growing plant derives its nutrients . . . from the soil through the medium of the soil solution . . . . The composition of the soil solution . . . may vary greatly."

Experiments with soil solution were especially helpful in delineating the role of Ca deficiency in acid soils. Calcium deficiency has often been considered to be synonymous with high soil acidity by those unaware of the true nature of soil acidity. However, Ca deficiency may also occur in nonacid soils, as illustrated by the soil solution data of Howard and Adams (1965). They used root growth to study Ca deficiency in soils and in culture solutions and found that it was definable in terms of the Ca:total-cations ratio in the solution, regardless of pH. More important, they showed that the critical Ca level for cotton root growth was the same in culture solutions as in soil solutions *in situ.* Subsequently, Adams (1966) and Bennett and Adams (1970b) found that Ca-deficient levels even in limed soils were defined by this same quantitative relationship. However, the Ca-plant interrelation required that soil solution cations be expressed as ionic activities rather than ionic concentrations. The critical Ca level in soil solution was thus defined as

$$a_{Ca}/\Sigma a_{cation\ i} \cong 0.15, \qquad\qquad [2]$$

where a_{Ca} is the molar activity of Ca^{2+} and $\Sigma a_{cation\ i}$ is the sum of cationic activities.

Since early experimental trials with $(NH_4)_2HPO_4$ as a potential fertilizer material, it has been observed that this compound frequently causes seedling injury or even death. The cause of injury was generally ascribed to NH_3 toxicity. By using soil solution data, however, Adams (1966) and Bennett and Adams (1970b) showed that injury was caused by an induced Ca deficiency for root growth. Soil solution Ca and exchangeable Ca were precipitated by the soluble phosphate, $(NH_4)_2HPO_4$, to the extent that Ca deficiency resulted. This conclusion was made possible by knowing the Ca and phosphate activities in soil solution.

The toxic nature of NH_3 has been recognized for years, but quantitative definitions of it were lacking until Bennett and Adams (1970a), with soil solution and culture solution data, defined the "critical" toxic concentration of solution NH_3 for cotton roots and sudangrass [*Sorghum bicolor* (Linn.) Moench] seedlings. They further showed that the "critical" NH_3 concentration in soil solution was identical to the value in culture solutions. Significantly, however, a quantitative evaluation of the "critical" toxic concentration of 0.2 mM NH_4OH, which applied to all systems, required that the electrolytic composition of the solutions be expressed as ionic activities and not as ionic concentrations.

Thus, the soil solution experiments of Adams and co-workers were the first to show that some ionic effects on root growth were the same in culture solutions as in soil solutions *in situ.* These experiments presented additional evidence that the chemical composition of the soil solution is indeed the chemical environment of root systems in soils. The experiments further demonstrated the advantage of expressing soil solution and culture solution compositions as ionic activities rather than as measured concentrations.

III. INTERRELATIONS OF SOIL PHASES

The chemical environment of roots in natural soil systems is so obviously complex that its precise definition has eluded plant physiologists and soil scientists alike. If this complex chemical system is to be defined quantitatively, the methods of thermodynamics must be used to evaluate the experimental data.

There have been two basic approaches to organizing chemical data to describe the root environment. One is the approach in which plants are grown in culture solutions of known salt concentrations. This approach has provided the basis for many present-day concepts of plant nutrition. However, it has failed to give precise definitions of nutrient-plant relationships because salts and ions have been interpreted in terms of concentrations instead of thermodynamic activities. The thermodynamics of dilute, simple electrolytic solutions have been known for decades; plant physiologists and soil scientists, however, have failed to apply them to soil-plant systems. It seems likely that a universal and accurate definition of a root's chemical environment awaits the proper application of thermodynamics to the root's ambient solution.

A second basic approach to defining a root's chemical environment is the empirical procedure of extracting soils with various reagents. This approach has been highly successful in elucidating such basic soil chemistry concepts as cation exchange reactions, soil P insolubility, soil salinity and alkali problems, soil acidity, and K fixation. Nevertheless it, too, fails to define the root's chemical environment adequately. This approach has two inherent weaknesses: (1) the extractants remove arbitrary and undetermined amounts of solid phase electrolytes and ions (or the extractants may cause precipitation of salts or ions from the soil solution); (2) the plant-soil interrelationship is defined in terms of the solid phase component of the soil, even though the solid phase is essentially inert except as it maintains thermodynamic equilibria with the solution phase. The emphasis on solid phase soil properties obviously results from the ease and convenience with which these data can be obtained and correlated with observed plant growth.

Because the chemical properties of the soil solution form the chemical environment of roots, the task of defining a root's chemical environment is a matter of fundamental soil solution chemistry. The other soil phases (solid and gas) affect a root's chemical environment only indirectly as they alter the chemical composition of the soil solution.

A. Solid Phase

Physically, the solid phase provides a skeletal structure of rigidity that anchors the root system and provides free space for the movement and retention of gases and water. Chemically, the solid phase is a heterogeneous mixture of compounds and minerals, some of which are in equilibrium with their component ions in the soil solution. The solid phase equilibrium ions may be the exchangeable cations or the sparingly soluble compounds of Ca, P, Mn, Zn, Al, Fe, and others. Thus, the role of the solid phase determines a root's chemical environment only as it affects the composition of the soil solution.

The solid phase components of a soil affect soil solution composition through two general thermodynamic reactions: (1) the solubility product principle of sparingly soluble salts and (2) cation exchange reactions.

The solubility product principle controls entirely the soil solution concentration of certain ions. For example, this is the reason that the concentration of soil solution phosphate is usually less than 1 ppm. Even the most soluble soil phosphate compound, $CaHPO_4 \cdot 2H_2O$, is only sparingly soluble. It dissolves according to the equation

$$CaHPO_4 \cdot 2H_2O(s) = Ca^{2+} + HPO_4^{2-} + 2H_2O. \qquad [3]$$

At equilibrium, its thermodynamic activity is defined by the equation

$$k_{sp} = (Ca^{2+})(HPO_4^{2-}), \qquad [4]$$

where parentheses denote activity. The solubilities of other soil phosphate compounds, e.g., octocalcium phosphate, hydroxyapatite, and aluminum phosphate, are similarly defined by thermodynamic constants of equilibria.

There have been numerous efforts in recent years to define the chemical behavior of soil P with the aid of thermodynamic solubility diagrams for Ca-, Al-, and Fe-phosphates. This approach was first suggested

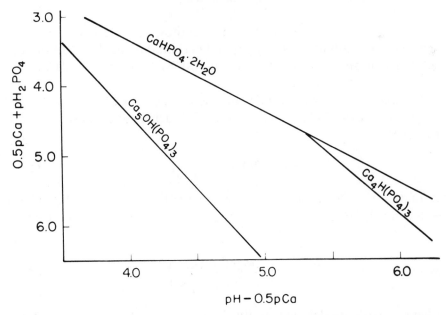

Fig. 15.1. Two-dimensional graph of the solubility of dicalcium phosphate, octocalcium phosphate, and hydroxyapatite (with molar activities expressed as negative logarithms)

by the comprehensive report of Aslyng (1954). He constructed two-dimensional solubility diagrams of the thermodynamic solubilities of $CaHPO_4 \cdot 2H_2O$, $Ca_4H(PO_4)_3 \cdot 3H_2O$, and $Ca_5OH(PO_4)_3 \cdot H_2O$, similar to those drawn in Figure 15.1. The assumptions were then made that (1) 0.01 M $CaCl_2$ typifies the ionic strength and Ca content of soil solutions, (2) extracting a soil with 0.01 M $CaCl_2$ does not alter soil solution P concentration, and (3) the phosphate compound controlling soil solution P concentration can be determined by analyzing the 0.01 M $CaCl_2$ extract for P. Because the solubility diagrams are based on actual ionic activities, Aslyng's procedure was obviously only qualitative with very limited applications because his assumptions of constant ionic composition and ionic strength were fallacious. From the examples of soil solution composition given in Table 15.9, it becomes apparent that Aslyng's assumptions were untenable and that the thermodynamic treatment of soil solution phosphate solubilities requires a more complete knowledge of ionic concentrations and activities in the soil solution than Aslyng proposed.

Analogous equilibria exist for other sparingly soluble compounds in soils. Of special interest are the solubilities of $CaCO_3$ in calcareous soils, $CaSO_4 \cdot 2H_2O$ in gypsiferous soils, $Al(OH)_3$ and MnO_2 in acid soils, and $Fe(OH)_2$ in anaerobic soils.

B. Gas Phase

The gas phase of soils influences the composition of all soil solutions through the solubilities of oxygen and carbon dioxide. In addition, the solubility of ammonia is of great interest because of the role of ammoniacal N as a plant nutrient.

The amount of dissolved CO_2 in a soil solution is determined by the partial pressure of $CO_2(g)$ in the soil air. It is defined thermodynamically by the equation

$$k_s = (H_2CO_3)/P_{CO_2(g)}. \qquad [5]$$

It follows from equation [5] that the partial pressure of soil air $CO_2(g)$ also controls HCO_3^- and CO_3^{2-} ionic concentrations through the equations

and
$$k_1 = (H^+)(HCO_3^-)/(H_2CO_3) \qquad [6]$$
$$k_2 = (H^+)(CO_3^{2-})/(HCO_3^-), \qquad [7]$$

where parentheses denote activity.

Table 15.9. Examples of soil solution composition

Soil no.	Ca	Mg	K	Na	NO$_3$ (mM)	Cl	SO$_4$	PO$_4$	HCO$_3$	pH
1	4.7	0.3	0.1	0.1	4.8	0.2	0.6	—	1.9	—
2	3.5	4.5	0.6	0.7	2.9	0.0	2.8	0.01	0.7	7.2
3	13.3	5.5	1.8	2.0	27.6	0.9	3.5	0.02	0.6	7.0
4	19.6	—	0.9	—	—	—	—	0.09	—	5.8
5	9.1	—	0.6	—	—	—	—	0.00	—	5.8
6	13.9	10.3	0.7	18.9	1.0	32.3	0.3	—	<0.1	7.2
7†	12.6	1.0	6.8	—	—	—	—	0.01	—	7.0
8†	1.7	3.5	0.8	—	—	—	—	0.01	—	6.8
9†	1.8	0.7	2.6	—	—	—	1.0	—	—	4.8

Sources: Data from (1) Schloesing, 1866, reviewed in Russell, 1950; (2 and 3) Burd and Martin, 1924; (4 and 5) Pierre, 1931; (6) Reitemier and Richards, 1944; (7 and 8) Bennett and Adams, 1970; (9) J. S. Richburg, 1969, Solubility and hydrolysis of aluminum in soil solutions and saturated-paste extracts, M.S. Thesis, Auburn University, Auburn, Ala.

† No. 7 also contained 6.1 mM NH$_4^+$. No. 8 contained 5.9 mM NH$_4^+$. No. 9 contained 5.0 mM NH$_4^+$, 2.4 mM Mn^{2+}, and 41.9μM of Al^{3+}.

The effect of $NH_3(g)$ on soil solution composition is defined by the thermodynamic equations

$$k_s = (NH_4OH)/P_{NH_3(g)}. \qquad [8]$$

and

$$k = (NH_4^+)(OH^-)/(NH_4OH). \qquad [9]$$

In addition, (NH_4^+) must satisfy the cation exchange equilibria of the soil system.

Effects of the solubility of $O_2(g)$ on soil solution composition are somewhat more complex and more difficult to measure than the effects of CO_2 and NH_3. Dissolved oxygen in water is defined by the equation

$$k_s = (O_2(aq))/P_{O_2(g)}. \qquad [10]$$

The activity of $O_2(aq)$ in turn affects soil solution composition through such important oxidation-reduction reactions as

$$2Fe^{2+} + 1/2O_2(aq) + 2H^+ = 2Fe^{3+} + H_2O \qquad [11]$$

and

$$Mn^{2+} + 1/2O_2(aq) + H_2O = MnO_2 + 2H^+. \qquad [12]$$

Thus, the soil solution activities of Fe^{3+}, Fe^{2+}, and Mn^{2+} are governed by the partial pressure of soil air $O_2(g)$ as well as by soil solution pH.

C. Solid Phase plus Gas Phase

The effect of the partial pressure of $CO_2(g)$ on the solubility of $CaCO_3$ has been of major interest to soil scientists for many years. This effect is described thermodynamically by the following equilibria equations. Solid $CaCO_3$ dissolves in water at 25C according to the equation

$$k = (Ca^{2+})(CO_3^{2-}) = 4.45 \times 10^{-9}. \qquad [13]$$

In pure water and normal atmospheric CO_2, it dissolves to the extent of 0.52 mM. The activity of Ca^{2+} in this solution is about 0.44 mM. Substitution of this value into equation [13] shows that

$$(CO_3^{2-}) = (4.45 \times 10^{-9})/(4.4 \times 10^{-4}) = 1.01 \times 10^{-5}M. \qquad [14]$$

Combining equations [5], [6], and [7] and solving for (H^+) gives

$$(H^+) = \{k_s k_1 k_2 P_{CO_2(g)}/(CO_3^{2-})\}^{1/2}. \qquad [15]$$

With P_{CO_2} at 0.00031 atm and (CO_3^{2-}) at $1.01 \times 10^{-5}M$, then (H^+) equals 4.6×10^{-9} (or a solution pH of 8.34). This is the basis for the soil scientists' assumption that the ideal pH of a calcareous soil is 8.3.

Soil solutions of calcareous soils are not pure $CaCO_3$ solutions, however, and therefore calcareous soils seldom have pH values of 8.3 (soil pH values this high are usually caused by the presence of Na_2CO_3). Since soil solutions contain more than 0.52 mM Ca^{2+}, as well as finite concentrations of other ions (see Table 15.9), the (CO_3^{2-}) of nonalkali, calcareous soils is less than the 10^{-5} M of equation [14], and soil solution pH is correspondingly lower.

IV. SOIL SOLUTION CHEMISTRY

The behavior of dilute electrolytic solutions obeys established laws of thermodynamics. Since soil solutions are dilute salt solutions, their properties can be treated with the methods of thermodynamics. This means, in practice, that measured ionic concentrations of soil solutions can be corrected to thermodynamic concentrations called *activities*. If soil solution composition is in terms of ionic activities, such diverse soil phenomena as ion uptake by plants, cation exchange reactions, and precipitation and dissolution of sparingly soluble compounds can be treated on a common basis.

The complexity of soil solution chemistry results from the numerous simultaneous equilibria that exist among solution phase ions, between solution and solid phase ions, and between dissolved and free gases. The Ca equilibria in a calcareous soil containing gypsum serve as an example of important simultaneous chemical equilibria that exist in a natural soil system.

$$CaCO_3(s) \rightleftharpoons CO_3^= \rightleftharpoons HCO_3^- \rightleftharpoons H_2CO_3 \rightleftharpoons CO_2(g)$$

$$CaHPO_4 \cdot 2H_2O(s)$$

$$SO_4^= \rightleftharpoons CaSO_4 \cdot 2H_2O(s) \rightleftharpoons Ca^{++} \qquad HPO_4^= \rightleftharpoons H_2PO_4^-$$

$$Ca_5OH(PO_4)_3(s)$$

$$Exch\text{-}Ca \rightleftharpoons Exch- \begin{cases} Mg \rightleftharpoons Mg^{++} \\ K \rightleftharpoons K^+ \\ Na \rightleftharpoons Na^+ \\ NH_4 \rightleftharpoons NH_4^+ \rightleftharpoons NH_4OH \rightleftharpoons NH_3(g) \end{cases}$$

This scheme shows that soil solution Ca^{2+} is simultaneously in equilibrium with solid phase components of $CaCO_3$, $CaSO_4 \cdot 2H_2O$, $CaHPO_4 \cdot 2H_2O$, $Ca_5OH(PO_4)_3$, and exchangeable Ca. Furthermore, these equilibria are affected by the solution activities of other electrolytes and ions, i.e., the exchangeable cations and soil solution ions of H^+, Mg^{2+}, and others.

   Interactions among the above equilibria are obviously complex. However, if activities of solution ions are known, all the indicated equilibria
are definable with the aid of known thermodynamic constants of equilibrium. Modern analytical procedures combined with computer techniques provide a ready means of determining solution activities. It is no
longer necessary to assume that ionic concentrations of soil solutions
equal ionic activities.

A. Principles

Adams (1971) has described in detail the procedure for correcting soil
solution ionic concentrations to ionic activities. It is outlined briefly
here.
   The principle of correcting measured ionic concentrations to ionic
activities depends upon the concepts of (1) ionic strengths of electrolytic
solutions, (2) single-ion activities, and (3) thermodynamic constants of
equilibrium.

1. Ionic Strength

The concept of ionic strength, μ, is basic to the thermodynamic treatment of electrolytic solutions. It is a measure of electrical field intensity
and is defined by the equation

$$\mu = 1/2\Sigma C_i Z_i^2 \qquad [16]$$

where C_i is the actual molar concentration of each ion in solution and
Z_i is its valence.

2. Single-Ion Activity

The single-ion activity is a highly pragmatic concept for soil solutions. It
is readily obtainable for dilute solutions of known ionic strength. Ionic
activity, a_i, is the product of ionic concentration, C_i, and its activity coefficient, f_i. Values of C_i are obtained by chemical analysis of the solution; f_i is obtained from the Debye-Hückel equation

$$- \log f_i = AZ_i^2(\mu)^{1/2}/[1 + Ba_i^1(\mu)^{1/2}], \qquad [17]$$

where A and B are temperature-dependent constants, μ is ionic strength
of the solution, and a_i' is an ion-size parameter. On the molar scale at
25C, $A = 0.509$ and $B = 0.329$ with a_i' in Angstrom units (see Table
15.10 for a_i' values). Thus, the activity coefficients of ions are readily calculable in soil solutions of known ionic strengths.

Table 15.10 Individual ion-size parameters, a_i', in Angstrom units for the Debye-Huckel equations

Ion	a_i'
H^+, Al^{3+}, Fe^{3+}	9
Mg^{2+}	8
Li^+, Ca^{2+}, Cu^{2+}, Zn^{2+}, Mn^{2+}, Fe^{2+}	6
Sr^{2+}, Ba^{2+}	5
Na^+, CO_3^{2-}, HCO_3^-, $H_2PO_4^-$	4.5
SO_4^{2-}, HPO_4^{2-}, PO_4^{3-}	4
OH^-, F^-	3.5
K^+, Cl^-, NO_3^-	3
Rb^+, Cs^+, NH_4^+	2.5

Source: Kielland, 1937.

3. Ion-Pair Concentrations

All strong electrolytes when dissolved in water do not completely dissociate into their component ions. The cations and anions of some are so strongly attracted to each other that a significant fraction behave as if they were un-ionized. Ions associated in this manner are called *ion-pairs* (Davies, 1962). The extent of ion-pairing in solution is expressed by the traditional method for expressing dissociation of weak electrolytes. For example, Ca^{2+} and SO_4^{2-} pair extensively as $CaSO_4^0$ in solution; the dissociation reaction is

$$CaSO_4^0 = Ca^{2+} + SO_4^{2-}, \qquad [18]$$

and the thermodynamic equilibrium expression is

$$k = (Ca^{2+})(SO_4^{2-})/(CaSO_4^0), \qquad [19]$$

where parentheses denote activity. In dilute solutions, e.g., soil solutions, the activity coefficient, f_i, of an ion-pair is assumed to be unity. Some thermodynamic equilibrium constants for ion-pairs expected to be present in soil solutions are listed in Table 15.11. In general, the soil solution anions that pair extensively with cations are SO_4^{2-}, HPO_4^{2-}, and HCO_3^-; pairing of NO_3^- with cations is only minor and can usually be ignored; Cl^- apparently does not pair with most cations.

As a result of ion-pairing, several soil solution ions will be present as different species. For example, soil solution Ca may be present as Ca^{2+}, $CaSO_4^0$, $CaHPO_4^0$, and $CaHCO_3^+$. Because the analytical procedure for determining solution Ca makes no distinction among these species, the "measured Ca" is the sum of all its solution species. However, the "actual" concentration of Ca^{2+} can be calculated.

Table 15.11.   Equilibrium constants for some common soil solution ion-pairs

Reaction	K
$AlSO_4^+ = Al^{3+} + SO_4^{2-}$	6.3×10^{-4}
$CaSO_4^\circ = Ca^{2+} + SO_4^{2-}$	5.25×10^{-3}
$CaHPO_4^\circ = Ca^{2+} + HPO_4^{2-}$	1.98×10^{-3}
$CaH_2PO_4^+ = Ca^{2+} + H_2PO_4^-$	8.3×10^{-2}
$CaCO_3^\circ = Ca^{2+} + CO_3^{2-}$	6.3×10^{-4}
$CaHCO_3^+ = Ca^{2+} + HCO_3^-$	5.5×10^{-2}
$CaNO_3^+ = Ca^{2+} + NO_3^-$	5.25×10^{-1}
$MgSO_4^\circ = Mg^{2+} + SO_4^{2-}$	5.88×10^{-3}
$MgHPO_4^\circ = Mg^{2+} + HPO_4^{2-}$	3.16×10^{-3}
$MgH_2PO_4^+ = Mg^{2+} + H_2PO_4^-$	0.1 (estimate)
$MgCO_3^\circ = Mg^{2+} + CO_3^{2-}$	4.0×10^{-4}
$MgHCO_3^+ = Mg^{2+} + HCO_3^-$	6.9×10^{-2}
$MnSO_4^\circ = Mn^{2+} + SO_4^{2-}$	5.25×10^{-3}
$NH_4SO_4^- = NH_4^+ + SO_4^{2-}$	7.93×10^{-2}
$NaSO_4^- = Na^+ + SO_4^{2-}$	2.4×10^{-1}
$NaCO_3^- = Na^+ + CO_3^{2-}$	5.35×10^{-2}
$NaHCO_3^\circ = Na^+ + HCO_3^-$	1.78
$KSO_4^- = K^+ + SO_4^{2-}$	1.1×10^{-1}

Sources: Davies, 1962; Sillen and Martell, 1964.

4. Weak Acids and Bases

Ions of weak acids and bases that exist in soil solutions as more than one species are the (1) phosphates, (2) carbonates, (3) ammonia, (4) aluminum, and (5) iron. Analytical procedures only measure the sum total of all ionic species in solution. However, the dissociation and hydrolytic reactions of these ions in water are defined by thermodynamic constants of equilibrium; thus the concentration of each species is calculable.

a. Phosphates. Unpaired soil solution phosphate ions are distributed between HPO_4^{2-} and $H_2PO_4^-$ at 25C according to the equation

$$6.23 \times 10^{-8} = (H^+)(HPO_4^{2-})/(H_2PO_4^-), \qquad [20]$$

where parentheses denote molar activity.

b. Carbonates. The H_2CO_3-HCO_3^--CO_3^{2-} distribution in solution is according to

$$3.38 \times 10^{-2} = (H_2CO_3)/P_{CO_2(g)} \qquad [21]$$

and

$$4.45 \times 10^{-7} = (H^+)(HCO_3^-)/(H_2CO_3) \qquad [22]$$

and

$$4.67 \times 10^{-11} = (H^+)(CO_3^{2-})/(HCO_3^-), \qquad [23]$$

where $P_{CO_2(g)}$ is in atmospheres.

c. Ammonia. Solution ammonia occurs as NH_4^+ and NH_4OH; their distribution is according to

$$58.9 = (NH_4OH)/P_{NH_3(g)} \qquad [24]$$

and

$$1.73 \times 10^{-5} = (NH_4^+)(OH^-)/(NH_4OH), \qquad [25]$$

where $P_{NH_3(g)}$ is in atmospheres.

5. Soil Solution Analysis

If a soil solution is to be treated thermodynamically, a rather complete knowledge of its composition must be available. The soil solution must be analyzed for (1) ions that affect ionic strength, (2) ions that pair significantly, and (3) ions of special interest. The dominant soil solution cations are usually Ca^{2+}, Mg^{2+}, and K^+. However, significant amounts of H^+, Na^+, NH_4^+, Mn^{2+}, Al^{3+}, or others may also be present. Soil solution anions consist primarily of Cl^-, NO_3^-, SO_4^{2-}, $H_2PO_4^-$, HPO_4^{2-}, and HCO_3^-. A few preliminary trials will show which ions should be determined analytically.

Soil solutions should probably always be analyzed for the cations Ca^{2+}, Mg^{2+}, K^+, and H^+. Similarly, they should always be analyzed for SO_4^{2-} because of its propensity for pairing with cations; phosphate and bicarbonate are sometimes needed. Concentrations of Cl^- and NO_3^- can usually be assumed to be those required for electrical neutrality of the soil solution.

B. Calculated Ionic Activities

Since the chemical analysis of a soil solution gives the total concentration of all species of a particular ion, the *actual* concentration of that ion is the *measured* concentration only if there is no ion-pairing and no hydrolysis or dissociation. Otherwise, the actual ionic concentration is less than that measured. For example, the actual ionic concentrations of a measured 10 mM $CaCl_2$ solution are 10 mM Ca^{2+} and 20 mM Cl^-. In contrast, the actual ionic concentrations of a measured 10 mM $CaSO_4$ solution are 7.17 mM Ca^{2+} and 7.17 mM SO_4 because 28.3% of the Ca^{2+} and SO_4^{2-} ions are paired as $CaSO_4^0$ (see Adams, 1971, for calculations).

Calculating ionic concentrations, ion-pair concentrations, and ionic activities in pure electrolytic solutions is rather tedious. It involves iterative estimations of (1) the ionic-strength equation, i.e., equation [16], (2) the Debye-Hückel (1923) equation for cation and anion activity coefficients, equation [17], and (3) the appropriate ion-pair equation, such as equation [19]. These equations are not solvable simultaneously but are conveniently solved by a method of successive approximations.

The principle of calculating ionic activities in pure electrolytic solutions is equally applicable to mixed electrolytic solutions, e.g., soil solutions. However, the number of equations to solve is more numerous for mixed solutions and the number of required iterations may be greater, but the basic procedure is the same.

The stepwise procedure for correcting measured ionic concentrations to actual concentrations and activities is illustrated below by using the analytical composition of actual displaced soil solutions.

1. Example No. 1: A Strongly Acid Soil Solution

Soil solutions of strongly acid soils characteristically contain low concentrations of Ca and Mg and toxic concentrations of Al and sometimes Mn. Analytical data of the displaced solution from such a soil are presented in Table 15.12, column 2.

The soil solution was analyzed for pH, Ca, Mg, Mn, K, NH_4, SO_4, and Al; phosphate concentration was too low to assess accurately; HCO_3^- was calculated from pH and $P_{CO_2(g)}$ values and found to be negligible. Thus, the cationic concentration was assumed to be balanced by the measured SO_4^{2-} concentration plus Cl^- and NO_3^- ions, neither of which was measured. This assumption is justified because soil solution anions consist primarily of Cl^-, NO_3^-, SO_4^{2-}, $H_2PO_4^-$, HPO_4^{2-}, and HCO_3^-. Only trace amounts of other anions are to be expected.

Step 1: Ionic strength. The measured ionic concentrations were assumed to be actual ionic concentrations. Ionic strength, μ, calculated according to equation [16], was 0.0234.

Table 15.12. Measured ionic concentrations and calculated ion-pair concentrations and ionic activities of a displaced strongly acid soil solution

Ion or ion-pair	1st approx.[†] Conc.	Activity	2nd approx. Conc.	Activity (mM)	Final approx. Conc.	Activity
pH	—	4.85	—	4.85	—	4.85
Ca^{2+}	1.76	1.01	1.65	0.96	1.68	0.98
Mg^{2+}	0.74	0.44	0.70	0.42	0.71	0.43
Mn^{2+}	2.44	1.41	2.29	1.34	2.33	1.35
K^+	2.61	2.23	2.60	2.23	2.60	2.23
NH_4^+	5.00	4.26	4.97	4.25	4.98	4.26
$Cl^- + NO_3^-$	15.49	13.25	15.49	13.30	15.49	13.29
SO_4^{2-}	1.00	0.55	0.66	0.37	0.75	0.42
Al^{3+}	0.0419	—	—[‡]	—	0.0255	0.0085
Ionic strength	0.0234 M	—	0.0221 M	—	0.0225 M	—
$CaSO_4^\circ$	0.11	0.11	—	—	0.08	0.08
$MgSO_4^\circ$	0.04	0.04	—	—	0.03	0.03
$MnSO_4^\circ$	0.15	0.15	—	—	0.11	0.11
KSO_4^-	0.01	0.01	—	—	0.01	0.01
$NH_4SO_4^-$	0.03	0.03	—	—	0.02	0.02
$AlSO_4^+$	—[‡]	—	—	—	0.0057	0.0057
$AlOH^{2+}$	—[‡]	—	—	—	0.0107	0.0063

Source: Original data from Richburg and Adams, 1970.
[†] The analytically measured concentrations are used as the first approximation of ionic concentrations.
[‡] Because Al concentration was too low to affect activity of other ions, its distribution among Al^{3+}, $AlOH^{2+}$, and $AlSO_4^+$ was not calculated until after all other ionic concentrations and activities were calculated.

Step 2: Activity coefficients. The activity coefficient, f_i, of each ion was calculated according to equation [17], using μ from step 1 and a_i' values from Table 15.10. Ionic activities, a_i, were then calculated by the equation

$$a_i = f_i C_i, \qquad [26]$$

where C_i is the molar concentration of each ion.

Step 3: Ion-pair concentrations. The concentrations of $CaSO_4^0$, $MgSO_4^0$, $MnSO_4^0$, KSO_4^-, and $NH_4SO_4^-$ ion-pairs were calculated by using the ionic activities from step 2 and ion-pair dissociation constants listed in

Table 15.11. For example, according to equation [19],

$$[CaSO_4^0] = k/(Ca^{2+})(SO_4^{2-})$$
$$= (5.25 \times 10^{-3})/(1.01 \times 10^{-3})(5.5 \times 10^{-4})$$
$$= 1.1 \times 10^{-4}M, \qquad [27]$$

where brackets denote concentration and parentheses denote activity.

Step 4: Revised ionic concentrations. A second estimate of ionic concentrations was made by subtracting the first estimate of ion-pair concentrations from the measured concentrations. For example, the second estimate of actual Ca^{2+} concentration was

$$[Ca^{2+}] = \text{measured } [Ca] - [CaSO_4^0]$$
$$= 1.76 - 0.11 = 1.65 \text{ mmole/liter.} \qquad [28]$$

The second estimate of actual SO_4^{2-} concentration was

$$[SO_4^{2-}] = \text{measured } [SO_4] - ([CaSO_4^0] + [MgSO_4^0]$$
$$+ [MnSO_4^0] + [KSO_4^-] + [NH_4SO_4^-])$$
$$= 1.00 - (0.11 + 0.04 + 0.15 + 0.01 + 0.03)$$
$$= 0.66mM. \qquad [29]$$

Step 5: Revised ionic strength. The ionic concentrations from step 4 were used to calculate a second estimate of μ, according to equation [16]; the new value was 0.0221.

Step 6: Revised ionic activities. New activity coefficients for all ions were calculated according to equation [17] and the revised μ value. Then new a_i values were calculated with equation [26], using the second estimates of C_i (step 4) and f_i.

Step 7: Revised ion-pair concentrations. Ion-pair concentrations were estimated a second time with new a_i values, according to equation [19] and its analogues, for each ion-pair.

Step 8: Iteration of equations. Steps 2, 3 and 4 were repeated using the latest values of μ, C_i, and a_i each time until C_i values remained unchanged with successive calculations. An alternative, more rapid solution of the problem can be made by averaging C_i values of step 1 and step 4, i.e., average the first two estimates of ionic concentrations to obtain the third estimate of C_i. Then repeat steps 2, 3, and 4 until C_i values are unchanged with successive iterations.

Step 9: Al distribution. Soil solution Al concentration was too low to affect μ or SO_4^{2-} concentration. Thus, solution Al was ignored until final values for μ and SO_4^{2-} activity were obtained. Then soil solution Al was distributed among its species. It was first assumed that Al^{3+} hydrolyzed in water according to the reaction

$$Al^{3+} + H_2O = AlOH^{2+} + H^+. \qquad [30]$$

It was further assumed that the hydrolytic k for this reaction was 9.5 $\times 10^{-6}$, or

$$9.5 \times 10^{-6} = (AlOH^{2+}) (H^+)/(Al^{3+}), \qquad [31]$$

where parentheses denote activity. Values of f for Al^{3+} and $AlOH^{2+}$ were calculated according to equation [15] with an a' value of 9 for Al^{3+} and 7 for $AlOH^{2+}$. Assuming that solution Al consists of the ionic species of Al^{3+}, $AlOH^{2+}$, and $AlSO_4^+$, then

$$\text{total } [Al] = \{(Al^{3+})/f_{Al^{3+}}\} + [AlSO_4^+]$$

$$+ \{(AlOH^{2+})/f_{AlOH^{2+}}\}. \qquad [32]$$

The concentration of $AlSO_4^+$ ion pair is defined by the equation

$$6.3 \times 10^{-4} = (Al^{3+}) (SO_4^{2-})/(AlSO_4^+). \qquad [33]$$

Substituting equations [31] and [33] into equation [32] yields

$$\text{total } [Al] = \{(Al^{3+})/f_{Al^{3+}}\} + \{(Al^{3+}) (SO_4^{2-})/6.3 \times 10^{-4}\}$$

$$+ \{9.5 \times 10^{-5}(Al^{3+})/f_{AlOH^{2+}}(H^+)\}, \qquad [34]$$

in which (Al^{3+}) is the only unknown. This scheme of Al distribution indicated that 61% of soil solution Al was present as Al^{3+}, 13% as $AlSO_4^+$, and 26% as $AlOH^{2+}$. Further, Al^{3+} activity was only 20% of the analytically measured Al concentration.

The above calculation of Al^{3+} activity is not presented as an accurate description of soil solution Al. Unfortunately it is only approximate because soil solution Al^{3+} does not hydrolyze simply to $AlOH^{2+}$; it also hydrolyzes to polymers, such as $Al_6(OH)_{15}^{3+}$. Soil solution Al^{3+} activity will remain imprecise until the hydrolytic reactions of Al^{3+} and H_2O have been more clearly defined. Nevertheless, this method provides a reasonable approximation of soil solution Al^{3+} activity.

2. Example No. 2: A Neutral, High-Salt Soil Solution

A highly fertilized soil will have a relatively high salt concentration in its soil solution. The composition of such a soil solution is given in Table 15.13, column 2. Unlike the soil solution in Table 15.12, this soil solution contained no measurable quantities of Mn and Al but did contain a measurable concentration of phosphate.

The procedure for determining ionic activities in this solution is the same as the previous solution except that measured $[PO_4]$ must be distributed among several ionic species just as measured $[Al]$ was distributed among its species. Consequently, each step will be only briefly explained.

Table 15.13. Calculated ionic and ion-pair concentrations and ionic activities of a displaced soil solution in which the second estimate of ionic concentrations resulted in a negative value for $[SO_4^{2-}]$

Ion or ion-pair	1st approx.[†] Conc.	1st approx.[†] Activity	2nd approx. Conc.	3rd approx.[‡] Conc. (mM)	3rd approx.[‡] Activity	Final approx. Conc.	Final approx. Activity
pH	—	6.26	—	—	6.26	—	6.26
Ca^{2+}	73.50	20.13	64.51	69.01	19.32	69.19	19.30
Mg^{2+}	10.15	3.39	8.80	9.48	3.21	9.51	3.22
NH_4^+	90.10	56.12	88.45	89.28	56.25	89.27	56.15
K^+	11.40	7.31	11.18	11.29	7.30	11.31	7.32
SO_4^{2-}	11.35	2.34	−0.83	5.26	1.12	5.50	1.17
Cl^-	246.10	157.84	246.10	246.10	159.36	246.10	159.26
P	0.050	—	0.021	0.036	—	0.031	—
$H_2PO_4^-$	—§	0.0257	—	0.029	0.0188	0.025	0.0159
HPO_4^{2-}	—§	0.0020	—	0.007	0.0015	0.006	0.0014
Ionic strength	0.364 M	—	—	0.339 M	—	0.342 M	—
$CaSO_4^\circ$	8.96	—	—	4.12	—	4.30	—
$MgSO_4^\circ$	1.35	—	—	0.61	—	0.64	—
$NH_4SO_4^-$	1.65	—	—	0.79	—	0.83	—
KSO_4^-	0.22	—	—	0.10	—	0.09	—
$CaHPO_4^\circ$	0.020	—	—	0.015	—	0.013	—
$MgHPO_4^\circ$	0.002	—	—	0.001	—	0.001	—
$CaH_2PO_4^+$	0.006	—	—	0.004	—	0.004	—
$MgH_2PO_4^+$	0.001	—	—	0.001	—	0.001	—

Source: M. A. A. Hashimi, 1969, Calcium phosphate compounds precipitated by addition of diammonium phosphate to soil, Ph.D. thesis, Auburn University, Auburn, Ala.
[†] Measured concentrations are the first approximated ionic concentrations.
[‡] The third approximated ionic concentrations are the average of the first and second approximations.
§ Initial estimates of $[H_2PO_4^-]$ and $[HPO_4^{2-}]$ were unnecessary because they were too low to affect significantly the initial estimate of ionic strength.

Step 1: Ionic strength. In solutions containing enough phosphate to affect ionic strength, the measured $[PO_4]$ may be initially divided arbitrarily between $H_2PO_4^-$ and HPO_4^{2-}. A convenient division is 50% each. In this particular example, however, phosphate concentration was ignored because it was too low to affect the initial estimate of ionic strength. (This is the usual case with soil solutions.) Ionic strength was then calculated as for Example 1.

Step 2: Activity coefficient. Values of f were calculated as in Example 1; a_i values were also calculated as before except for phosphate. Activities of $H_2PO_4^-$ and HPO_4^{2-} were calculated by combining the equation

$$6.23 \times 10^{-8} = (H^+)(HPO_4^{2-})/(H_2PO_4^-) \qquad [35]$$

with

$$\text{total } [PO_4] = [H_2PO_4^-] + [HPO_4^{2-}]$$
$$= \{(H_2PO_4^-)/f_{H_2PO_4^-}\} + \{(HPO_4^{2-})/f_{HPO_4^{2-}}\} \quad [36]$$

to give

$$(HPO_4^{2-}) = \frac{[H_2PO_4^-] + [HPO_4^{2-}]}{\{(H^+)/6.3 \times 10^{-8}f_{H_2PO_4^-}\} + 1/f_{HPO_4^{2-}}}, \qquad [37]$$

where (HPO_4^{2-}) is the only unknown; $(H_2PO_4^-)$ was then calculated from equation [35].

Step 3: Ion-pair concentration. Calculations were made as before but including phosphate ion-pairs this time.

Step 4: Revised ionic concentrations. A second estimate of ionic concentrations was made as before. For ionic phosphate, this was

$$[H_2PO_4^-] + [HPO_4^{2-}] = \text{measured } [PO_4] - \{[CaHPO_4^0]$$
$$+ [MgHPO_4^0] + [CaH_2PO_4^+]$$
$$+ [MgH_2PO_4^+]\}. \qquad [38]$$

Step 5: Revised ionic concentrations. Because the first estimate of ionic concentrations was too high and the second estimate was too low, a third estimate of ionic concentrations was made by averaging values for the first two estimates (including a negative $[SO_4^{2-}]$ value in the second estimate).

Step 6: Revised μ, f_i, a_i. The procedure for calculating new values of μ, f_i, a_i, and ion-pair concentrations was repeated until C_i remained unchanged with successive iterations.

Finally, with the activity of each soil solution ion known, the soil solution composition is in a form that can be related to such diverse phenomena as soil-plant interrelations, ion exchange reactions, and precipitation and dissolution reactions.

3. Corroborative Experimental Data

The above procedure for calculating ionic activities of soil solutions requires validation by experimentation. Adams and co-workers used

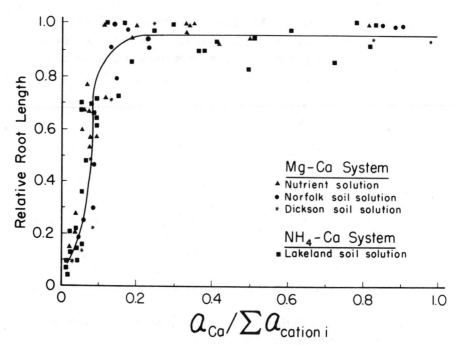

Fig. 15.2. Effect of the molar activity ratio of Ca^{2+}/Σ cations in culture solutions and in soil solutions *in situ* on the relative length of 3-day-old cotton seedling radicles. (Howard and Adams, 1965; Adams, 1966)

two approaches in their confirmatory experiments: (1) ionic activities for incipient inhibition of root growth in soils were defined for Ca, NH_3, and Al; (2) $CaSO_4$ activity in soil solutions saturated with gypsum was verified as being the same as in pure solutions.

The first reports on the relationships between soil solution ionic activities and root growth were based on the assumption that measured concentrations were actual ionic concentrations; ionic activities were calculated from measured concentrations according to the Debye-Hückel equation (Howard and Adams, 1965; Adams, 1966; Adams and Lund, 1966). Although this was an incomplete thermodynamic treatment of the soil solution, it allowed root growth to be expressed as a single function of the solution ion, whether it be in culture solution or in soil solution. This is illustrated by the data plotted in Figure 15.2, in which a single curve fits the root growth data for numerous culture solutions and displaced soil solutions *in situ* with a wide range in Ca concentrations as well as in concentrations of other ions.

Subsequently, Bennett and Adams (1970b) made a more complete thermodynamic treatment of similar soil solution data. They corrected measured ionic concentrations to actual concentrations by calculating ion-pair concentrations, as described previously. This resulted in lower Ca^{2+} activities and a slightly lower "critical" ratio of $a_{Ca}/\Sigma a_{cation\ i}$ for

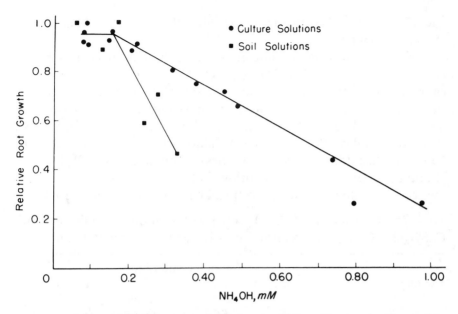

Fig. 15.3. Effect of NH$_4$OH on the relative length of 3-day-old cotton seedling radicles in culture solutions and in soil solutions *in situ*. (Bennett and Adams, 1970b)

root growth than previously reported where ion-pairing had been ignored.

The more complete thermodynamic treatment was also applied to the composition of diverse culture solutions and soil solutions containing NH$_3$ in which cotton roots were grown (Bennett and Adams, 1970a). These data, plotted in Figure 15.3, showed the NH$_3$ activity for incipient toxicity to be 0.2 mM NH$_4$OH in both kinds of solutions. Slopes of the two curves in the toxic NH$_3$ range are different because soil solutions containing toxic concentrations of NH$_4$OH were also deficient in Ca for root growth.

Showing root growth to be a function of ionic activities in the ambient soil solution is indirect evidence of the validity of calculated ionic activities and ion-pair concentrations in displaced soil solutions. To obtain more direct experimental evidence of their validity, Bennett and Adams (unpublished data) thoroughly mixed excess CaSO$_4$·2H$_2$O with a number of soil samples and equilibrated the samples for several days. Then the displaced soil solutions were analyzed, and CaSO$_4$ activity was calculated for each. The CaSO$_4$ activity of a saturated solution of CaSO$_4$·2H$_2$O water at 25C is about 2.4×10^{-5} M (the value varies with method and researcher). Measuring the solubility of CaSO$_4$·2H$_2$O in water and correcting measured Ca and SO$_4$ concentrations to ionic activities, as described in this chapter, Bennett and Adams found CaSO$_4$ activity in pure water to be 2.56×10^{-5} M. Applying the corrective calculations to displaced soil solutions, CaSO$_4$ activity ranged between 2.56×10^{-5}

and 2.75×10^{-5} M, with an average value of 2.64×10^{-5} M (see Table 15.14 for examples).

In view of the possible accumulative analytical errors from the separate determinations of Ca, Mg, Mn, K, NH_3, PO_4, and SO_4 (including the capriciousness of the sulfate determination), the calculated solution activity of $CaSO_4$ in the various displaced soil solutions agreed remarkably well with that in pure water. The soil solution data on $CaSO_4$ activity serve two significant roles: (1) they validate the procedure for correcting ionic concentrations to ionic activities in soil solutions; (2) they add additional support to the contention that the displaced soil solution is the true soil solution and the chemical environment of plant roots.

Table 15.14. Calculated $CaSO_4$ activity of solution at equilibrium with $CaSO_4 \cdot 2H_2O$ in pure water and in a Lucedale fine sandy loam

Medium	Solution composition				
	pH	Ca	SO_4	Ionic strength	$CaSO_4$ activity
		(mM)			
Water	6.75	15.3	15.3	0.061	2.56×10^{-5}
Soil	4.39	23.8	15.4	0.104	2.75×10^{-5}
Soil	4.67	15.6	19.4	0.083	2.62×10^{-5}
Soil	4.81	15.6	18.7	0.079	2.62×10^{-5}
Soil	5.32	15.3	18.2	0.079	2.56×10^{-5}

References

Adams, Fred. 1966. Calcium deficiency as a causal agent of ammonium phosphate injury to cotton seedlings. *Soil Sci. Soc. Amer. Proc.* 30: 485–88.

——. 1971. Ionic concentrations and activities in soil solutions. *Soil Sci. Soc. Amer. Proc.* 35: 420–26.

Adams, Fred, and Z. F. Lund. 1966. Effect of chemical activity of soil solution aluminum on cotton root penetration of acid subsoils. *Soil Sci.* 101: 193–98.

Adams, Fred, R. W. Pearson, and B. D. Doss. 1967. Relative effects of acid subsoils on cotton yields in field experiments and on cotton roots in growth-chamber experiments. *Agron. J.* 59: 453–56.

Aslying, H. E. 1954. The lime and phosphate potentials of soils: The solubility and availability of phosphates. *Roy. Vet. Agr. Coll. Yearbook* (Copenhagen), pp. 1–50.

Beckett, P. H. T., and J. B. Craig. 1964. The determination of potassium potentials. *Trans. 8th Int. Congr. Soil Sci.* 3: 249–55.

Bennett, A. C., and Fred Adams. 1970a. Concentration of NH_3(aq) required for incipient NH_3 toxicity to seedlings. *Soil Sci. Soc. Amer. Proc.* 34: 259–63.

——. 1970b. Calcium deficiency and ammonia toxicity as separate causal factors of $(NH_4)_2HPO_4$-injury to seedlings. *Soil Sci. Soc. Amer. Proc.* 34: 255–59.

Burd, J. S., and J. C. Martin. 1923. Water displacement of soils and the soil solution. *J. Agr. Sci.* 13: 265–95.

——. 1924. Secular and seasonal changes in the soil solution. *Soil Sci.* 18: 151–67.

——. 1931. Secular and seasonal changes in soils. *Hilgardia* 5: 455–509.

Burgess, P. S. 1922. The soil solution, extracted by Lipman's direct pressure method, compared with 1:5 water extracts. *Soil Sci.* 14: 191–216.

Conrad, J. P., E. L. Proebsting, and L. R. McKinnon. 1930. Equipment and procedure for obtaining the displaced soil solution. *Soil Sci.* 29: 323–29.

Davies, C. W. 1962. *Ion Association.* Washington, D.C.: Butterworth. 190 pp.

Debye, P., and E. Hückel, 1923. Zur theories der electrolyte. *Phy. Z.* 24: 185–206, 305–25.

Evans, C. E., and E. J. Kamprath. 1970. Lime response as related to percent Al saturation, solution Al, and organic matter content. *Soil Sci. Soc. Amer. Proc.* 34: 893–96.

Fudge, J. F. 1928. Influence of various nitrogenous fertilizers on availability of phosphate. *J. Amer. Soc. Agron.* 20: 280–93.

Howard, D. D., and Fred Adams. 1965. Calcium requirement for penetration of subsoils by primary cotton roots. *Soil Sci. Soc. Amer. Proc.* 29: 558–62.

Kemper, W. D. 1959. Estimation of osmotic stress in soil water from the electrical resistance of finely porous ceramic units. *Soil Sci.* 87: 345–52.

Khasawneh, F. E., and Fred Adams. 1967. Effect of dilution on calcium and potassium contents of soil solutions. *Soil Sci. Soc. Amer. Proc.* 31: 172–76.

Kielland, J. 1937. Individual activity coefficients of ions in aqueous solutions. *J. Amer. Chem. Soc.* 59: 1675–78.

Lauritzen, C. W. 1934. Displacement of soil solubles through plant roots by means of air pressure as a method of studying soil fertility problems. *J. Amer. Soc. Agron.* 26: 807–19.

Magistad, O. C. 1925. The aluminum content of the soil solution and its relation to soil reaction and plant growth. *Soil Sci.* 20: 181–226.

McGeorge, W. T. 1924. Iron, aluminum, and manganese in the soil solution of Hawaiian soils. *Soil Sci.* 18: 1–11.

Moss, P. 1969. A comparison of potassium activity ratios derived from equilibration procedures and from measurements on displaced soil solution. *J. Soil Sci.* 20: 297–306.

Parker, F. W. 1921. Methods of studying the concentration and composition of the soil solution. *Soil Sci.* 12: 209–32.

Pierre, W. H. 1925. The H-ion concentration of soils as affected by carbonic acid and the soil-water ratio, and the nature of soil acidity as revealed by these studies. *Soil Sci.* 20: 285–305.

———. 1931. Hydrogen-ion concentration, aluminum concentration in the soil solution, and percentage base saturation as factors affecting plant growth on acid soils. *Soil Sci.* 31: 183–207.

Pierre, W. H., and F. W. Parker. 1927. Soil phosphorus studies, II. The concentration of organic and inorganic phosphorus in the soil solution and the availability of organic phosphorus to plants. *Soil Sci.* 24: 119–28.

Pierre, W. H., G. G. Pohlman, and T. C. McIlvaine. 1932. Soluble aluminum studies: I. The concentration of aluminum in the displaced solution of naturally acid soils. *Soil Sci.* 34: 145–60.

Proebsting, E. L. 1929. Changes in the nitrate and sulfate content of the soil solution under orchard conditions. *Hilgardia* 4: 57–76.

———. 1930. Concentration of certain constituents of the soil solution under orchard conditions. *Hilgardia* 5: 35–59.

———. 1933. Effect of covercrops on the soil solution at different depths under orchard conditions. *Hilgardia* 7: 553–84.

Reeve, R. C., and E. J. Doering. 1965. Sampling the soil solution for salinity appraisal. *Soil Sci.* 99: 339–44.

Reitemeier, R. F. 1945. Effect of moisture content on the dissolved and exchangeable ions of soils in arid regions. *Soil Sci.* 61: 195–214.

Reitemeier, R. F., and L. A. Richards. 1944. Reliability of the pressure-membrane method for extraction of soil solution. *Soil Sci.* 57: 119–35.

Richards, L. A. 1941. A pressure-membrane extraction apparatus for soil solutions. *Soil Sci.* 51: 377–86.

Richburg, J. S., and Fred Adams. 1970. Solubility and hydrolysis of aluminum in soil solutions and saturated-paste extracts. *Soil Sci. Soc. Amer. Proc.* 34: 728–34.

Russell, E. W. 1950. *Soil Conditions and Plant Growth.* 8th ed. New York: Longmans, Green and Co. 635 pp.

Schofield, R. K. 1947. A ratio law governing the equilibrium of cations in solution. *Proc. 11th Int. Congr. Pure Appl. Chem.* (London) 3: 257–61.

——. 1952. Soil Colloids. *Chem. and Industry* (London), no. 4, pp. 476–78.

Shimshi, D. 1966. Use of ceramic points for the sampling of soil solution. *Soil Sci.* 101: 98–103.

Sillen, L. G., and A. E. Martell. 1964. *Stability Constants of Metal-ion Complexes.* London: The Chemical Society. 754 pp.

Soil Science Society of America, Committee Report. 1965. Glossary of soil science terms. *Soil Sci. Soc. Amer. Proc.* 29: 330–51.

Tidmore, J. W. 1930. The phosphorus content of the soil solution and its relation to plant growth. *J. Amer. Soc. Agron.* 22: 481–88.

Wagner, G. H. 1962. Use of porous ceramic cups to sample soil water within the profile. *Soil Sci.* 94: 379–86.

White, L. M., and W. H. Ross. 1936. Influence of fertilizers on the concentration of the soil solution. *Soil Sci. Soc. Amer. Proc.* 1: 181–86.

16. Chemical Reactions Controlling Soil Solution Electrolyte Concentration

Grant W. Thomas

ONE of the most important long-term goals of soil chemistry has been to improve the nutrient status of soils. To do this efficiently requires that we be able to determine what the addition of nutrients will do to the composition of the soil solution from which the plants absorb nutrients. Many important nutrients are held rather tightly by the solid portion of the soil. Among these are calcium, potassium, and phosphorus. Two metal cations that affect many reactions in soils by their tendency to form insoluble compounds are iron and aluminum.

A number of different approaches will be presented here to describe the quantities of these ions in the soil solution. Equations for description of cation exchange and the solubility product are given, and examples are worked out in some detail. It is my hope that this information will enable the reader to make a number of useful calculations without having to go back to original source material. If this chapter serves as a recipe book for cation exchange and solubility calculations, it will have achieved its purpose.

I. CATION EXCHANGE REACTIONS

A. Types of Cation Exchange Reactions in Soils

1. Stoichiometric Replacement

The earliest cation exchange experiments of Thomas Way (1850) were carried out on neutral soils using ammonium sulfate. The resulting exchange was given by

$$2NH_4^+ + SO_4^{2-} + \text{Ca-Soil} \rightarrow CaSO_4 + (NH_4)_2\text{-Soil}. \quad [1]$$

Way correctly surmised that the calcium sulfate resulted from replacement of "exchangeable" calcium by ammonium. Sodium and potassium
gave similar reactions, whereas added calcium came through the soil
essentially unchanged. Way's finding that there was a relatively constant
capacity to hold ammonium and subsequent experiments in the early
1900's gave rise to the idea that cation exchange is basically a stoichiometric reaction. Therefore, most of the experimental work done on
selectivity coefficients has been done on clays having a definite cation
exchange capacity, using salts that did not cause precipitation reactions.

The assumptions that the cation exchange capacity is a constant and
that equivalent quantities of cations are adsorbed and replaced are
essentially correct in neutral soils. Reactions such as replacement of
calcium by ammonium, potassium, sodium, or magnesium can be
described adequately by the chemical reaction discovered by Way.
However, in both acid and calcareous soils other reactions must be
considered as well.

2. Neutralization and Precipitation Reactions

In acid soils the addition of calcium carbonate (limestone) to neutralize
acidity that is largely from aluminum results in the formation of an
insoluble salt as follows:

$$CaCO_3 + 2H_2O \rightarrow Ca(OH)_2 + H_2CO_3, \qquad [2]$$

$$3/2\ Ca(OH)_2 + Al\text{-}Soil \rightarrow Al(OH)_3 + Ca_{3/2}\text{-}Soil. \qquad [3]$$

In this case, the reaction will proceed to completion, limited only by
the formation of hydroxy-aluminum polymers of the composition
$[Al(OH)_{3-x}]_y^{xy+}$, which are not truly exchangeable. In any case, the
reaction is not in the same category as

$$3/2\ Ca^{2+} + 3Cl^- + Al\text{-}Soil \rightarrow Ca\ 3/2\text{-}Soil + AlCl_3. \qquad [4]$$

In reaction [3], replacement of aluminum is virtually complete because
of the low solubility of aluminum hydroxide. In reaction [4], aluminum
strongly competes for the exchange sites, making calcium chloride only a
mediocre replacer of aluminum, as will be discussed later.

In calcareous soils, the cation exchange equilibrium is dominated by
the solubility of calcium carbonate. The largest factor affecting the
amount of calcium in solution is the partial pressure of carbon dioxide in
the soil atmosphere. Bradfield (1941) presented the data that are given in
Table 16.1.

At a partial pressure of 0.00031 atm (the value found in the earth's
atmosphere), the pH of a soil dominated by calcium carbonate should be
8.3, and the soil solution concentration of calcium should be 0.5 mM if

Table 16.1. Effect of partial pressure of CO_2 on pH and on calcium concentrations in solution, in a suspension of $CaCO_3$ and water

P_{CO_2} (atm)	pH	Ca^{2+} (mM)
0.00031	8.30	0.53
0.00332	7.62	1.17
0.0160	7.18	2.01
0.0432	6.90	2.87
0.1116	6.64	4.03
0.9684	6.05	8.91

Source: Bradfield, 1941.

no other salts are present. However, at a partial pressure of 0.00332 atm, which is common in soils, the pH will be 7.62 and the calcium concentration 1.17 mM. Obviously, the amount of calcium in solution depends strongly on biological reactions that control carbon dioxide production. Changes in temperature and moisture can cause drastic fluctuations in pH values, much to the consternation of soil-testing laboratories.

In summary, the cation exchange reactions in soils where aluminum is reacting with limestone or where calcium carbonate is present in excess are controlled by the solubilities of the aluminum hydroxides and of calcium carbonate, respectively. For this reason, the reactions will be discussed under the section on solubility product principles.

B. The Origin of Cation Exchange Equilibrium Equations

Excellent reviews of this subject are given by Kelley (1948) and by Marshall (1964), so that only the barest outline will be given here.

1. The Freundlich Equation

By its mathematical nature, the Freundlich equation should be adaptable to most adsorption reactions over a narrow range. The equation is

$$q = kc^{1/n},$$

where q = meq of cation adsorbed per gram of adsorbent,
  c = concentration of cation in moles per liter,
  k = the "distribution" constant,
  n = a correction factor.

When K is being adsorbed from an ambient solution the equation is

$$(K \text{ ads})/[K^+] \, 1/n = k.$$

If $n = 1$, the equation reduces to the distribution coefficient

$$(K \text{ ads})/[K^+] = k,$$

which is valid for cation or anion absorption over narrow ranges. The two main drawbacks of the Freundlich k are its extreme variability with concentration and the lack of an adsorption maximum. Nevertheless as a proportionality coefficient it has some value.

2. The Langmuir Equation

Unlike the Freundlich equation, the Langmuir equation does contain a capacity term, but it does not consider that any ion is replaced by the adsorbed ion.

The equation is

$$q/q^0 = kc/(1 + kc),$$

where q and c have the same meaning as in the Freundlich equation, k is an affinity coefficient, and q^0 is total exchange capacity. Multiplying both sides by q^0 and then by $(1 + kc)$, we get

$$q + kcq = kcq^0$$
$$q = kcq^0 - kcq$$
$$q = kc(q^0 - q)$$
$$k = q/c(q^0 - q). \qquad [5]$$

In the reaction

$$NH_3 + H\text{-Soil} \rightarrow NH_4\text{-Soil},$$
$$q = (NH_4 \text{ ads})$$
$$q^0 = (NH_4 \text{ ads}) + (H \text{ ads})$$
$$c = (NH_3).$$

Using these quantities in equation [5], we obtain

$$k = (NH_4 \text{ ads})/(NH_3)(H \text{ ads}).$$

This equation describes the reaction of ammonia with a hydrogen-saturated soil very well, but it would not describe the reaction of a normal stoichiometric cation exchange reaction such as sodium for hydrogen. Since the Langmuir equation describes adsorption where replacement does not occur, it also has been used to determine the adsorption capacity of soils for phosphorus (Woodruff and Kamprath, 1965). When the equation is used in this way, it generally is arranged for plotting as

$$c/q = (1/kq^0) + (c/q^0), \qquad [6]$$

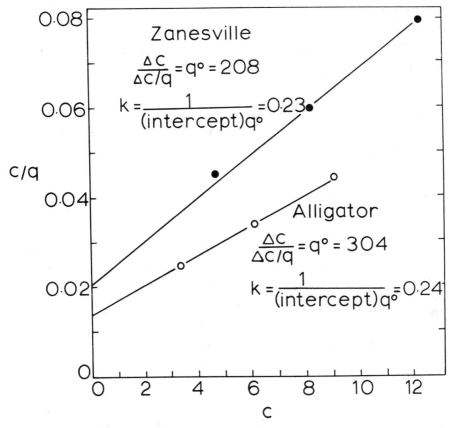

$$\frac{\Delta c}{\Delta c/q} = q^0 = 208$$

$$k = \frac{1}{(\text{intercept})q^0} = 0.23$$

$$\frac{\Delta c}{\Delta c/q} = q^0 = 304$$

$$k = \frac{1}{(\text{intercept})q^0} = 0.24$$

Fig. 16.1. Langmuir plot of phosphorus adsorption data for two soils, showing how q^0 and k are calculated. (J. C. Ballaux, unpublished data, University of Kentucky, 1971)

and c/q is plotted against values of c. In a plot of this type, the reciprocal of the slope gives the value of q^0, the adsorption capacity, and the intercept is equal to the reciprocal of the product of the adsorption capacity, q^0, and the affinity constant, k.

Some data of J. C. Ballaux (unpublished data, University of Kentucky, 1971) for phosphorus adsorption on Alligator and Zanesville soils are shown in Figure 16.1. The calculations for q^0 and k are also presented there. Very similar calculations for P adsorption can be made for any soil where the adsorption data are available.

3. Mass Action Equations

Since the reaction discovered by Way (1850) is superficially similar to the double decomposition of a salt, Kerr (1928) used the familiar mass action equation to describe the equilibrium attained. For sodium-potassium exchange,

$$Na^+ + K\text{-Soil} \rightarrow Na\text{-Soil} + K^+,$$

and

$$(Na\ ads)\,[K^+]/(K\ ads)\,[Na^+] = k. \tag{7}$$

The mass action equation for heterovalent exchange is given by Kerr (1928) as

$$2Na + Ca\text{-Soil} \rightarrow Na_2\text{-Soil} + Ca^{2+},$$

and

$$k = (Na\ ads)^2\,[Ca^{2+}]/(Ca\ ads)\,[Na^+]^2. \tag{8}$$

The value of k from this equation was found to fluctuate rather wildly, and in the early days constancy was highly valued. Vanselow (1932) suggested that the true activity of adsorbed cations on a clay was obtained from the mole fraction occupied by a given cation, i.e.,

$$\text{activity of (Na ads)} = (Na\ ads)/(Na\ ads + 1/2\ Ca\ ads).$$

Using this value in Kerr's equation, one gets

$$\{(Na\ ads)^2/(Na\ ads + 1/2\ Ca\ ads)\}\,[Ca^{2+}]/(Ca\ ads)\,[Na^+]^2 = k,$$

or

$$(Na\ ads)^2\,[Ca^{2+}]/(Ca\ ads)\,[Na^+]^2 = k\,(Na\ ads + 1/2\ Ca\ ads). \tag{9}$$

Thus, the Vanselow equation for heterovalent exchange differs from the Kerr equation by the factor $(Na\ ads + 1/2\ Ca\ ads)$.

4. Kinetic Equations

A number of investigators have used the kinetic approach to cation exchange. In homovalent exchange, this gives the same equation as the mass action approach (Jenny, 1936). In heterovalent exchange, equations have been developed that have the form used by Gapon (1933) or Davis (1945), i.e.,

$$Na + Ca_{1/2}\text{-Clay} \rightarrow Na\text{-Clay} + 1/2\ Ca,$$

and

$$(Na\ ads)\,[Ca^{2+}]^{1/2}/(Ca\ ads)\,[Na^+] = k_g, \tag{10}$$

where the cations adsorbed are expressed in millequivalents per gram and the cations in solution are expressed in millimoles per milliliter. Other workers (Krishnamoorthy and Overstreet, 1949) have used a kinetic theory to come up with an equation similar to that of Vanselow,

$$2Na + Ca\text{-}Clay \rightarrow Na_2\text{-}Clay + Ca^{2+},$$

and

$$(Na\ ads)^2\,[Ca^{2+}]/(Ca\ ads)\,[Na^+]^2 = k\,(Na\ ads + 0.75\ Ca\ ads). \quad [11]$$

5. Thermodynamic Equation

A very similar equation to [11] was developed from thermodynamic theory by Gaines and Thomas (1955) and has the form

$$(Na\ ads)^2\,[Ca^{2+}]/(Ca\ ads)\,[Na^+]^2 = k2(Na\ ads + Ca\ ads), \quad [12]$$

where (Na ads + Ca ads) equals the cation exchange capacity.

6. The Double-Layer Theory

Eriksson (1952) developed an equation from the Gouy (double layer) theory which has the potential of predicting the cation exchange equilibrium between sodium and calcium, but which does not work very well for potassium and calcium (Lagerwerff and Bolt, 1959). The equation is

$$\frac{S\Gamma^+}{S\Gamma^{++}} = \frac{\dfrac{1}{\beta}\ \text{arg sinh}\ \dfrac{(\beta)^{1/2}\Gamma}{R + 4Vc(a^{++})^{1/2}}}{\Gamma - \dfrac{R}{(\beta)^{1/2}}\ \text{arg sinh}\ \dfrac{(\beta)^{1/2}\Gamma}{R + 4Vc(a^{++})^{1/2}}} \cdot R. \quad [13]$$

For sodium-calcium exchange,

$$\frac{Na\ ads}{Ca\ ads} = \frac{\dfrac{1}{\beta}\ \text{arg sinh}\ \dfrac{(\beta)^{1/2}\Gamma}{\dfrac{[Na^+]}{[Ca^{2+}]^{1/2}} + 4Vc[Ca^{2+}]^{1/2}}}{\Gamma - \dfrac{\dfrac{[Na^+]}{[Ca^{2+}]^{1/2}}}{(\beta)^{1/2}}\ \text{arg sinh}\ \dfrac{(\beta)^{1/2}\Gamma}{\dfrac{[Na^+]}{[Ca^{2+}]^{1/2}} + 4Vc[Ca^{2+}]^{1/2}}} \cdot \frac{[Na^+]}{[Ca^{2+}]^{1/2}}.$$

The symbols have the meaning

S = surface area of exchanger in cm^2

Γ = charge density in meq/cm^2

Γ^+ = meq monovalent cation/cm^2

Table 16.2. Comparison of several equilibrium equations for mono-divalent exchange

---

For the reaction

$$2Na^+ + Ca\text{-}Clay \rightarrow Na_2\text{-}Clay + Ca^{2+}$$

1. Kerr (mass action)
   $$(Na\ ads)^2\ [Ca^{2+}]/(Ca\ ads)\ [Na^+]^2 = k \qquad\qquad [8]$$
2. Vanselow
   $$(Na\ ads)^2\ [Ca^{2+}]/(Ca\ ads)\ [Na^+]^2 = k\ (Na\ ads + 0.5\ Ca\ ads) \quad [9]$$
3. Krishnamoorthy and Overstreet
   $$(Na\ ads)^2\ [Ca^{2+}]/(Ca\ ads)\ [Na^+]^2 = k\ (Na\ ads + 0.75\ Ca\ ads)\ [11]$$
4. Gaines and Thomas
   $$(Na\ ads)^2\ [Ca^{2+}]/(Ca\ ads)\ [Na^+]^2 = k\ (Na\ ads + Ca\ ads) \qquad [12]$$
5. Gapon (square-root form)
   $$(Na\ ads)\ [Ca^{2+}]^{1/2}/(Ca\ ads)\ [Na^+] = k_g \qquad\qquad [10]$$

Note: Adsorbed ions are in meq; solution ions in mM.

$$\Gamma^{++} = meq\ divalent\ cation/cm^2$$
$$\beta = 8/100\ RT;\ at\ 30C = 10^{15}cm/mM$$
$$Vc = 1$$
$$arg\ sinh = sinh^{-1}$$
$$R = a^+/(a^{++})^{1/2}.$$

When this equation is reduced, it has the same form as equation [10]. The U.S. Salinity Laboratory (1954) also has empirically developed an equation identical to [10].

On the basis of use, equation [10] has gained general acceptance for mono-divalent exchange such as sodium-calcium or potassium-calcium. For aluminum-potassium or aluminum-calcium exchange, equation [12] has been used by Nye *et al.* (1961). For homovalent exchange (sodium-potassium or calcium-magnesium), equation [7] is always used. Table 16.2 gives the equations developed for heterovalent exchange and compares them with the Kerr equation.

C. The Use of Cation Exchange Equations

1. Predicting Soil Solution Concentrations

In this section, the Gapon equation [10] will be used for making most of the calculations. This will be done for two reasons: first, it is the most widely used equation; and second, the square-root form is readily adaptable to calculations of the ratio of ions in solution (Schofield, 1947; Beckett, 1964a, 1964b).

a. Sodium-Calcium Exchange. Using data from the U.S.O.A. Handbook no. 60 prepared by the U.S. Salinity Laboratory (1954) we can calculate an average kilogram value for a number of California soils. The value obtained from Figure 9 in Handbook no. 60 is

$$k_g = [Ca^{2+}]^{1/2} (Na\ ads)/[Na^+] (Ca\ ads) = 0.01475.$$

At an exchangeable sodium ratio of 0.05, one can calculate the amounts of sodium in solution at different concentrations of calcium. For example, if calcium is present at a concentration of 5 millimoles per liter in the soil solution, the calculations are

$$(Na\ ads)/(Ca\ ads) = 0.05;$$

$$k_g = 0.01475;$$

$$[Ca^{2+}] = 5; [Ca^{2+}]^{1/2} = 2.24.$$

Then

$$(0.05)\ (2.24)/0.01475 = [Na^+] = 7.6\ mM.$$

At a calcium concentration of 10 mM, $[Ca^{2+}]^{1/2} = 3.16$, and the value would be

$$(0.05)\ (3.16)/0.01475 = [Na^+] = 10.7\ mM.$$

Table 16.3 lists values of sodium in solution for different values of calcium in solution and different exchangeable sodium ratios.

b. Potassium-Calcium Exchange. The values in Table 16.3 are much more simply calculated than most soil solution concentration values since it was assumed that the value of k_g was constant over exchangeable sodium ratios varying between 0.05 and 0.20. A more general case, where k_g varies as the value (K ads)/(Ca ads) changes, can next be calculated from the data of W. G. J. Knibbe (Ph.D. Thesis, Texas A & M University, 1968), where the value of k_g on a Houston Black soil varied from 0.697 at 2% potassium saturation down to 0.158 at 20% potassium saturation. In this case, the potassium in solution actually was determined at a constant

Table 16.3. Calculated values of $[Na^+]$ using a k_g value of 0.01475 [†]

Exchangeable sodium ratio (Na ads/Ca ads)	[Na$^+$] values (mM)			
	1 mM [Ca^{2+}]	5 mM [Ca^{2+}]	10 mM [Ca^{2+}]	20 mM [Ca^{2+}]
0.05	3.4	7.6	10.7	15.2
0.10	6.8	15.2	21.4	30.3
0.15	10.2	22.8	32.1	45.4
0.20	13.6	30.3	42.8	60.6

[†] k_g taken from U.S. Salinity Laboratory, 1954, fig. 9.

Table 16.4. Calculated values of $[K^+]$ using a $[Ca^{2+}]$ concentration of 2 mM with variable (K ads)/(Ca ads) and k_g

	$[K^+]$ values (mM)			
	Exchangeable potassium ratio (K ads/Ca ads)			
k_g	0.02	0.04	0.08	0.20
0.7	0.040	0.081	0.162	0.562
0.5	0.057	0.113	0.227	0.800
0.3	0.094	0.188	0.378	1.332
0.1	0.283	0.566	0.132	4.000

calcium concentration of about 0.05 N, which is considerably stronger than found in the soil solution. The calculations appearing below are made on the basis that k_g is variable with the percentage of potassium saturation, but not with ionic strength.

Suppose that at an exchangeable potassium ratio of 0.02 (about 2% exchangeable potassium saturation) the calcium in solution is 2 mM (a realistic value) and the value of $k_g = 0.7$. Then

$$[Ca^{2+}] = 2, \text{ and } [Ca^{2+}]^{1/2} = 1.414; \text{ therefore,}$$
$$[K^+] = (0.02)(1.414)/(0.7) = 0.0404 \text{ mM}.$$

At an exchangeable potassium ratio of 0.08 with a k_g of 0.253,

$$[K^+] = (0.08)(1.414)/0.253 = 0.447.$$

Other values calculated using the above procedure at a $[Ca^{2+}]$ concentration of 2 mM are given in Table 16.4. Intermediate values can be obtained by simply making a plot of the data in Table 16.4.

In all these calculations, the assumption has been made that calcium is the only divalent ion present. In neutral soils or calcareous soils this is not a bad assumption since the only other cation that is likely to be present in large amounts is magnesium. Several workers (Beckett, 1964b; U.S. Salinity Lab., 1954) have shown that calcium and magnesium act very much alike when compared to sodium and potassium.

But suppose that magnesium is the ion of interest in a predominantly calcium-saturated soil to which different amounts of potassium are added. In this case, for rough estimation, it can be assumed that magnesium acts like an isotope of calcium. That would mean that the exchange constant for calcium-magnesium exchange has a value of 1.0. In a situation where there are 20 meq of calcium on the exchange sites and 2.0 meq of magnesium and where the calcium concentration is 5 mM, the magnesium concentration in the soil solution should be given by

$$(\text{Mg ads}) [Ca^{2+}]/(\text{Ca ads}) = [Mg^{2+}].$$

Or in this case

$$[Mg^{2+}] = (2.0)\,(5)/(20) = 0.5 \text{ mM}.$$

Using this approach, a "minimum" magnesium concentration can be calculated from the calcium solution concentration by knowing the ratio of exchangeable magnesium to calcium. Usually the amount of magnesium actually in solution will be slightly larger than calculated.

c. Aluminum-Potassium Reactions. The very high affinity of exchangeable aluminum for clays makes the probability of its being in solution in high concentrations rather remote. However, as predicted by equation [10], the concentration of the replacing solution is very important. Potassium is unexpectedly an exceptional replacer of aluminum, whereas other cations present in the soil are not. This effect is shown graphically in Figure 16.2, taken from the paper of Nye *et al.* (1961). The solid diagonal line represents equal affinities between aluminum and a competing cation. Points above the line favor aluminum and those below the line favor the competing cation, which is potassium.

Nye *et al.* (1961) used equation [12] and calculated that the value for k varied from 1,010 down to 11 for kaolinite and 200 down to 5.9 for

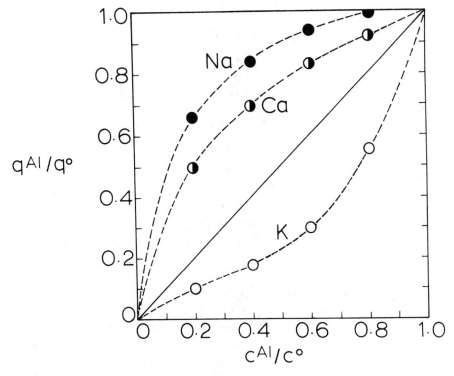

Fig. 16.2. Affinity diagram for sodium, calcium, and potassium versus aluminum in clay. The solid line with a slope of one represents equal affinities.

montmorillonite as the fraction of KCl/KCl + $AlCl_3$ varied from 0.05 to 0.90. This indicates a strong affinity for a few potassium ions even as compared to a cation as strongly held as aluminum. It also indicates that addition of potassium fertilizer could have lethal effects on plants in an acid soil. This is especially true in the neighborhood of a fertilizer band of potassium chloride. However, the effect is a temporary one, and aluminum in the soil solution is most often governed by the solubility of aluminum hydroxide and other rather insoluble compounds.

2. The Ratio Law

The ratio law, as enunciated by Schofield (1947), stated that the ratio of a pair of ions in the soil solution from a given soil is given by $[K^+]/[Ca^{2+}]^{1/2}$ and is a constant as long as the values of the exchangeable cations on the soil are not markedly changed. This ratio will be recognized immediately as being equivalent to the solution phase of equation [10]. Notice how the values change when the soil solution is concentrated. Assume that the original concentrations are 1 mM calcium and 0.1 mM potassium. Then,

$$0.1/1.0 = k' = 0.1.$$

Now suppose enough evaporation occurs that $[Ca^{2+}]$ equals 10 mM. Since k' is a constant,

$$[K^+] = k'(10)^{1/2}$$
$$= (0.1)(3.16) = 0.316 \text{ mM}.$$

Increasing [Ca] to 20 mM gives

$$[K^+] = (0.1)(4.47) = 0.447 \text{ mM}.$$

In general then, the ratio law predicts that as soil solutions are concentrated, calcium will be displaced from the soil and potassium (or any other monovalent cation) will be adsorbed. The amount of calcium in the original solution was 10 times that of potassium. When the amount of calcium was increased by a factor of 10, potassium increased only by the square root of 10, i.e., 3.16. This law predicts that in dry soils the relative amount of potassium in the soil solution will be low. Moss (1963) has shown that there is a relationship between $[K^+]/[Ca + Mg]^{1/2}$ in the soil solution and $[K^+]/[Ca + Mg]^{1/2}$ in the plant. Likewise, this law predicts that the concentration of any polyvalent cation in solution will rise proportionately faster than the concentration of a monovalent cation. In the case of aluminum, the effect is even greater:

$$[K^+]/[Al]^{1/3} = k.$$

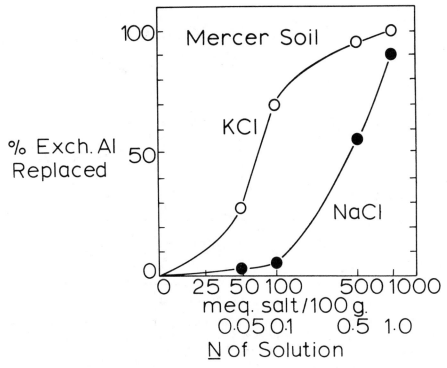

Fig. 16.3. Replacement of aluminum from Mercer soil by KCl and NaCl solutions of different concentrations. (D. E. Kissel, Ph.D. Thesis, University of Kentucky, 1969)

This suggests that the higher the concentration of potassium, the more effective the replacement of aluminum. This is amply borne out by some data of Kissel (Ph.D. Thesis, University of Kentucky, 1969), shown in Figure 16.3. However, it is even more striking in the case of sodium chloride. For example, 50 meq removed only 2.6% of the exchangeable aluminum, whereas 500 meq removed 55%. Thus, increasing the sodium concentration by a factor of 10 caused an increase in the removal of aluminum by a factor of 20.

The ratio law has been used especially in the area of potassium availability and in defining the soil acidity. Both of these uses will be outlined fully below.

If it is remembered that the ratio law can be derived from the Gapon equation [10], then it will be seen that the solution phase is an indirect measurement of (1) the amount of potassium present on the exchangeable sites and (2) the affinity with which it is held by the soil. These relationships are shown by

$$[K^+]/[Ca^{2+}] = \{(K \text{ ads})/(Ca \text{ ads})\} \{1/k_g\}. \qquad [14]$$

In comparing a number of soils, Beckett (1964a, 1964b) discovered that there was a clever way of plotting the $[K^+]/[Ca^{2+} + Mg^{2+}]^{1/2}$ values

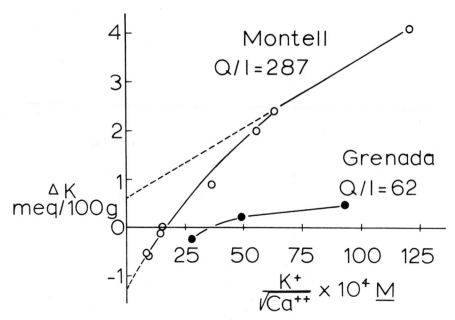

Fig. 16.4. "Beckett" plot of potassium equilibrium data from two soils. (Montell: from W. G. J. Knibbe, Ph.D. Thesis, Texas A & M University, 1968; Grenada: from S. Yimprasert, M. S. Thesis, University of Kentucky, 1969)

which gave clues to the potassium behavior. Figure 16.4 is taken from the work of Knibbe (Ph.D. Thesis, Texas A & M University, 1968), and Yimprasert (M.S. Thesis, University of Kentucky, 1969). The upper curve was obtained from the $<2\mu$ clay fraction of a Montell clay soil, the same one for which k_g values are shown in Table 16.4. The first observation that can be made from the "Beckett" plot is that the slope of the curve ΔK exch/$[K^+]/[Ca]^{1/2}$ (Q/I) equals 287. The second observation is that the curve diverges from linearity and the intercepts are 1.87 meq apart. The original potassium content of the soil is 2.1 meq per 100 g of clay, and the plot indicates that essentially all of it is held very tightly. The actual curve indicates that the solution potassium would be zero when exchangeable potassium is reduced to 1.25 meq per 100 g. The curve also indicates that to double the activity ratio would require the addition of 0.80 meq of potassium per 100 g. In actual fact, this soil and a number of others like it showed in practice that the "Beckett" plot is quite accurate. The native potassium was high but quite unavailable to plants. It was almost impossible to deplete the soils severely by cropping, and addition of a very substantial quantity of potassium to the soil was required to show any real difference.

Many soils common to the Southeast, in contrast, have curves that lay almost horizontally. In such soils, represented by the Grenada soil (also Figure 16.4), potassium is so available that it is taken up by plants in large amounts until the exchangeable potassium present is almost depleted. Usually, at a value of 0.05 meq per 100 g or so, the exchangeable potassium becomes very tightly held, and further removal is very slow. In soils like the Grenada with nearly horizontal "Beckett" curves, frequent fertilization at moderate rates is a useful practice, whereas with soils like the Montell, only occasional large applications are needed.

The "Beckett" type curve also can be applied to phosphorus in the soil solution by plotting the change in "labile" phosphorus against a function of phosphorus in solution, $1/2$ pCa + pH_2PO_4. As in the potassium plots, the more horizontal the curve, the more frequently fertilization should be practiced. There will be a tendency for the Q/I slope to increase with an increase in clay content.

II. SOLUBILITY PRODUCT CONSTANTS

A. Theory behind Their Use in Soils

With the discovery that many clay minerals, aluminum and iron oxides, and phosphate-containing minerals have well-defined morphology, crystalline structure, and composition came a number of studies designed to show that the principle of the solubility product was usable in soils. As a review, it may be useful to go through the reasoning behind the solubility product constant.

When sparingly soluble salts are placed in aqueous solution, they dissociate to some degree. The reaction can be illustrated using gibbsite,

$$Al(OH)_3 \rightarrow Al^{3+} + 3\ OH^-.$$

At equilibrium, i.e., when the amount of Al going into solution as the trivalent ion just equals the amount of trivalent aluminum precipitating as the hydroxide, we can write

$$K_{sp} = [Al^{3+}]\,[OH]^3/[Al(OH)_3], \qquad [15a]$$

where the brackets denote activities and K_{sp} is the solubility product constant. Since the activity of aluminum hydroxide (or any sparingly soluble salt) can be taken as 1.0, we can remove it from the denominator and simply write

$$K_{sp} = [Al^{3+}]\,[OH]^3. \qquad [15b]$$

Negative logarithms (similar to pH) often are used to make calculations more simple. Transforming equation [15b] by this means we get

$$pK_{sp} = pAl^{3+} + 3pOH \qquad [15c]$$

Since pOH + pH = 14.02, pOH generally is expressed as 14 − pH.

The solubility product constant of gibbsite is 1.9×10^{-33}. At a pH of 4, the activity of OH^- is 1×10^{-10}, so that

$$1.9 \times 10^{-33} = [Al] [10^{-10}]^3,$$

and

$$[Al^{3+}] = (1.9 \times 10^{-33})/(1 \times 10^{-30}) = 1.9 \times 10^{-3},$$

or the activity of aluminum is 1.9×10^{-3} M.

However, at pH 5, the activity of OH^- is 1×10^{-9}, and

$$[Al^{3+}] (1.9 \times 10^{-33})/(1 \times 10^{-27}) = 1.9 \times 10^{-6},$$

so that increasing the pH by one unit decreases aluminum activity by 1,000 times. Performing the same operation using negative logs, we get at pH 6,

$$32.72 = pAl + 3(14 - 6)$$

$$pAl = 32.72 - 24 = 8.72 \text{ or } 1.9 \times 10^{-9} \text{ M Al.}$$

As applied to soils, the foregoing equations would describe the activity of trivalent aluminum in solution if gibbsite were controlling the aluminum present in the soil solution. In most soils, there are more soluble, poorly organized hydroxy-aluminum compounds than gibbsite, but the solubility product principle is still useful. In this case it tells us that as the pH increases, the amount of trivalent aluminum in solution decreases at a very rapid rate.

Many workers correct the concentration of ions (other than H^+ and OH^-) to activities using the Debye-Huckel equation

$$-\log f = 0.5Z^2(\mu)^{1/2}/[1 + (\mu)^{1/2}], \qquad [16]$$

where f = activity coefficient,

$\quad Z$ = valence,

$\quad \mu$ = ionic strength = $1/2 \Sigma cZ^2$ (c = molal concentration).

Although this correction procedure may have merit in simple salt solutions, the activities of ions in solutions as complex as the soil solution have never been defined by the Debye-Hückel equation or any other equation. Furthermore, as will be seen from phosphorus data, the solubility product constants are so inconstant that it does not appear to be worth the trouble to make a lot of extra calculations so that the data will fit better. Therefore, for the balance of this discussion concentrations will be used instead of activities except for H^+ and OH^-, where activities are measured directly by the electrode.

B. Behavior of Aluminum and Iron Compounds of Phosphorus

In studying aluminum and iron concentrations in the soil solution, there are several problems to contend with. The first is the solubility of relatively insoluble compounds such as gibbsite, $Al(OH)_3$, or hematite, Fe_2O_3. A second problem is the strong tendency of trivalent aluminum or iron to hydrolyze in aqueous solution, giving

$$Al^{3+} + H_2O \rightarrow AlOH + H^+.$$

When we add to that the easy reducibility of iron to the ferrous (Fe^{2+}) ion and the fact that both aluminum and iron are complexed somewhat by organic matter in a poorly defined way, it appears that the system is not completely describable.

However, for the purposes of determining the amounts of iron and aluminum in soils where they are not abundantly present as exchangeable cations, the solubility product constants of ferric hydroxide and aluminum hydroxide give reasonable values that can be used for calculating availability of phosphorus in solution.

Most of the data on phosphorus in aluminum and iron compounds have been collected at relatively low pH because concentrations of both cations become very low at higher pH values. However, from extensive fractionation data it appears that both iron and aluminum phosphates are fairly stable at pH values from 4 to 6.5.

Following the methods outlined by Cole and Jackson (1951), which are very straightforward, the solubility product constant of variscite, $Al(OH)_2H_2PO_4$, has been calculated by a number of workers. Supposing that variscite dissociates as

$$Al(OH)_2H_2PO_4 \rightarrow Al^{3+} + 2\,OH^- + H_2PO_4^-.$$

Then

$$K_{sp} = [Al^{3+}]\,[OH^-]^2\,[H_2PO_4^-]. \qquad [17]$$

Using values from Cole and Jackson,

$$H_2PO_4 \times 10^4\ M = 2.50,\ pH_2PO_4 = 3.60,$$
$$Al^{3+} \times 10^4\ M = 0.07,\ pAl^{3+} = 5.15,$$
$$pH = 4.10,\ [H^+] = 0.795 \times 10^{-4},$$
$$pOH = 9.90,\ [OH^-] = 1.26 \times 10^{-10},$$

$$K_{sp} = (7 \times 10^{-6})\,(1.26 \times 10^{-10})^2\,(2.50 \times 10^{-4}) = 2.8 \times 10^{-29}.$$

Using negative logarithms,

$$pK_{sp} = 5.15 + 2(9.90) + 3.60 = 28.55.$$

To predict the amount of phosphorus in the soil solution at other values, we can assume that gibbsite controls the aluminum concentration. Then, using pK_{sp} gibbsite $= 32.72$, we can make a table of calculations:

$$pK_{sp} = pAl^{3+} + 3\ pOH^- = 32.72;$$
$$\text{at pH 4, } pAl^{3+} = 2.72;$$
$$\text{at pH 5, } pAl = 5.72.$$

For variscite at pH 4,

$$pK_{sp} - 2.72 + 2(10) = pH_2PO_4.$$

Using $pK_{sp} = 28.55$,

$$28.55 - (2.72 + 20) = 5.83.$$

Similar calculations can be made for iron phosphate if ferric hydroxide and strengite are the sparingly soluble salts present (Chang and Jackson, 1957):

$$pK_{sp}\ Fe(OH)_3 = 37;$$
$$pK_{sp}\ Fe(OH)_2\ H_2PO_4 = 33.5;$$
$$pFe = pK_{sp} - 3\ pOH;$$
$$\text{at pH 4, } pFe^{3+} = 37 - 30 = 7.$$

Table 16.5, taken from Chang and Jackson (1957), shows the ranges of values obtained for both iron and aluminum phosphates using the above assumptions.

It remains to determine whether soil solution values of H_2PO_4 approach those predicted from solubility product data. In Figure 16.5 the values for pK_{sp} 28 and 30 are plotted against pAl, and data from different sources are superimposed. As can be seen, the data show di-

Table 16.5. Calculated values for Al^{3+}, Fe^{3+}, and $H_2PO_4^-$ using solubility product constant

pH	2 pOH	pFe^{3+}	pAl^{3+}	$Fe(OH)_2H_2PO_4$†	$Al(OH)_2H_2PO_4$‡
				\multicolumn{2}{c}{$pH_2PO_4^-$}
4.0	20	7.0	2.72	6.0–8.0	5.3–9.3
4.5	19	8.5	4.22	5.5–7.5	4.8–8.8
5.0	18	10.0	5.72	5.0–7.0	4.3–8.3
5.5	17	11.5	7.72	4.5–6.5	3.8–7.8
6.0	16	13.0	8.72	4.0–6.0	3.3–7.3
6.5	15	14.5	10.22	3.5–5.5	2.8–6.8
7.0	14	16.0	11.72	3.0–5.0	2.3–6.3
7.5	13	17.5	13.22	2.5–4.5	1.8–5.8
8.0	12	19.0	14.72	2.0–4.0	1.3–5.3

† $pK_{sp} = 33$–35.
‡ $pK_{sp} = 28$–32.

hydrogen phosphate to be much more soluble than predicted by the solubility product constant. The main source of error appears to be the high solubility of aluminum predicted by the solubility product constant of gibbsite at lower pH values. Since there obviously are other sinks for aluminum than variscite, all the data are thrown off. However, the general trends of the data follow the solubility product principle; i.e., as aluminum decreases, P increases in solution.

C. Calcium Phosphate Compounds

Table 16.6 shows the relative concentrations of the various ions of phosphate at several pH values. Similar values can be calculated from the equation of Aslyng (1954), in which

$$pH_2PO_4 = pP + p[H/(K_2 + H)],$$

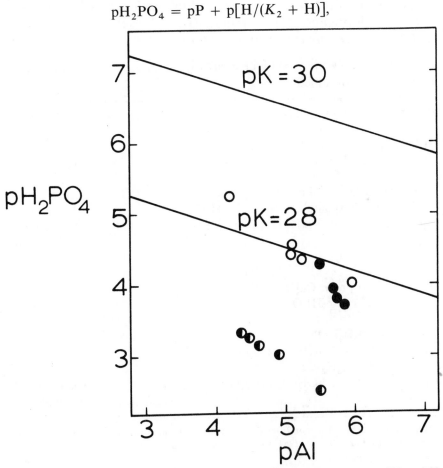

Fig. 16.5. Plot of pH_2PO_4 vs. pAl for variscite with assumed p*K* values of 28 and 30. (*Open circles* from Hemwall, 1957; *dark circles* from Kittrick and Jackson, 1955; and *shaded circles* from Coleman *et al.*, 1960)

Table 16.6 Percentages of phosphate forms present in solutions of different pH values

pH	% of each form present			
	H_3PO_4	$H_2PO_4^-$	HPO_4^{2-}	PO_4^{3-}
4	0.9	98.9	0.2	7×10^{-10}
5	0.1	98.0	2	7×10^{-8}
6	8×10^{-3}	82	18	6×10^{-6}
7	3×10^{-4}	33	67	2×10^{-4}
8	4×10^{-6}	3	97	2×10^{-3}
9	5×10^{-8}	0.5	99.5	4×10^{-2}

and K_2 equals the second ionization constant of phosphoric acid (about 10^{-7}). At pH values above 6, $H_2PO_4^-$ no longer is the most prevalent ionic form present, and for most calculations the concentration of HPO_4^{2-} is more convenient to use. At a pH of 7, using the expression

Table 16.7. Solubility product expressions for important phosphates and constants obtained

Phosphates	pK	Notes
Hydroxyapatite, $Ca_5(PO_4)_3OH \cdot H_2O$		
$\quad pK = 5\,pCa^{2+} + 3\,p(HPO_4^{2-}) - 4\,pH$	7.5	Dissolution[†]
	6.5	Precipitation[†]
$\quad pK = 10\,pCa^{2+} + 6\,p(PO_4^{3-}) + 2\,pOH^-$	113.7	Calc.[‡]
Octocalcium phosphate, $Ca_4H(PO_4)_3 \cdot 3H_2O$		
$\quad pK = 4\,pCa^{2+} + 3\,p(HPO_4^{2-}) - 2\,pH$	11.8	Dissolution[†]
	9.3	Precipitation[†]
	9.93[‡]	
$\quad pK = 4\,pCa^{2+} + pH + 3\,p(PO_4^{3-})$	46.91[‡]	
Dicalcium phosphate, $CaHPO_4 \cdot 2H_2O$		
$\quad pK = pCa^{2+} + p(HPO_4^{2-})$	6.57	Precipitation[†]
	6.56[‡]	
Variscite, $Al(OH)_2H_2PO_4$		
$\quad pK = pAl^{3+} + 2\,pOH + p(H_2PO_4^-)$	30.5[§]	
Strengite, $Fe(OH)_2H_2PO_4$		
$\quad pK = pFe^{3+} + 2\,pOH + p(H_2PO_4^-)$	35.1–33.6	Dissolution[‖]
	33.0–33.2	Precipitation[‖]

Source: Reprinted by permission from C. E. Marshall, *The Physical Chemistry and Mineralogy of Soils.* I: *Soil Materials,* © 1964, John Wiley & Sons, Inc.
† Data from Aslyng, 1954.
‡ Data from Lindsay and Moreno, 1950.
§ Data from Lindsay *et al.*, 1959.
‖ Data from Chang and Jackson, 1957.

of Aslyng (1954) for hydroxyapatite, a molarity for calcium of 2×10^{-3}, and a pK_{sp} of 7,

$$Ca_5 (PO_4)_3 O_4 \cdot H_2O + 4 H^+ \rightarrow 5 Ca^{3+} + 3 HPO_4, \text{ and}$$

$$pK_{sp} = 7 = 5 \, pCa + 3 \, p(H \, PO_4^{2-}) - 4 \, pH, \qquad [18]$$

$$3pHPO_4 = 7 - 5 (2.7) + 28$$
$$= 35 - 13.5 = 21.5, \text{ and}$$

$$pHPO_4^{2-} = \frac{21.5}{3} = 7.17, \text{ or}$$

$$[HPO_4^{2-}] = 6.76 \times 10^{-8} \text{ M.}$$

At pH 8, with the same calcium concentration,

$$3 \, pHPO_4^{2-} = 7 - 5(2.7) + 32$$
$$= 39 - 13.5 = 25.5, \text{ and}$$

$$pHPO_4^{2-} = \frac{25.5}{3} = 8.5, \text{ or}$$

$$[HPO_4^{2-}] = 3.16 \times 10^{-9} \text{ M,}$$

showing that unlike phosphate bound by aluminum or iron, the calcium-bound phosphate becomes less soluble with increasing pH.

Marshall (1964) has presented a summary of solubility product expressions together with values for the solubility product constants. These are presented in Table 16.7. Lindsay and Moreno (1960), using similar data, have calculated the solubility of H_2PO_4 in several systems. Their plot is shown in Figure 16.6.

It has been mentioned previously that calcium in the solution of calcareous soils varies with the partial pressure of carbon dioxide in the atmosphere. Turner and Clark (1956) have given the equation

$$1/2 \, pCa = pH + 1/2 \log P_{CO_2} - 4.93, \qquad [19]$$

where 4.93 is a constant and P_{CO_2} = partial pressure of CO_2. The data of Bradfield (1941) previously given in Table 16.1 fit this equation fairly well. Therefore, using either equation [19] or the data from Table 16.1, we can determine the values for calcium phosphates in calcareous soils in the same way as the values for iron and aluminum phosphate. Thus, by using these calculations, we can calculate the general trends for calcium, phosphorus, iron, and aluminum in the soil solution.

III. SUMMARY

Equations have been presented to calculate quantities of several nutrients in the soil solution. A number of examples have been worked out to

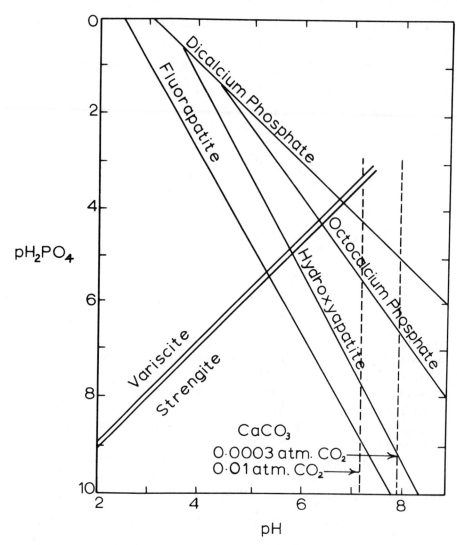

Fig. 16.6.   Solubility diagram for several important phosphate compounds. (Lindsay and Moreno, 1960)

indicate the methods used in calculation. It should be possible for the reader to calculate values for other situations simply by using the same methods used here and substituting other numbers. For further information, the interested reader is directed to the original source material cited in the list of references.

References

Aslyng, H. C. 1954. The lime and phosphate potential of soils; the solubility and availability of phosphates. *Roy. Vet. Agr. Coll. Yearbook* (Copenhagen), pp. 1–50.

Beckett, P. H. T. 1964a. Studies on soil potassium I. Confirmation of the ratio law: Measurement of potassium potentials. *J. Soil Sci.* 15: 1–8.

——. 1964b. Studies on soil potassium II. The "immediate" Q/I relations of labile potassium in the soil. *J. Soil Sci.* 15: 9–23.

Bradfield, R. 1941. Calcium in the soil I: Physico-chemical relations. *Soil Sci. Soc. Amer. Proc.* 6: 8–15.

Chang, S. C., and M. L. Jackson. 1957. Solubility product of iron phosphate. *Soil Sci. Soc. Amer. Proc.* 21: 265–69.

Cole, C. V., and M. L. Jackson. 1951. Solubility equilibrium constant of dihydroxy aluminum dihydrogen phosphate relating to a mechanism of phosphate fixation in soils. *Soil Sci. Soc. Amer. Proc.* 15: 84–89.

Coleman, N. T., J. T. Thorup, and W. A. Jackson. 1960. Phosphate sorption reactions that involve exchangeable Al. *Soil Sci.* 90: 1–7.

Davis, L. E. 1945. Simple kinetic theory of ionic exchange for ions of unequal charge. *J. Phys. Chem.* 49: 473–79.

Eriksson, E. 1952. Cation-exchange equilibria on clay minerals. *Soil Sci.* 74: 103–13.

Gaines, G. L., and H. C. Thomas. 1955. Adsorption studies on clay minerals. II. A formulation of the thermodynamics of exchange absorption. *J. Chem. Phys.* 21: 714–18.

Gapon, E. N. 1933. Theory of exchange adsorption in soils. *J. Gen. Chem.* (USSR) 3(2): 144–52.

Hemwall, John B. 1957. The role of soil clay minerals in phosphorus fixation. *Soil Sci.* 83: 101–8.

Jenny, H. 1936. Simple kinetic theory of ionic exchange. I: Ions of equal valency. *J. Phys. Chem.* 40: 501–17.

Kelley, W. P. 1948. *Cation Exchange in Soils.* A.C.S. Monograph no. 109. New York: Reinhold.

Kerr, H. W. 1928. The identification and composition of the soil aluminosilicate active in base exchange and soil acidity. *Soil Sci.* 26: 385–98.

Kittrick, J. A., and M. L. Jackson. 1955. Application of solubility product principles to the variscite-kaolinite system. *Soil Sci. Soc. Amer. Proc.* 19: 455–57.

Krishnamoorthy, C., and R. Overstreet. 1949. Theory of ion exchange relationships. *Soil Sci.* 68: 307–15.

Lagerwerff, J. V., and G. H. Bolt. 1959. Theoretical and experimental analysis of Gapon's equation for ion exchange. *Soil Sci.* 87: 127–222.

Lindsay, W. L., and E. C. Moreno. 1960. Phosphate phase equilibria in soils. *Soil Sci. Soc. Amer. Proc.* 24: 177–82.

Lindsay, W. L., M. Peech, and J. S. Clark. 1959. Solubility criteria for the existence of variscite in soils. *Soil Sci. Soc. Amer. Proc.* 23:357–60.

Marshall, C. E. 1964. *The Physical Chemistry and Mineralogy of Soils.* I: *Soil Materials.* New York: Wiley.

Moss, P. 1963. Some aspects of the cation status of soil moisture. Part I: The ratio law and soil moisture content. *Plant Soil* 18: 99–113.

Nye, P., D. Craig, N. T. Coleman, and J. L. Ragland. 1961. Ion exchange equilibria involving aluminum. *Soil Sci. Soc. Amer. Proc.* 25: 14–17.

Schofield, R. K. 1947. A ratio law governing the equilibrium of cations in the soil solution. *Proc. 11th Int. Congr. Pure Appl. Chem.* 3: 257–61.

Turner, R. C., and J. S. Clark. 1956. The pH of calcareous soils. *Soil Sci.* 82:337–41.

U.S. Salinity Laboratory. 1954. *Diagnosis and Improvement of Saline and Alkali Soils.* U.S.D.A. Agriculture Handbook no. 60.

Vanselow, A. P. 1932. Equilibria of the base-exchange reactions of bentonites, permutities, soil colloids, and zeolites. *Soil Sci.* 33: 95–113.

Way, J. T. 1850. On the power of soils to absorb manure. *J. Roy. Agr. Soc.* 11: 313–79.

Woodruff, J. R., and E. J. Kamprath. 1965. Phosphorus adsorption maximum as measured by the Langmuir isothern and its relationship to phosphorus availability. *Soil Sci. Soc. Amer. Proc.* 29: 148–50.

17.   Role of Chelation in Micronutrient Availability

W. L. Lindsay

CHELATION of metal ions occurring in the vicinity of plant roots plays an important role in supplying nutrients to plants. Were it not for this important phenomenon, most soils would be devoid of plant growth because iron and, in some cases, zinc, copper, and manganese are too insoluble in soils to maintain adequate levels of soluble nutrients.

Some of the questions that arise regarding chelates in soils are these: Where do they come from? What reactions do they undergo in soils? How are they able to assist in the transport processes of moving nutrients to plant roots? This chapter discusses these topics.

The hypothesis proposed here is that chelating agents—either natural or synthetic—combine with micronutrient cations to increase the total level of these nutrients in solution; these chelated metals increase nutrient availability by increasing both mass-flow and diffusion of nutrients to roots.

I. THE PROBLEM OF MICRONUTRIENT SUPPLY

The problem of supplying micronutrients to plant roots can be readily demonstrated for Fe. A recent survey of 37 Colorado soils shows an average content of about 2% total Fe or 20,000 ppm (Follett and Lindsay, 1970). Most agricultural crops require less than 0.5 ppm Fe from the plow layer. This means that many soils contain over 40,000 times more Fe in the plow layer than most crops require annually. Obviously, the problem of Fe deficiency is not one of total supply.

The solubility of inorganic Fe in soils is highly pH dependent and is governed largely by the solubility of iron oxides (Norvell, 1970). The manner in which Fe^{3+} and its hydrolysis species govern Fe solubility in the pH range of soils is known (Bohn, 1967; Lindsay, 1972).

Figure 17.1, which is similar to that of O'Connor *et al.* (1971), shows that total soluble Fe in soils must be at least 10^{-6} M to enable mass-flow of water to the roots to transport sufficient Fe to the plant. If soluble Fe in soils drops to 10^{-8} M, only 1% of that required by plants can arrive at the root by mass-flow (500 g water per gram of plant dry matter). The obvious conclusion is that precipitation of iron oxides in soils and in nutrient solutions lowers the soluble inorganic Fe far below that required by plants. The Fe^{3+} and Fe^{2+} in equilibrium with iron oxides and 0.2 atm oxygen contribute only slightly to the total Fe in solution. In the pH range of 7 to 8, total inorganic Fe drops to approxi-

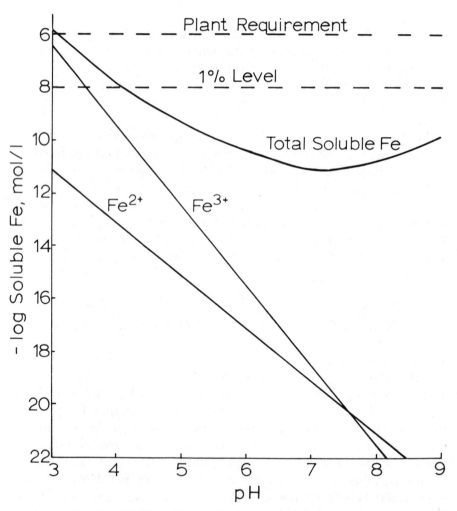

Fig. 17.1. Solubility of Fe^{2+}, Fe^{3+}, and total Fe in equilibrium with iron oxide and 0.2 atm O_2 showing the levels of soluble Fe required for mass-flow to supply plants with adequate Fe

mately 10^{-11} M, or 10^5 times lower than that required for mass-flow transport. As plant roots lower the concentration of Fe in their immediate vicinity, diffusion as well as mass-flow are operative. Simple estimates based on diffusive fluxes (O'Connor *et al.*, 1972) show that the diffusion of inorganic Fe at these low solubility levels is far less than that needed to supply plants with adequate Fe.

One way in which soluble Fe can be increased in soils is by the formation of soluble Fe complexes or chelates. Chelating agents may originate as root exudates, as substances released from decaying organic matter, as products of microbiological synthesis, or as chelated fertilizers added to soils. Chelate concentrations as low as 10^{-8} to 10^{-7} M are estimated to be adequate when both diffusion and mass-flow participate in the transport process. Similar problems are encountered with Zn, Cu, and Mn in soils as the pH rises to neutral or above. The presence of micronutrient complexes in soils has been clearly demonstrated (Hodgson *et al.*, 1966; Geering *et al.*, 1969).

II. CHELATE EQUILIBRIA IN SOILS

Numerous studies have been made during the past 20 years to test the effectiveness of various chelates in supplying micronutrients to plants. Most of these studies have been empirical; that is, the chelates have been tested in different soils and nutrient solutions to see what would happen. Only recently have successful attempts been made to develop theoretical models from which the behavior of various chelates can be predicted and explained (Lindsay *et al.*, 1967; Lindsay and Norvell, 1969; Halvorson and Lindsay, 1972).

Figure 17.2, taken from Lindsay *et al.* (1967), illustrates the initial development of stability-pH diagrams to predict the behavior of Fe chelates in soils. Included are the three synthetic chelating agents EDTA (ethylenediaminetetraacetic acid), DTPA (diethylenetriamine pentaacetic acid), and EDDHA [ethylenediamine di (o-hydroxyphenylacetic acid)]. All three chelates are saturated with Fe at low pH values. As pH increases above 6, Fe^{3+} is displaced by Ca^{2+} to give $CaEDTA^{2-}$ and precipitated iron oxide. Above pH 7, $FeDTPA^{-2}$ undergoes a similar exchange, whereas $FeEDDHA^-$ remains stable throughout the pH range of 4 to 10. These theoretical relationships were tested experimentally by reacting Fe chelates and chelating agents with soils for various periods. The results show a close correspondence between theoretical predictions and experimental findings.

Norvell and Lindsay (1969) conducted further experimental studies of the reactions of EDTA and DTPA metal complexes with soils. Figure 17.3, taken from their work, demonstrates this correspondence.

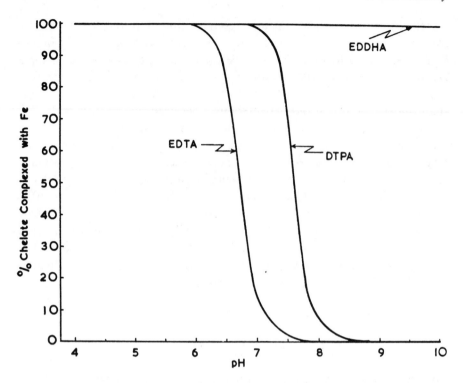

Fig. 17.2. Stability of various Fe chelates in soils when Ca^{2+} is 2.5×10^{-3} M. (Lindsay *et al.*, 1967)

During the 30-day reactions, EDTA picked up Fe at low pH values and at high pH values lost it to the soil. Similar studies were also carried out with DTPA and will be published in the near future. The experimental results corresponded closely to those predicted by the chelate stability diagrams. These diagrams were developed from a knowledge of the formation constants of the chelated species and the activity of the competing metal ions expected in soils.

The question that naturally arises is this: Can the availability of chelated plant nutrients be predicted from chelate stability diagrams? Figure 17.4 summarizes the greenhouse results reported by Lindsay *et al.* (1967) for an Fe-deficient calcareous soil. For the first crop, FeEDTA and FeDTPA gave equal responses, but Fe deficiency was only partially corrected. From Figure 17.2 one might not expect FeEDTA to be sufficiently stable to correct Fe deficiency in a calcareous soil. Norvell and Lindsay (1969) pointed out that metal chelates often require several weeks to obtain equilibrium with other cations in the soil. Apparently during this period in which Ca displaces Fe from the chelate, the released Fe is partially available. By the second crop the exchange was apparently complete, since the FeEDTA treatment showed no beneficial effect. The FeDTPA continued to show slight

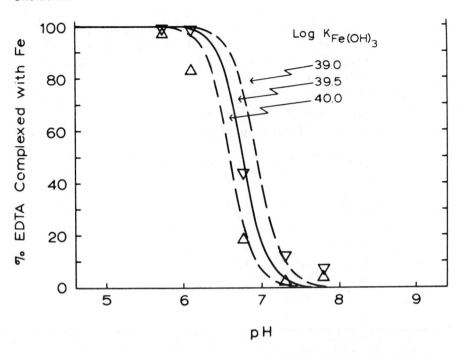

Fig. 17.3. Comparison of calculated and measured FeEDTA stability after 30 days' re-action of NaEDTA (Δ) and FeEDTA (∇) with five soils. (Norvell and Lindsay, 1969)

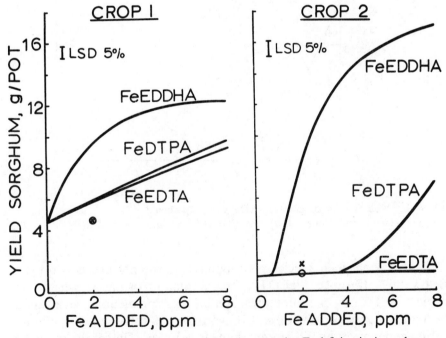

Fig. 17.4. Effectiveness of various Fe chelates in correcting Fe deficiencies in a calcareous soil for two croppings of sorghum. (Data from Lindsay *et al.*, 1967)

correction of Fe deficiency in the second crop. As predicted, the stable
FeEDDHA was highly available to both crops. Thus, chelate stability
diagrams are useful in predicting plant response to these Fe chelates
if kinetic as well as thermodynamic factors are considered.

Lindsay and Norvell (1969) later extended their chelate stability
diagrams to include the competition among Zn^{2+}, Fe^{3+}, Ca^{2+}, and
H^+. Figure 17.5, taken from their paper, illustrates the fact that the

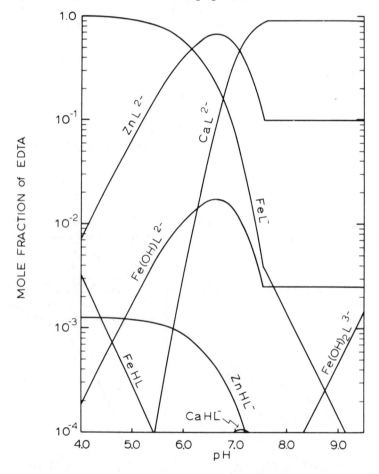

Fig. 17.5. Mole fraction diagram for EDTA in soils when Zn^{2+}, Fe^{3+}, Ca^{2+}, and H^+
are the competing metal ions at 0.003 atm CO_2. (Lindsay and Norvell, 1969)

$ZnEDTA^{2-}$ chelated species predominates in the pH range of 6 to 7.
Below this pH range, $FeEDTA^-$ predominates, while at higher pH
values $CaEDTA^{2-}$ predominates. Again, these predictions were tested
by reacting ZnEDTA with soil of various pH values over a 30-day
period and measuring the Zn remaining in solution. The results reported
in Figure 17.6 (Norvell and Lindsay, 1969) confirm the close correspon-

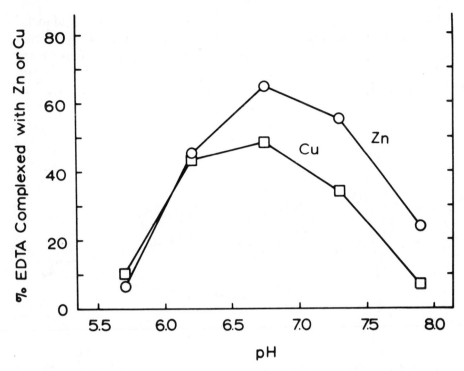

Fig. 17.6. Stability of ZnEDTA and CuEDTA in five soils after 30 days' reaction. (Norvell and Lindsay, 1969)

dence between the predicted and experimental behavior of ZnEDTA. At low pH values Fe^{3+} displaced Zn from EDTA, and at high pH values Ca^{2+} displaced it. Similar reactions with CuEDTA showed almost identical behavior to that of Zn.

Lindsay and Norvell (1969) compared the theoretical stabilities of ZnEDTA and ZnDTPA complexes. They found them to be equivalent below pH 7.0, but above this pH, ZnDTPA was more stable. Anderson (1964) compared the availability of these two Zn sources in a calcareous soil using corn (*Zea mays*) and found that ZnDTPA was slightly more available than ZnEDTA. Again, the predicted stability and measured availability of these Zn chelates show a close correspondence.

Reactions of MnEDTA with soils (Norvell and Lindsay, 1969) showed that Mn was displaced from the chelate complex by Fe at low pH values and by Ca at high pH values. The significance of these results was demonstrated in recent studies in Michigan (Knezek and Greinert, 1971). They found that MnEDTA did not correct Mn deficiencies; in fact, it intensified them. The explanation is that Fe dissolves from the soil to displace Mn from the chelate and the released Mn precipitates in an insoluble form. Aggravated Fe-induced Mn deficiency resulted from the increased availability of FeEDTA.

Another example of interactions that can result from the use of metal chelates in soils was reported by Lindsay *et al.* (1967). They reacted ZnEDDHA with soils for an 11-day period and periodically measured the soluble cations and ^{14}C-labeled chelating agent in solution. Figure 17.7, taken from their work, summarizes their findings. Zinc was

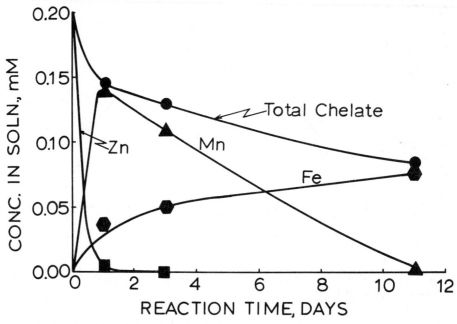

Fig. 17.7. Reaction of 0.2 mM ZnEDDHA with a Platner loam (pH 7.0) during an 11-day period. (Lindsay *et al.*, 1967)

rapidly displaced from the EDDHA chelate by Mn dissolved from the soil. Later the Mn was precipitated from solution as dissolved soil Fe saturated the chelate. It is interesting to speculate how these displacement reactions might affect the nutrition of plants growing in a soil fertilized with ZnEDDHA. Initially, Zn would be available, but its availability would decrease rather rapidly. The availability of Mn is also expected to increase for several days while Fe dissolves to saturate the chelate. In fact, during this period the increased availability of Mn and a high ratio of free $EDDHA^{4-}$: Fe^{3+} in the soil solution are expected to depress Fe activity and reduce its availability. Eventually, however, as most of the chelate becomes saturated with Fe, the availability of Fe would increase and that of Mn would subside. These examples point out the complexity of nutrient interactions that may be encountered when using metal chelates in soils and in plant nutritional studies. Chelate stability diagrams provide a very useful tool for interpreting such interactions.

III. ROLE OF CHELATES NEAR PLANT ROOTS

The greatest benefit from chelates in increasing nutrient availability in soils is believed to be due to their effect on diffusion and mass-flow of nutrients in the immediate vicinity of plant roots. Elgawhary *et al.* (1970b) very convincingly demonstrated that chelating agents can increase the diffusion of nutrients by using a "simulated root" technique. Their laboratory setup is depicted in Figure 17.8, and some of their

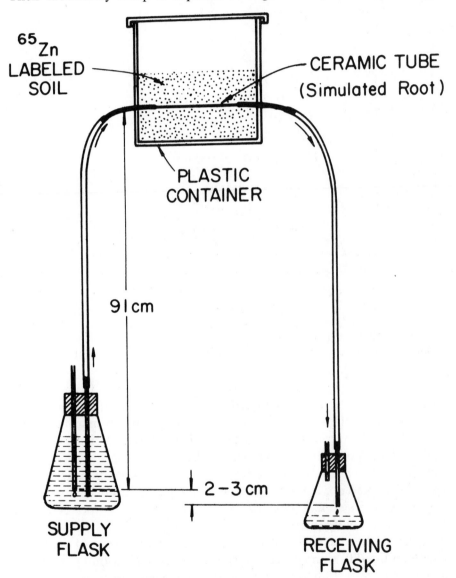

Fig. 17.8. Experimental setup used to study the effect of complexing agents and acids on the diffusion of Zn to a simulated root. (Elgawhary *et al.*, 1970b)

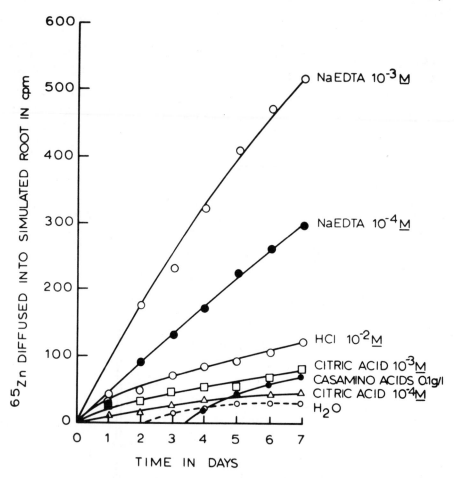

Fig. 17.9. Effect of various complexing agents and acids on the accumulative diffusion of Zn into a simulated root over a 7-day period. (Elgawhary *et al.*, 1970b)

results are given in Figure 17.9. Mass-flow of solution from the soil to the porous tube was eliminated by equalizing the water tension in the soil and in the solution within the tube. Diffusion of complexing agents from the tube into the adjacent soil and diffusion of the chelated ^{65}Zn into the porous tube demonstrate the striking effect that chelating agents can have on the solubilization and diffusion of nutrients to roots. The inference is made that root exudates, natural chelates in the soil, or added synthetic chelates provide a mechanism for increasing nutrient availability to roots.

Other studies have also demonstrated the importance of chelating agents in the diffusion of micronutrient cations. Hodgson *et al.* (1967) showed that 2×10^{-3} M citrate acting as a chelating agent increased the transport of Zn through agar gel from a $ZnCO_3$ source by 100-fold. The results compare favorably with a theoretical treatment of the solubility and diffusional process involved. Elgawhary *et al.* (1970a) showed that addition of either ZnEDTA or EDTA chelating agents increased the self-diffusion of Zn in soils. O'Connor *et al.* (1971) showed a similar effect from the addition of FeEDDHA or CaEDDHA to soils. They also showed that the uptake of Fe by sorghum roots was directly related to total soluble Fe in solution. The evidence is rapidly accumulating that chelating agents provide a means of solubilizing micronutrient cations and that the increased concentration levels provide greater diffusion gradients for the transport of Fe in soils. The presence of adequate chelates becomes critical when nutrient concentrations are low and depletion zones develop in the immediate vicinity of plant roots (O'Connor *et al.*, 1972). Chelating agents also contribute directly to the supply of nutrients arriving at roots through mass-flow of water to the roots.

The role of chelates in the immediate vicinity of plant roots is illustrated in Figure 17.10. Depicted here is a root hair absorbing a

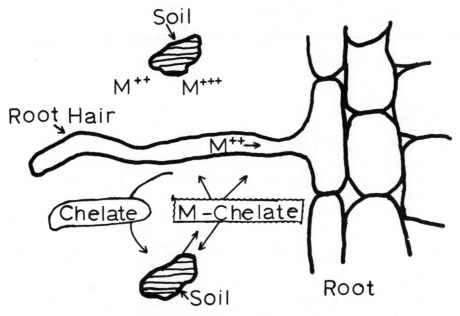

Fig. 17.10. Diagramatical representation of how chelating agents transport metal ions to plant roots

metal ion. In the process, the activity of that ion in the immediate vicinity of the root is lowered. Metal chelates next to the root dissociate to some extent to replenish the metal ion taken into the plant. Removal of the chelated metal thus establishes a diffusion gradient to transport more chelated metal to the root. At the same time the chelating ligand released at the root establishes a diffusion gradient for the free ligand to move away from the root where it can resaturate with metal cations from the soil. The diffusion gradient for the newly formed metal chelate is back toward the root. Thus chelating agents play an important role in transporting insoluble nutrients to root surfaces. Hodgson (1968, 1969) has developed a theoretical approach for considering the contribution of chelates to the movement of Fe to roots. This model brings out many important factors that certainly deserve further critical examination and testing.

The question of whether chelated metals enter the root intact or whether the complex dissociates at the root surface is often asked. Wallace (1963) and more recently Hill-Cottingham and Lloyd-Jones (1965) reviewed this subject and concluded that both mechanisms are operative. The differential uptake of chelating agent and metal varied among plant species, the kind of chelating agent, and other parameters such as pH.

The rate at which metal ions get into and out of chelating ligands is undoubtedly an important factor. Recently Chaney *et al.* (1971) proposed that reduction of Fe^{3+} at the root surface is important in making chelated Fe available for absorption. Further investigations in this area are needed.

IV. ROLE OF CHELATES IN NUTRIENT SOLUTIONS

Iron is the most difficult nutrient to keep in solution in hydroponic cultures. Organic acids such as citrate, tartrate, and various chelating agents have been used to prevent precipitation; yet the use of such chelates often produces side reactions and nutrient interactions that are difficult to explain.

Halvorson and Lindsay (1972) developed stability diagrams to predict equilibrium relationships among the chelated species present in nutrient solutions. An example of their calculations when FeEDTA was used to supply Fe in a modified 2/5-strength Hoagland nutrient solution is shown in Figure 17.11. Not only Fe^{3+} but also Zn^{2+}, Cu^{2+}, Mn^{2+}, Ca^{2+}, and Mg^{2+} react to form soluble chelate complexes. Changes in the concentration of any of these cations in the nutrient media change the equilibrium concentrations of all chelated species.

Halvorson (1971) conducted several nutrient solution experiments to test the usefulness of such diagrams to predict and interpret plant nutrient responses. One such experiment is summarized in Figure 17.12. All four Fe chelates used in this study were effective in supplying available Fe to corn at pH 5.3, but at pH 7.5 only FeEDTA was effective. The poor response from FeEGTA (ferric ethyleneglycol-bisaminoethylether tetraacetate) at pH 7.5 was caused by Fe deficiency.

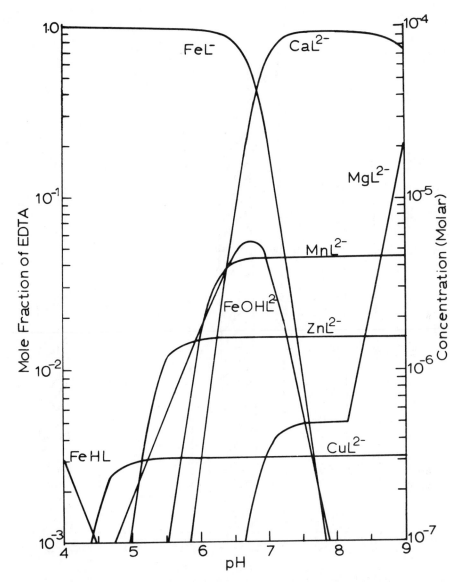

Fig. 17.11. Mole fraction diagram for EDTA in a nutrient solution in equilibrium with hydrous iron oxide and 0.0003 atm CO_2. (Halvorson and Lindsay, 1972)

At this pH, Ca^{2+} displaced Fe^{3+} from the chelate, giving $CaEGTA^{2-}$ and insoluble iron oxide. If the EGTA stability curve were plotted in Figure 17.2, it would lie to the left of the EDTA curve, showing that Fe is readily displaced from FeEGTA as pH increases from 5 to 6.

The poor response with FeDTPA at pH 7.5 was caused by Zn deficiency and not Fe. Spraying the plants with Zn or increasing Zn in the nutrient solution corrected the deficiency. Theoretical calculations confirm that DTPA combines with Zn so strongly at pH 7.5 that

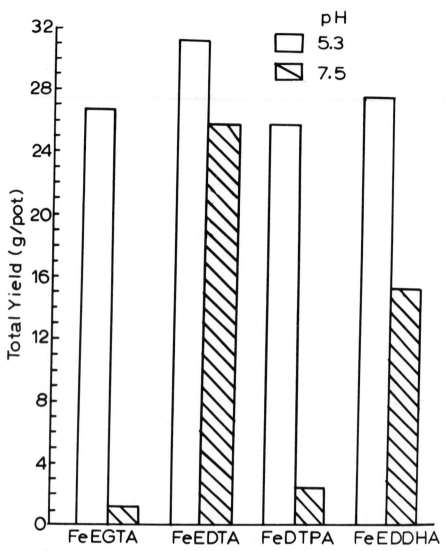

Fig. 17.12. Growth of corn (*Zea mays*) tops in nutrient solutions as affected by different Fe chelates and pH. (Data from Halvorson, 1971)

free Zn^{2+} in the nutrient solution drops from the added $10^{-5.8}$ M to $10^{-11.6}$ M. A somewhat smaller depression of Zn^{2+} activity accompanied the FeEDTA treatment where corn growth was normal.

When Zn in the FeDTPA treatment was raised, Zn displaced sufficient Fe from the chelate above 20 mM to give Fe deficiency and reduced growth. Again, the chelate stability diagrams were most useful in interpreting these plant responses and nutrient interactions.

The poor response from FeEDDHA in nutrient solution at pH 7.5 (Fig. 17.12) may seem surprising at first, since FeEDDHA is stable at pH 7.5 (Fig. 17.2) and is usually effective in correcting Fe deficiency in calcareous soil (Fig. 17.4). How might this apparent discrepancy be resolved? In the presence of soil, removal of Fe from the chelate at the root surface releases free chelating agent that diffuses away and becomes resaturated with soil Fe (Fig. 17.10). In solution cultures removal of Fe at the root leaves behind free chelating agent that accumulates in solution and depresses the activity of free Fe^{3+}. Theoretical calculations (Halvorson, 1971) clearly show that increasing the molar ratios of EDDHA to Fe above unity depresses Fe^{3+} activity in solution. These nutrient solution experiments also show that increasing the ratio of EDDHA to Fe in solution cultures causes Fe deficiency and reduces growth. Similar plant growth depressions with excessive chelating agents were demonstrated several years ago by Brown *et al.* (1960).

Most significantly, the studies of Halvorson (1971) and Halvorson and Lindsay (1972) pointed out the great advantage of theoretical stability diagrams as guides for predicting and interpreting metal chelate behavior in nutrient solutions. Even when molar EDDHA exceeds that of Fe by only 0.5%, free Fe^{3+} in the nutrient solution decreased 10^6-fold. Unsatisfactory responses to FeEDDHA obtained in nutrient solution experiments can often be traced to this cause. The problem can be avoided by adding excess soluble Fe with FeEDDHA. Although the nonchelated Fe readily precipitates, it can dissolve to resaturate the chelate when Fe is removed by the plant. The chelate competition is much less severe for chelates like EGTA, EDTA, and DTPA, where other metal ions more easily replace Fe so that only slight increases in activity of free chelating agents result.

V. SUMMARY

Chelation of micronutrient cations is an extremely important mechanism by which insoluble micronutrient cations are made available to plants. Such chelates increase the diffusion and mass-flow of nutrients to roots by replenishing those taken up by the plant. In the immediate vicinity of roots, chelates provide the carrier mechanism by which depleted nutrients at the root surface are replenished.

Recently, chelate stability diagrams have been developed to predict metal chelate equilibria in soils and in nutrient solutions. For soils, the availability of chelated Fe, Zn, and Mn was shown to correspond to the predicted stability of these chelated metals. In nutrient solution, examples were cited to demonstrate that (1) Fe chelates of low stability allow Fe to precipitate from solution, causing Fe deficiency; (2) use of FeDTPA causes Zn deficiency unless the normal Zn level is raised, but raising the Zn too much gives Fe deficiency; and (3) removal of Fe from solutions containing equimolar Fe and EDDHA leaves behind a free chelating agent that lowers the activity of the Fe^{3+} in solution and causes Fe deficiency.

Considerable progress has been made in understanding the behavior of chelates in soils and the important role they have on the availability of nutrients to roots. Further research in this important area of the root and its environment is to be encouraged.

References

Anderson, W. B. 1964. Effectiveness of synthetic chelating agents as sources of zinc for calcareous soils. Ph.D. Thesis. Colorado State University.

Bohn, H. L. 1967. The (Fe) $(OH)^3$ ion product in suspensions of acid soils. *Soil Sci. Soc. Amer. Proc.* 31: 641–44.

Brown, J. C., L. O. Tiffin, and R. S. Holmes. 1960. Competition between chelating agents and roots as factors affecting absorption of iron and other ions by plant species. *Plant Physiol.* 35: 878–86.

Chaney, R. L., J. C. Brown, and L. O. Tiffin. 1972. Obligatory reduction of ferric chelates in iron uptake by soybeans. *Plant Physiol.* 50: 208–13.

Elgawhary, S. M., W. L. Lindsay, and W. D. Kemper. 1970a. Effect of EDTA on the self-diffusion of Zn in aqueous solution and in soil. *Soil Sci. Soc. Amer. Proc.* 34: 66–70.

——. 1970b. Effect of complexing agents and acids on the diffusion of Zn to a simulated root. *Soil Sci. Soc. Amer. Proc.* 34: 211–14.

Follett, R. H., and W. L. Lindsay. 1970. *Profile Distribution of Zinc, Iron, Manganese, and Copper in Colorado Soils*. Colo. Exp. Sta. Tech. Bull. 110.

Geering, H. R., J. F. Hodgson, and Caroline Sdano. 1969. Micronutrient cation complexes in soil solution: IV. The chemical state of manganese in soil solution. *Soil Sci. Soc. Amer. Proc.* 33: 81–85.

Halvorson, A. D. 1971. Chelation and availability of metal ions in nutrient solutions. Ph.D. Thesis, Colorado State University.

Halvorson, A. D., and W. L. Lindsay. 1972. Equilibrium relations of metal chelates in hydroponic solutions. *Soil Sci. Soc. Amer. Proc.* 36: 755–61.

Hill-Cottingham, D. G., and C. P. Lloyd-Jones. 1965. The behavior of iron chelating agents with plants. *J. Exp. Bot.* 16: 233–42.

Hodgson, J. F. 1968. Theoretical approach for the contribution of chelates to the movement of iron to roots. *Trans. 9th Int. Congr. of Soil Sci.* (Adelaide) 2: 229–41.

——. 1969. Contribution of metal-organic complexing agents to the transport of metals to roots. *Soil Sci. Soc. Amer. Proc.* 33: 68–75.

Hodgson, J. F., W. L. Lindsay, and W. D. Kemper. 1967. Contributions of fixed charge and mobile complexing agents to the diffusion of Zn. *Soil Sci. Soc. Amer. Proc.* 31: 410–13.

Hodgson, J. F., W. L. Lindsay, and J. F. Trierweiler. 1966. Micronutrient cation complexing in soil solution: II. Complexing of zinc and copper in displaced solution from calcareous soils. *Soil Sci. Soc. Amer. Proc.* 30: 723–26.

Knezek, B. D., and H. Greinert. 1971. Influence of soil Fe and MnEDTA interactions upon the iron and manganese nutrition of bean plants. *Agron J.* 63: 617–27.

Lindsay, W. L. 1972. Inorganic phase equilibria of micronutrients in soils. In *Micronutrients in Agriculture*, ed. J. J. Mortvedt, P. M. Giovdano, and W. L. Lindsay, pp. 41–57, Madison, Wis.: Soil Sci. Soc. Amer.

Lindsay, W. L., J. F. Hodgson, and W. A. Norvell. 1967. The physico-chemical equilibrium of metal chelates in soils and their influence on the availability of micronutrient cations. *Trans. Comm. II & IV Int. Soc. Soil Sci.* (Aberdeen, 1966), pp. 305–16.

Lindsay, W. L., and W. A. Norvell. 1969. Equilibrium relationships of Zn^{2+}, Fe^{3+}, Ca^{2+}, and H^+ with EDTA and DTPA in soils. *Soil Sci. Soc. Amer. Proc.* 33: 62–68.

Norvell, W. A. 1970. Solubility of iron in soils. Ph.D. Thesis. Colorado State University.

Norvell, W. A., and W. L. Lindsay. 1969. Reactions of EDTA complexes of Fe, Zn, Mn, and Cu with soils. *Soil Sci. Soc. Amer. Proc.* 33: 86–91.

O'Connor, G. A., W. L. Lindsay, and S. R. Olsen. 1971a. Diffusion of iron and iron chelates in soil. *Soil Sci. Soc. Amer. Proc.* 35: 407–10.

——. 1973. Diffusion of iron to plant roots. *Soil Sci. Soc. Amer. Proc.* (in press).

Wallace, A. 1963. Review of chelation in plant nutrition. *Agr. Food Chem.* 11: 103–7.

18. Influence of the Plant Root on Ion Movement in Soil

Stanley A. Barber

THIS chapter discusses both the movement of ions through the soil to the root surface and the effect of the root on its soil environment. The dynamic nature of conditions at the root-soil interface is in contrast with the usual equilibrium measurements made to evaluate soil nutrient availability. Although the primary concern here is with nutrient flux to the root, nonnutrient ions follow the same supply mechanisms and may affect root growth and nutrient availability.

Earlier chapters have been concerned with the effect of the soil on root growth. In this chapter we are more concerned with the influence of the root on the soil. The influence of the root may make the soil immediately adjacent to it either more favorable or less favorable for root growth.

I. EFFECT OF THE ROOT ON THE PHYSICAL PROPERTIES OF THE SOIL

A. Changes in Soil Bulk Density

Changes in soil bulk density affect rates of ion diffusion in the soil (see section III B2). Hence, it is important to recognize the change caused by the root. As the root grows into the soil it occupies space that was previously occupied by soil pore space and soil particles. Since root diameters are usually larger than soil pores (roots may be 0.1 to 3 mm in diameter, while soil pores are of the order of .002 to 0.2 mm), soil particles are pushed aside, and the bulk density of the soil near the root increases. There is undoubtedly a wide range in effects depending on root diameter and the nature of the soil. The effect of roots in deforming the soil has been discussed by Barley (1968) and Greacen *et al.* (1968). The latter show increases in bulk density of the soil next to the root to 1.6 and 1.7 g/cm^3 from an initial level of 1.5 g/cm^3.

When ions diffuse toward a root along a concentration gradient, the concentrations are ordinarily given per unit volume of soil. Hence, as the soil is compacted about the root, the concentration per unit volume is increased, and steeper concentration gradients may be obtained.

Changes in soil density about cotton (*Gossypium*) roots can be observed in the image of the thin section of soil and roots published by Lund and Beals (1965). It appears that the proportion of clay and organic matter is higher in the volume near the root, due to differential packing of soil particles, than at some distance from the root. Where this occurs, increases in concentration of nutrients per volume of soil may be greater than increases in bulk density.

B. Amount of Roots in the Soil

The amount of physical change that roots exert on a soil varies with the density of roots in the soil and with the root morphology. The amount of roots present per unit volume of soil can be measured in terms of root surface area, root length, or root volume. The amount of displacement of soil is measured by root volume. The nature of the displacement is influenced by the average diameter of the roots, which is a function of root length per unit of root volume. The root surface available for ion absorption is a function of surface area.

Dittmer (1940) found that roots of soybeans (*Glycine max*), oats (*Avena sativa*), rye (*Secale cereale*), and Kentucky bluegrass (*Poa pratensis*) occupied 0.91%, 0.55%, 0.85%, and 2.8%, respectively, of the soil volume of the 0- to 15-cm soil layer. Barber (1971) found that corn (*Zea mays* L.) growing in field plots occupied from 0.19% to 1.06% of the volume of the 0- to 15-cm layer. Values varied with year and tillage practice. Barber (unpublished data, 1971) measured the root volumes of crops that had reached full growth on a Chalmers silt loam, Typic Argiaquoll, and obtained the following average fractions for the 0- to 15-cm layer: wheat (*Triticum aestivum*), 0.67%; alfalfa (*Medicago sativa*) and bromegrass (*Bromus inermis*) mixture, 1.10%.

The length of roots present in the soil is also significant, particularly when nutrient flux to the root is considered. Barber (1971) found corn root lengths varying from 0.74 to 3.40 cm/cm^3 when the 0- to 60-cm layer of soil was sampled by 5-cm depths. Root density varied with tillage treatment and decreased with depth of sampling. Pavylechenko and Harrington (1934) measured the length of roots in a 0- to 68.5-cm soil layer in Saskatchewan, Canada, and obtained average values of 0.95, 0.51, and 0.26 cm/cm^3 for barley (*Hordeum vulgare*), wheat, and oats, respectively. Dittmer sampled oats, rye, and bluegrass in Iowa and obtained values of 6.7, 9.3, and 28 cm/cm^3, respectively, in the 0- to 15-cm layer. Barber (unpublished data) found values of 3.1 and 4.5 cm/cm^3 in the 0- to 15-cm layer for wheat and alfalfa and bromegrass, respectively, in Indiana. Root length per cm^3 of soil varies widely with species and soil conditions.

II. EFFECT OF PLANT ROOTS ON ION DISTRIBUTION IN THE SOIL

The plant root affects the distribution of ions in the soil because of its physical presence and because of its absorption of nutrients and water.

A. Mechanisms Affecting Ion Distribution

1. Root Interception

In the previous section we indicated the relative volume of roots in the soil and their influence on soil bulk density. Nutrients are measured by concentration per unit volume of soil when diffusive flow to plant roots is determined. Hence, when the bulk density of soil is increased near the root, the concentration of available nutrients per unit soil volume is also increased. This increase results in a greater concentration gradient for nutrients diffusing to the root. The root may also bump into some nutrients and absorb them as it forces its way through the soil.

Oliver and Barber (1966) calculated the ions displaced by the roots as a supply mechanism, since although a part may be pushed away and return to the root by mass-flow and diffusion, the amount returning would be in addition to the amount for mass-flow and diffusion calculated on the assumption that the root did not influence soil bulk density near its surface.

2. Mass-Flow

Plant roots absorb water, causing a flow of water from the soil to the root surface. This water contains inorganic anions and cations as well as soluble organic molecules. These are mass-transported to the root in the convective flow of the water; hence the term *mass-flow*. (Some use *convective flow*.) The amount of ions reaching the root depends upon the rate of water flow to the root and the average ion content of this water. The level of a particular ion in the soil near the root may be increased, stay the same, or be decreased depending on the balance between the rate of supply to the root by mass-flow and the rate of absorption into the root. Each ion behaves in an independent manner.

Calculation of the supply of nutrients to the root by mass-flow assumes that the concentration of the solution flowing to the root and the amount of water absorbed can be measured. Total water uptake can be determined by measuring the water transpired plus that contained in the plant. Measurement of the average concentration in solution by analyzing displaced solutions is reasonably accurate where the solution concentration does not change appreciably during plant growth. Use of soils of high water-holding capacity and use of a large soil: plant ratio so that only a portion of the total soil water originally present is absorbed allows measurement errors to be kept within narrow limits.

3. Diffusion

When the supply of a particular ion to the root by the initial root interception and subsequent mass-flow does not equal or exceed the amounts absorbed by the root, the concentration of this ion at the root surface is reduced and a concentration gradient normal to the root is established. Because a concentration gradient exists, the ion diffuses toward the root due to the thermal motion of the particles. The supply to the root and the rate of uptake is largely regulated by the rate of diffusion.

Diffusion follows Fick's law, which is

$$F = - DA(\partial c/\partial x), \tag{1}$$

where F is the amount diffusing per unit of time, t; D is the diffusion coefficient; A is the area for diffusion; and $\partial c/\partial x$ is the concentration gradient. Since diffusion to a root is in radial coordinates, the appropriate equation becomes more complex. A simplified version of this equation, as given by Passioura (1963), is

$$F = - A[(C - C_o)Dk/r_o], \tag{2}$$

where C is the initial total concentration of the nutrient in the soil; C_o is the concentration at the root surface; r_o is the root radius; and k is a monotonically decreasing function of Dt/r_o^2, t being the time the sink has been operating.

The soil reacts chemically with many of the nutrients that diffuse to the root and also physically makes the diffusion pathway more tortuous.

The value of D in a soil reflects the reduction in rate of diffusion because of the chemical reaction and the increase in tortuosity because of physical factors. The effect of the various soil properties on D are discussed in section III B since those properties are important in determining the availability of nutrients that reach the root mainly by diffusion. The combined effects of mass-flow and diffusion are discussed in section II B.

B. Relation between Mass-Flow and Diffusion and Nutrient Uptake: Theoretical Investigations

1. Equations

Equations expressing the theoretical flow of ions by mass-flow and diffusion to a model consisting of a cylindrical sink in soil have been developed or modified by Bouldin (1961), Olsen *et al.* (1962), Passioura (1963), Nye and Spiers (1964), Gardner (1965), Nye (1966), Lewis and Quirk (1967), Passioura and Frere (1967), Gardner (1968), Kautsky *et al.* (1968), Olsen and Kemper (1968), and Nye and Marriott (1969). The discussion in this section is restricted to general relationships and a discussion of the results that have been predicted from the equations developed. The reader is referred to the listed references for a more complete mathematical treatment.

Uptake by plant roots is related to the concentration in the soil at the root surface. Relationships between uptake rate and concentration that have been developed from experiments with stirred or flowing culture solutions may be expressed by the relation

$$dM/dt = \alpha C_l, \qquad [3]$$

where dM/dt is the amount of ion M absorbed per unit area per unit time (moles cm^2 sec^{-1}), C_l is the concentration at the root surface in moles/cm^3, and α is a proportionality coefficient in cm^2/sec which has sometimes been called the *root absorption coefficient*. In equation [3] the value of α remains approximately constant over a limited range at low concentrations and decreases as the concentration in solution is increased.

The equation for the flow of a nutrient to a root of radius r cm growing in a soil from which it can extract both nutrients and water is

$$F = -D_e(\partial C/\partial r) + vC_l, \qquad [4]$$

where C is the total concentration of diffusible solute in moles/cm^3 of soil, C_1 is the concentration in the soil solution in moles per liter, r is the radial distance from the root axis in cm, v is the inward flux of water into the root in cm^3 cm^{-2} sec^{-1}, F is the inward radial flux of diffusible solute in moles cm^{-2} sec^{-1}, and D_e is the effective diffusion coefficient in cm^2/sec. The first term on the right side of the equation is the contribution from diffusion and the second term is from mass-flow.

The effect of diffusion and mass-flow on the concentration of a particular ion can be evaluated by integrating equation [4] with respect to the change in concentration at the root surface with time. The resulting equation (Nye and Marriott, 1969) is

$$\partial C_l/\partial t = (1/r)\,(\partial/\partial r)\,[rD_e(\partial C_l/\partial r) + (v_o r_o C_l/b)], \qquad [5]$$

where v_o is the inward flux of water at r_o, the radius of the root; and b is the buffering power, defined as dC/dC_l.

Several boundary conditions have been used in solving the equations for flow of nutrients to the plant root. These consist of (1) a constant concentration at the root surface (Olsen *et al.*, 1962), (2) a constant rate of absorption per unit area of root (Olsen *et al.*, 1962), (3) a concentration at the root decreasing exponentially with time (Lewis and Quirk, 1967), and (4) ion absorption proportional to the concentration in solution (Passioura, 1963; Nye, 1966; Olsen and Kemper, 1968). The most appropriate boundary condition is the latter, since with mass-flow contributing to supply, the concentration at the root could be either greater or less than that originally present.

The boundary conditions used by Nye and Marriott (1969) in solving the equation were

$$(1)\ t = 0, r > r_o, C_l = C_{lo},$$

where C_{lo} is the concentration in solution at the root surface, and

$$(2)\ t > 0,\ r = r_o,$$

$$D_e b(\partial C_e/\partial r) + v_o C_{lo} = \alpha C_{lo}/[1 + (\alpha C_{lo}/F\max)].$$

In the second boundary condition the rate of uptake is related by the Michaelis-type relation, which is

$$F = \alpha C_{lo}/[1 + (\alpha C_{lo}/F\max)]. \qquad [6]$$

Passioura and Frere (1967) have obtained solutions for a similar equation; however, in their boundary conditions they kept uptake equal to zero and evaluated the accumulation that may occur at the root when mass-flow is supplying at a greater rate than absorption.

2. Predictions

The effect of varying different parameters on the computer solution of equation [5] has been investigated (Passioura and Frere, 1967; Nye and Marriott, 1969). A number of assumptions were used since experimental data were not available to show the functional relationships. These include:

    a. $F_{\max}$ and α are independent of v_o.
    b. D_e is independent of v_o.
    c. The root is a uniform sink of constant radius.
    d. Neither α nor v_o vary with root age.
    e. D_e and b are independent of C.

The range of values for each of the parameters was based on the limited observations that have been reported. They were α, 0 to 2 ×

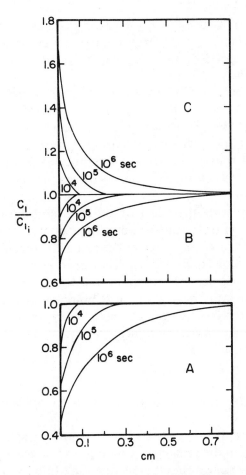

Fig. 18.1. Effect of rate of soil solution flow on the relative concentration near the root surface. $\alpha = 2 \times 10^{-7}$ cm/sec; $D = 10^{-7}$ cm^2/sec; $b = 0.2$; F_{max} high; $r_o = 0.05$ cm. *A*, $v_o = 0$, diffusion only; *B*, $v_o = 10^{-7}$ cm/sec; *C*, $v_o = 4 \times 10^{-7}$ cm/sec. The $Cl_i = Cl_o$ of text. (Nye and Marriott, 1969)

10^{-5} cm/sec; v_o, 0 to 2×10^{-6} cm^3 cm^{-2} sec^{-1}; and D_e, 2×10^{-9} to 2×10^{-7} cm^2/sec.

Calculations were made of C_l/C_{lo} with distance normal to the root surface.

An example of the effect of varying the rate of solution flow on the theoretical change in the relative ion concentration near the root surface with time is shown in Figure 18.1, taken from Nye and Marriott (1969). Values of v_o increasing from 0 to 4×10^{-7} are shown for varying times. The constant values used for the other parameters were α, 2×10^{-7}; D_e, 10^{-7}; b, 0.2; F_{max}, high; r_o, 0.05 cm.

The concentration of the solution at the root surface, relative to the initial soil solution, approached a limiting value of v_o/α with time.

Increasing v_o increased C_l/C_{lo} in a linear manner. Increasing α decreased C_l/C_{lo}. Where v_o was large enough to cause accumulation, increasing D_e reduced C_l/C_{lo} and caused the accumulated zone at the root to be more diffuse.

Passioura and Frere (1967) treat the case where no nutrient absorption occurs. They show that the C_l/C_{lo} level obtained at the root is a function of $r_o v_o/D_e$. Hence, increasing r_o or v_o causes an increase in the accumulation at the root, whereas increasing D_e reduces it. They predict that the level at the root may increase to as much as 20 times the initial level under conditions of a small D_e, such as occur at low moisture contents.

The equations relating the effect of diffusion and mass-flow to ion uptake and to the ion distribution about the root are useful in making theoretical predictions of the effect of changing the magnitude of each of the parameters involved. Experimental evidence is necessary to determine if the predictions are valid. Because the conditions present at the surface of a root growing in soil are more complex than those used in the equations referred to here, it is important to obtain quantitative information for each significant parameter so that it can be included in future mathematical models.

C. Predicting the General Significance of the Supply Mechanisms

The effect of the supply mechanisms can be approximated by calculating each separately and comparing it with the total absorption by the plant. This can be done to a root segment per unit time, or it can be done by evaluating the whole system at harvest. This latter averages the effects that may have varied during growth. To obtain a more detailed evaluation it would be necessary to measure the total plant parameters at a number of growth stages.

1. Calculation of Supply

a. Root Interception. Although the ions displaced by root interception may diffuse to the root, as indicated previously, we prefer to handle this effect separately since it will be an effect in addition to normally calculated diffusion and a portion may be absorbed as the root bumps into them. The amount intercepted is the amount of nutrient in a volume of soil equal to the root volume. As indicated earlier, this varies with species and soil conditions from 0.1% to 2.0% of the 0- to 15-cm soil layer. Hence, although roots only intercept a small portion of the total available nutrients in the soil, the significance depends upon a comparison of the amount intercepted with the amount required by the plant.

b. Mass-flow. The amount of nutrients moving by mass-flow toward the root at a point beyond where a concentration gradient is caused by diffusive flux is equal to the amount of water flowing to the root multiplied by the average concentration of the nutrients in this water. The quantity is independent of the extent of the root system providing the water uptake is not influenced by the extent of the roots. When diffusion also occurs (i.e., mass-flow does not supply ions to the root at a rate as great as absorption), there is an interaction between diffusion and mass-flow in the movement of the nutrients through the rhizosphere zone to the root, and the significance of mass-flow can be evaluated in two ways. The first and simplest is to determine the movement to the root rhizocylinder (the root plus that soil influenced by the root) by multiplying water use by its average nutrient content and comparing this with uptake. This is the approach used in the evaluation in section II C2. An alternative approach requiring much more information is to compare the amount supplied by diffusion in the absence of mass-flow with that supplied to the root by diffusion plus mass-flow in the presence of mass-flow. In this case, providing diffusion toward the root occurs, the extent of the root system would influence the results because an increase in the extent of the roots increases the relative supply by diffusion, and as a result as root extent increases the importance of mass-flow can be said to decrease.

In both approaches the assumption is made that uptake of ions and water occurs at the same locations on the root. This probably occurs on younger roots and may be less true the older the roots become. When a plant is actively growing, the rate of root growth increases exponentially; hence most of the roots are young.

The amount of water transpired by a crop varies but is in the range of 2 to 4 million kg/ha. The concentration in soil solution also varies. Measurement of the amount in a saturation extract or a displaced soil solution gives a measure of this quantity. Hydrodynamic dispersion may occur when ions move through soil with the flow of water, and this may alter the average concentration reaching the root. It is assumed that the same effect occurs during the slow displacement of the soil solution, so that the concentration found is a reasonable estimate of that which reaches the root by mass-flow. Barber *et al.* (1963) reported values for calcium, magnesium, potassium, and phosphorus obtained from analysis of the displaced extracts from 135 north central United States soils. They obtained median values for Ca, Mg, K, and P of 30, 25, 4, and 0.04 ppm, respectively. Multiplying these median values by three gives a reasonable idea of the kilograms per hectare that may be transported to the root by mass-flow on these soils.

c. Diffusion. In order to calculate supply by diffusion, we need to know the values for the concentration gradient about the root, the diffusion coefficient for the system, and the area of active roots. It is difficult to evaluate these for a crop growing in the field. In this evaluation we will calculate the diffusive supply by determining the difference between uptake and the supply by root interception and mass-flow.

2. Estimated Supply of Calcium, Magnesium, Potassium, and Phosphorus

The estimated supply of Ca, Mg, K, and P to a corn crop growing on a fertile silt loam soil is shown in Table 18.1. The soil was assumed to have 4,000, 800, 300, and 100 kg/ha of available Ca, Mg, K, and P in the top 15 cm of soil. The roots were all assumed to have grown in this layer. The uptake values are for a corn crop yielding 9,000 kg grain/ha and having a root volume equal to 1 % of the 0- to 15-cm soil layer.

The data in Table 18.1 indicate that the major portion of P and K supply reached the root by diffusion, whereas mass-flow could supply more Ca and Mg to the root than was required. Using soils with different levels of available and solution Ca, Mg, K, and P would alter the amounts estimated. However, for many soils the same general conclusions would be obtained.

D. Relation between Uptake and Measured Supply

1. Mass-flow

The amount of nutrient reaching a root by mass-flow can be increased either by increasing the average content maintained in solution or by increasing the amount of water transpired per gram of plant produced. The increase in solution content of a cation is usually increased when a salt is added, even though the salt does not contain the particular cation, since equilibration with the exchangeable cations in the soil occurs.

Table 18.1. Estimated supply of Ca, Mg, K, and P to corn roots by root interception, mass-flow, and diffusion

Ion	Crop uptake (kg/ha)	Supply (kg/ha) by Root interception	Mass-flow	Diffusion
Ca	45	40	90	—
Mg	35	8	75	—
K	110	3	12	95
P	30	1	.12	28.9

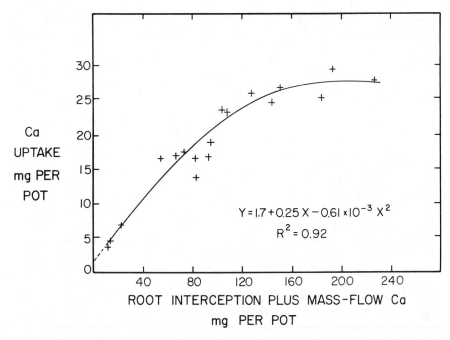

Fig. 18.2. Relation between Ca uptake by soybeans and the sum of the Ca intercepted by the root and the amount moving to the root by mass-flow

a. Altering the Calcium Content of Solution. The reliability of using mass-flow calculations to predict availability of a nutrient to a plant is best tested using a variety of soils varying in quantities of the particular ion in solution yet at a level in which mass-flow is the dominant supply mechanism. Al Abbas and Barber (1964) conducted greenhouse experiments with soybeans in six different soils, three of which had additional treatments of $CaCO_3$, CaO, MgO, and $CaSiO_4$. The results in Figure 18.2 show the correlation between the amount supplied by mass-flow plus root interception and uptake. When mass-flow was used alone, only five of the soils showed a correlation. Uptake from Chalmers silt loam, a soil much higher in exchangeable Ca, had approximately 50% higher Ca uptake values than the other soils at the same level of supply by mass-flow. This shows that either root interception or diffusion was having an effect in this experiment. However, increasing the calculated supply by mass-flow where root interception did not vary appreciably increased uptake as expected.

b. Altering the Rate of Transpiration. If the rate of transpiration per gram of plant is increased, the quantity of ions moved to the root by mass-flow is increased. This creates a higher concentration near the root surface and influences uptake by the plant. Under high-salt conditions increasing the transpiration increases translocation within the plant with

Table 18.2.   Effect of increasing transpiration on the average uptake of Ca and Mn by four plant species as related to changes in calculated mass flow

	Effects of rate of transpiration	
	106 ml/g	444 ml/g
Plant weight (g/pot)	2.28	2.38
Ca uptake (meq/pot)	0.91	2.22
Mass-flow Ca (meq/pot)	0.47	1.94
Mn uptake (μeq/pot)	21.8	28.9
Mass-flow Mn (μeq/pot)	4.7	19.5

Note: All values are the means of three replications of four species, tomatoes (*Lycopersicon esculentum*), soybeans (*Glycine max*), lettuce (*Lactuca sativa*), and wheat (*Triticum aestivum*).

a subsequent increase in uptake of ions; however, with low-salt conditions the transpiration rate has little influence on the uptake of nutrients (Russell and Shorrocks, 1959).

The effects of increasing mass-flow on the relative uptake of calcium and manganese are shown in Table 18.2. Increasing mass-flow of Ca increased uptake greatly since mass-flow supplied a large portion of the Ca reaching the root. Increasing mass-flow had a smaller effect on Mn uptake, presumably because diffusion supplied a greater portion of Mn to the root. Oliver and Barber (1966) have shown similar results for Ca and K where increasing transpiration, and consequently mass-flow, increased Ca uptake by a factor of 2 but had little effect on K in a situation where most of the K reached the root by diffusion.

2. Diffusion

The diffusive flux of an ion to the plant root depends upon (1) the diffusivity of the ion in the soil, (2) the concentration gradient along which diffusion occurs, and (3) the area through which diffusion occurs. Soil and root properties greatly influence the magnitude of all these factors in any particular soil-crop situation.

In this section I discuss the relation of each parameter to uptake with selected supporting experimental evidence. The effect of soil and root parameters on the magnitude of the diffusion coefficient is discussed in section III B.

a. Effect of the Diffusion Coefficient on Uptake. There should be a rather direct relation between the size of D_e and the rate of uptake, as illustrated by equation [4]. The apparent or effective diffusion coefficient, D_e, in a soil can be determined in the transient-state system by measuring the movement of the ion in question from one block of soil across a known area into a second untreated block of soil. Place and Barber (1964) measured D_e for rubidium at three levels of exchangeable Rb each at four levels of soil moisture. Increases in both moisture content and Rb concentration caused an increase in D_e. The uptake of Rb from each of these soil systems was measured by growing a corn root through a 3-cm band of soil in order to have the same area of root contact for each treatment. Moisture and Rb level also increased uptake, as they did D_e. When they compared D_e and Rb uptake, they obtained the relation shown in Figure 18.3, indicating a very high linear relation between D_e and uptake.

Effect of D_e on supply by diffusion can also be shown by comparing D_e for Rb diffusion on a number of soils with uptake from these soils by corn roots, as Evans and Barber (1964) have done. They obtained a high

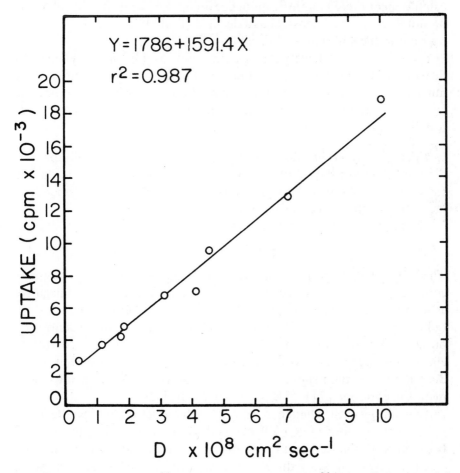

Fig. 18.3. Relation between ^{86}Rb uptake by corn and D_e for ^{86}Rb in soil. D_e was varied by varying moisture and Rb levels.

degree of correlation ($r^2 = 0.97$) for the relation between D_e and uptake, as would be anticipated.

b. Effect of Root Area on Uptake of Ions Supplied by Diffusion. The effect of area should be directly proportional to uptake, providing the nutrient is in limited supply. Evans and Barber (1964), in an experiment on the effect of area of root exposed to a labeled soil, found that after the first day doubling the root area essentially doubled uptake.

A question that has not been answered is, What effect does reducing the total root system by one half have on the uptake of ions diffusing to the root? It is likely that the reduction in uptake will not be proportional to the reduction in root length unless the nutrient demand per unit root length by the top growth is very large. Barley (1970) presented a theoretical evaluation of the effects and used the data of Cornforth (1968) to illustrate that for the mobile nitrate ion, root density within the range usually encountered has little influence on uptake, while for the relatively immobile P ion, root density has a decided effect on the proportion of the nutrient in the soil that is utilized.

The effect of root density on nutrient uptake is influenced by the plant demand for the nutrient, the nutrient level in the soil, the mobility of the nutrient in the soil, and the root radius. There is very little experimental evidence evaluating the significance of root density under conditions usually found in the field.

c. Effect of Concentration Gradient. The concentration gradient is directly related to the supply by diffusion, as indicated by equation [4]. In addition, the diffusion coefficient is usually concentration dependent, increasing as the concentration increases.

The soil parameters affecting D_e are

$$D_e = D_l \theta f_l \gamma (\Delta C_l / \Delta C), \qquad [7]$$

where D_l is the diffusion coefficient in water, θ is the volumetric moisture content, f_l is the tortuosity factor, γ is a factor accounting for negative adsorption and water viscosity effects, and $\Delta C_l / \Delta C$ accounts for the buffering-capacity effect. The principal effect of changes in concentration on D_e is due to changes in $\Delta C_l / \Delta C$ with changes in C. Concentration effects may occur as C is depleted near the root in a situation where diffusion is the main supply mechanism. The effect of the various parameters in equation [7] is discussed in Section III B.

An example of the change in $\Delta C_l / \Delta C$ with reduction in C can be found in the investigations on K by Beckett (1964). He determined the relation between K in solution and the K on the exchange sites and found that when K in solution was reduced below 1×10^{-4} M, the relation between K in solution and K on the exchange sites became curvilinear. This curvilinearity was such that $\Delta C_l / \Delta C$ became smaller, and hence the effective rate of diffusion is less than predicted from a linear extrapolation. For ions that equilibrate with the soil in this manner, a reduction of concentration near the root would cause a reduction in D_e and consequent reduction in flux. This would create a problem in calculation of diffusive flux to the root unless the appropriate adjustments are added to the model.

d. Calculated via Measured Diffusion of Nutrients to the Plant Root. As indicated in section II B1, equations have been developed relating flux of nutrients to the plant root; hence, if all of the necessary parameters are known, the flux to the root can be calculated. Brewster and Tinker (1970) measured the flux of potassium, calcium, magnesium, and sodium to leek (*Allium porrum*) roots growing in large covered pots set in the field. They calculated the theoretical change in concentration of each nutrient at the root surface with time. Since it is difficult to measure the actual concentration at the root surface, calculated changes in concentration cannot be compared with actual changes to determine how well values predicted from the equation agree with observation.

The relation between nutrient flow to the plant root and the accumulation or depletion that actually occurs has been shown by using autoradiographic procedures. For example, Lavy and Barber (1964) calculated the supply of molybdenum by mass-flow to the root and measured its uptake by the plant. Their calculations of depletion or accumulation about the root corresponded with observations on the autoradiographs.

E. Autoradiographic Evidence of Plant Root Effects

Autoradiographic procedures to study the effects of plant roots on ion distribution in the soil were first used by Walker and Barber (1962) to show the effect of corn roots on the distribution of ^{86}Rb in the soil. The procedure involved growing plants in a uniformly labeled soil in a special container. The container was made with one side covered with thin mylar film and supported by a hinged Plexiglas door. The autoradiograph was made by placing X-ray film between the hinged door and the mylar film while the container was in complete darkness. The results obtained for several nutrients illustrate the effect of the root on ion distribution about the root due to diffusion and mass-flow mechanisms.

1. Mass-Flow

a. Sulfate. Barber *et al.* (1963) used sulfate labeled with ^{35}S to demonstrate the movement of this ion to corn roots. The autoradiograph in Figure

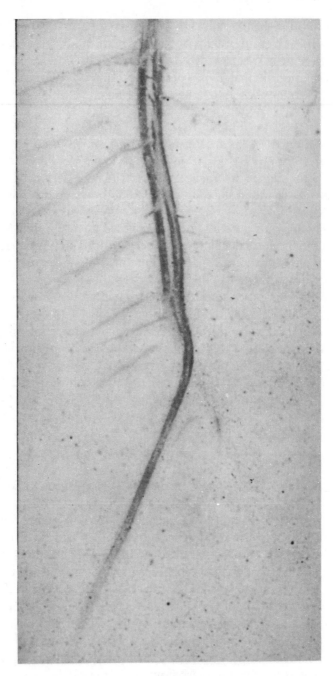

Fig. 18.4. Autoradiograph showing accumulation of ^{35}S-labeled sulfate about a corn root as a result of mass-flow. (Reprinted by permission from S. A. Barber, J. W. Walker, and E. H. Vasey, *J. Agr. Food Chem.* 11: 204–7, © 1963, American Chemical Society)

18.4 shows the effect of the corn root on ^{35}S distribution. The dark area represents ^{35}S accumulation about the corn root. The large amount but narrow zone of accumulation at the root surface area appears to be precipitation of $CaSO_4$. A comparable experiment with wheat, using a soil solution strength of sulfate of 1×10^{-4} M, gave similar results. Cutting the root did not cause the ^{35}S to diffuse rapidly away from the root, indicating that it was bound tightly.

b. Strontium. Corn roots were grown in agar in a system containing labeled ^{90}Sr (Barber, 1962). The ^{90}Sr accumulated about the root in the same manner as ^{35}S had in the experiments reported in section II E1*a* (Barber, 1962). In this case the solution contained 2×10^{-3}M sulfate, 5×10^{-3}M Ca, and a level of Sr determined by the contamination present in the salts used. When the root was cut, Sr did diffuse away, but the rate was much slower than what would occur if the Sr was present as ions in solution.

c. Calcium. Movement of Ca to plant roots by mass-flow causes accumulation to occur when mass-flow exceeds uptake rate. Barber and Ozanne (1970) showed that this occurred when they grew ryegrass (*Lolium rigidum*) and subterranean clover (*Trifolium subterraneum* L.) in a soil uniformly labeled with ^{45}Ca. They grew the plants from seed in a sandy soil with an exchangeable Ca content of 0.62 meq/100 g. The soil solution contained 2.5×10^{-3}M Ca. The calculated supply of Ca to the root by mass-flow was greater than uptake for both ryegrass and subterranean clover. The autoradiograph obtained is shown in Figure 18.5. The observed accumulation on the autoradiograph agreed with the accumulation calculated with transpiration and soil solution data.

Barber and Ozanne (1970) also grew lupine (*Lupinus digitatus*) and capeweed (*Arctotheca calendula*). The autoradiograph showing the effect of these species on Ca distribution in the same soil is shown in Figure 18.6. Lupine absorbed a large amount of Ca, contained a large amount within its roots, and caused a depletion pattern in the soil near some of its roots. This depletion indicates that Ca uptake exceeded supply mass-flow and diffusion occurred. The calculated results indicated that Ca supplied by mass-flow was less than uptake and hence agreed with the observations in the autoradiograph. The autoradiograph indicated capeweed caused little effect on the distribution. The calculated supply by mass-flow was only slightly greater than uptake; hence these results are also in agreement.

The autoradiograph results for these four species agreed with the calculated supply by mass-flow. The order of accumulation evident on the autoradiographs was ryegrass > subterranean clover > capeweed > lupine. This was the same order for the values of calculated mass-flow

Fig. 18.5. Autoradiograph showing accumulation of [45]Ca-labeled Ca about ryegrass (*left*) and subterranean clover (*right*) roots as a result of mass-flow

minus uptake. These data give evidence to support the use of calculations to determine ion distribution about the plant root.

2. Diffusion

When mass-flow does not supply as much of an ion to the root as the root absorbs, a reduction in concentration at the root surface occurs. This should be evident on autoradiographs.

a. Rubidium. Autoradiographs with ^{86}Rb show depleted areas about corn roots resulting from diffusion (Walker and Barber, 1962). In the auto-radiograph shown in Figure 18.7, the light areas are the depleted areas

Fig. 18.6. Autoradiograph showing depletion of ^{45}Ca-labeled Ca about lupine roots (*left*), where mass-flow was less than uptake, and little change about capeweed roots (*right*), where mass-flow approximately equaled uptake

about the corn roots growing through the soil. The width of the depleted area corresponds to the width predicted for an apparent diffusion coefficient of 2×10^{-8} cm^2/sec. The change in the concentration in the soil can be followed by taking sequential autoradiographs, as illustrated by Walker and Barber (1962). The concentration near the root surface can be estimated by including standards, made up of the same soil mixed with varying levels of radioactivity, in the same box. The linearity between concentration and density of darkening of the film depends upon the amount of exposure. For the areas in Figure 18.7, with the exception of dark areas of the root tips, we have found that the relation is approximately linear. This indicates that the concentration level at the root in this case was less than one-tenth the original.

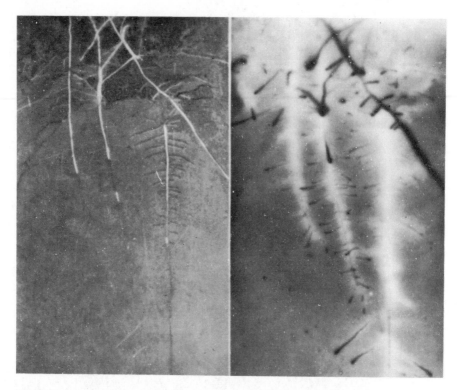

Fig. 18.7. Autoradiograph showing depletion of ^{86}Rb-labeled Rb about corn roots resulting from diffusion of Rb in soil. (Reprinted by permission from S. A. Barber, J. W. Walker, and E. H. Vasey, *J. Agr. Food Chem.* 11:204–7, © 1963, American Chemical Society)

b. Phosphorus. Lewis and Quirk (1967) and Barber *et al.* (1963) have obtained autoradiographs of ^{32}P depletion about plant roots. Phosphorus diffuses more slowly in soil because of the higher buffer capacity of the soil for P. Lewis and Quirk used a very sandy soil and got depletion throughout the root hair zone. On a silt loam soil Barber *et al.* observed a much narrower depletion zone.

c. Molybdenum. Lavy and Barber (1964) published autoradiographs of both accumulation of [99]Mo about the root due to mass-flow and depletion with a concentration gradient showing supply by diffusion. When the soil solution Mo content was less than 4×10^{-8}M, a diffusion pattern developed; when the content in solution was larger than this concentration, accumulation due to mass-flow occurred. In one instance a diffusion pattern occurred in an area of low moisture and accumulation occurred at the zone of watering on the same root, indicating the variation within the root zone that differences in moisture content may cause.

d. Calcium. When uptake exceeds the supply by mass-flow, depletion may occur near the root. This was shown for ^{45}Ca in Figure 18.6, where a diffusion pattern of Ca developed about lupine roots.

e. Zinc. Wilkinson *et al.* (1968) published autoradiographs of ^{65}Zn diffusion to wheat. They got similar autoradiographs whether they used high or low transpiration, indicating that even at high transpiration, mass-flow was not able to supply the Zn needs of the plant.

F. Visual Evidence of Plant Root Effects

It is possible to observe visually the effects of the plant root on the ion distribution in the soil. Bidwell *et al.* (1968) observed ferro-manganese pedotubules around bromegrass and big bluestem (*Andropogon gerardii*) in Kansas soils. The pedotubules were 3 to 15 mm in diameter. They contained 12.6% and 15.5% Fe and 2.02% and 1.68% Mn on brome-grass and big bluestem, respectively. These pedotubules apparently were formed by the movement of Ca, Fe, and Mn to the root by mass-flow in larger quantities than were absorbed by the roots and by precipitation of these ions at that location.

Barber (unpublished data, 1968) has observed pedotubules in sandy Australian soils near the coast line. The pedotubules were not around roots when observed but showed patterns that indicated they had once been around roots. Figure 18.8 indicates a pedotubule.

There may be several ways in which these pedotubules develop. We postulate that they can form when mass-flow moves ions in excess of up-take to the root surface. As we shall see in section IV, the root may increase the pH and supply carbonate that could cause precipitation of $CaCO_3$.

III. SOIL PROPERTIES INFLUENCING NUTRIENT SUPPLY RATES TO THE ROOT

A. Mass-Flow

Two factors influence the amount supplied by mass-flow: the rate of transpiration of the plant and the ion content of the solution flowing to the root. The ion content of the solution may increase because of addition of soluble anions such as nitrate or chloride to the soil. The cations balancing these anions are not necessarily the ones added since they equilibrate with the exchangeable cations. Hence, mass-flow of many cations is increased when soluble salts are added to the soil.

Fig. 18.8. Pedotubules of CaCO₃ believed to have formed about roots by Ca flowing to the root by mass-flow faster than Ca was absorbed by the root

B. Diffusion

The rate of diffusion of an ion from the soil to the root surface is dependent upon a number of chemical and physical properties of the soil. These include volumetric moisture, soil bulk density, soil buffering capacity, concentration, the counterdiffusing ion, and the diffusion mechanism that is operating.

1. Volumetric Water Content

Increasing the water content reduces the tortuosity of the diffusion path and increases the cross-sectional area for diffusion. Hence increasing the moisture level in the soil usually increases the rate of diffusion. Increasing the moisture may also influence the distribution of ions between the solid and the solution phases. Highly soluble nonadsorbed ions decrease in concentration in solution, whereas solution ions in equilibrium with adsorbed ions may stay the same.

The factors affecting the rate of diffusion in soil were given in equation [7], repeated here:

$$D_e = D_l \theta f_l \gamma (\Delta C_l / \Delta C). \qquad [7]$$

Increasing θ should directly increase the rate of diffusion providing it does not alter factors such as $\Delta C_l / \Delta C$ or γ. An example of the effect of increasing θ or D_e for Zn from data of Warncke and Barber (1971a) is shown in Figure 18.9. On four of the six soils there was an increase in D_e with increasing moisture, but on two soils there was very little effect. The latter soils were high in solution Zn, and increasing θ apparently reduced $\Delta C_l / \Delta C$ so that D_e remained approximately constant.

2. Soil Bulk Density

The solid particles of soil create a tortuous path along which an ion must diffuse to reach the root; hence, this reduces the rate of diffusion. Changing the soil bulk density affects the tortuosity of the diffusion path as well as influencing the effect of buffering capacity and moisture content.

The rate of diffusion usually increases when bulk density is increased up to a maximum value; beyond this the diffusion rate decreases rapidly with further increases in bulk density. Warncke and Barber (1971b) found in a study with six silt loam soils that the tortuosity as measured by ^{36}Cl diffusion was least at a bulk density of 1.3 g/cm^3. At a constant moisture content, measured by weight, increasing the soil compaction increased the volumetric moisture content, and this increased the cross-sectional

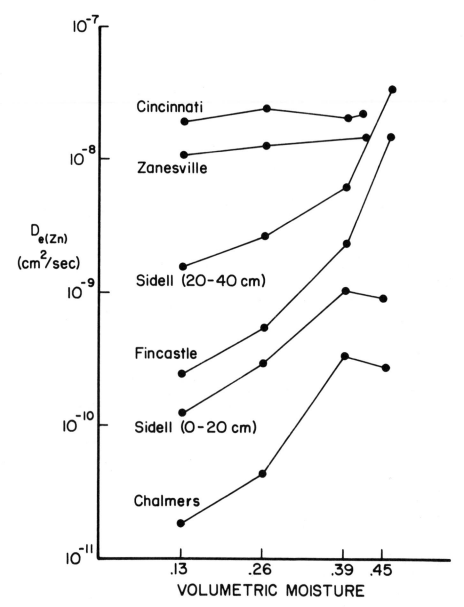

Fig. 18.9. Effect of increasing the volumetric water content on Zn diffusion in six soils

area for diffusion. It also reduced the tortuosity because it increased the continuity of the water and physically reduced the length of the diffusion path. However, above a bulk density of 1.3 g/cm³, the solid particles came close enough together to begin to make the diffusion path more tortuous.

The effect of bulk density on the diffusion coefficient for Zn in six soils, obtained by Warncke and Barber (1971b), is shown in Figure 18.10. The

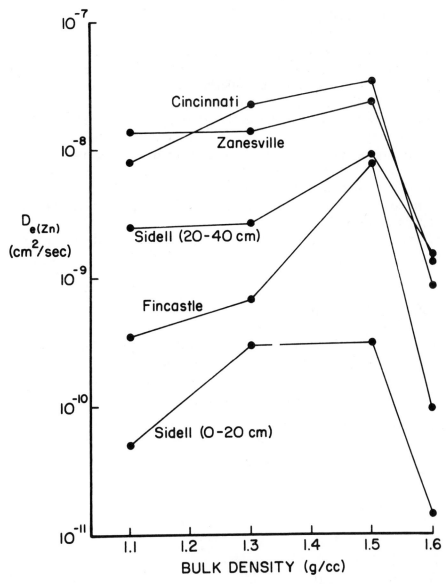

Fig. 18.10.   Effect of increasing soil bulk density on Zn diffusion in six soils

values are for a moisture content of 20% (w/w). The greatest diffusion coefficient occurred at a bulk density of 1.5 g/cm³. However, it was very much less when the bulk density was increased to 1.6 g/cm³. Since tortuosity, as indicated by the measurement of ^{36}Cl diffusion, was greatest at 1.3, the increase in Zn diffusion when the bulk density was increased from 1.3 to 1.5 was apparently due to a reduction in the interaction of Zn with the soil.

3. Buffering Capacity

The buffering capacity of the soil reduces the diffusion coefficient as compared to that in water. It is frequently the factor that has the greatest effect on D_e. The amount that buffering capacity reduces the diffusion coefficient is $\Delta C_l/\Delta C$. This ratio may change with C; so it is necessary to determine the relation and use the appropriate value. The value of $\Delta C_l/\Delta C$ may be as small as 0.001 for a strongly adsorbed ion such as P and as large as 1.0 for a relatively nonadsorbed anion such as chlorine. The value of $\Delta C_l/\Delta C$ varies with soil and for exchangeable cations. It also varies with the soluble anion level of the soil solution. Moisture level in the soil may also affect $\Delta C_l/\Delta C$, depending on how it affects the equilibrium between C_l and C.

An increase in concentration of an ion in the soil may increase $\Delta C_l/\Delta C$ and make D_e larger.

4. Effect of Concentration

Increasing the concentration of an ion in the soil usually increases D_e. Figure 18.11 shows how an increase in ^{86}Rb increased the ^{86}Rb diffusion

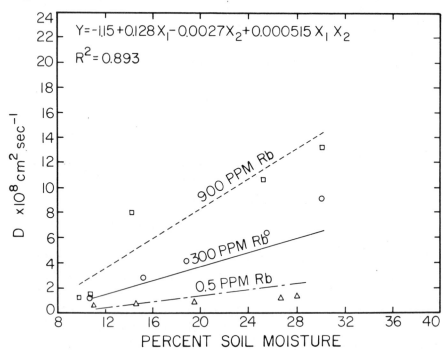

Fig. 18.11. Effect of Rb concentration and moisture content on the Rb diffusion coefficient in soil

coefficient. The increase that occurs may be attributed to changes in $\Delta C_l / \Delta C$. Concentration does influence D_l in pure water systems, but to a much smaller degree than a change in concentration in the soil.

Parsons (1959) gives values of D_l at various concentrations. For $CaCl_2$ at 1.05 mM, D_l was 1.248×10^{-5} cm^2/sec, and at 5.01 mM D_l was 1.179×10^{-5} cm^2/sec.

The amount of ion supplied by diffusion increases as the concentration gradient is increased. Hence adding an ion to the soil directly increases the amount diffusing to a sink such as the root because of both an increase in the concentration gradient and an increase in D_e.

5. Effect of the Counterdiffusing Ion

When ions diffuse they either diffuse as a cation-anion pair (salt diffusion) or as counterdiffusion. In counterdiffusion, an ion such as H^+ may diffuse in one direction while Ca diffuses in the opposite direction. The rate of diffusion of an ion depends upon the rate of diffusion of the counter ion. With a slowly diffusing counter ion, D_e is less than it would be with a more rapidly diffusing counter ion. Mathematical relationships for the effect of counter ions have been developed (Low, 1962).

Table 18.3. Values of D_e for diffusion in soils

Ion	D_e(cm^2/sec)	References
Na$^+$	1×10^{-5}	Vaidyanathan and Nye (1968)
NH$_4^+$	$0.4 - 3 \times 10^{-7}$	Clarke and Barley (1968)
K$^+$	2.3×10^{-7}	Vaidyanathan and Nye (1968)
Rb$^+$	$2 - 1.0 \times 10^{-7}$	Place and Barber (1964)
	$0.3 - 64.5 \times 10^{-7}$	Evans and Barber (1964)
	$3.9 - 24.3 \times 10^{-7}$	Phillips and Brown (1964)
Ca^{2+}	3.28×10^{-7}	Vaidyanathan and Nye (1968)
	$3.2 - 7.4 \times 10^{-8}$	Elgawhary (unpublished data, 1971)
Zn^{2+}	$3.1 \times 10^{-10} - 2.66 \times 10^{-8}$	Warncke and Barber (1971a)
	$1 \times 10^{-8} - 2 \times 10^{-7}$	Clarke and Graham (1968)
Mn^{2+}	$3.3 - 22.0 \times 10^{-8}$	Halstead and Barber (1968)
NO$_3^-$	$0.5 - 5.0 \times 10^{-6}$	Clarke and Barley (1968)
Cl$^-$	$0.27 - 13.5 \times 10^{-6}$	Warncke and Barber (1971b)
	$0.5 - 5.7 \times 10^{-6}$	Porter *et al.* (1960)
H$_2$PO$_4^-$	$1 - 3.8 \times 10^{-9}$	Olsen and Watanabe (1963)
	$2 - 4 \times 10^{-11}$	Vasey and Barber (1963)
	9.8×10^{-11}	Vaidyanathan and Nye (1968)
	$2 \times 10^{-10} - 4 \times 10^{-8}$	Lewis and Quirk (1967)
MoO$_4^{2-}$	$4.6 - 51.0 \times 10^{-8}$	Lavy and Barber (1964)

6. Effect of the Ion

The effect of the soil on D_e varies with each ion. The tortuosity factor is similar for most ions; however, the degree of reaction of the ion with the soil or $\Delta C_l/\Delta C$ varies widely with the ions. Values for D_e in soils that have been reported in the literature are given in Table 18.3. Many values have also been reported for clay minerals. They are usually larger than those reported here for soils.

7. Exchange Diffusion vs. Solution Diffusion

Soils have two phases, the solution phase containing soluble salts and the solid phase, which has exchangeable cations associated with it. Although the exchangeable cations are in the solution, they are balanced by negative charges on the solid phase so that they are only in the solution near the solid phase surface. Diffusion in soil is frequently calculated as diffusion in the solution phase only. The ions in the solution phase equilibrate with those associated with the solid phase, and this is the buffering capacity that influences diffusion. We can also calculate the diffusion as being in both the solution and the solid phases. When there is a relatively high salt level, most diffusion probably occurs through solution. However, when the ion content in solution is very low, diffusion of cations on the solid phase, sometimes called *exchange diffusion*, may occur.

Bole and Barber (1971) used the Sr : Ca ratio to determine the phase in which diffusion was occurring to plant roots. They used both a soil and an exchange resin as media. In each the Sr : Ca ratio on the exchange phase was different from the ratio in solution due to differential adsorption of Sr and Ca. When diffusion supplied most of the Sr and Ca to the root, uptake was in the ratio of the exchangeable Sr and Ca. This indicated that exchange diffusion was important in supplying Sr and Ca in this system. When mass-flow supplied most of the Sr and Ca, uptake was similar to the ratio in solution. This would be expected where the Sr and Ca were moving to the root in solution.

IV. EFFECT OF ROOT EXUDATES ON THE SOIL AND ON ION ABSORPTION

A. Inorganic Exudates

The inorganic ions released by the root are in exchange for the cations and anions absorbed into the root. Significant release occurs when cation and anion absorption are unequal. When more cations are absorbed than anions, H^+ is released by the root to balance the charge. When more anions are adsorbed than cations, OH^- or HCO_3^- are released to balance the charge. The amount of release is proportional to the disparity between cation and anion uptake.

The degree of disparity between cation and anion uptake varies with nitrogen nutrition, plant species, and stage of plant growth.

1. Effect of Nitrogen Nutrition

Nitrogen can be supplied to the plant root in either the ammonium or the nitrate form. Since the ammonium form is the cation form, supplying all the N in the ammonium form causes cation uptake to greatly exceed anion uptake and H to be released. When this is done in soils, the soil pH near the root is reduced. If it is done in nutrient culture, the pH of the media drops. When N is supplied in the nitrate form, anion uptake usually exceeds cation uptake, and HCO_3^- is released, which increases the pH.

The degree to which the form of N alters cation vs. anion absorption is shown in the unpublished data of Ozanne and Barber in Table 18.4.

The ryegrass was grown in nutrient culture, and the solution was adjusted to pH 6.5 daily by addition of either HCl or KOH. The amounts

Table 18.4. Effect of form of nitrogen on the uptake of cations and anions by ryegrass

	Uptake (meq/pot)	
Ion	NH_4^+	NO_3^-
Ca^{2+}	0.20	0.46
Mg^{2+}	0.30	0.52
K^+	1.76	1.87
Na^+	0.03	0.03
NH_4^+	5.59	—
Total cations	7.88	2.88
SO_4^{2-}	0.97	0.33
Cl^-	0.45	0.17
$H_2PO_4^-$	0.37	0.34
NO_3^-	—	4.72
Total anions	1.79	5.56

Note: Values are means of three replicates.

of HCl or KOH added agreed reasonably well with the calculated differences in uptake of cations vs. anions.

2. Effect of Plant Species

The nutrient composition of plants varies with species. Hence, plants growing in the same media differ in their relative uptake of cations vs. anions and in the degree to which they influence the media in which they are growing.

   The results obtained by Ozanne and Barber (unpublished data) can be used to compare uptake by capeweed with that by ryegrass. For plants grown in nutrient solution with nitrate as the N source, the cations and anions taken up per pot by ryegrass were 3.05 and 6.16 meq, respectively. The comparable values for capeweed were 3.73 and 5.66 meq. Theoretically ryegrass would raise the pH about the roots more than capeweed, and this occurred.

3. Changes in Soil pH from Differential Plant Cation-to-Anion Absorption

Riley and Barber (1971) have shown that soybeans supplied with ammonium N can reduce the pH of the rhizocylinder (root plus rhizosphere soil) by 0.5 pH units in a silt loam soil which had an initial pH of 5.2. The same system with nitrate as the N source raised the pH of the rhizocylinder by 1.4 pH units. The pH of the rhizocylinder changes, while the pH of the soil not influenced by H^+ or HCO_3^- released by the root remains constant. Because of this change, plant roots growing in soil may have a pH at their interface which is very different from the average pH of the soil.

B. Organic Exudates

Plant roots may release organic materials as well as inorganic ions. These are discussed in Chapter 6. The effect of organic ions released by roots on diffusion or mass-flow is related to their effect on ion solubility. Organic molecules may complex metal ions such as iron, manganese, copper, and zinc. This complexing may increase the amount of the ion in the solution phase relative to that associated with the solid phase. This increases the amount reaching the root by both mass-flow and diffusion. The larger complex may diffuse more slowly in water; however, the $\Delta C_l/\Delta C$ factor in soil is much larger, so that the rate of diffusion from the soil to the root will be increased. Hodgson *et al.* (1966) have shown that most of the Cu in

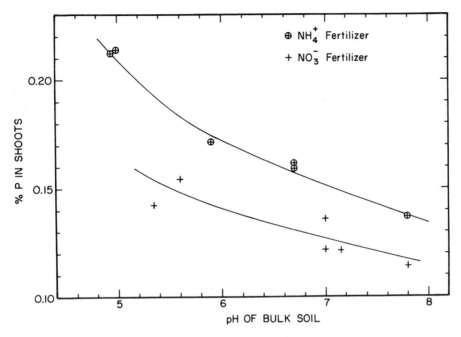

Fig. 18.12. Relation between the percentage P in soybean plants and the pH of the bulk soil and the form of N used

soil solution was in organic complexes; hence, the natural complexes in the soil are important in influencing Cu availability. Elgawhary *et al.* (1970) have shown that adding chelates to the soil system increases the diffusive flow of chelated metals to a cylindrical sink.

Chaney *et al.* (1972) have shown that Fe-stressed plants release a reducing agent which reduces Fe at the root surface so that it can be absorbed into the plants.

C. Influence of Rhizocylinder pH Changes on Soil Nutrient Availability

1. Phosphorus

Riley and Barber (1971) have demonstrated the effect of rhizocylinder pH on P availability for soybeans growing in a Chalmers silt loam soil. They grew soybeans on soil limed to four pH levels and added two forms of N, NH_4^+ vs. NO_3^-. On this soil, reducing the pH increased P solubility. After growing the soybeans for three weeks, they measured the P content of the plant, the pH of the rhizocylinder, and the pH of the bulk soil. Yield was not affected significantly by soil pH or form of N.

The P content of the soybean top vs. the bulk soil pH is plotted in Figure 18.12. There are separate correlations for NH_4^+ and NO_3^--N. At a

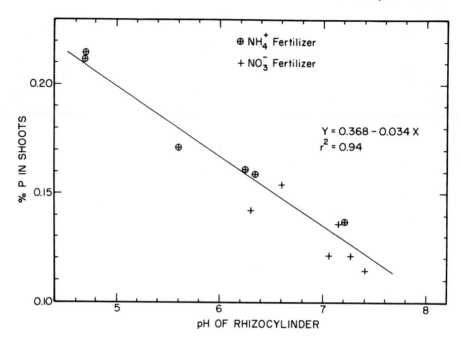

Fig. 18.13. Relation between the percentage of P in soybean plants and the pH of the rhizocylinder and the form of N used

similar bulk soil pH, the NH_4^+ form of N increased P content as compared to NO_3^-. This effect has been observed by many investigators. However, plots of the P content of the soybean top vs. rhizocylinder pH (Fig. 18.13), revealed only one relation. The effects of NH_4^+ and NO_3^- treatments were similar. These results indicate that the soybean root altered P availability in the soil by changing the pH at the root-soil interface.

2. Boron

In the experiment of Riley and Barber just referred to, boron content of the soybean plant was determined (Barber, 1971). There was a variation in B which was related to soil pH. When B content was plotted vs. the pH of the bulk soil (Fig. 18.14), there was a separate relation for each of the NH_4^+ and NO_3^- treatments. Ammonium N increased B uptake as compared with NO_3^--N. However, when the pH of the rhizocylinder was plotted vs. the B content of the plant (Figure 18.15), the data for NH_4^+ and NO_3^- fitted the same regression line. Again, the data indicate that the soybean plant modified B availability in the soil by changing the pH. The changes in pH used in this experiment were due to the form of N supplied to the plant.

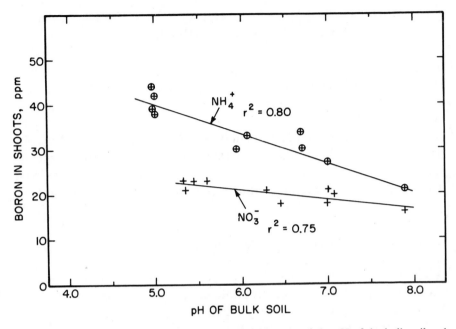

Fig. 18.14. Relation between the B content of soybeans and the pH of the bulk soil and the form of N used

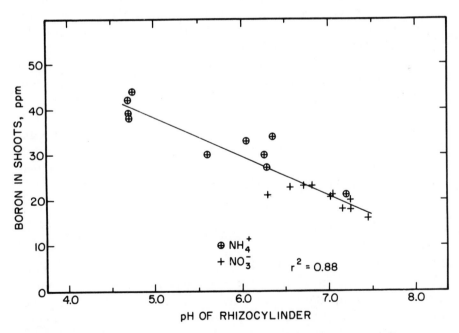

Fig. 18.15. Relation between the B content of soybeans and the pH of the rhizocylinder and the form of N used

V. EFFECT OF ION ACCUMULATION AT THE ROOT ON NUTRIENT AVAILABILITY

When mass-flow supplies more nutrients to the root than the root absorbs, these nutrients accumulate. Their increased concentration may influence the availability of other nutrients that reach the root mainly by diffusion.

The effect of accumulated salts may be due to (1) a change in the concentration of an ion in solution where the ion's concentration depends upon the solubility of the solid phase; (2) exchange of adsorbed ions from the soil; and (3) an effect on the rate of absorption of an ion into the root.

A. Effect of Calcium Accumulation on Phosphorus Uptake

Calcium frequently reaches the root in excessive amounts, especially when the plant absorbs small amounts so that its Ca uptake rate is low. As indicated earlier, this Ca may remain as solution Ca near the root, or it may precipitate as calcium sulfate or carbonate.

Hoffmann and Barber (1971) investigated the possibility that Ca accumulation may reduce P uptake by the plant. They investigated four different soils and compared P uptake under conditions of accumulation with P uptake where little Ca accumulation occurred. The amount of Ca reaching the root was changed by changing the transpiration rate and by changing the amount of Ca in the soil solution.

The results of their experiments, shown in Table 18.5, indicate that in soils containing soluble calcium phosphate as the source of soluble P to the plant, increasing the soluble Ca reduced P uptake significantly at high transpiration rates. However, in the acid soils, where the available P was in the form of iron and aluminum phosphates, increasing the Ca did not influence P uptake. Low transpiration rates that caused less Ca accumulation also resulted in smaller differences in plant P content between the treatments receiving and those not receiving $CaSO_4$.

B. Effect of Calcium Accumulation on Manganese Uptake

Barber (1968) reported on the influence of Ca accumulation on Mn uptake. In an experiment with four different species, the species varied in their rate of Ca uptake but were similar in their rate of Mn uptake. Because the species differed in their rate of Ca uptake, Ca accumulated outside the root on the low Ca absorber, wheat, and was depleted about the roots of the high Ca absorber, tomato (*Lycopersicon esculentum*). As a result he was able to investigate the relation between this Ca accumula-

Table 18.5. Effect of $CaSO_4$ and transpiration on yield and P content of the shoot plus root of wheat grown in soil cultures

	Yield (g/pot)		% P	
	Low tran-spiration	High tran-spiration	Low tran-spiration	High tran-spiration
Raub − $CaSO_4$	1.42	1.25	0.57	0.59
+ $CaSO_4$	1.37	1.25	0.56	0.55
Genesee − $CaSO_4$	1.68	1.48	0.51	0.54
+ $CaSO_4$	1.70	1.47	0.49	0.49
Fincastle − $CaSO_4$	1.45	1.35	0.28	0.32
+ $CaSO_4$	1.44	1.33	0.27	0.31
Limed Fincastle − $CaSO_4$	1.62	1.48	0.31	0.31
+ $CaSO_4$	1.52	1.52	0.30	0.30
LSD (.05)	—	0.105	—	0.026

tion and the accumulation or depletion of Mn. The data in Figure 18.16 show a direct correlation between the two. The results are interpreted to indicate that when Ca accumulates outside the root, it reduces Mn solubility, and as a result Mn also accumulates. The ability of the species used to absorb Mn actively over a very wide concentration range indicates that the Mn accumulation is not due to the inability of the root to absorb the Mn that is at its surface.

VI. SUMMARY AND CONCLUSIONS

The plant root can greatly alter the physical, chemical, and biological nature of the soil adjacent to the root. This soil represents only a small percentage of the soil in the root media. However, it is the soil that determines the rate of supply of nutrients as well as toxic ions to the root surface. Measurement of the average conditions in the total soil present does not reflect the changes occurring at the root.

A large number of factors are involved in the changes that occur in this rhizosphere zone. Plant species, soil, climate, and fertilization are the major factors. However, each of these has a variety of effects. Species differ in root morphology and extent, in amount of nutrients absorbed, and in the amount of H^+ or HCO_3^- released. Soils differ in the level of nutrients, physical nature, buffering capacity, and moisture-holding capacity, all of which influence the supply of nutrients to the root, as well as affecting root morphology and extent. Climate affects the relative amount of root growth, the rate of nutrient absorption, and the rate of

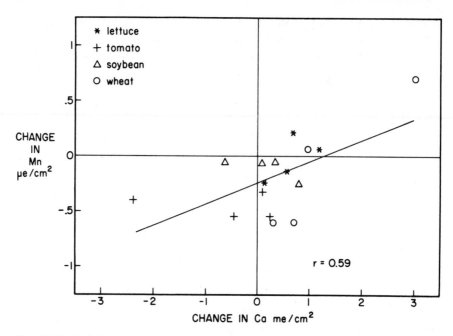

Fig. 18.16.   Relation between the calculated change in Ca concentration at the root surface and the calculated change in Mn concentration for four species growing in soil

transpiration per unit of nutrient absorbed. Fertilization affects the root environment by altering the salt content of the soil solution, changing the concentration of nutrients, and altering the balance between cation and anion absorption, which affects the H^+ or HCO_3^- released. In addition there is undoubtedly an interaction of all these factors with the little-known effects of the biological nature of the root environment.

In order to get a more accurate measure of the soil as a suitable environment for root growth, we need to investigate fully the micro-environment of the plant root. Measurement of equilibrium values on the total soil may not reflect the dynamic conditions occurring around the plant root.

References

Al Abbas, A. H., and S. A. Barber. 1964. The effect of root growth and mass-flow on the availability of soil calcium and magnesium to soybeans in a greenhouse experiment. *Soil Sci.* 97: 103–7.

Barber, S. A. 1962. A diffusion and mass-flow concept of soil nutrient availability. *Soil Sci.* 93: 39–49.

——. 1968. On the mechanisms governing nutrient supply to plant roots growing in soil. *Trans. 9th Int. Congr. Soil Sci.* 2: 243–50.

——. 1971. The influence of the plant root system in the evaluation of soil fertility. *Proc. Int. Symp. Soil Fertility Evaluation* (New Delhi) 1 : 249–56.

Barber, S. A., and P. G. Ozanne. 1970. Autoradiographic evidence for the differential effect of four plant species in altering the Ca content of the rhizosphere soil. *Soil Sci. Soc. Amer. Proc.* 34: 635–37.

Barber, S. A., J. W. Walker, and E. H. Vasey. 1963. Mechanisms for the movement of plant nutrients from the soil and fertilizer to the plant root. *J. Agr. Food Chem.* 11: 204–7.

Barley, K. P. 1968. Deformation of the soil by the growth of plants. *Trans. 9th Int. Congr. Soil Sci.* 1: 759–68.

——. 1970. The configuration of the root system in relation to nutrient uptake. In *Advances in Agronomy*, ed. N. C. Brady, pp. 159–201. New York: Academic Press.

Beckett, P. 1964. Potassium-calcium exchange equilibria in soils: specific adsorption sites for potassium. *Soil Sci.* 97: 376–83.

Bidwell, O. W., D. A. Gier, and J. E. Cipra. 1968. Ferromanganese pedotubules on roots of *Bromus inermis* and *Andropogon Gerardii. Trans. 9th Int. Congr. Soil Sci.* 4: 683–92.

Bole, J. B., and S. A. Barber. 1971. Differentiation of Sr-Ca supply mechanisms to roots growing in soil, clay and exchange resin cultures. *Soil Sci. Soc. Amer. Proc.* 35 : 768–72.

Bouldin, D. R. 1961. Mathematical description of diffusion processes in the soil-plant system. *Soil Sci. Soc. Amer. Proc.* 25: 476–80.

Brewster, J. L., and P. B. Tinker. 1970. Nutrient cation flows in soil around plant roots. *Soil Sci. Soc. Amer. Proc.* 34: 421–26.

Chaney, R. L., J. C. Brown, and L. O. Tiffin. 1972. Obligatory reduction of ferric chelates in iron uptake by soybeans. *Plant Physiol.* 50 : 208–13.

Clarke, A. L., and K. P. Barley. 1968. The uptake of nitrogen from soils in relation to solute diffusion. *Aust. J. Soil Res.* 6: 75–92.

Clarke, A. L., and E. R. Graham. 1968. Zinc diffusion and distribution coefficients in soil as affected by soil texture, zinc concentration and pH. *Soil Sci.* 105: 409–18.

Cornforth, J. S. 1968. Relationships between soil volume used by roots and nutrient accessibility. *J. Soil Sci.* 19: 291–301.

Dittmer, H. J. 1940. A quantitative study of the subterranean members of soybean. *Soil Conserv.* 6: 33–34.

Elgawhary, S. M., W. L. Lindsay, and W. D. Kemper. 1970. Effect of complexing agent and acids on the diffusion of zinc to a simulated root. *Soil Sci. Soc. Amer. Proc.* 34: 211–14.

Evans, S. D., and S. A. Barber. 1964. The effect of rubidium-86 diffusion on the uptake of rubidium-86 by corn. *Soil Sci. Soc. Amer. Proc.* 28: 56–57.

Gardner, W. R. 1965. Movement of nitrogen in soil. In *Soil Nitrogen*, ed. W. V. Bartholomew and F. E. Clark, pp. 550–72. Agronomy 10. Madison, Wis.: Amer. Soc. Agron.

———. 1968. Nutrient transport to plant roots. *Trans. 9th Int. Congr. Soil Sci.* 1: 135–41.

Greacen, E. L., D. A. Farrell, and B. Cockroft. 1968. Soil resistance to metal probes and plant roots. *Trans. 9th Int. Congr. Soil Sci.* 1: 769–79.

Halstead, E. H., and S. A. Barber. 1968. Manganese uptake attributed to diffusion from soil. *Soil Sci. Soc. Amer. Proc.* 32: 540–42.

Hodgson, J. F., W. L. Lindsay, and J. F. Trierweiler. 1966. Micronutrient cation complexing in soil: II. Complexing of zinc and copper in displaced solution from calcareous soils. *Soil Sci. Soc. Amer. Proc.* 30: 723–26.

Hoffmann, W. F., and S. A. Barber. 1971. Phosphorus uptake by wheat (*Triticum aestivum*) as influenced by ion accumulation in the rhizocylinder. *Soil Sci.* 112: 256–62.

Kautsky, J., K. P. Barley, and D. K. Fiddaman. 1968. Ion uptake from soils by plant roots, subject to the Epstein-Hagen relation. *Aust. J. Soil Res.* 6: 159–67.

Lavy, T. L., and S. A. Barber. 1964. Movement of molybdenum in the soil and its effect on availability to the plant. *Soil Sci. Soc. Amer. Proc.* 28: 9–97.

Lewis, D. G., and J. P. Quirk. 1967. Phosphate diffusion in soil and uptake by plants. IV. Computed uptake by model roots as a result of diffusive flow. *Plant Soil* 26: 454–68.

Low, P. F. 1962. Effect of quasicrystalline water on rate processes involved in plant nutrition. *Soil Sci.* 93: 6–15.

Lund, Z. F., and H. O. Beals. 1965. A technique for making thin sections of soil with roots in place. *Soil Sci. Soc. Amer. Proc.* 29: 633–34.

Nye, P. H. 1966. The effect of nutrient intensity and buffering power of a soil, and the absorbing power, size and root hairs of a root, on nutrient absorption by diffusion. *Plant Soil* 25: 81–105.

Nye, P. H., and F. H. C. Marriott. 1969. A theoretical study of the distribution of substances around roots resulting from simultaneous diffusion and mass-flow. *Plant Soil* 30: 459–72.

Nye, P. H., and J. A. Spiers. 1964. Simultaneous diffusion and mass-flow to plant roots. *Trans. 8th Int. Congr. Soil Sci.* (Bucharest) 3: 535–44.

Oliver, S., and S. A. Barber. 1966. An evaluation of mechanisms governing the supply of Ca, Mg, K, and Na to soybean roots (*Glycine max*). *Soil Sci. Soc. Amer. Proc.* 30: 82–86.

Olsen, S. R., and W. D. Kemper. 1968. Movement of nutrients to plant roots. In *Advances in Agronomy*, ed. A. G. Norman, pp. 91–149. New York: Academic Press.

Olsen, S. R., W. D. Kemper, and R. D. Jackson. 1962. Phosphorus absorption by corn roots as affected by moisture and phosphorus concentration. *Soil Sci. Soc. Amer. Proc.* 25: 289–94.

Olsen, S. R., and F. S. Watanabe. 1963. Diffusion of phosphorus as related to soil texture and plant uptake. *Soil Sci. Soc. Amer. Proc.* 27: 648–53.

Parsons, R. L. 1959. *Handbook of Electrochemical Constants*. New York: Academic Press.

Passioura, J. B. 1963. A mathematical model for the uptake of ions from the soil solution. *Plant Soil* 18: 225–38.

Passioura, J. B., and M. H. Frere. 1967. Numerical analysis of the convection and diffusion of solutes to roots. *Aust. J. Soil Res.* 5: 149–59.

Pavylechenko, T. K., and J. B. Harrington. 1934. Competitive efficiency of weeds and cereal crops. *Can. J. Res.* 10: 77–94.

Phillips, R. E., and D. A. Brown. 1964. Ion diffusion: II. Comparison of apparent self and counter diffusion coefficients. *Soil Sci. Soc. Amer. Proc.* 28: 758–63.

Place, G. A., and S. A. Barber. 1964. The effect of soil moisture and rubidium concentration on diffusion and uptake of rubidium-86. *Soil Sci. Soc. Amer. Proc.* 28: 239–43.

Porter, L. K., W. D. Kemper, R. D. Jackson, and B. A. Stewart. 1960. Chloride diffusion in soils as influenced by moisture content. *Soil Sci. Soc. Amer. Proc.* 24: 460–63.

Riley, D., and S. A. Barber. 1971. Effect of ammonium and nitrate fertilization on phosphorus uptake as related to root-induced pH changes at the root-soil interface. *Soil Sci. Soc. Amer. Proc.* 35: 301–6.

Russell, R. S., and V. M. Shorrocks. 1959. The relationship between transpiration and the absorption or inorganic ions by intact plants. *J. Exp. Bot.* 10: 301–16.

Vaidyanathan, L. V., and P. H. Nye. 1968. The measurement and mechanism of ion diffusion in soils. II. An exchange resin paper method for measurement of the diffusive flux and diffusion coefficient of nutrient ions in soils. *J. Soil Sci.* 17: 175–83.

Vasey, E. H., and S. A. Barber. 1963. Effect of placement on absorption of ^{86}Rb and ^{32}P from soil by corn roots. *Soil Sci. Soc. Amer. Proc.* 27: 193–97.

Walker, J. M., and S. A. Barber. 1962. Uptake of rubidium and potassium from soil by corn roots. *Plant Soil* 17: 243–59.

Warncke, D. D., and S. A. Barber. 1971a. Diffusion of Zn in soils: I. The influence of soil moisture. *Soil Sci. Soc. Amer. Proc.* 36: 39–42.

——. 1971b. Diffusion of Zn in soils: II. The influence of soil bulk density and its interaction with soil moisture. *Soil Sci. Soc. Amer. Proc.* 36: 42–46.

Wilkinson, H. F., J. F. Loneragan, and J. P. Quirk. 1968. The movement of zinc to plant roots. *Soil Sci. Soc. Amer. Proc.* 32: 831–33.

19.   Effects of Soil Calcium Availability on Plant Growth

Charles D. Foy

I. EFFECTS OF CALCIUM ON PLANT STRUCTURE AND FUNCTION

A. Calcium Deficiency Symptoms

ALTHOUGH this volume is emphasizing the effects of various factors on root growth, the interdependence of roots and tops makes it necessary to include some discussion of top growth and the translocation of various ions from root to top. Calcium deficiency in plant tops has been variously characterized by marginal chlorosis, blackening, curling and necrosis of apical leaves, petiole collapse, flower abscission, ovule collapse, and poorly developed seeds (Hewitt, 1963; Millikan and Hanger, 1964; Jackson, 1967; Nightingale and Smith, 1968). Specific physiological disorders associated with Ca deficiency include tipburn of lettuce (*Latuca sativa* L.) and cabbage (*Brassica oleracea*), dark plumule of peanuts (*Arachis hypogaea* L.), petiole collapse in soybeans (*Glycine max* L., Merrill), blossom-end rot (BER) in tomatoes (*Lycopersicon esculentum* Mill) and peppers (*Capsicum* spp.), blackheart of celery (*Apicum graveolens*), and rolling of terminal leaves in wheat (*Triticum aestivum* L.) and barley (*Hordeum vulgare* L.) (Walker *et al.*, 1961; Hewitt, 1963; Maynard *et al.*, 1965; Harris and Brolmann, 1966; Foy *et al.*, 1969a; Geraldson, 1970; Long and Foy, 1970).

Because Ca tends to become immobilized in older plant tissues (Loneragan and Snowball, 1969a, 1969b), the symptoms of deficiency generally appear first in the young, meristematic parts of plants, but in some instances the petioles of older leaves collapse before those nearer the apex of the plant. In such cases the newly absorbed Ca apparently

Contribution from the Northeastern Region, Agricultural Research Center, Agricultural Research Service, U.S. Department of Agriculture, Beltsville, Md. 20705.

bypasses the older tissue and translocates to younger plant parts (Millikan and Hanger, 1964).

Calcium deficiency reduces growth of root tips, root laterals, and root hairs (Jackson, 1967). In general, roots grown for 4 to 5 days without Ca develop a swollen, stubby, and spatulate appearance. Hypocotyl necrosis in snapbean (*Phaseolus vulgaris*) seedlings in artificial growth media and in acid, low Ca soil has been attributed to Ca deficiency (Shannon *et al.*, 1967). Although Ca is often considered especially important in root growth, Ca deficiency sometimes reduces the growth of plant tops more than that of roots (Jackson and Evans, 1962; Clark, 1970).

B. Effects of Calcium on Cell Division and Extension

Calcium deficiency in meristematic roots of peas (*Pisum sativum*) has been associated with disturbed cell division (as reflected by polyploid and constricted nuclei), binucleate cells, and susceptibility to invasion by fungi (Sorokin and Sommer, 1940). Calcium-deficient roots in these studies were further characterized by the appearance of translucent extensions of the main root tips, destruction of cytoplasm, resulting in cell vacuolation, and reduced cell elongation and differentiation.

The process of cell division apparently has a small but specific requirement for Ca which is associated with the differentiation of organelles in the cytoplasm (Burstrom, 1968). In wheat roots Ca is required at 0.04 ppm for cell division, 0.4 ppm for cell elongation, and still higher concentrations for the detoxification of the hydrogen ion. Burstrom and Tullin (1957) found that in wheat roots the addition of 10^{-5} M EDTA, which removed Ca by chelation, inhibited mitotic activity but that the subsequent addition of Ca at the same concentration restored mitosis to normal levels. Iron and manganese could not replace Ca in this role. Cell elongation, which had been inhibited by a Ca-deficient growth medium, was also restored by a solution containing 0.4 ppm Ca (Burstrom, 1952). Calcium deficiency has also been associated with chromosome abnormality (Steffenson, 1958; Hewitt, 1963).

The role of Ca in cell wall stability and auxin action is still debated (Jones and Lunt, 1967; Burstrom, 1968). Ginzburg (1961) concluded that the cell wall is cemented by pectins and cross-linked macromolecules, with Ca as a stabilizing agent, but iron and copper can apparently replace Ca in this role. One basis for questioning the Ca pectate theory of cell wall structure is the recent observation that apparently healthy roots contain extremely low levels of Ca (Jones and Lunt, 1967). Bennet-Clark (1956) proposed that auxin promoted the extension of cell walls by removing Ca from Ca pectate. However, more recent evidence indicates that indoleacetic acid (IAA) does not affect the binding, distribution, or loss of Ca by cell walls of oat (*Avena sativa*) coleoptiles (Cleland, 1960; Burling and Jackson, 1965). The actions of Ca and auxin appear to be independent (Ray and Baker, 1965; Burstrom, 1968).

C. Effects of Calcium on Membrane Structure and Function

Many investigators have noted the beneficial effects of Ca on the integrity of plant membrane structures such as the nuclear envelope, plasmalemma, mitochondria, and Golgi apparatus (Marinos, 1962; Marschner and Overstreet, 1966; Galey *et al.*, 1968). For example, Ca protects plant cells against leakage of potassium and other ions induced by excess H ions (Marschner and Overstreet, 1966; Mengel, 1968) and reverses membrane damage brought about by the removal of Ca (and RNA) with EDTA (Hanson, 1960; Hermann, 1964; Van Steveninck, 1965). However, other ions such as strontium, manganese, aluminum, magnesium and barium can at least partially substitute for Ca in its membrane building and/or stabilizing role (Hermann, 1964; Van Steveninck, 1965). It has been suggested that Ca binds negative charges of the plasmalemma to the cell wall (Van Steveninck, 1965), stabilizes RNA protein in cytoplasmic membranes (Foote and Hansen, 1964), and binds to nucleoproteins (Masuda, 1959).

D. Effects of Calcium on Ion Uptake

Under certain conditions Ca increases K uptake by excised roots (Viets, 1944), counteracts the inhibitory effects of the H ion on K uptake (Jacobson *et al.*, 1961), and reduces the leakage of K and other ions from roots (Marschner and Overstreet, 1966). However, Hooymans (1966) reported that Ca may inhibit, stimulate, or have no effect on K uptake, depending upon the K concentration and time. He found that initial K uptake was reduced by Ca but steady-state K uptake was unaffected. Mengel (1968) found that Ca additions lowered the exchangeable K level of oat roots but did not affect the nonexchangeable K fraction. Leggett and Gilbert (1967) observed that soybean roots generally accumulate more Mg than K from solutions containing equivalent concentrations of the two ions but the addition of Ca to the medium reversed this accumulation ratio. Epstein (1961) found that 0.04 ppm Ca was required for roots to maintain their selective uptake of K over sodium. Foote and Hanson (1964) suggested that a Ca-nucleoprotein complex is involved in K uptake.

Calcium also affects the uptake of anions. For example, Foote and Hanson (1964) showed that Ca deficiency was associated with reduced uptake of phosphorus, nitrate, chlorine, and bromine, and Leggett and Epstein (1956) found that Ca increased the uptake of sulfate. Calcium and P uptake by corn (*Zea mays*) mitochondria are mutually dependent (Hodges and Hanson, 1965); however, Sr and Mg will partially substitute for Ca in promoting such P uptake (Truelove and Hanson; 1966; Kenefick and Hanson, 1966). Calcium promotes inorganic P accumulation by roots as well as by mitochondria. In this role Ca seems to act by diverting high-energy P from ATP formation into inorganic P uptake. Kenefick and Hanson (1966) postulated that Ca reacts with an intermediate to form an unstable CaXP complex which degrades, delivering Ca and inorganic P into the matrix and releasing X for subsequent coupled electron flow. Hodges and Elzam (1967) provided evidence that in corn mitochondria a common high-energy intermediate of oxidative phosphorylation provides the energy for either Ca ion transport or ATP formation in both plant and animal mitochondria. Leggett *et al.* (1965) attributed Ca-stimulated P uptake in excised barley roots to an increased turnover rather than a change in concentration of the rate-limiting intermediate. For more specific information on the role of Ca in mitochondrial structure and function, see recent reviews by Lehninger (1970) and Green and Young (1971).

Robson *et al.* (1970) found that in flowing culture solutions containing Ca and P concentrations within the range found in soils, increasing the Ca concentrations markedly increased P absorption by several annual legumes. This effect was greater at low than at high P concentrations and also was greater for two *Medicago* species than for two cultivars of subterranean clover (*Trifolium subterraneum* L.). Such differential Ca-induced P uptake was suggested as a partial explanation for the greater sensitivity of the *Medicago* species to soil acidity and their greater tolerance to alkalinity when compared with subterranean clover (Robson and Loneragan, 1970; Trumble and Donald, 1938). In the studies of Robson *et al.* (1970), pretreatment with different Ca levels did not affect P absorption, but immediate increases were obtained by transferring plants to solutions containing different Ca concentrations. These investigators suggested that Ca increased P absorption directly by screening negative charges on roots, thus increasing the accessibility of absorption sites to P. Franklin (1969) drew a similar conclusion regarding the Ca-stimulated P uptake by excised roots of barley, corn, and soybeans; however, he observed that pretreatment of roots with Ca was effective in promoting P uptake and that the Ca-enhanced P uptake was reversed by the addition of K.

The older literature contains some suggestion of a link between calcium and boron in plant nutrition. For example, Smith (1944) found that about 50 % of the B and Ca in squash (*Cucurbita pepo*) leaf cells was immobilized in cell walls or intercellular substances. Both elements were readily extracted with dilute acid but not with dilute alkali. The distribution of B in cells of deficient, normal, and high-B leaves suggested that this element plays a role in the cytoplasm and cell wall but not in the chloroplasts or vacuoles. Boron and Ca contents and uptake were positively correlated in squash plants. However, Neales and Hinde (1962) concluded that B and Ca metabolism were not associated in the growth of broad bean (*Vicia faba*) roots. Addition of boric acid did not reduce the restriction of root growth imposed by Ca deficiency, and root growth restriction by B deficiency was not associated with reduced Ca uptake by the whole root or root tip. More recently Harris and Brolmann (1966) concluded that Ca deficiency is distinctly different from B deficiency in peanuts. Boron deficiency changed foliage characteristics, flowering pattern, shoot and root yields, and produced hollow heart in the fruit. In contrast, Ca deficiency seemed to affect only fruit yield and quality, particularly the vascular system at the base of the plumule, producing dark plumule. However, these investigators did postulate an interaction of B and Ca in flower production. Shear and Faust (1970) found that increasing the B supply in nutrient solutions or foliar sprays increased Ca accumulation in apple (*Pyrus malus*) trees, particularly when the Ca supply was low. Millikan and Hanger (1965) concluded that B deficiency impaired xylem function in subterranean clover. Parkhash and Subbiah (1964) reported that ^{45}Ca uptake was reduced by Cu and B deficiencies in citrus seedlings.

The role of metabolism in the uptake of Ca by plants is debatable. Moore *et al.* (1965) found that exudates from barley roots contained Ca concentrations 58 times those in the external medium and considered this to be evidence of a metabolically mediated process. At ambient Ca concentrations of 100 ppm, the rate of Ca accumulation in the exudate was 35 times that in the bulk of the excised roots. They concluded that most root cells were not active in Ca absorption and that the endodermis was the barrier against ion accumulation. Leggett and Gilbert (1967) found that a major portion of the Ca absorbed by excised soybean roots was associated with less than 10 % of the root volume and was located almost entirely in the epidermal cell layer. They concluded that the active accumulation of Ca by these roots was negligible. Mengel (1961) showed that 95 % of the Ca absorbed by oat seedlings (from a solution containing 16 ppm Ca) was exchangeable and stated that binding sites for Ca were nonspecific and not connected with active cationic uptake. Drew and Biddulph (1969) concluded that Ca accumulation by kidney bean (*Phaseolus vulgaris* L.) roots was not directly controlled by metabolic processes but that the movement of Ca to plant tops was metabolic in nature.

E. Calcium in Relation to Enzyme Activities

Jones and Lunt (1967) presented a table of enzymes whose activities have been associated with Ca. Chrispeels and Varner (1967) found that Ca at 800 ppm was required for the activity of α-amylase in barley aleurone and suggested that Ca was involved in the maintenance of membrane functions and cytoplasmic organization. Jeffries *et al.* (1969) found that Ca ions changed the activity of malic dehydrogenase in liverwort (*Lemma minor*). They believed that the effect resulted from a change in configuration of the enzyme protein. Calcium appeared to stabilize the higher molecular aggregates that were associated with enzyme activity, but Na and K were ineffective in this regard. They further suggested that such effects could explain the adaptability of organisms to growth media varying in Ca concentration. They pointed out that the observed Ca-induced changes in enzyme activity could result from the production of a range of genetically distinct iso-enzymes having different responses to Ca or from the production of one enzyme having different sensitivities to Ca depending on the ionic environment under which it was synthesized. The pectinesterase enzymes have recently been reported to be activated by Ca and Na in both microorganisms and higher plants (Mayorga and Rolz, 1971). Dodds and Ellis (1966) found that adenosine triphosphates activity in plant cell walls, which is stimulated by K, is also dependent upon Ca and Mg.

F. Effects of Calcium on Rhizobia

Root infection or nodule initiation in subterranean clover apparently has a higher Ca requirement than either nodule development or growth of the host plant supplied with fixed N (Lowther and Loneragan, 1968). This high Ca requirement did not seem to be related to the survival or growth of the rhizobia or the effects of Ca on taproot length, root hair development, or lateral root initiation. Increasing the Ca concentration from 9.8 to 28.8 ppm had no effect on the growth of the host plant but increased the number of nodules 7% to 24%. Decreasing the Ca from 9.8 to 0.16 ppm progressively decreased both plant growth and nodule numbers. In later work with subterranean clover, Lowther and Loneragan (1970) obtained evidence of adequate root hair infection at low Ca concentrations, but the majority of the infections formed did not produce nodules except when transferred to a high Ca concentration. The Ca-sensitive stage was completed as early as 3 days after inoculation. These investigators also found an acid-sensitive state of 2 days' duration preceding the Ca-sensitive stage in subterranean clover. This stage could be modified by Ca concentrations in solution. For example, decreasing the Ca from 30 to 10 ppm at pH 5.0 increased the acid-sensitive stage to 3 days. However, even 400 ppm Ca did not affect the inhibition from acidity at pH 4.0, where no nodulation occurred (Loneragan and Dowling, 1958). At a Ca concentration of 0.4 ppm no nodules formed at any pH. Above these critical levels the nodulation could be increased to optimum levels by increasing either pH or Ca level. At pH 4.5 increasing the Ca level from 4 to 280 ppm increased numbers of root nodules from 0 to 10. At the high Ca concentration, maximum nodulation was obtained at pH 5.0, but with the lower Ca concentration a pH of 6.0 was required. These results help to explain the increased nodulation of clover obtained by pelleting seeds with lime and rhizobia mixtures (Loneragan *et al.*, 1955). In addition to the effects mentioned previously, Benath *et al.* (1966) suggested that Ca deficiency interferes with the rate of N reduction in the nodules of subclover, rather than with the export of reduced N.

II. SOIL FACTORS AFFECTING CALCIUM AVAILABILITY TO PLANTS

A. Calcium Contents of Soils

Although soils vary widely in Ca content (Lane and Sartor, 1966), absolute Ca deficiency is not generally a primary growth-limiting factor except in very sandy, acid soils having low cation exchange capacities (Adams and Pearson, 1967; Evans and Kamprath, 1970). In the absence of toxic factors, most acid soils could probably supply adequate Ca for most plants. For example, Howard and Adams (1965) concluded that Norfolk subsoil at pH 5.0 and Dickson subsoil at pH 4.6, as obtained from the field, contained adequate Ca for normal growth of primary cotton (*Gossypium* sp.) roots. The addition of soluble salts of Ca to acid soils either has no effect (Table 19.1) or may actually decrease plant growth (Fig. 19.1). Such detrimental effects are believed to result from the displacement of Al from cation exchange sites and a lowering of soil solution pH, which in turn increases the solubility and toxicity of Al in the Bladen soil shown. In other acid soils the decrease in soil pH, induced by the addition of Ca salts, can increase Mn toxicity (Foy, 1964). Calcium deficiency symptoms, only rarely observed in the field, are more likely due to Al-Ca or other ionic antagonisms than to a low absolute Ca level in the soil (Melsted, 1953; Foy *et al.*, 1969a; Long and Foy, 1970). However, peanuts and sugarcane (*Saccharum* sp.) appear to give more direct yield responses to Ca as a nutrient than to pH change (Colwell and Brady, 1945; Ayers, 1963). In both instances yields were closely related to the level of exchangeable Ca and not to changes in soil pH.

Table 19.1. Yield and composition of Chief soybean tops grown on Bladen soil treated with various fertilizer and liming materials

Treatment no.	Soil treatment†	Final soil pH	Yield of plant tops (g/pot dry wt)	Composition of leaves and petioles (meq/100 g)		
				Ca	Al	H_2PO_4
1	No lime	4.4	7.66 b‡	10.7 b	1.86 a	5.42 a b
2	$Ca(NO_3)_2 \cdot 4H_2O$	4.5	7.84 b	26.0 b	1.64 b	6.02 a
3	$CaSO_4 \cdot 2H_2O$	4.2	7.91 b	18.1 b	1.66 b	4.08 b
4	$CaCO_3$	5.3	13.82 a	61.0 a	1.19 c	5.66 a
5	$MgCO_3$	5.0	12.66 a	24.0 b	1.13 d	6.24 a

† *Key to treatments:* 1. Basal fertilizer only: 100, 109, and 137 ppm N, P, and K, respectively, added as NH_4NO_3 and KH_2PO_4. 2. Basal fertilizer plus $Ca(NO_3)_2 \cdot 4H_2O$ to supply 100 ppm N and 143 ppm Ca. 3. Basal fertilizer plus $CaSO_4 \cdot 2H_2O$ to supply 143 ppm Ca. 4. $CaCO_3$ at 3,000 ppm. 5. $MgCO_3$ equivalent in neutralizing value to 3,000 ppm $CaCO_3$.

‡ Within any vertical column any two values having a letter in common are not significantly different at the 5% level by the Duncan Multiple Range Test.

B. Soil pH

Direct effects of the H ion on Ca uptake and plant growth are difficult to study because at soil pH levels where the H ion is considered harmful to higher plants, Al, Mn, and perhaps other elements may be soluble in toxic concentrations. In addition, these elements are more harmful to plants at higher soil pH levels than is the H ion itself. In most acid soils (pH 4.2 or above), the harmful effects of low pH on Ca uptake and the growth of higher plants are therefore largely indirect. However, Lund (1970) found that soybean taproots in the nutrient solution portion of a split medium required higher Ca levels for optimum growth at pH 4.5 than at pH 5.6. Maas (1969) reported that Ca uptake by excised corn roots in nutrient solution was markedly reduced by H ion in the pH range of 3.0 to 5.0.

C. Soil Colloid Type and Calcium Saturation

The Ca saturation associated with good plant growth varies widely, depending upon the type of soil colloid and the plant. Allaway (1945)

Fig. 19.1. Effects of $CaSO_4$ and $CaCO_3$ on the growth of Hudson barley in acid Bladen soil. *Left to right*: 0, 100, 200, and 400 ppm Ca as $CaSO_4$ and 3,000 ppm $CaCO_3$. Final pH values (1:1 soil-water) were (*left to right*): 4.3, 4.1, 3.9, 3.8, and 5.5. Corresponding KCl-extractable Al levels were: 3.6, 3.5, 3.5, 3.4, and .18 meq/100 g. The soil initially had an NH_4OAc CEC of 12.0 meq/100 g and an exchangeable Ca level of 1.5 meq/100 g. (C. D. Foy, unpublished data)

reported that the availability of colloid-bound Ca to soybeans decreased in the order: peat, kaolin, illite, Wyoming bentonite, and Mississippi bentonite. Aluminum-coated montmorillonite seems to have a higher preference for binding Ca over Mg than does the original clay (Hunsaker and Pratt, 1971). In the same studies, allophane, an organic soil, and soils containing amorphous minerals or kaolinite plus gibbsite also showed higher Ca-binding preferences than expected in montmorillonitic clays and soils. Mehlich and Colwell (1944) concluded that in acid mineral soils the percentage of base saturation was more important than the total Ca present in determining the availability of Ca to plants.

The percentage of Ca saturation of soils at which plant growth is decreased has been reported to vary from 40 % to 4 %. Adams *et al.* (1967) found that in acid Norfolk, Magnolia, and Greenville subsoils, cotton yields were reduced at Ca saturations in the range of 30 % to 40 % and at Al saturations of 40 % to 60 %. Facteau and Eck (1970) concluded that a Ca saturation greater than 35 % was necessary for adequate growth of blueberries (*Vaccinium* sp.) in a bentonite-sand substrate. Henderson (1969) reported Ca deficiency in soybeans on Norfolk soil at a Ca saturation of 20 % or below, an Al saturation of 68 % or greater, and a Ca concentration in plant tops of 0.58 % or less. A Ca saturation of 12 % produced near-normal growth of sugarcane in volcanic soils of Hawaii (Mahilum *et al.*, 1970). Martin and Page (1969) found that a Ca saturation of only 4 % did not produce foliar symptoms or reduce growth of citrus in a sandy loam soil at pH 5.4.

Gonzalez-Erico (1968) reported that for acid soils containing a low exchange acidity, liming appeared to be more important for supplying Ca than for neutralizing exchangeable acidity. Calcitic lime applied at 500 ppm above the level required to neutralize twice the exchangeable acidity was optimum for supplying Ca to plants. Plant growth in these soils was closely related to exchangeable acidity and Ca saturation.

D. Ratios between Calcium and Other Cations in Soil Solutions

Howard and Adams (1965) found that in either nutrient solutions or displaced soil solutions from acid Norfolk and Dickson subsoils, a Ca: total-cation ratio of 0.10 to 0.15 was required for optimum growth of cotton roots. Bennett and Adams (1970) showed that this relationship was also valid for limed soils when soil solution cations were expressed as ionic activities rather than ionic concentrations. Lund (1970) reported that in nutrient cultures, soybean root elongation rate was highest when the Ca: Ca + Mg + K ratios were between 0.10 and 0.20. Low ratios of Ca: total cations were less detrimental when K was substituted for half of the Mg. Increasing the Ca level from 10 to 40 ppm decreased the injury caused by Al added at 0.5, 1.0, or 2.0 ppm. According to Gerald-son (1957), the ratio of Ca: total soluble salts in the soil solution must be maintained in the range of 0.16 to 0.20 to prevent BER in tomatoes. More recently Geraldson (1970) showed that in nutrient solutions Ca deficiency can be induced either by decreasing the Ca: total-soluble-salt ratio, at a given salt level, or by increasing the level of soluble salts at a given Ca: total-salt ratio. For example, 150 ppm Ca: 1,000 ppm total salts (15% Ca) produced normal celery plants, but increasing the total-salt concentration at the same equivalent proportion of Ca (15%) increased the incidence and severity of blackheart (Ca deficiency). Reducing the Ca: total-salt ratio to 5% at a fixed salt level (50 ppm Ca: 1,000 ppm total salts) also produced Ca deficiency in celery. Effects of K, Na, NH_4, and Mg ions on tomatoes at a given salt level, and with a Ca: total-salt ratio of 5%, are shown in Table 19.2. The NH_4 ion was

Table 19.2. Effect of partially replacing Ca with K, Na, NH_4, or Mg (on an equivalent basis) in the nutrient solution on tomato yield, incidence of blossom-end rot (BER), and Ca concentrations of leaves

Calcium ratio (ppm Ca/ppm salt)	Relative fruit yield (%)	% fruit with BER	% Ca in leaves
150/1,000	100.0	0	1.35
50/1,000 (K)	10.1	40	0.80
50/1,000 (Na)	57.4	16	0.95
50/1,000 (NH_4)	2.7	75	0.47
50/1,000 (Mg)	32.9	22	0.67

Source: Geraldson, 1970.

the most effective in decreasing fruit yields, increasing BER, and decreasing the percentage of Ca in the leaves. The K treatment decreased yields and increased BER more than Mg, and Mg more than Na. Gerard (1971) reported that increasing salinity levels increased the Ca requirement for elongation of cotton roots. Increased Ca levels reduced the detrimental effects of salinity, high temperature (32C–38C), and low temperature (21C) on root growth.

Magnesium toxicity (Ca-Mg imbalance) is apparently a growth-limiting factor in some serpentine soils, and the problem is reduced by the addition of Ca (Proctor, 1970). Magnesium saturations above 80% in a peat-sand mixture severely reduced the rooting of chrysanthemums (Paul and Thornhill, 1969), and Vlamis (1949) reported unbalanced Ca nutrition in barley and lettuce when soils were above 80% Mg-saturated. Lund (1970) found that low Ca: Mg ratios in nutrient solution were detrimental to soybean roots. In acid Nason and Tatum soils, containing high levels of exchangeable Al and extremely low levels of exchangeable

Fig. 19.2. Effects of CaSO$_4$, MgCO$_3$, and CaCO$_3$ on the growth of Hudson barley in acid Nason (*top*) and Tatum (*bottom*) soils. *Left to right*: No lime, 402 ppm Ca as CaSO$_4$, 840 ppm MgCO$_3$, and 750 ppm CaCO$_3$. Final pH values (1:1 soil-water) for Nason soil were (*left to right*): 3.8, 3.7, 4.1, and 4.2. Corresponding pH values for Tatum soil were: 3.8, 3.7, 4.2, and 4.2. Nason soil initially had a pH of 4.4, NH$_4$OAc CEC of 8.8 meq/100 g, 0.20 meq of NH$_4$OAc exchangeable Ca per 100 g, and 3.54 meq KCl extractable Al per 100 g. Corresponding values for Tatum soil were: pH 4.3, CEC-12.6, exch. Ca 0.05 meq/100 g, and 5.93 meq/100 g KCl extractable Al. Growth is limited by high Al, extremely low Ca, and the interactions of these two factors. (C. D. Foy, unpublished data)

Ca, the addition of $MgCO_3$ improved growth somewhat but also induced a Ca deficiency (leaf rolling) in Hudson barley (Fig. 19.2). Adding $CaCO_3$, instead of $MgCO_3$, gave greater increases in growth and prevented the Mg-induced Ca deficiency symptoms (Fig. 19.2). However, in an acid Bladen soil, which also contains a high level of exchangeable Al but much more exchangeable Ca than Nason or Tatum, $MgCO_3$ was as effective as $CaCO_3$ in preventing Al-induced Ca deficiency symptoms and increasing the growth of Chief soybeans (Fig. 19.3 and Table 19.1; Foy *et al.*, 1969a).

Fig. 19.3. Chief soybeans on acid Bladen soil. Aluminum-induced Ca deficiency (petiole collapse) is prevented and yield is increased by increasing the soil pH from 4.4 to 5.5 with either $CaCO_3$ or $MgCO_3$. Primary function of lime is to reduce solubility of Al and create a more favorable Ca:Al ratio. Absolute Ca level is not limiting. Original NH_4OAc CEC was 12 meq/100 g, and exchangeable Ca was 1.5 meq/100 g. (Foy *et al.*, 1969a)

Potassium is often antagonistic to Ca. Chaudhry *et al.* (1964) reported that a soil K saturation greater than 10 % reduced Ca uptake. Bussler (1962) recognized the importance of K:Ca ratios in cytoplasmic permeability. Gammon (1957) pointed out that high ratios of Na and K to Ca were more detrimental to root development than equivalent ratios of H to Ca. Nelson and Brady (1953) found that adding K to one portion of a ladino (*Trifolium repens*) root system decreased the Ca uptake from another portion of the same system. Jackson (1967) has noted the possible significance of this to plants growing in heterogeneous soil media.

Increased Mn uptake and toxicity (crinkle leaf) in Pima S-2 cotton was associated with reduced Ca concentrations in plant tops (Foy *et al.*, 1969b). Robson and Loneragan (1970) found that increasing the Ca concentration from 10 to 100 ppm in the nutrient solution decreased Mn toxicity in two annual *Medicago* species, primarily by decreasing Mn uptake.

In acid soils (below pH 5.5), Al-Ca antagonism is probably the most important factor affecting Ca uptake by plants. The drastic effects of Al on Ca are illustrated in studies of Lance and Pearson (1969), who showed that reduction in Ca uptake was the first externally observable symptom of Al damage in cotton seedling roots; the effect was noted within one hour after root exposure to a solution containing only 0.3 ppm Al. The inhibition of Ca uptake was avoided by increasing the Ca concentration of the nutrient solution to 600 ppm.

III. PLANT FACTORS AFFECTING CALCIUM AVAILABILITY AND USE

A. Calcium Movement in Plants

Many Ca deficiencies are actually due to poor Ca transport (Loneragan and Snowball, 1969a, 1969b). Calcium immobility is generally more pronounced in older plant tissues, but Wiebe and Kramer (1954) showed that Ca absorbed by the terminal 4 mm of barley roots (consisting of root cap and meristematic cells) was largely immobilized in those regions. Rios and Pearson (1964) found that downward transport of Ca in cotton plants was inadequate to support root growth in the Ca-deficient nutrient solution portion of a split medium.

Fong and Ulrich (1970) reported that Ca concentrations in White Rose potato (*Solanum tuberosum*) leaves increased with age, even under severe Ca deficiency, indicating that Ca was poorly transported. Calcium concentrations in the roots of Ca-deficient plants were twice those of Ca-deficient petioles. These investigators emphasized the importance of maintaining a continuous supply of Ca in the medium for good growth. In this connection, Millikan *et al.* (1969) concluded that the main factor affecting [45]Ca distribution in subterranean clover was the Ca concentration in the substrate. The percentage of Ca retained in the roots was greatest at the low Ca level and greater in plants supplied with NH_4-N than in those supplied with NO_3-N. Millikan and Hanger (1969) observed that [45]Ca applied to the middle leaves of brussels sprouts (*Brassica oleracia gemifera*) remained essentially immobile, even when applied in doses containing extra Ca, plus water, citric acid, malic acid, diphenylamine, dimethyl sulfoxide, or oxalic acid. However, injecting a combination of ^{45}Ca and citric acid or malic acid into the midrib of corresponding leaves promoted both acropetal and basipetal movement of the isotope. They reported movement in both xylem and phloem of plant stems.

Calcium deficiency (tipburn) in lettuce is associated with a temporary localized shortage of Ca (Thibodeau and Minotti, 1969). Foliar sprays of $Ca(NO_3)_2$ or $CaCl_2$ controlled the disorder and increased the Ca concentration of leaves fivefold. Foliar sprays of organic acid salts, particularly oxalate (which precipitates Ca), increased the rate of development and severity of tipburn.

Many factors affect the transport of Ca in plants. Copper deficiency in wheat is characterized by Ca deficiency in the youngest leaves and an accumulation of Ca in the older leaves immediately below them (Brown and Foy, 1964). Thus, Cu seems to play a part in the normal transport of Ca to the growing point. Brown (1965) found that P accentuated Cu deficiency in barley and wheat and decreased the translocation of Ca into the upper leaves of Cu-deficient plants. Copper-sufficient plants maintained normal Ca levels in their upper leaves, even in the presence of high P.

Jacoby (1966) found that Ca absorbed by the intact roots of beans was distributed throughout the plant but Ca absorbed by cut surfaces of similar but derooted plants was retained in the basal stem segments. Under the same conditions Cl, SO_4, K, and an acid dye (fuchsin) were absorbed by derooted plants and distributed throughout the plant. Additional studies with fuchsin indicated that passing Ca ions through plant roots increased their transport in the xylem vessels. He first suggested that this resulted from binding of these ions with organic chelating agents in the roots but from a later study (Jacoby, 1967) concluded that bean roots promoted the transport of Ca by supplying previously accumulated cations that ascended with the plant sap and displaced Ca bound on sites in the stem. Shear and Faust (1969) noted that Ca transport in apple stems was brought about by a nonspecific ion exchange; any divalent cation could free Ca for movement. They later (1970) concluded that Ca moves by exchange on lignin to areas of high metabolic activity in growing tissues and accumulates in vascular tissues of older plant parts. Ammonium-N increased Ca translocation to young leaves, but nitrate-N increased movement to mature leaves. Spraying the foliage with NH_4 salt solutions had similar effects. A Ca supply sufficient to prevent leaf symptoms was not sufficient to prevent fruit symptoms, especially in rapidly growing fruit.

Wiersum (1965) found that forcing the growth rate of fruit and reducing transpiration of fruits increased BER in tomatoes and raised the K: Ca ratio. Reducing the growth rate of fruit increased the Ca content and reduced the K: Ca ratio. He believed the explanation lay in the fact that the fruits received water and assimilates by means of sieve tubes, which do not transport Ca. He concluded that extra water must be supplied through the xylem to transport adequate Ca for the fruit. However, Bible *et al.* (1968) found that decreasing the transpiration rate of entire tomato plants by mist irrigation increased fruit yields. Walker[1] noted that high soil temperature induced a Ca deficiency in corn. Increasing the relative humidity of air in the growth chamber from 74% to 83% decreased transpiration and the severity of the Ca deficiency. This treatment also increased the soil temperature at which maximum growth was obtained from 23C to 26C.

Al-Ani and Koontz (1969) reported that Ca uptake and distribution in beans is affected by location and age of roots. Applications of ^{45}Ca to the terminal 5 cm of a specific lateral root for 6 hours translocated preferentially to specific areas of the plant, and the amount transported was greatly decreased in roots older than 20 days. The reduced rate of ^{45}Ca uptake was associated with a reduced root growth rate. Continued Ca uptake, therefore, depends upon continuing development of new roots. If Al toxicity or some other factor reduces this process, Ca uptake and growth will be reduced accordingly.

Calcium transport in plants is affected by the temperature of both air and soil. Chang *et al.* (1968) reported a severe Ca deficiency in tobacco (*Nicotiana tabacum*) at growth chamber temperatures of 29C or 30C. This was associated with increased accumulation of Ca in stems. Resnik *et al.* (1969) reported that Ca accumulation by roots of bush beans and barley appeared to be independent of temperature but that Ca transport to tops was temperature dependent. Walker (1969, 1970) found that soil temperatures of 27C to 35C induced Ca deficiency in the young leaves of corn on a Benevola soil having a CEC of 18.6 meq/100 g and a Ca saturation of 75%

B. Calcium Requirements of Plants

1. General Requirements

Leaves of certain legumes, tobacco, tomatoes, and other dicots accumulate rather high Ca concentrations (2% to 4%) on a dry-weight basis, and this led to the conclusion that they have a high Ca requirement. However, there is considerable recent evidence that the level of Ca

---

[1] John M. Walker, 1973, A small change in aerial relative humidity can benefit plant growth and nutrition (manuscript in preparation), U.S. Soils Laboratory, USDA, ARS, SWC, Beltsville, Md. 20705.

essential for growth is very small and actually approaches that of a micronutrient (Wallace *et al.*, 1966; Jones and Lunt, 1967; Burstrom, 1968). A primary role of the additional Ca in the growth medium and in the plant is believed to be that of detoxifying excesses of other cations. Wallace *et al.* (1966) found that corn and tobacco plants grew well in nutrient cultures at Ca levels lower than those in 1/50 Hoagland solution if levels of Cu, Fe, Mn, Zn, and Mg were in proper balance. Adequate Ca concentrations in leaves were only 0.01 % to 0.02 % for corn and 0.08 % for tobacco. These investigators concluded that the published critical Ca levels of plants do not reflect a direct need for growth. Wallace *et al.* (1970) found that Ca (at concentrations up to 400 ppm) was more effective than EDTA in reducing chromium toxicity of bush beans. Figure 19.4 shows how Ca can reduce Al toxicity in two soybean varieties. Lund (1970) found that the absolute Ca requirement of soybean roots is extremely low if other essential ions are in balance and toxic ions are absent (Table 19.3).

Calcium requirement has been defined in terms of optimum concentrations in the growth medium and concentrations in various plant parts. For example, Neales and Hinde (1962) reported that Ca was required at 10^{-3} M to 10^{-4} M $CaCl_2$ (0.4 to 4.0 ppm Ca) for unrestricted growth of broad bean radicles. However, as discussed earlier, the Ca concentration required in the growth medium depends to a large extent on the ratios of Ca to other cations. Ratios between Ca and other cations within the plant are also important, and this may help to explain why Ca deficiency symptoms occur over a wide range of Ca concentrations in plants tops. For example, Nightingale and Smith (1968) found that for alfalfa (*Medicago sativa*) plants grown in complete nutrient solutions, petiole collapse occurred most frequently when the Ca content of shoots was between 0.22 % and 0.32 % but it occasionally occurred at Ca contents of 0.64 %. These investigators concluded that 0.64 % to 1.3 % was sufficient for normal development and that growth was reduced at values below 0.22 %. Soybeans showed collapse of apical meristems when whole plant tops contained less than 0.84 % Ca (Foy *et al.*, 1969a). Melsted (1953) reported that corn grown on acid soils below pH 4.5 showed Ca deficiency symptoms when the whole plant tops contained less than 0.2 % Ca. Fong and Ulrich (1970) reported a critical level of 0.15 % for petiole and blade tissue of potatoes. Under severe Ca stress the petioles of recently matured leaves were higher in Na, Mg, and P than normal petioles, lower in NO_3-N, and about the same in K concentrations.

Loneragan *et al.* (1970) recently reexamined the question of Ca requirement in plants and defined the following terms: (1) *solution Ca requirement*—the minimal Ca concentration permitted in solution for maximum growth rate; (2) *functional Ca requirement*—the minimum Ca

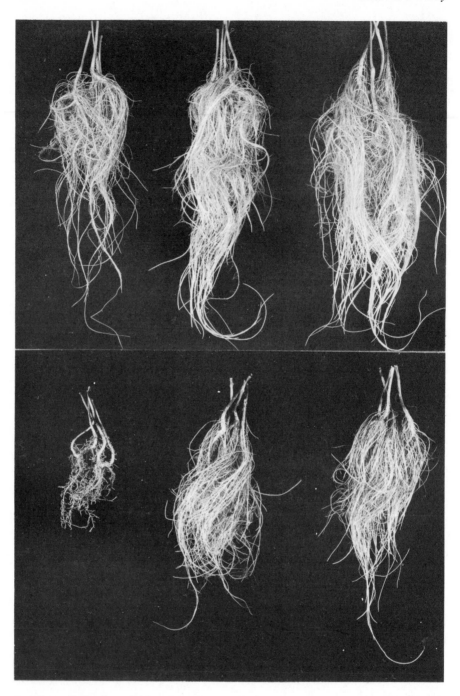

Fig. 19.4. Reduction of Al toxicity by Ca in Al-tolerant Perry soybean (*top*) and Al-sensitive Chief (*bottom*). *Left to right*: 2, 8, and 16 ppm Ca added with 8 ppm Al at initial pH 4.6. (Foy *et al.*, 1969a)

Table 19.3. Effect of Ca concentration in the nutrient solution portion of a split medium on the length and unit weight of soybean taproots

Ca added (ppm)	Taproot elongation rate† (mm/hr)	Harvest length of taproot‡ (mm)	Oven-dry wt/mm (mg)
0.00	0.57	63	0.32
0.05	2.29	395	0.21
0.25	2.90	490	0.27
1.25	2.95	474	0.33
6.25	3.04	455	0.39
31.25	2.97	452	0.42
156.25	2.96	467	0.44

Source: Lund, 1970.
Note: LSD at 5% = 37 and at 1% = 49.
† Elongation rate during first 48 hr in solution.
‡ Harvested 7½ days after entering solution.

concentration required at functional sites within plant tissues to sustain maximum growth rate; and (3) *critical Ca concentration*—the Ca concentration actually present in the plant or its organs at the time Ca becomes deficient for growth. These investigators emphasized that because Ca has low mobility between the organs of plants, the "critical" concentration can vary greatly with the conditions under which the deficiency is produced and may have little relation to the functional Ca requirement. They regarded growth rate and Ca uptake rate as important factors determining Ca requirement in solution and critical Ca requirement in plants. When plants were grown in dilute flowing solutions with constant Ca concentrations, estimates of critical and functional Ca requirements were lower than published values for most species. However, some species appear to differ in critical Ca concentration and in functional Ca requirement, as will be discussed later. The difficulty in determining the critical internal Ca requirement of plants is illustrated in Table 19.1, where both good and poor growth of soybean tops was associated with Ca concentrations of 0.48% to 0.52% in plant tops. Petiole collapse occurred when plant tops contained about 0.22% Ca.

2. Differential Calcium Requirements of Plant Species and Varieties

a. Occurrence and Magnitude. Ecologists have classified many plant species into two broad categories: those occurring on acid, low Ca soils (calcifuges) and those occurring on calcareous soils (calcicoles) (Bur-

strom, 1968). Hou and Merkle (1950) found that 14 calcifuges contained
0.48% to 1.08% Ca and 7 calcicoles contained 1.56% to 2.27%. In
general, the calcifuges have been characterized by tolerance to low Ca
levels in soils and nutrient solutions, low growth responses to increased
Ca levels (or even growth reduction at levels below those considered
optimum for calcicoles), low Ca uptake rates, high K:Ca ratios in
plant tissues, slow growth rates, tolerance to Al, and susceptibility to
Fe deficiency chlorosis in calcareous soils. However, there is no sharp
line of demarcation between the two groups. Rather, there appears to
be a continuous scale of sensitivity to high or low Ca soils among plant
species. Strains or varieties within species also differ widely in this regard.

A major problem in using the calcifuge-calcicole classification is that
of determining which of the physiological factors correlated with
tolerance to soil conditions are causally related to plant adaptation. It
is now recognized that factors other than absolute Ca requirement are
involved. Aluminum tolerance is a major factor in determining the
growth of some plant species and varieties on some acid soils. Sensitivity
to excess Al is often characterized by reduced uptake and utilization of
Ca. This relationship is discussed more fully in Chapter 20. In some
cases tolerance to excess Mn may explain plant adaptation to acid soils.

The evidence indicates that Al and Mn toxicities are more important
than absolute Ca levels in limiting plant growth on acid soils and that
the differential Ca uptake by plant species on acid soils may be merely
the result of differential Al-Ca, Mn-Ca, or other ion antagonisms in
the growth medium or in the plant. Nevertheless, plant species also
apparently differ in specific response to Ca in the absence of toxic
factors. For example, Bradshaw *et al.* (1958) showed that plants classified
as calcicoles (*Lolium perenne*, *Agrostis stolonifera*, and *Cynosurus
cristatus*) showed marked yield increases with increasing Ca concentrations in sand cultures, but a calcifuge species (*Nardus stricta*) was
decreased in yield by Ca concentrations above 15 ppm. Species classified
as intermediate (*Agrostis tenuis* and *A. canina*), which grow on soils
with a wide pH range, gave small yield increases with increased Ca
supply. Snaydon and Bradshaw (1962) reported that the Ca requirement
of crimson clover (*Trifolium incarnatum*) was higher than that of alsike
clover (*T. hybridum*); those of red clover (*T. pratense*) and white clover
(*T. repens*) were intermediate. Jeffries and Willis (1964b) found that
grasses in England differ widely in Ca requirement. Tomatoes require
more Ca than wheat, and legumes (with the exception of the *Lupinus*
species) and herbs require more than grasses and cereals (Loneragan
and Snowball, 1969a, 1969b). Gladstones and Loneragan (1970) show
Ca contents of a wide range of plants grown on an infertile lateritic
gravelly sand of western Australia (pH 5.0 in .01 M $CaCl_2$). Clarkson

(1965) concluded that *Agrostis stolonifera* has a higher Ca requirement than *A. setacea*.

Varietal differences in Ca requirement, uptake, or response within a plant species have also been reported for *Festuca ovina* (Snaydon and Bradshaw, 1961), tobacco (McEvoy, 1963), barley (Young and Rasmussen, 1966), cabbage (Maynard *et al.*, 1965), corn (Bradford *et al.*, 1966), tomatoes (Greenleaf and Adams, 1969), lespedeza (Morris and Pierre, 1949), white clover (Snaydon, 1962), bermudagrass (*Cynodon dactylon*) (Ramakrishman and Singh, 1966), apples (Shear and Faust, 1970), and snapbeans (Shannon *et al.*, 1967).

Soybean varieties differ in nodulating abilities and in their nodulation response to gypsum. Mammouth, an easy nodulating variety, gave optimum nodulation with 100 ppm gypsum, but two strains which are more difficult to nodulate, L-571 and L-2006, required 200 ppm (Dobereiner and Arruda, 1967).

b. Physiological Characterization. Many of the species and varietal differences in Ca requirement or response that have been reported involve differences in Ca uptake and translocation, but some differences in internal Ca requirement have also been found. Jeffries and Willis (1964a) found that the calcicole *Origanum vulgare* absorbed less Ca in relation to the amount of Ca in the soil than did the calcifuge *Juncus squarrous*. These investigators also noted that the calcicole showed a preferential absorption of K over Ca and mentioned that calcareous soils often have low K:Ca ratios. Thus, efficiency in K uptake may be a factor in the adaptation of calcicoles. Root growth of *J. squarrous*, a calcifuge, reached a maximum at Ca concentrations of 10 ppm in solution and decreased above this level. *Sieglingia decumbens*, a calcicole, required Ca at 40 ppm for maximum growth (Jeffries and Willis, 1964b).

Walker *et al.* (1955) reported that yields of crop plants—common sunflower (*Helianthus annuus*), tomato, and buckwheat (*Fagopyrum esculentum*)—were reduced at Ca saturations below 20% on serpentine soils but that the yields of plant species adapted to such soils—serpentine sunflower (*Helianthus bolanderi exilis* Heisner) and two species of *Strepanthus*—were not decreased until the Ca saturations were below 3%. The better growth of the serpentine species on the serpentine soil was attributed to more efficient absorption of Ca and exclusion of excess Mg rather than to a lower Ca requirement. Kruckeberg (1954) found that under laboratory conditions serpentine strains of *Phacelia* tolerated low Ca levels better than nonserpentine strains. He concluded that serpentine plants did not require the high magnesium, iron, nickel, or chromium concentrations often found in serpentine but rather tolerated these elements. However, Madhok and Walker (1969) reported that the serpentine sunflower species not only tolerated higher Mg concentrations in nutrient solutions but actually required higher concentrations within the plant for maximum growth than did the common cultivated sunflower species.

Clarkson (1965) reported that *Agrostis setacea*, which grows on acid soils having low Ca and high Al levels, made healthy growth at lower Ca concentrations in its shoots than did *A. stolonifera*, which grows on high Ca soils and is more sensitive to Al. The two species bound Ca in their roots to about the same extent. Clarkson concluded that the Ca transport mechanism in *A. setacea* becomes saturated at lower Ca concentrations in the external medium. In an acid podzolic soil (pH 4.0), $CaCO_3$ or $CaSO_4$ increased growth in *A. stolonifera*, but there was no increase in growth or Ca content of *A. setacea*. Loneragan and Snowball (1969b) found that herbs and legumes absorb Ca more rapidly than grasses and cereals in ordinary nutrient solutions and also appear to have a higher functional requirement for Ca; however, in flowing cultures, with constant Ca concentration, some herbs and legumes grew at lower Ca concentrations than many grasses and cereals. Tomatoes depleted standard nutrient solutions of Ca 10 to 15 times as rapidly as Gabo wheat. Legumes and herbs appeared to have higher functional requirements for Ca in their tissues than grasses and cereals. The *Lupinus* species were the only legumes having functional Ca requirements as low as grasses and cereals. Higher solution Ca requirement, higher critical Ca requirement, and higher internal Ca requirement for maximum yield in the tropical legume *Medicago truncatula*, compared with subterranean clover, have been associated with less effective distribution of Ca in plant tops, a higher Ca concentration in apical tissue, and a higher relative growth rate (Loneragan *et al.*, 1970). Barber and Ozanne (1970) observed that lupines caused a Ca depletion in their root zones, but Ca accumulated in the root zones of *Lolium rigidum* Gaud., subterranean clover, and capeweed (*Arctotheca calendula* L., Leoyns). Differences in the uptake of Ca by some plant species have been related to naturally occurring differences in root CEC values (Asher and Ozanne, 1961; Morita and Aoki, 1961) and also to root CEC differences induced by N fertilization of a single species (White *et al.*, 1965). Bradford *et al.*, (1966) found that corn hybrid IV accumulated lower Ca concentrations at a soil pH of 5.0 but higher concentrations at pH 6.0 than did corn hybrid II. Potassium added at 400 ppm decreased Ca accumulation to a greater degree in corn hybrid IV than in hybrid II.

Snaydon and Bradshaw (1961) found that ecotypes of *Festuca ovina* differed in Ca response. Maximum growth occurred at 20 ppm Ca for calcifuge ecotypes and at 100 ppm Ca for calcicoles. Differences in Ca response among populations of this species were attributed to differences in Ca-absorbing ability rather than differences in Ca metabolism at low internal Ca concentrations in plants. Greenleaf and Adams (1969) found that tomato lines that were most resistant to BER (Ca deficiency) absorbed and accumulated Ca more effectively in the fruit than did the most susceptible lines. However, one moderately resistant line appeared to require a lower Ca level in the fruit to prevent BER. Susceptibility to internal tipburn in certain cabbage varieties (also a Ca deficiency) is related to a lower efficiency in Ca uptake and transport from basal and wrapper leaves to head leaves (Maynard *et al.*, 1965). Ramakrishman and Singh (1966) found that a bermudagrass population from a non-calcareous soil absorbed more Ca than a population from a highly calcareous soil when both were grown on the same soil. The upper leaves and buds of burley tobacco varieties that were susceptible to Ca deficiency contained lower Ca concentrations than those of non-susceptible varieties; however, the Ca contents of whole tops were not related to susceptibility to Ca deficiency (Brumagen and Hiatt, 1966). It was concluded that high levels of oxalic acid in the upper stalks of susceptible varieties interfered with the transport and use of Ca. Brown (1965) found that Ca deficiency induced by low Cu in the growth medium and aggravated by high P uptake was more severe in Atlas 66 wheat than in the Monon variety.

Shannon *et al.* (1967) reported that snapbean varieties differed in susceptibility to a hypocotyl necrosis which was attributed to Ca deficiency. The varieties Tenderwhite, Earligreen, and Processer were quite resistant, White Seeded Tendercrop was moderately susceptible, and Tendercrop was quite susceptible. The disorder occurred within 2 to 4 days after germination and was observed in both acid, low Ca soils and in artificial media. It was corrected by adding Ca salts at 10 meq/l to the germinating medium. Magnesium salts were about half as effective as Ca, and even KNO_3 was quite effective at 40 meq/l. Phosphate salts increased the Ca deficiency at the same concentration. Incidence of the disorder in the susceptible varieties was inversely related to the Ca contents of the seeds. Plants of normally resistant varieties grown in Ca-deficient nutrient solutions produced seeds with a higher incidence of hypocotyl necrosis. The seeds of some resistant varieties had the same Ca content as the seeds of moderately susceptible varieties, but their K and Mg contents were higher. The researchers suggested that these other elements increased the transport of Ca from seeds and thus prevented hypocotyl necrosis at lower levels of Ca in the seeds.

Shear and Faust (1970) reported genetic differences in Ca uptake and translocation by apple seedlings. Calcium (^{45}Ca) applied to roots moved readily to developing leaves. Varieties showing juvenile leaf characteristics were most effective in translocating Ca into mature leaves.

Perry soybean variety, which is Al tolerant, generally produces a higher percentage of its maximum top yield and accumulates a higher Ca concentration in its tops, when grown at low Ca levels, than does the Al-sensitive Chief variety (Table 19.4; Foy *et al.*, 1969a). Under these conditions the apices of Perry collapsed before those of Chief.

Table 19.4.   Growth and Ca concentrations in Perry and Chief soybean tops grown in 1/5 Steinberg solution modified to contain various Ca concentrations

Ca added (ppm)	Relative top yield (%)		Ca concentrations in tops (meq/100 dry wt)	
	Perry	Chief	Perry	Chief
0	15	14	7.2	2.1
1	82	66	8.6	7.2
2	86	68	10.6	9.8
4	96	75	12.2	18.7
8	63	79	17.1	29.4
16	94	86	42.8	55.5
32	100	100	93.2	85.7

Source: Foy *et al.*, 1969a.

This was probably caused by a more rapid depletion of Ca from the solution by Perry. The Perry variety is less susceptible than Chief to Al-induced Ca deficiency, which is manifested as a collapse of petioles of older first, second, or third trifoliate leaves.

Loneragan and Snowball (1969b) have emphasized the importance of Ca uptake rate in the development of differential Ca deficiencies in plant species. They pointed out that Ca sensitivity rankings developed by a continuous low Ca level in the solution and a low uptake rate are quite different from those obtained by an initially high but declining Ca level in the medium and corresponding Ca uptake rate by the plant. Thus, rankings of plant species according to their abilities to absorb nutrients from solutions at high concentrations had no relationship to their abilities to absorb from solutions at low concentrations. For example, *Erodium botrys*, which had the highest Ca uptake rate at solution concentrations of 1000μM (40 ppm), had the lowest at 2.5μM (0.1 ppm). Since Loneragan and Snowball have obtained good agreement between Ca accumulation in soil and in solution for a wide range of plant species, these relationships should be valid in field soils that differ in initial or continuous Ca-supplying abilities.

For phloem mobile elements like P, excesses absorbed during early stages of growth can be used for later growth when the nutrient medium has been depleted. But Ca is considered essentially nonmobile in phloem, and excesses absorbed early are not generally available for later growth. Loneragan and Snowball (1969a, 1969b) emphasized this inability of plants to utilize excess Ca accumulated in their tissues. When plants were first exposed to luxury levels of Ca (40 ppm) and then transferred to a low Ca medium (0.012 ppm), deficiency symptoms developed rapidly. These investigators found that even during the development of Ca deficiency, the remobilization of Ca from cotyledons and leaves of *Medicago truncatula* and subterranean clover was negligible. The leaves of the former species had much higher Ca concentrations than the apices.

c. Breeding Calcium-efficient Plant Varieties. The problem of Ca deficiency can be divided into two parts: first, the inability of plants to absorb sufficient Ca from soils containing low absolute Ca levels, low Ca: other-cation ratios, or high salt levels; and second, the inability of plants to distribute previously absorbed Ca to growing zones in times of reduced Ca uptake, a condition which may be imposed by drought, unfavorable temperature, or other factors. The first situation can occur in acid surface soils or subsoils where Al-Ca antagonisms greatly reduce Ca uptake, in heavily fertilized soils where low Ca: Mg, Ca: K, or Ca: NH_4 ratios may result (particularly where fertilizers are band-placed in sandy, low CEC soils), and in soils where brackish water is used for irrigation and in which low Ca : total-soluble-salt ratios reduce Ca availability to plants. High P fertilization may also induce Ca deficiency. The second situation, inadequate Ca distribution within the plant, can occur when Ca is available in adequate quantities during the early part of the growing season (such as would be the case with fertilizer bands containing Ca) but is greatly reduced as the fruiting stage proceeds. Because Ca absorbed earlier is generally rather immobile within the older tissues, the growing point may die of Ca deficiency while the leaf below it contains abundant Ca. Copper deficiency in wheat, a rolling of terminal leaves, is manifested as such a Ca deficiency (reduced Ca transport), and the problem is magnified by high P levels in the plant.

The situations mentioned above, which are associated with Ca deficiency, may be difficult or impractical to control by the usual soil fertilization, liming, and management practices. In such cases breeding plant varieties better adapted to these soil conditions would appear worthy of considerable research effort. Plants are needed that can obtain adequate Ca in the presence of excess Al, Mn, Mg, NH_4, and K or total salts or low levels of available Cu, particularly in the sandy, low CEC soils of the South. There is an even greater need for plants that can more effectively transport previously absorbed Ca to growing tips and fruits in times of stress. In the case of legumes, rhizobial strains that can tolerate low levels of available Ca may also be worth investigating, particularly for use on hostile sites, such as strongly acid, Al-toxic mine spoils.

Considerable progress has been made in selecting and breeding tomatoes for resistance to BER. Walter (1957) selected four BER-resistant varieties on the acid, sandy, Ca-deficient soils of Florida. Several breeding lines, differing widely in BER resistance, have been developed at Auburn University (Table 19.5; Greenleaf and Adams, 1969). The exact genetic nature of BER resistance is not known, but the

Table 19.5. Blossom-end rot (BER) incidence in two tomato lines as affected by four levels of Ca in the watering solutions

Breeding line	Ca added (ppm)	No. of fruit per 4 plants	No. of fruit with BER	% fruit with BER †
Au-2	0	98	41	41.8 a
	22	85	35	41.2 a b
	60	94	34	36.2 a b
	161	98	17	17.3 b
Au-1	0	85	1	1.2 ns
	22	67	0	0.0 ns
	60	94	0	0.0 ns
	161	104	0	0.0 ns

Source: Greenleaf and Adams, 1969.
† Means having a letter in common are not significantly different at the 1% level by the Duncan Multiple Range Test.

low frequency of resistant plants in segregating populations suggested to Greenleaf and Adams (1969) that it is a recessive trait. These investigators suggested that the use of BER-resistant varieties offers insurance against the sudden occurrence of this disorder. Geraldson (1957) recommended the incorporation of lime and gypsum into soils before planting and supplementary biweekly sprays with 0.04 M $CaCl_2$ as controls for BER, but the problem apparently still occurs under some conditions of soil and climate (Wilson, 1963; Riggleman, 1964; Gerard, 1966). This suggests that additional preventive measures are needed. Breeding plants to resist BER and other physiological disorders associated with Ca deficiency or inefficient use appears to be a worthy endeavor. In view of the close association between Ca, Al, and Cu, varieties selected for greater Ca efficiency may also be more tolerant to excess Al and low Cu availability in soils. Gorsline *et al.* (1968) postulated that three major genes regulated Ca-Sr accumulation in certain corn inbred lines.

Plant breeding offers a "natural" solution to some of the more difficult problems in soil fertility that may not be corrected by applying more fertilizer and lime.

References

Adams, F., and R. W. Pearson. 1967. Crop response to lime in the Southern United States and Puerto Rico. In *Soil Acidity and Liming*, ed. R. W. Pearson and Fred Adams, pp. 161–206. Agronomy 12. Madison, Wis.: Amer. Soc. Agron.

Adams, F., R. W. Pearson, and B. D. Doss. 1967. Relative effects of acid subsoils on cotton yields in field experiments and on cotton roots in growth chamber experiments. *Agron. J.* 59: 453–56.

Al-Ani, Tariq A., and H. V. Koontz. 1969. Distribution of calcium absorbed by all or part of the root system of beans. *Plant Physiol.* 44: 711–16.

Allaway, W. H. 1945. Availability of replaceable calcium from different types of colloids as affected by degree of Ca saturation. *Soil Sci.* 59: 207–17.

Asher, C. J., and P. G. Ozanne. 1961. The cation exchange capacity of plant roots and its relationship to the uptake of soluble nutrients. *Aust. J. Agr. Res.* 12: 755–60.

Ayers, A. S. 1963. The utility of soil analysis in determining the need for applying calcium to sugarcane. *Proc. 11th Congr. Int. Soc. Sugarcane Technol.* (1962), pp. 162–70.

Barber, S. A., and P. G. Ozanne. 1970. Autoradiographic evidence for the differential effect of four plant species in altering the calcium content of the rhizosphere soil. *Soil Sci. Soc. Amer. Proc.* 34: 635–37.

Benath, C. L., E. A. N. Greenwood, and J. F. Loneragan. 1966. Effects of calcium deficiency on symbiotic nitrogen fixation. *Plant Physiol.* 41: 760–63.

Bennet-Clark, T. A. 1956. Salt accumulation and mode of action of auxin. A preliminary hypothesis. In *The Chemistry and Mode of Plant Growth Substances*, ed. R. L. Wain and F. Wightman, pp. 284–91. Proceedings of a Symposium Held at Wye College (University of London) July 1955. New York: Academic Press.

Bennett, A. C., and Fred Adams. 1970. Calcium deficiency and ammonia toxicity as separate causal factors of $(NH_4)_2SO_4$ injury to seedlings. *Soil Sci. Soc. Amer. Proc.* 34: 255–59.

Bible, B. B., R. L. Cuthbert, and R. L. Carolus. 1968. Response of some vegetable crops to atmospheric modifications under field conditions. *Proc. Amer. Soc. Hort. Sci.* 92: 590–94.

Bradford, R. R., Dale E. Baker, and W. I. Thomas. 1966. Effect of soil treatments on chemical element accumulation of four corn inbred lines. *Agron J.* 58: 614–17.

Bradshaw, A. D., R. W. Lodge, D. Jowett, and M. J. Chadwick. 1958. Experimental investigations into the mineral nutrition of several grass species. I. Calcium level. *J. Ecol.* 46: 749–57.

Brown, J. C. 1965. Calcium movement in barley and wheat as affected by copper and phosphorus. *Agron. J.* 57: 617–21.

Brown, J. C., and C. D. Foy. 1964. Effect of Cu on the distribution of P, Ca, and Fe in barley plants. *Soil Sci.* 98: 362–70.

Brumagen, D. M., and A. J. Hiatt. 1966. The relationship of oxalic acid to the translocation and utilization of calcium in *Nicotiana tobacum*. *Plant Soil* 24: 239–49.

Burling, Edwin, and William T. Jackson. 1965. Changes in calcium levels in cell walls during elongation of oat coleoptile sections. *Plant Physiol.* 40: 138–41.

Burstrom, H. G. 1952. Studies on growth and metabolism of roots. VIII. Calcium as a growth factor. *Physiol. Plant.* 5: 391–402.

———. 1968. Calcium and plant growth. *Biol. Rev.* 43: 287–316.

Burstrom, H., and V. Tullin. 1957. Observations on chelates and roots growth. *Physiol. Plant.* 17: 207–19.

Bussler, W. 1962. Ca-mangelsymptoms bei sonnenblumen A. *Z. Pflanzenernähr. Düng. Bodenk.* 99: 207–15.

Chang, S. Y., R. H. Lowe, and A. J. Hiatt. 1968. Relationship of temperature to the development of calcium deficiency symptoms in *Nicotiana tabacum*. *Agron. J.* 60: 435–36.

Chaudhry, M. S., E. O. McLean, and R. E. Franklin. 1964. Effects of N, Ca:K saturation ratio, and electrolyte concentration on uptake of Ca and K by rice plants. *Agron. J.* 56: 304–7.

Chrispeels, M. J., and J. E. Varner. 1967. Gibberellic acid enhanced synthesis and release of α-amylase and ribonuclease by isolated barley aleurone layers. *Plant Physiol.* 42: 398–406.

Clark, R. B. 1970. *Effects of mineral nutrient levels on the inorganic composition and growth of corn* (Zea mays *L.*). Ohio Agr. Res. and Develop. Center Research Circular 151. 21 pp.

Clarkson, D. T. 1965. Calcium uptake by calcicole and calcifuge species in the genus *Agrostis*. L. *J. Ecol.* 53: 427–35.

Cleland, Robert. 1960. Effect of auxin upon loss of calcium from cell walls. *Plant Physiol.* 35: 581–84.

Colwell, W. E., and N. C. Brady. 1945. The effect of calcium on yield and quality of large seeded type peanuts. *J. Amer. Soc. Agron.* 37: 413–28.

Dobereiner, Johanna, and Norma Bergallo De Arruda. 1967. Interrelacoes entre variedades e nutricao na nodulacao e simbiose da soja (*Glycine max* L., Merrill). *Pesq. Agropec Bras.* 2: 475–86 (English abstr.).

Dodds, J. A. A., and R. J. Ellis. 1966. Cation stimulated ATPase activity in plant cell walls. *Biochem. J.* 101: 31.

Drew, M. C., and O. Biddulph. 1969. Non-metabolic uptake of calcium by intact bean roots. *Plant Physiol.* 44, supplement no. 94 : 20.

Epstein, E. 1961. The essential role of calcium in selective cation transport by plant cells. *Plant Physiol.* 36: 437–44.

Evans, C. E., and E. J. Kamprath. 1970. Lime response as related to percent Al saturation, solution Al and organic matter content. *Soil Sci. Soc. Amer. Proc.* 34: 893–96.

Facteau, T. J., and Paul Eck. 1970. Effects of base saturation and cation level of a sand-clay substrate upon growth and composition of highbush blueberry. *Soil Sci.* 110: 244–52.

Fong, Kwok H., and Albert Ulrich. 1970. Calcium nutrition of White Rose potato in relation to growth and leaf minerals. *Soil Sci. Plant Anal.* 1: 43–55.

Foote, B. D., and J. B. Hanson. 1964. Ion uptake by soybean root tissue depleted of calcium by ethylenediaminetetraacetic acid. *Plant Physiol.* 39: 450–60.

Foy, C. D. 1964. *Toxic Factors in Acid Soils of the Southeastern United States as Related to the Response of Alfalfa to Lime.* USDA-ARS Prod. Res. Rep. 80.

Foy, C. D., A. L. Fleming, and W. H. Armiger. 1969a. Aluminum tolerance of soybean varieties in relation to calcium nutrition. *Agron. J.* 61: 505–11.

——. 1969b. Differential tolerance of cotton varieties to excess manganese. *Agron. J.* 61: 690–94.

Franklin, R. E. 1969. Effect of adsorbed cations on phosphorus uptake by excised roots. *Plant Physiol.* 44: 697–700.

Galey, F., R. G. W. Jones, and O. R. Lunt. 1968. Microscopic and histochemical studies on calcium deficient root apices of *Zea mays.* In preparation, Jones and Lunt, 1967.

Gammon, J., Jr. 1957. Root growth responses to soil pH adjustments made with carbonates of calcium, sodium or potassium. *Soil Crop Sci. Soc. Fla. Proc.* 17: 249–54.

Geraldson, C. M. 1957. Control of blossom end rot of tomatoes. *Proc. Amer. Soc. Hort. Sci.* 69: 309–17.

——. 1970. Intensity and balance concept of an approach to optimum vegetable production. *Soil Sci. Plant. Anal.* 1: 187–96.

Gerard, C. J. 1966. Blossom end rot of pear shaped tomatoes. *J. Rio Grande Valley Hort. Soc.* 20: 134–41.

——. 1971. Influence of osmotic potential, temperature, and calcium on growth of plant roots. *Agron. J.* 63: 555–58.

Ginzburg, B. Z. 1961. Evidence for a protein gel structure cross linked by metal cations in the intercellular cement of plant tissue. *J. Exp. Bot.* 12: 85–107.

Gladstones, J. S., and J. F. Loneragan. 1970. Nutrient elements in herbage plants in relation to soil adaptation and animal nutrition. Proc. *11th Int. Grasslands Congr.* (Surfers Paradise, Aust.), pp. 350–54.

Gonzalez-Erico, Enrique. 1968. Desirable calcium levels in relation to aluminum in some soils of the tropics. M. S. Thesis. North Carolina State University.

Gorsline, G. W., W. I. Thomas, and D. E. Baker. 1968. *Major Gene Inheritance of Sr-Ca, Mg, K, P, Zn, Cu, B, Al-Fe and Mn Concentrations in Corn (*Zea mays *L.).* Pa. Agr. Exp. Sta. Bull. 746.

Green, David E., and John H. Young. 1971. Energy transduction in membrane systems. *Amer. Sci.* 59: 92–100.

Greenleaf, W. H., and Fred Adams. 1969. Genetic control of blossom end rot disease in tomatoes. *J. Amer. Soc. Hort. Sci.* 94: 248–50.

Hanson, J. B. 1960. Impairment of respiration, ion accumulation, and ion retention in root tissue treated with ribonuclease and ethylenediaminetetraacetic acid. *Plant Physiol.* 35: 372–79.

Harris, Henry C., and John B. Brolmann. 1966. Comparison of Ca and B deficiencies of the peanut. II. Seed quality in relation to histology and viability. *Agron. J.* 58: 578–82.

Henderson, J. B. 1969. The calcium to magnesium ratio of liming materials for efficient magnesium retention in soils and utilization by plants. Ph.D. Thesis. North Carolina State University, University Microfilms, Ann Arbor, Mich., *Diss. Abstr.*, B, 30 (11): 4871–72.

Hermann, R. 1964. Die Wirkungen des Oxalats und der Ethylenediaminetetraessigsäure (AeDTE) auf die Ausbildung des Plasmalemmas bei Zwiebelinnenpidermiszellen von *Allium cepa*. *Protoplasma* 58: 172–89.

Hewitt, E. J. 1963. The essential nutrient elements: Requirements and interactions in plants. In *Plant Physiology: A Treatise*. III. *Inorganic Nutrition of Plants*, ed. F. C. Steward, pp. 137–329. New York: Academic Press.

Hodges, T. K., and O. E. Elzam. 1967. Effect of azide and oligomycin on the transport of calcium ions in corn mitochondria. *Nature* 215: 970–72.

Hodges, T. K., and J. B. Hanson. 1965. Calcium accumulation by maize mitochondria. *Plant Physiol.* 40: 101–9.

Hooymans, J. J. 1966. The role of calcium in the absorption of anions and cations by excised barley roots. *Acta Bot. Neerl.* 13: 507–40 (*Biol. Abstr.* 47: 38936.)

Hou, Hsioh-Yu, and F. G. Merkle. 1950. Chemical composition of certain calcicolous plants. *Soil Sci.* 69: 474–86.

Howard, D. D., and Fred Adams. 1965. Calcium requirement for penetration of subsoils by primary cotton roots. *Soil Sci. Soc. Amer. Proc.* 29: 558–61.

Hunsaker, V. E., and P. F. Pratt. 1971. Calcium and magnesium exchange equilibria in soils. *Soil Sci. Soc. Amer. Proc.* 35: 151–52.

Jackson, William A. 1967. Physiological effects of soil acidity. In *Soil Acidity and Liming*, ed. R. W. Pearson and Fred Adams, pp. 43–124. Agronomy 12. Madison, Wis.: Amer. Soc. Agron.

Jackson, W. A., and H. J. Evans. 1962. Effect of Ca supply on the development and composition of soybean seedlings. *Soil Sci.* 94: 180–86.

Jacoby, B. 1966. Ascent of calcium in intact and de-rooted bean plants. *Nature* 211: 212.

———. 1967. The effects of the roots on calcium ascent in bean stems. *Ann. Bot.* 31: 725–30.

Jacobson, L. R., J. Hannapel, D. P. Moore, and M. Schaedle. 1961. Influence of calcium on selectivity of ion absorption process. *Plant Physiol.* 36: 58–61.

Jeffries, R. L., D. Laycock, G. R. Stewart, and A. P. Sims. 1969. The development of mechanisms involved in the uptake and utilization of calcium and potassium by plants in relation to an understanding of plant distribution. In *Ecological Aspects of the Mineral Nutrition of Plants*, ed. I. H. Rorison, pp. 281–307. Symp. Brit. Ecol. Soc. no. 9, Sheffield, 1968. Oxford: Blackwell Scientific Publications.

Jeffries, R. L., and A. J. Willis. 1964a. Studies on the calcicole, calcifuge habit. I. Method of analysis of soil and plant tissue and some results of investigations on four species. *J. Ecol.* 52: 121–38.

——. 1964b. Studies on the calicole, calcifuge habit. II. The influence of calcium on the growth and establishment of four species in soil and sand cultures. *J. Ecol.* 52: 691–707.

Jones, R. G. W., and O. R. Lunt. 1967. The function of calcium in plants. *Bot. Rev.* 33: 407–26.

Kenefick, D. G. and J. B. Hanson. 1966. Contracted state as an energy source for Ca binding and Ca and inorganic P accumulation by corn mitochondria. *Plant Physiol.* 41: 1601–9.

Kruckeberg, A. R. 1954. The ecology of serpentine soils. II. Plant species in relation to serpentine soils. *Ecology* 35: 267–74.

Lance, J. C., and R. W. Pearson. 1969. Effect of low concentration of aluminum on growth and water and nutrient uptake by cotton roots. *Soil Sci. Soc. Amer. Proc.* 33: 95–98.

Lane, William B., and James D. Sartor. 1966. Exchangeable calcium content of United States soils. *Soil Sci.* 101: 390–91.

Leggett, J. E., and E. Epstein. 1956. Kinetics of sulfate absorption by barley roots. *Plant Physiol.* 31: 222–60.

Leggett, J. E., and R. A. Galloway and H. G. Gauch. 1965. Calcium activation of orthophosphate absorption by barley roots. *Plant Physiol.* 40: 897–902.

Leggett, J. E., and W. A. Gilbert. 1967. Localization of the Ca-mediated apparent ion selectivity in the cross sectional volume of soybean roots. *Plant Physiol.* 42: 1658–64.

Lehninger, Albert S. 1970. Mitochondria and calcium ion transport. *Biochem. J.* 119: 129–38.

Loneragan, J. F., and E. J. Dowling. 1958. The interaction of calcium and hydrogen ions in the nodulation of subterranean clover. *Aust. J. Agr. Res.* 9: 464–72.

Loneragan, J. F., D. Meyer, R. G. Fawcett, and A. J. Anderson. 1955. Lime pelleted clover seeds for nodulation on acid soils. *J. Aust. Inst. Agr. Sci.* 21: 264–65.

Loneragan, J. F., I. C. Rowland, A. D. Robson, and K. Snowball. 1970. The calcium nutrition of plants. *Proc. 11th Int. Grasslands Congr.* (Surfers Paradise, Aust.), pp. 358–67.

Loneragan, J. F., and K. Snowball. 1969a. Calcium requirements of plants. *Aust. J. Agr. Res.* 20: 465–78.

Loneragan, J. F. and K. Snowball. 1969b. Rate of calcium absorption by plant roots and its relation to growth. *Aust. J. Agri. Res.* 20: 479–90.

Long, F. Leslie, and C. D. Foy. 1970. Plant varieties as indicators of aluminum toxicity in the A$_2$ horizon of a Norfolk soil. *Agron. J.* 62: 679–81.

Lowther, W. L., and J. F. Loneragan. 1968. Calcium and nodulation in sub-terranean clover. *Plant Physiol.* 42: 1362–66.

——. 1970. Calcium in the nodulation of legumes. *Proc. 11th Int. Grasslands Congr.* (Surfers Paradise, Aust.), pp. 446–50.

Lund, Z. F. 1970. The effect of calcium and its relation to some other cations on soybean root growth. *Soil Sci. Soc. Amer. Proc.* 34: 456–59.

Maas, E. V. 1969. Calcium uptake by excised maize roots and interactions with alkali cations. *Plant Physiol.* 44: 985–89.

McEvoy, E. T. 1963. Varietal differences in calcium level in leaves of flue cured tobacco. *Can. J. Plant Sci.* 43: 141–45.

Madhok, O. P., and R. B. Walker. 1969. Magnesium nutrition of two species of sunflower. *Plant Physiol.* 44: 1022.

Mahilum, B. C., R. L. Fox, and J. A. Silva. 1970. Residual effects of liming volcanic ash soils in the humid tropics. *Soil Sci.* 109: 102–9.

Marinos, N. G. 1962. Studies on submicroscopic aspects of mineral deficiencies. I. Calcium deficiency in the shoot apex of barley. *Amer. J. Bot.* 49: 834–41.

Marschner, H. R., and R. Overstreet. 1966. Potassium loss and change in the fine structure of corn root tips induced by H ion. *Plant Physiol.* 41: 1725–35.

Martin, J. P., and A. L. Page. 1969. Influence of exchangeable calcium and magnesium and of percentage base saturation on growth of citrus plants. *Soil Sci.* 107: 39–46.

Masuda, Y. 1959. Role of cellular ribonucleic acid in the growth response of *Avena* coleoptile to auxin. *Physiol. Plant.* 12: 324–35.

Maynard, D. N., B. Gersten, and H. F. Vernell. 1965. The distribution of calcium as related to internal tipburn, variety and calcium nutrition of cabbage. *Proc. Amer. Soc. Hort. Sci.* 86: 392–96.

Mayorga, Hector, and Carlos Rolz. 1971. Pectinesterase activity as a function of pH, enzyme and cation concentrations. *J. Agr. Food Chem.* 19: 179–81.

Mehlich, A., and W. E. Colwell. 1944. Influence of nature of soil colloids and degree of base saturation on growth and nutrient uptake by cotton and soybeans. *Soil Sci. Soc. Amer. Proc.* 8: 179–84.

Melsted, W. W. 1953. Some observed calcium deficiencies in corn under field conditions. *Soil Sci. Soc. Amer. Proc.* 17: 52–54.

Mengel, K. 1961. Die Donnan-verteilung der kationen im freien raum der planzenwurzel und ihr dodentung fur die aktivekationen aufnahme. *Z. Pflanzenernähr. Dung. Bodenk.* 95: 240–59.

——. 1968. Exchangeable cations of plant roots and potassium absorption by the plant. *The Role of Potassium in Agriculture*, ed. V. J. Kilmer, S. E. Younts, and N. C. Brady, pp. 311–19. Madison, Wis.: ASA, CSSA, SSSA.

Millikan, C. R., and B. C. Hanger. 1964. Effect of calcium level in the substrate on the distribution of Ca-45 in subterranean clover (*Trifolium subterraneum* L.). *Aust. J. Biol. Sci.* 17: 823–44.

——. 1965. Transport of Ca-45 and Zn-65 through collapsed petioles of calcium deficient subterranean clover. (*Trifolium subterraneum* L.). *Aust. J. Biol. Sci.* 18: 1083–97.

———. 1969. Movement of foliar applied Ca-45 in brussels sprouts. *Aust. J. Biol. Sci.* 22: 545–58.

Millikan, C. R., B. C. Hanger, and E. N. Bjarnason. 1969. Interactions between calcium level and nitrogen source on growth and Ca-45 distribution of subterranean clover. *Aust. J. Biol. Sci.* 22: 535–44.

Moore, D. P., B. J. Mason, and E. V. Maas. 1965. Accumulation of Ca in the exudate of individual barley roots. *Plant Physiol.* 40: 641–44.

Morita, S., and A. Aoki. 1961. Studies on nutrient uptake by the roots of fruit trees. I. Cation exchange capacity of roots and cation absorption from solution. *J. Sci. Soil Tokyo.* 31: 234–36 (Jap.).

Morris, H. D., and W. H. Pierre. 1949. Minimum concentration of manganese necessary for injury to various legumes in solution culture. *Agron. J.* 41: 107–13.

Neales, T. F., and N. R. W. Hinde. 1962. A test of the Ca-B interaction hypothesis using the growth of the bean radicle. *Physiol. Plant.* 15: 217–28.

Nelson, L. E., and N. C. Brady. 1953. Some greenhouse studies of cation interactions in Ladino clover using split root techniques. *Soil Sci. Soc. Amer. Proc.* 17: 274–78.

Nightingale, H. I., and R. L. Smith. 1968. Collapse of alfalfa petioles and their calcium content. *Agron. J.* 60: 475–77.

Parkhash, V., and B. V. Subbiah. 1964. Distribution pattern of radioactive Ca-45 and radiosulfur (S-35) in citrus seedlings under normal and deficient conditions of different micronutrient elements. *Soil Sci. Plant Nutrition* 10: 24–30.

Paul, J. L., and W. H. Thornhill. 1969. Effects of magnesium on rooting of chrysanthemums. *J. Amer. Soc. Hort. Sci.* 94: 280–82.

Proctor, J. 1970. Magnesium as a toxic element. *Nature* 227: 742–43.

Ramakrishman, P. S., and V. K. Singh. 1966. Differential responses of the edaphic ecotypes in *Cynodon dactylon* L. Pers. to soil calcium. *New Phytol.* 65: 100–108.

Ray, P. M., and D. B. Baker. 1965. The effect of auxin on synthesis of oat coleoptile cell wall constituents. *Plant Physiol.* 40: 353–60.

Resnik, M. C., O. R. Lunt, and A. Wallace. 1969. Cesium, potassium, strontium and calcium transport in two different plant species. *Soil Sci.* 108: 64–73.

Riggleman, J. D. 1964. Blossom end rot of tomatoes as influenced by certain cultural practices. Ph.D. Thesis. University of Maryland. (Hort. Abstr. 1965, no. 5937.).

Rios, M. A., and R. W. Pearson. 1964. The effect of some chemical environmental factors on cotton root behavior. *Soil Sci. Soc. Amer. Proc.* 28: 232–35.

Robson, A. D., D. G. Edwards, and J. F. Loneragan. 1970. Calcium stimulation of phosphate absorption by annual legumes. *Aust. J. Agr. Res.* 21: 601–12.

Robson, A. D., and J. F. Loneragan. 1970. Effect of calcium carbonate and inoculation level on the nodulation of *Medicado truncatula* on a moderately acid soil. *Aust. J. Agr. Res.* 21: 427–34.

Shannon, S. J., J. Nath, and J. D. Atkin. 1967. Relation of calcium nutrition to hypocotyl necrosis of snapbean (*Phaseolus vulgaris* L.). *Proc. Amer. Soc. Hort. Sci.* 90: 180–90.

Shear, C. B. and Miklos Faust. 1969. Calcium transport in apple trees. *Plant Physiol.* 45: 670–74.

———. 1970. Value of various tissue analyses in determining calcium status of apple tree and fruit. In *Plant Analysis and Fertilizer Problems*. Abstracts 6th Int. Colloqium, Tel Aviv., Israel, March 13–17, 1970. Geneva, New York: W. F. Humphrey Press.

Smith, Malcom E. 1944. The role of boron in plant metabolism. 1. Boron in relation to the absorption and solubility of calcium. *Aust. J. Exp. Biol.* 22: 256–63.

Snaydon, R. W. 1962. The response of certain *Trifolium* species to calcium in sand culture. *Plant Soil* 16: 381–88.

Snaydon, R. W., and A. D. Bradshaw. 1961. Differential response to calcium within the species *Festuca ovina* L. *New Phytol.* 60: 219–34.

———. 1962. Differences between natural populations of *Trifolium repens* (white clover) in response to mineral nutrients. I. Phosphate. *J. Exp. Bot.* 13: 422–34.

Sorokin, H., and A. L. Sommer. 1940. Effects of calcium deficiency upon the roots of *Pisum sativum*. *Amer. J. Bot.* 27: 308–18.

Steffenson, D. 1958. Chromosome aberrations in calcium deficient *Transdescantia* produced by irradiation. *Nature* 182: 1750–51.

Thibodeau, P. O., and P. L. Minotti. 1969. The influence of calcium on the development of lettuce tipburn. *J. Amer. Soc. Hort. Sci.* 94: 372–76.

Truelove, B., and J. B. Hanson. 1966. Calcium activated phosphate uptake in corn mitochondria. *Plant Physiol.* 41: 1004–13.

Trumble, H. C., and C. M. Donald. 1938. Soil factors in relation to the distribution of subterranean clover and some alternatives legumes. *J. Aust. Inst. Agr. Sci.* 4: 206–8.

Van Steveninck, R. F. M. 1965. The significance of calcium on the apparent premeability of cell membranes and the effects of substitution with other divalent ions. *Physiol. Plant.* 18: 54–69.

Viets, F., Jr. 1944. Calcium and other polyvalent cations as accelerators of ion accumulation by excised barley roots. *Plant Physiol.* 19: 466–79.

Vlamis, J. 1949. Growth of lettuce and barley as influenced by degree of calcium saturation of soil. *Soil Sci.* 67: 453–66.

Walker, J. C., L. V. Edington, and M. V. Nayudu. 1961. *Tipburn of Cabbage, Nature and Control*. Wis. Agr. Exp. Sta. Bull. 230.

Walker, John M. 1969. Effects of alternating versus constant soil temperatures on maize seedlings growth. *Soil Sci. Soc. Amer. Proc.* 34: 889–92.

———. 1970. One degree increments in soil temperatures affect maize seedling behavior. *Soil Sci. Soc. Amer. Proc.* 33: 729–36.

Walker, R. B., H. M. Walker, and P. R. Ashworth. 1955. Calcium-magnesium nutrition with special reference to serpentine soils. *Plant Physiol.* 30: 214–25.

Wallace, A., E. Frohlich, and O. R. Lunt. 1966. Calcium requirements of higher plants. *Nature* 209: 634.

Wallace, A., M. Sadek, M. Sufi, and Evan M. Pomney. 1970. Regulation of heavy metal uptake and responses in plants. In *Plant Analysis and Fertilizer Problems.* Abstracts of 6th Int. Colloquium, Tel Aviv, Israel, March 13–17, 1970. Geneva, New York: W. F. Humphrey Press.

Walter, J. M. 1957. Hereditary resistance to disease in tomato. *Ann. Rev. Phytopathol.* 5: 131–62.

Weibe, H. H., and P. J. Kramer. 1954. Translocation of various radioactive isotopes from various regions of roots of barley seedlings. *Plant Physiol.* 29: 342–48.

White, R. P., M. Drake, and J. H. Barker. 1965. Effect of induced changes in root CEC on Ca adsorption from bentomite systems by excised barley roots. *Soil Sci.* 99: 267–71.

Wilson, J. D. 1963. Closer planting of tomatoes favors blossom end rot. *Plant Dis. Reptr.* 47: 729–31.

Wiersum, L. K. 1965. Effect of growth and evaporation of the fruits on the occurrence of blossom end rot in tomato. *Overdruk Mededel. Dir. Tumb.* 28: 264–67 (English summary).

Young, W. I., and D. C. Rasmussen. 1966. Variety differences in strontium and calcium accumulation in seedlings of barley. *Agron. J.* 58: 481–83.

20. Effects of Aluminum on Plant Growth

Charles D. Foy

I. ALUMINUM TOXICITY IN SOILS

A. Occurrence and Importance

ALUMINUM toxicity is an important growth-limiting factor in many acid soils (Adams and Pearson, 1967). The problem is particularly serious in strongly acid subsoils that are difficult to lime (Adams, 1968, 1969; Adams and Lund, 1966), and it is being intensified by the heavy use of acid-forming nitrogen fertilizers (Abruna *et al.*, 1958; Pearson *et al.*, 1962; Wolcott *et al.*, 1965; Pierre *et al.*, 1971). Aluminum toxicity is also increased by the addition of nonnitrogenous fertilizers that displace exchangeable Al into the soil solution and lower soil pH even more (Ragland and Coleman, 1962). Strong subsoil acidity, with Al at toxic levels, reduces root penetration and increases the probability of injury by drought, a frequent growth-limiting factor for crops, even in the humid eastern United States. In acid soils Al is toxic as a cation, but aluminate (anion) toxicity has also been reported in alkaline fly ash deposits of England (Jones, 1961).

B. Soil Factors Affecting Aluminum Toxicity

The solubility of Al and the severity of its toxicity to plants are affected by many soil factors, including soil pH, type of predominant clay mineral, concentrations of other cations, total-salt concentrations, and organic matter content.

In general, Al toxicity does not occur in soils above pH 5.5 (McCart and Kamprath, 1965), but it is common at lower pH values, and parti-

Contribution from the Northeastern Region, Agricultural Research Center, Agricultural Research Service, U.S. Department of Agriculture, Beltsville, Md. 20705.

cularly severe below pH 5.0, where the solubility of Al increases sharply (Magistad, 1925) and more than half the cation exchange sites may be occupied by Al (Evans and Kamprath, 1970). For example, Adams (1968) reported that cotton (*Gossypium* sp.) roots failed to proliferate in subsoil of pH 5.0 or below, growth was stunted, and plants wilted during midseason within 3 to 4 days after a rain. On the other hand, when the subsoils were within the pH range of 5.2 to 5.5, subsoil rooting occurred, yields were not reduced, and plants could withstand drought periods of 10 to 14 days without wilting.

For a given acid soil, lime responses of crops are often well correlated with the KCl-exchangeable Al levels (Table 20.1; Moschler *et al.*, 1960; Abruna-Rodriguez *et al.*, 1970; Kamprath, 1970; Foy *et al.*, 1965a), but the soil pH at which Al becomes soluble in toxic concentrations is different in different soils. Hester (1935) found that the pH at which plant growth was markedly inhibited in Coastal Plains soils coincided with that at which Al appeared in soil drainage water. Soils having colloids of low silica : sesquioxide ratios (Norfolk) produced Al toxicity at higher pH values than those with higher ratios (Bladen). These critical pH values were 4.8 for Portsmouth, 4.9 for Bladen, and 5.5 for Norfolk. In support of Hester's conclusions, Adams and Lund (1966) reported that the displaced solution of a Norfolk subsoil contained toxic levels of Al for cotton at a soil pH of 5.4, but Bladen subsoil did not contain toxic levels above pH 4.9. The critical soil pH for primary cotton root penetration was about 5.5 for Norfolk subsoil and less than 5.0 for Dickson and Bladen subsoils. In recent field studies with cotton and corn (*Zea mays*), Adams (1968, 1969) also demonstrated that certain soils of the Coastal Plains differ in "critical" pH, the maximum pH at which a given crop responds to lime.

Table 20.1. Effects of lime on the top growth of Hudson barley and on the pH and level of KCL–extractable Al in Tatum surface soil

CaCO$_3$ added (ppm)	Yield of barley tops (g/pot)†	Soil properties‡	
		pH	KCl-extractable Al (meq/100 g)
0	0.29 e	4.1	5.75
375	0.91 d	4.3	4.81
750	2.72 c	4.5	4.33
1,500	4.29 b	4.8	2.75
3,000	5.07 a	5.5	0.37

Source: Foy *et al.*, 1965a.
† Values having different letters are significantly different at the 5% level by Duncan's Multiple Range Test.
‡ Averages of three replicates.

Adams and Lund (1966) found that the levels of exchangeable Al required for toxicity were also quite different in different soils, being 0.1 meq/100 g for Norfolk, 1.5 for Dickson, and 2.5 for Bladen. Thus, Al is toxic at a higher soil pH level and at a lower level of exchangeable Al in Norfolk subsoil, whose predominant clay mineral is kaolinite, than in Dickson and Bladen, whose major clay mineral components are vermiculite and montmorillonite, respectively. These investigators concluded that the soil pH, exchangeable Al, and degree of Al saturation were not satisfactory indicators of root growth inhibition in several Coastal Plains soils. In addition, they found that no single critical Al concentration in displaced soil solutions would apply to all the soils studied. For example, an Al concentration of 0.013 meq/l in Bladen subsoil solution was more toxic to roots than was 0.042 meq/l in Dickson subsoil solution, even at lower pH levels in Dickson. However, placing soil solution Al on a molar activity basis resulted in similar critical values for all soils. Cotton root penetration was restricted at values above 0.15×10^{-5} for both displaced soil solutions and the nutrient solution portion of a split medium. They attributed the lower toxicity of Dickson soil solution to a higher salt concentration which reduced the activity coefficient of the Al. Increasing the calcium concentration of a nutrient solution, at the same initial pH, is known to reduce the toxicity of Al and other excess cations (Table 20.2; Foy *et al.*, 1969; Wallace *et al.*, 1970; Lund, 1970).

Richburg and Adams (1970) concluded that differences in critical pH values of soil were caused both by differences in $Al(OH)_3$ activity and by differences in the relationship between the pH of soil-water suspensions and the pH of displaced soil solutions. For example, Lucedale soil, which had a critical pH of 5.1, had higher $pAl(OH)_3$ values than those of Norfolk and Magnolia (critical pH 5.5). The Lucedale subsoil also had the highest displaced solution pH; a soil suspension of pH 5.1 in all three soils produced displaced soil solution pH values of 5.1 in Norfolk, 5.2 in Magnolia, and 5.5 in Lucedale.

Organic soils are known to have lower critical pH values for good crop growth than mineral soils (Welch and Nelson, 1950). The addition of humic acid lowers the pH at which plants are injured in certain acid soils (Mattson and Hester, 1933) and prevents Al toxicity of alfalfa (*Medicago sativa*) in nutrient solutions (Brogan, 1967). Hester (1935) found that the detoxification of Al by the addition of organic matter to acid soils was associated with decreased Al solubility. Evans (1968) reported that Al solubility is very low at pH 5.0 in organic Coastal Plains soils. Bhumbla and McLean (1965) suggested that the exchange acidity displaced at higher soil pH values in some soils is due to organic complexed Al or hydroxy Al polymers. The evidence indicates that the lower critical pH values for plant growth in organic soils compared

Table 20.2. Effects of Al on soybean taproot growth as modified by Ca concentration in the subsurface nutrient solution portion of a split medium

Al added (ppm)	Root elong. rate (mm/hr)†		Harvested root length (mm)		Dry root wt (mg/mm)	
	10 ppm Ca‡	40 ppm Ca‡	10 ppm Ca	40 ppm Ca	10 ppm Ca	40 ppm Ca
0	2.71	2.86	376	445	0.28	0.38
0.5	1.88	2.72	295	443	0.27	0.36
1.0	0.44	2.61	59	427	0.61	0.31
2.0	0.24	1.57	40	245	0.67	0.40
LSD 5%	0.38	0.38	115	56		
1%	0.57	0.57	174	84		

Source: Adapted from Lund, 1970.
† Elongation rate during first 48 hours in solution.
‡ Calcium levels were studied in two separate experiments.

with mineral soils is due, at least in part, to the formation of Al-organic matter complexes of lower solubilities (Schnitzer and Skinner, 1963; Greene, 1963; Evans, 1968). However, there is also the possibility that Al is detoxified by chelation in water-soluble forms. For details on the chemistry of Al in soils, see reviews by Coleman and Thomas (1967), and Coulter (1969), and Chapters 15 and 16 of this volume.

II. EFFECTS OF ALUMINUM ON PLANT GROWTH

A. Beneficial Effects of Aluminum

Although Al is generally regarded as a nonessential element, during the past 60 years various claims have been made for its beneficial effects on plants when used at low concentration. Stoklasa (1911) found small quantities of Al beneficial and believed it to be a catalytic agent in photosynthesis; however, higher concentrations coagulated plant proteins and caused a loss of calcium and potassium from the injured cells. Varvarro (1912) reported that aluminum oxide accelerated the germination of corn. McLean and Gilbert (1928) found that 3 to 13 ppm Al stimulated plant growth but higher concentrations were toxic.

MacLeod and Jackson (1965) found that Al concentrations of 0.1 to 0.2 ppm in nutrient solutions increased the growth of alfalfa and red clover (*Trifolium pratense*) seedlings. Aluminum at 5 ppm stimulated the root growth of *Deschampsia flexuosa*, *Alopecurus pratensis*, *Festuca pratensis*, and *Lolium perenne* (Hackett, 1962, 1967). Dios and Broyer (1962) reported that Al stimulated magnesium uptake by corn. Paterson (1965) found that Al at 0.25 to 0.50 ppm stimulated the growth of young corn plants in a Hoagland's solution containing 5 ppm phosphorus at pH 4.1. Bertrand and deWolf (1968) concluded that Al is actually required by corn and specified the optimum dose as 0.25 to 0.30 ppm in nutrient cultures. This corresponds very closely with the findings of Paterson (1965). Growth of the tea (*Thea sinensis*) plant is said to be increased by Al (Goletiani, 1965). Lee (1971b) found that Al at 1 to 5 ppm (pH 3.7) stimulated the vegetative growth and in some cases the uptake of Mg and K by Irish potatoes (*Solanum tuberosum*). In another study Lee (1971a) found that the addition of 20 ppm Al (pH 3.5) decreased the overall yield of potato tubers. All the decreased yield was in the small and knobby potatoes; yield of the larger tubers and the specific gravity of tubers were increased. The mangold (*Beta vulgaris*) plant (an Al-tolerant species) showed increased yields when treated with alkaline fly ash, which contains toxic levels of Al as aluminate (Rees and Sidrak, 1956). Much smaller quantities of ash reduced yields of Al-sensitive barley (*Hordeum vulgare*). Aluminum added at 2.5 ppm (pH 3.5) stimulated root growth of Al-tolerant cranberry (*Vaccinium macrocarpon*) (Medappa and Dana, 1970). Flower color in *Hydrangea macrophylla* is related to the Al content of the floral tissue (Asen *et al.*, 1963). Blue flowers contain higher concentrations of Al than pink flowers.

The mechanisms by which small quantities of Al benefit plant growth are not clear. A possible explanation is increased iron solubility and availability in the growth medium, resulting from Al hydrolysis and a lower pH. Grime and Hodgson (1969) have presented evidence that the positive Al response of *Scabiosa columbaria* is due to the displacement of Fe (by Al) from bound sites within the plant, thereby relieving an Fe distribution or utilization problem which had caused deficiency.

B. Inhibitory Effects of Aluminum

1. Gross Symptoms of Aluminum Toxicity

Aluminum toxicity in plant tops is often characterized by symptoms resembling those of P deficiency (overall stunting; small, abnormally dark green leaves; purpling of stems, leaves, and veins; and yellowing and death of leaf tips), or of Ca deficiency (cupping or rolling of young leaves and collapse of plant apex or petioles). Recent evidence indicates that bronzing, an important physiological disease of rice (*Oryza sativa*), attributed by some investigators to Fe toxicity (Tanaka and Yoshida, 1970), is actually caused mainly by excess Al in certain soils of Ceylon (Ota, 1968). Velasco *et al.* (1959) noted that high Al levels (which reduce water absorption), plus low copper levels in soils, are closely associated with another physiological disease, cadang-cadang, found in coconut (*Cocos nucifera*) trees in the Philippine Islands.

Aluminum-injured roots are characteristically stubby and spatulate in appearance. Root tips are inhibited and turn brown. The root system as a whole is coralloid in appearance, having many inhibited and thickened lateral roots but lacking in fine branching (Fleming and Foy, 1968; Clarkson, 1969; Reid *et al.*, 1971). Aluminum injury also appears to predispose plant roots to fungal infection (Ota, 1968). Cate and Sukhai (1964) reported that 1 ppm Al had no effect on the germination of rice in Petri dishes but did stunt lateral root growth. Dessureaux (1969) found that up to 100 ppm Al in nutrient solutions did not inhibit the germination of alfalfa but that toxicity began shortly after emergence of the radicle.

2. Cytological Effects of Aluminum

Levan (1945) reported that Al and other salts caused severe cytological abnormalities in the dividing cells of onion (*Allium cepa*) roots, including the formation of "sticky chromosomes" and anaphase bridges. Clarkson (1965) found that Al at 5.4 to 54 ppm, added as $Al_2(SO_4)_3$ (pH not given), completely inhibited elongation in onion root after 6 to 8 hours. Cessation of root elongation was closely correlated with the disappearance of mitotic figures. However, other trivalent ions, such as gallium, indium, and lanthanum, produced similar results. Rios and Pearson (1964) observed that Al concentrations above 0.5 ppm (pH not given) prevented growth of cotton seedling roots and that these roots did not recover when placed in Al-free solutions. The appearance of binucleate cells in meristematic regions of root tips indicated that cell division was inhibited. Huck (1972) found that exposing cotton roots to 1 ppm Al (pH 4.3) for 12 hours produced a breakdown of cells in the pericycle and a high frequency of binucleate cells. The Al-damaged tissue was soft and necrotic.

3. Physiological and Biochemical Effects of Aluminum

a. Effects on Protoplasm, Enzyme Activity, and General Metabolism. Many investigators have shown that Al increases the viscosity of protoplasm in plant root cells and decreases overall permeability to salts, dyes, and water (Stoklasa, 1911; Szucs, 1912; McLean and Gilbert, 1927; Hofler, 1958; Bohm-Tuchy, 1960; Aimi and Murakami, 1964). Clarkson and Sanderson (1969) suggested that these effects are due to cross-linking between adjacent protein molecules. Rorison (1958) concluded that Al decreases the extensibility of cell walls by cross-linking pectins in the middle lamella. Aluminum has been shown to accumulate in the nuclei of some Al-injured plant cells (McLean and Gilbert, 1927; Aimi and Murakami, 1964). Avdonin *et al.* (1957) noted that Al decreased the sugar content, increased the ratio of non-protein to protein N, and decreased P contents of leaves from several plants grown on acid podzolic soils. Aluminum toxicity (bronzing) in rice has been associated with interference in the synthesis of starch and protein, reduced translocation of sugars, and increases in the peroxidase and decreases in the cytochrome oxidase activities of leaves (Ota, 1968). Huck (1972) concluded that Al reduced the abilities of cotton roots to utilize sucrose in building cell-wall polysaccharides. Anderson and Evans (1956) found that Fe or Al inhibits isocitric dehydrogenase and malic enzyme activities in bean plants.

Aluminum has been reported to reduce sugar phosphorylation (Rorison, 1965; Clarkson, 1966b), respiration (Norton, 1966–67), and DNA synthesis (Sampson *et al.*, 1965; Clarkson, 1969). Eichhorn (1962) provided evidence that metal cations, which are bound to DNA in vitro, increased the stability of the double helix. On the basis of this evidence, Clarkson and Sanderson (1969) have suggested that the observed interference in DNA replication induced by Al (Clarkson, 1969) takes place by cross-linking of polymers, which increases the rigidity of the DNA double helix.

Progress in discovering the mechanism of Al toxicity has been greatly hindered by the unavailability of a suitable radioactive isotope of Al. Clarkson and Sanderson (1969) noted that at equal concentrations, aluminum and scandium are very similar in their inhibitions of root elongation and cell division and have concluded that ^{46}Sc can reliably indicate the behavior of Al in root meristems. They suggested that these two elements injure plants through a common mechanism which may be more closely related to their polyvalency than to their other properties. However, Clarkson (1969) has also observed that although other trivalent metal ions such as gallium, scandium, yttrium, and lanthanum may produce similar effects on cell division, resistance to Al injury in one species of bentgrass (*Agrostis setacea*) and in rye (*Secale cereale*) appears to be specific and does not necessarily coincide with tolerance to these other ions.

b. Nutrient Uptake and Utilization. Many investigators have associated Al toxicity with reduced uptake of several nutrient elements by plants, particularly Ca and P. However, the question remains: Are these effects intimately involved in the actual mechanism of Al toxicity or merely consequences of an earlier disturbance by Al at the cellular or subcellular level?

Reduced Ca uptake is commonly observed in Al-injured plants. For example, Johnson and Jackson (1964) found that Al reduced both the absorption and accumulation of Ca by wheat (*Triticum aestivum*). Paterson (1965) concluded that Al reduced Ca uptake by corn but did not appear to inhibit Ca transport to plant tops. Lance and Pearson (1969) found that Al at only 0.3 ppm in solution reduced Ca uptake by cotton seedling roots within one hour. This inhibition was prevented by increasing the solution Ca concentration to 600 ppm.

Clarkson (1970) found that Al reduced both the Ca associated with the free space of barley roots and the rate at which Ca previously absorbed by cells moved outward across the plasma membrane. He suggested that Al blocks, neutralizes, or reverses the negative charge on the pores of the free space and thereby reduces the abilities of such pores to bind Ca. From this it was postulated that Al should increase anion binding by the free space and total anion uptake by plants. Data obtained from ^{36}Cl uptake supported the hypothesis.

DeWard and Sutton (1960) associated Al toxicity in black pepper (*Piper nigrum* L.) vines with reduced uptake of Ca and Mg, plant deficiency symptoms of these elements, death of roots, wilting of foliage, and increased uptake of K and Al. They suggested that the ratio of K:Ca plus Mg in plant tops can be used as an index of Al injury. Pepper vines were slightly affected at a ratio of 1.29 and severely affected at a ratio of 3.89.

Aluminum toxicity in rice has been characterized by lower concentrations of Ca, Mg, K, Mn, and Si and higher concentrations of N and P in plant tops (Ota, 1968). Plant injury in rice was positively correlated with Al contents of roots. Lee (1971b) found that Al inhibited the transport of P to potato plant tops; decreased the absorption of Ca, Mg, and Zn by roots; and caused the accumulation of P, Al, Mn, Cu, and Fe in plant roots. Potassium absorption was stimulated by 1 to 2 ppm Al but inhibited by 5 to 10 ppm. Otsuka (1968a) reported that Al induced an Fe deficiency chlorosis in acid-soil-sensitive wheat and barley varieties.

Aluminum reduced the concentrations of Ca, Mg, and K in mature leaves of lettuce (*Lactuca sativa*) and Ca concentrations in all of the leaves (Harward *et al.*, 1955). Total Cu uptake by excised wheat roots was reduced by 0.1 ppm Al in nutrient solutions (Hiatt *et al.*, 1963). The latter investigators suggested that Al was competing with Cu for common binding sites at or near the root surface but not for a common ion carrier involved in active accumulation.

c. Phosphorus Uptake and Utilization. Aluminum toxicity often appears as a P deficiency in plants grown on acid soils or in nutrient solutions (Foy and Brown, 1963, 1964; Chiasson, 1964). Excess Al may reduce the solubility of P in the growth medium and its uptake and utilization by plants. For example, Suchting (1948) found that removing the soluble Al from acid soils increased P recovery of pine trees from 2.5% to 40% of applied P. Increasing the P supply in the growth medium can precipitate and thus detoxify Al, increase P uptake, and prevent P deficiency symptoms. There is also evidence that in the presence of excess soluble Al, a higher level of soluble P is required to prevent P deficiency in cotton (Foy and Brown, 1963).

The detrimental effects of high Al and low P are often extremely difficult to separate in acid soils. However, Reeve and Sumner (1970) concluded that on eight oxisols Al toxicity, P deficiency, and P fixation were primary but independent growth-limiting factors. They attributed the beneficial effects of P fertilization to the elimination of Al toxicity and the resulting increased ability of plants to absorb P, rather than to increased P availability in soils. Munns (1965a, 1965b) concluded that P treatments reduced the detrimental effects of Al only when they caused Al to precipitate in the medium. He noted that Al decreased P concentrations in both roots and tops of alfalfa but that Al-induced P deficiency symptoms were not corrected by increasing the P level in the growth medium, even when this restored the P concentrations in the plants to high levels. Chiasson (1964) found that in an acid Canadian soil (pH 5.0), banding 20 ppm N and 22 ppm P directly with barley seed prevented leaf-yellowing (P deficiency) symptoms and doubled yields but that lime and $MgSO_4$ (250 ppm) similarly applied did not influence symptoms or yield. One possible cause of such a beneficial effect is the precipitation of toxic Al by P in the root zone.

In some instances Al toxicity has been associated with reduced P concentrations in both roots and tops (Munns, 1965b). In others Al was believed to precipitate P inside plant roots and thereby cause P deficiency in the tops (Wright, 1937, 1948, 1952). In Wright's studies the P bound by Al-injured roots was largely inorganic and had a low water solubility. He attributed the beneficial effects of large applications of superphosphate on acid soils to precipitation of Al and P inside the plant with sufficient P remaining for metabolic purposes. MacLeod and Jackson (1965) reported that P appeared to be immobilized by Al in alfalfa roots. In contrast to the evidence cited above, Ruschel *et al.* (1968) suggested that the harmful effects of Al on *Phaseolus vulgaris* were not due to an Al-induced P deficiency. Wallihan (1958) concluded that Al did not interfere with P metabolism in ladino clover. Jones (1961) found that S-100 white clover, which he classified as Al-tolerant, accumulated considerable amounts of Al in the leaves but showed little evidence of P deficiency.

Although Al generally reduces the uptake of P by plants, exceptions have been reported. Randall and Vose (1963) found that 5.0 ppm Al in solution culture increased the total P and the P concentrations in both roots and tops of 8-week-old perennial ryegrass plants. Low Al concentrations, given as a pretreatment or during a 4-hour absorption period, greatly increased P uptake. Because this Al-stimulated P uptake was reduced by KCN, they judged it to be connected with metabolism, but they did not rule out precipitation of Al and P on root surfaces. Higher levels of Al (50 ppm) also increased P concentrations in plants but depressed total P uptake by reducing growth. They concluded that Al bound P within the plant and that this produced the P deficiency that they observed. Aluminum-stimulated P uptake has also been found in cranberry (Medappa and Dana, 1968), in excised snapbean roots (Ragland and Coleman, 1962), and in barley seedlings (Clarkson, 1966b). In Clarkson's work the P associated with Al-treated roots was inorganic and almost entirely exchangeable. However, in contrast to the results of Randall and Vose (1963) with perennial ryegrass, the Al-stimulated P uptake was not reduced by DNP or low temperature, suggesting that the process was not directly dependent upon metabolism.

Rasmussen (1968), using electron microprobe analysis, reported that Al absorbed from acid solutions (as a cation) precipitated on cell walls of corn roots, with no penetration of the cortex as long as root surfaces were intact, but openings made by emerging laterals did provide channels for Al entry. The observation that the distribution of P was exactly the same as that of Al led them to suggest that aluminum phosphate was the precipitate. On the other hand, Waisel *et al.* (1970), using anionic Al (pH 9.5) and X-ray microanalysis, found no correlation between the distribution of Al and P in bean and barley roots. Aluminum appeared to be located inside the cell lumen (with its distribution overlapping that of N), with only trace amounts found in cell walls. The absence of Al precipitates in cell walls suggested to these investigators that the Al in the protoplast was mobile. Differences in the results of Rasmussen (1968) and Waisel *et al.* (1970) could be due to different plant species or varieties, different electron microprobe techniques, pH of the medium, or perhaps a difference in the behavior of cationic and anionic Al in plants. However, Jones (1961) found that both forms produced P deficiency in some plants.

Rios and Pearson (1964) found that cotton grown with 2.5 ppm Al (but no P) in the subsurface solutions of a soil-nutrient solution split medium had a higher P concentration than those grown without Al. This suggested that P absorbed from the soil above moved downward and was precipitated by Al absorbed from the nutrient solution. This could have implications regarding P fertilizer efficiency in soils having Al toxic subsurface layers.

Aluminum is known to reduce hexose phosphorylation (Rorison, 1965; Clarkson, 1966b, 1969), the process by which P from ATP combines with sugars to give sugar phosphates and ADP (catalyzed by hexokinases). Clarkson (1966b) showed further that Al citrate inhibited the activity of a purified yeast hexokinase, reduced the rate of sugar phosphorylation by crude mitochondrial extracts from barley roots, and increased concentrations of ATP in barley roots. Woolhouse (1969) has noted that Al inhibits adenosine triphosphatases in plants, and this led Clarkson (1969) to conclude that Al acts directly or indirectly to prevent the utilization of ATP in glucose phosphorylation.

In summary, the biochemical evidence indicates that Al binds P on root surfaces and cell walls and in the free space of plant roots (Rorison, 1965; Clarkson, 1966b, 1967a), making P less available to metabolic sites within the cells. Some of the Al absorbed by plant roots also appears to penetrate the cells of meristematic tissue and to interfere with cell division, respiration, DNA synthesis, and sugar phosphorylation (Rorison, 1965; Clarkson, 1966b, 1969; Norton, 1966–67; Sampson *et al.*, 1965).

4. *Effects of Aluminum on Rhizobia*

Very little information is available concerning the direct effects of the Al ion on rhizobia. MacLeod and Jackson (1965) cited the work of Kliewer (1961), who reported that 0.4 to 0.7 ppm Al in solution greatly reduced the effective nodulation of birdsfoot trefoil (*Lotus corniculatus*). No other references were found on the subject.

III. DIFFERENTIAL ALUMINUM TOLERANCE OF PLANT SPECIES AND VARIETIES

A. *Range of Tolerance*

Plant species differ widely in their tolerance to excess soluble or exchangeable Al in acid soils or nutrient solutions (McLean and Gilbert, 1927; Lignon and Pierre, 1932; Hewitt, 1948; Jones, 1961; Foy and Brown, 1964; Clarkson, 1966a; Jackson, 1967; Grime and Hodgson, 1969; Adams and Pearson, 1970). McLean and Gilbert (1927) classified lettuce, beets, timothy (*Phleum pratense*), and barley as Al sensitive; radishes, sorghum, cabbage, oats, and rye as medium sensitive; and corn, turnips, and redtop as resistant. Aimi and Murakami (1964) found that lettuce was more sensitive to Al in solution than turnips and radishes were; these in turn were more sensitive than maize, rice, cucumber (*Cucumis sativus*), and squash (*Cucumis* sp.). Aluminum tolerance in nutrient solution was well correlated with acid soil tolerance. Jones (1961) rated barley as Al sensitive; brussels sprouts (*Brassica oleracea*) and peas (*Pisum* sp.) as semitolerant; and S-100 white clover, mangold, mustard, and *Atriplex hastata* as tolerant. In general, plants classified as calcifuges (acid-soil plants), such as *Deschampsis flexuosa* and *Carex demissa*, are more tolerant to Al than those classified as calcicoles (calcareous-soil plants), such as *C. lepidocarpa* (Clymo, 1962; Grime and Hodgson, 1969).

The tomato (*Lycopersicon esculentum*) plant is very sensitive to Al and has been suggested as an indicator of Al availability in soils (Rees and Sidrak, 1956; Mercado and Velasco, 1961). The datura plant is much more tolerant than tomato or eggplant (Otsuka, 1968b). Alfalfa and cotton are injured by as little as 0.5 ppm Al in nutrient solutions or displaced soil solutions (Rios and Pearson, 1964; MacLeod and

Jackson, 1965). Buckwheat (*Fagopyrum esculentum*) is much more tolerant than barley or cotton (Foy and Brown, 1964). Cotton root growth was practically stopped in Al-toxic subsoils, which had no effect on peanut roots (Adams and Pearson, 1970). Perennial ryegrass, white clover, and subclover (*Trifolium subterraneum*) are reportedly unaffected by Al toxicity at soil pH values where alfalfa is severely injured.

The cranberry plant appears extremely tolerant to Al, requiring 150 ppm Al added to solutions (at pH 3.5) for reduction of shoot growth (Medappa and Dana, 1970). Root length tended to be reduced at Al concentrations above 2.5 ppm, but root weight was not seriously decreased by 25 ppm Al.

In acid soils Kamprath (1970) found that corn made good growth at Al saturations of up to 44% but soybeans (*Glycine max*) would not tolerate above 20%. Evans and Kamprath (1970) reported that corn responded to liming when the soils contained above 0.4 meq Al/100 g and soybeans above 0.2 meq/100 g, corresponding to Al concentrations of 3.6 and 1.8 ppm in displaced soil solutions. Clarkson (1966a) found that species of the *Agrostis* (bentgrass) genus differed widely in Al tolerance. Roots of *A. stolonifera* showed injury at 5.4 ppm Al and *A. canina* at 10.8 ppm, but those of *A. setacea* and *A. tenuis* showed no root damage at 21.6 ppm Al in nutrient solution. Furthermore, *A. setacea* grew at 43.2 ppm Al, which inhibited root growth of *A. tenuis*.

The Al tolerance ratings of the species given above can only be considered estimates of the range of tolerance available in plants. The tolerance of a specific plant to a given amount of Al added to a culture is determined by many factors (pH, Ca, P, plant variety, and so on), which probably varied in the experiments cited.

Plant varieties within species also differ widely in Al tolerance, and this is partly responsible for the different species rankings in the older literature. Such varietal differences have been found in wheat and barley (Neenan, 1960; Foy *et al.*, 1965a, 1965b, and 1967; MacLean and Chiasson, 1966; Kerridge *et al.*, 1971), alfalfa (Ouellette and Dessureaux, 1958), perennial ryegrass (Vose and Randall, 1962), soybeans (Foy *et al.*, 1969), rice (Ota, 1968), Irish potatoes (Lee, 1971a, 1971b), and peanuts (Adams and Pearson, 1970). Lee (1971b) rated the Al tolerance of potato varieties as: Netted Gem (Russet Burbank) > Katahdin > Sebago > Green Mountain. Netted Gem actually seemed to require 1 to 5 ppm Al in the nutrient solution for best vegetative growth. Katahdin benefited from 1 ppm Al but was seriously injured by 5 ppm. Sebago was not affected by 1 to 5 ppm Al, and 10 ppm were required for injury. The Green Mountain variety was sensitive to 1 ppm Al. For additional details on species and varietal differences in Al tolerance, see tables in Jackson (1967) and Adams and Pearson (1967). Varietal differences in tolerance to acid, Al-toxic soils are illustrated for wheat (Table 20.3), barley (Fig. 20.1), and soybeans (Table 20.4). Differential Al tolerance of two barley varieties in nutrient solution is shown in Figure 20.2.

B. Characterization of Differential Aluminum Tolerances among Plant Species and Varieties

The exact biochemical nature of Al toxicity is still debated. Morphological, physiological, and biochemical properties, which have been associated with differential Al tolerance, are summarized here.

Table 20.3. Differential lime response of three wheat varieties on Al-toxic Tatum soil

CaCO$_3$ added (ppm)	Average soil pH	Yield of plant tops (g/pot)†		
		Atlas 66	Monon	Thatcher
0	4.2	1.50 a	0.49 c	0.23 d
1500	5.1	4.23 a	3.66 a	3.71 a
3000	5.8	4.25 b	4.66 a	4.76 a
4500	6.7	3.67 a	3.95 a	3.99 a
6000	7.2	3.16 a	2.99 a b	2.81 a b c

Source: Foy *et al.*, 1965a.
† Within a given lime level yield values having a letter in common are not significantly different at the 5% level by the Duncan Multiple Range Test. Origins of varieties are: Atlas 66, N.C.; Monon, Ind.; and Thatcher, Minn.

Table 20.4. Yields of tops of four soybean varieties on Al-toxic Bladen surface soil at different lime levels

CaCO$_3$ added (ppm)	Final average soil pH	Yield of plant tops (g/pot)†				
		Wayne	Clark	PI 85666	Perry	Chief
0	4.7	4.37 b c	3.63 e f	2.91 g	4.28 b c	3.26 f g
1500	5.0	5.39 b c	5.39 b c	4.84 c d e	5.38 a b c	5.31 b c d
3000	5.5	5.29 b c	5.32 b c	5.38 a b c	5.51 a b c	5.11 b c d
6000	6.4	4.67 c d	4.79 b c d	5.31 b c d	5.39 a b c	5.09 b c d

Source: Armiger *et al.*, 1968.
† Within a given lime level any two yields having a letter in common are not significantly different at the 5% level by the Duncan Multiple Range Test.

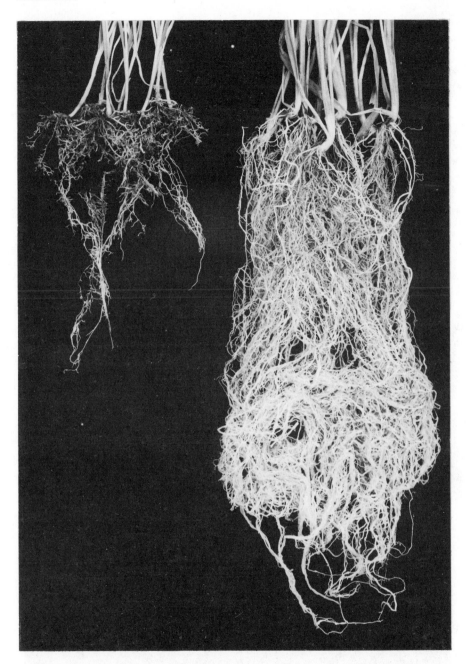

Fig. 20.1. Roots of Kearney (*left*) and Dayton (*right*) barley varieties grown in acid, Al-toxic Bladen soil (initial pH 4.6). (C. D. Foy, unpublished data)

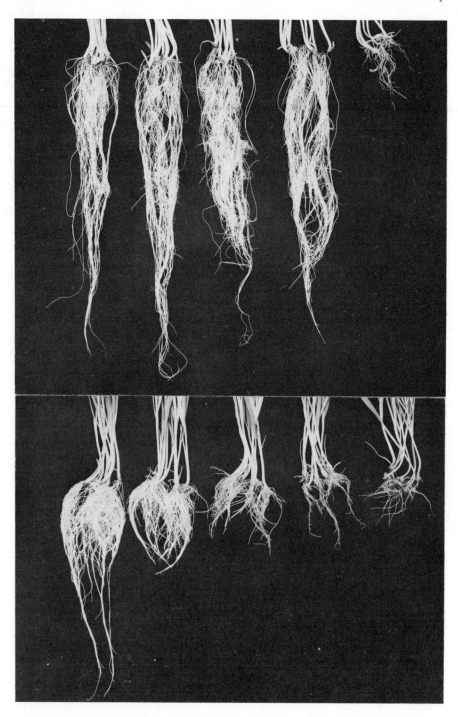

Fig. 20.2. Differential Al tolerance of Dayton (*top*) and Kearney (*bottom*) barley varieties in nutrient solutions. *Left to right*: 0, .75, 1.5, 3.0, and 6.0 ppm Al added at initial pH 4.8. Final pH values were 6.8, 7.0, 7.2, 6.7, and 4.4 for Dayton and 7.2, 6.6, 6.2, 4.7, and 4.4 for Kearney. (Foy *et al.*, 1967)

1. Root Morphology

Fleming and Foy (1968) found that greater Al tolerance in Atlas 66 wheat compared with Monon was associated with greater ability to continue root elongation and resist morphological damage to root tips and lateral roots when under Al stress and to initiate new lateral roots when the stress was removed. Differential Al tolerance of Atlas 66 and Monon wheat roots are shown in Figures 20.3 and 20.4. In addition, the greater Al injury in the more sensitive Monon variety was characterized by greater internal disorganization and the appearance of binucleate cells in roots. Reid *et al.* (1971) showed that Al injury in barley varieties is characterized by an increase in the numbers of roots, a decrease in root length and root weight per plant, and inhibition and darkening of root tips and laterals (Table 20.5).

2. Plant-induced pH Changes in Root Zones

Differential Al tolerance among some plant species and varieties is associated with differential ability to alter the pH of their root zones. For example, Foy *et al.* (1965b, 1967) showed that Al-sensitive Monon wheat and Kearney barley varieties induced lower pH values in their growth media than did the Al-tolerant Atlas 66 wheat and Dayton

Table 20.5. Differential effects of aluminum on root characteristics of barley varieties in nutrient solutions and in Al-toxic Tatum soil.

Variety	Nutrient solution† (4 ppm Al/no Al%)			Tatum soil (low lime/high lime %)
	No. of roots per plant	Max. root length	Root wt. per plant	Root wt. per pot
Al-tolerant				
Colonial 2	117	62	159	77
Calhoun	115	63	168	89
Dayton	110	70	149	75
Barsoy	109	74	143	108
Al-sensitive				
Cordova	207	17	67	53
Kenbar	190	17	79	62
Kearney	165	9	44	29

Sources: Nutrient solution data adapted from Reid *et al.*, 1971; Tatum soil data from Reid *et al.*, 1969.
†Nutrient solutions contained 3 ppm P; pH was adjusted initially to 4.8 and twice daily thereafter for the growth period of 18 days.

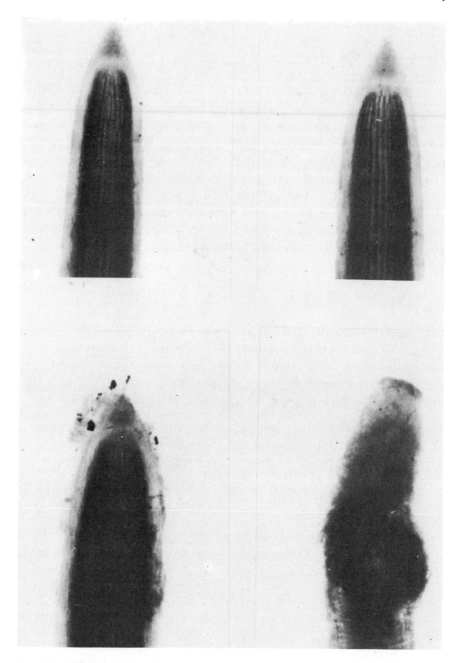

Fig. 20.3.   Effects of Al on primary root tip structure in Atlas 66 (*left*) and Monon (*right*)
wheat varieties, with no Al (*top*) and 9 ppm Al (*bottom*) added. (Fleming and Foy, 1968)

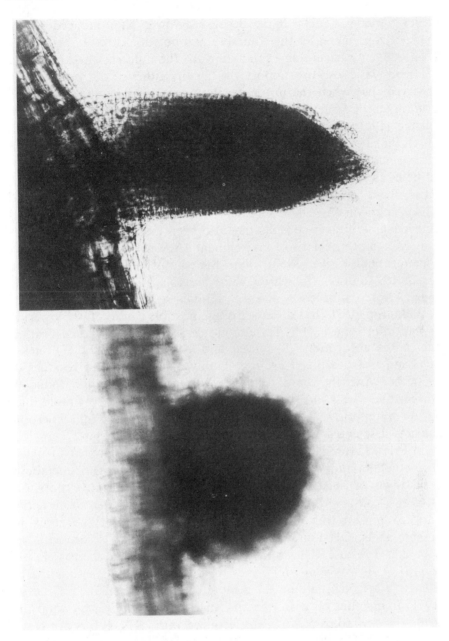

Fig. 20.4. Differential effects of 9 ppm Al on lateral root structure in Atlas 66 (*top*) and Monon (*bottom*) wheat varieties. (Fleming and Foy, 1968)

barley varieties (Table 20.6). Because a lower plant-induced pH in the root zone increases the solubility and potential toxicity of Al, this is an adequate beginning explanation for the differential Al tolerance of these varieties. In support of this hypothesis, Clarkson (1970) observed that when the nutrient solution pH was maintained at 4.2, Dayton and Kearney barley varieties appeared equally sensitive to Al. Otsuka (1968c) reported that an Al-tolerant wheat variety (Hiracki) raised the pH of its growth medium, but an Al-sensitive variety (Norin 25) lowered it. Rice seeds from acid-soil-tolerant (probably Al-tolerant) varieties raised the soil pH from 3.2 to 5.5 during germination, but those from acid-soil-sensitive varieties did not (Subramoney and Saukaranarayanan, 1964). Greater Al sensitivity of cotton compared with peanuts in acid subsoils has also been associated with the creation of a more acid root environment (Adams and Pearson, 1970). Chamura and Koike (1960) found that the pH of plant roots grown in culture solutions increased with increasing acid resistance in the order: rice > tobacco > maize > mint > sorghum.

Wilkinson (1970) has discussed some possible causes of pH lowering in plant root zones: (1) hydrogen ion release as a result of excess cation over anion absorption; (2) release and hydrolysis of CO_2; (3) release of H ion from carboxyl groups of polygalacturonic acid residues of pectic acid (which would result in the higher CEC values found in Al-sensitive wheat and barley varieties, which lower the pH more than Al-tolerant varieties); and (4) excretion of protons from microorganisms associated with roots. Pitman (1970) noted that H ion loss from excised, low-salt barley roots appeared to be coupled to phase II salt uptake and was not due to increased respiration or metabolic breakdown of sugars. Because of the large net negative potential (-60 mV) observed, he suggested that H ion loss is due to an active outward proton transport. He also pointed out that the efflux of organic acids, OH ions, or HCO_3 ions will increase or decrease the apparent H release. Riley and Barber (1969) attributed increase in pH and HCO_3 in soybean root zones to greater uptake of anions (particularly NO_3) than cations. In a more recent study Riley and Barber (1971) found that on a soil having an initial pH of 5.2, the final pH in the rhizocylinders of soybeans fertilized with NH_4^+-N was decreased to 4.7 but that in the root zone of plants fertilized with NO_3^--N, the pH was increased to 6.6.

Differential Al tolerances between Perry and Chief soybean varieties (Foy *et al.*, 1969) and Dade and Romano snapbean varieties (Foy *et al.*, 1972) do not appear to be associated with differential pH changes in nutrient cultures. This may mean that in some plants differential pH changes are merely results, rather than primary causes, of differential Al

Table 20.6. Differential Al tolerance of Dayton and Kearney barley varieties in relation to pH changes in nutrient solutions and composition of plant tops

Barley variety	Al added (ppm)	Final solution pH†	Plant yield (g/pot)‡		Plant composition (meq/100 g)				
			Tops	Roots	Tops			Roots	
					Al	H$_2$PO$_4$	Ca	Al	H$_2$PO$_4$
Dayton	0	6.8	2.41	0.81	4.5	20.2	28.2	2.4	18.7
Kearney	0	7.2	3.19	1.07	4.4	20.7	48.8	3.8	20.0
Dayton	3	6.7	2.20	0.84	5.0	10.0	27.4	20.6	13.8
Kearney	3	4.7	2.40	0.22	4.5	15.8	12.4	35.3	21.5

Source: Foy et al., 1967.
† pH initially adjusted to 4.8 and left unadjusted thereafter.
‡ Yields of 16 plants grown for 20 days in 9 l of solution.

tolerance. Because pH affects so many aspects of soil chemistry and plant nutrition, this subject deserves much more exhaustive study.

Small and Jackson (1949) reported that the root sap of oats (rather tolerant to acid soils and Al) had a higher buffer capacity (within the pH ranges of 4.6 to 4.8 and 6.4 to 6.6) than barley, which is more sensitive to these conditions. Ikeda (1965) also associated the higher Al tolerance in certain wheat varieties with higher buffer indexes of root sap. Chenery (1948) noted that the pH of cell sap from Al-accumulating plants was generally below 5.3. The significance of such information is not immediately apparent in explaining differential Al tolerance, but the subject should receive further study in relation to Al mobility in plant sap and the interactions of Al with root exudates in soils.

3. Aluminum Uptake and Transport

Aluminum tends to accumulate in the roots but not in the tops of some Al-injured plants (MacLeod and Jackson, 1965; Foy *et al.*, 1967). Rorison (1958) found that the inhibition of alfalfa root elongation by Al was associated with a rapid Al uptake and suggested that tolerant plants might take up Al more slowly. Aluminum concentrations in plant roots are often negatively correlated with crop yields on acid soils (Chamura, 1962; Ota, 1968). Rorison (1965) found that Al movement into excised roots of sanfoin (*Onobrychis sativa* Lam.) followed a normal curve which suggested little uptake into nonfree space. Clarkson (1967a) found most of the Al in Al-injured barley roots to be associated with the cell walls.

Differences in Al tolerance among wheat, barley, and soybean varieties have not been associated with differential Al concentrations in plant tops (Table 20.6; Foy *et al.*, 1967, 1969a). Aluminum-tolerant varieties of perennial ryegrass, wheat, and barley have lower root cation exchange capacities and accumulate lower concentrations of Al in their roots than do sensitive varieties (Vose and Randall, 1962; Foy *et al.*, 1967). However, the Al-tolerant Perry soybean variety did not accumulate any less Al in its roots than did the Al-sensitive Chief variety (Foy *et al.*, 1969).

Some investigators have reported that Al concentrations in plant tops correlate with Al injury. Otsuka (1968a) observed that Al-tolerant Hiracki wheat contained a lower Al concentration in its shoots than did the Al-sensitive Norin 25 variety. Ouellette and Dessureaux (1958) found that Al-tolerant alfalfa clones contained lower concentrations of Al in their tops and higher concentrations of Al in their roots than did Al-sensitive clones. Lunt and Kofranek (1970) observed that the azalea (*Rhododendron*) plant, which is quite tolerant to Al, retains 90% of its total Al content in the roots. Other plants that appear to tolerate Al by excluding it from their tops are cranberry (Medappa and Dana, 1970) and pangolagrass (*Digitaria decumbens* Stent.) (Moomaw *et al.*, 1959).

Hoyt and Nyborg (1971) found that Al concentrations in alfalfa tops grown on acid Canadian soils (pH 4.0 to 5.6) were highly correlated with Al extracted from soils by 0.01 M $CaCl_2$ and with Al activity calculated by the Debye-Huckel method but less well correlated with that extracted by 0.002 *N* HCl or ammonium acetate at pH 4.8. Salt-extractable Al was also significantly correlated, but to a lesser extent, with alfalfa yield response to lime. In these studies the Al contents of alfalfa were only 44 to 134 ppm, which are well below the 200 to 325-ppm critical range established by Ouellette and Dessureaux (1958) for alfalfa injury.

In some plants Al tolerance is associated with accumulation of Al in the tops, rather than exclusion from tops and/or roots. Hu *et al.* (1957) reported that plants which are indicative of acid soils in China usually had lower concentrations of Ca, P, Na, K, and Fe and higher concentrations of Mn and Al than calcareous-soil plants. Jones (1961) found that Al-tolerant plant species, such as *Atriplex hastata*, took up more Al than Al-sensitive species, such as barley. Suchting (1948) noted that pine trees, which are noted for acid soil and Al tolerance, accumulated 1,000 to 1,500 ppm Al in the tops when grown on an acid soil and only 200 to 450 ppm when grown on a sand-humus medium. Mangrove (*Rhizophora horisonii*) is also reported to be an Al accumulator (Hesse, 1963). Moomaw *et al.* (1959) found that 13 of 23 Hawaiian plant species contained more than 1,000 ppm Al in their tops and were classified as Al accumulators. Staghorn fern (*Gleichenia linearis* Burm, C. B. Clarke) appeared to be an obligate accumulator of Al and a fair indicator of bauxite deposits. In such plants Al may be prevented from reaching critical metabolic sites within the cells (trapped in cell walls), or enzymes may be altered to tolerate Al (Turner, 1969).

4. Calcium Uptake and Use

Differential Al tolerance among varieties of several plant species and varieties appears to be closely associated with the differential uptake and transport of Ca. Aluminum-sensitive wheat and barley varieties are more susceptible to Al-induced Ca deficiency than are tolerant varieties (Foy *et al.*, 1967, 1969; Long and Foy, 1970). Table 20.6 shows that Al inhibits Ca uptake to a greater extent in Al-sensitive

Kearney barley tops than in those of Al-tolerant Dayton. Perry soybeans are more tolerant to Al in acid Bladen soil and in nutrient solution than is the Chief variety. Greater Al sensitivity of the Chief variety is associated with greater susceptibility to a petiole collapse which was identified as an Al-induced Ca deficiency in the actual zone of collapse. Chief is also more susceptible to a cupping symptom in young leaves which has been observed in plants grown on Bladen subsoil (Armiger *et al.*, 1968). Aluminum treatment reduced Ca concentration in roots and tops of Al-sensitive Chief to a greater degree than in those of Al-tolerant Perry. Differential Al tolerance between the two varieties was greater with 2 ppm Ca (Fig. 20.5) than with 50 ppm (Fig. 20.6) in the solutions. Note in Figure 20.5 that with low Ca (2 ppm), adding Al induces a coralloid root structure, which is quite different from that at higher Ca levels (50 ppm) shown in Figure 20.6. Note also in Figure 20.5 that Perry can produce a fibrous root at higher Al levels than Chief.

Ouellette and Dessureaux (1958) found that Al-tolerant alfalfa clones contained lower concentrations of Al in their tops and higher concentrations of Al and Ca in their roots than did Al-sensitive clones. They suggested that Ca reduced Al toxicity by reducing Al uptake and transport to plant tops.

5. Magnesium, Potassium, and Silicon Uptake

Greater Al sensitivity in sorghum compared with corn has been associated with a greater reduction in K uptake (Chamura and Hoshi, 1960). Effects of Al on K uptake were greater in young than in old plants. Lee (1971b) suggested that the greater Al tolerance in certain varieties of Irish potatoes is associated with the abilities of plant roots to absorb Mg and K. Ota (1968) found that rice varieties having the greatest resistance to bronzing (Al toxicity) have much higher levels of Si in the epiderminal cells of leaves than do susceptible varieties. Silicon is known to reduce the internal toxicity of Mn in barley leaves (Williams and Vlamis, 1957) and may play a similar role in detoxifying Al. Silicon may also complex Al in soils and reduce its toxicity. Mattson and Hester (1933) showed that the addition of silicates to strongly acid soils lowered the soil pH at which plants were injured.

6. Phosphorus Uptake and Metabolism

Differential Al tolerance among certain plant species and varieties has been correlated with differential abilities to absorb and utilize P in the presence of Al. For example, Al-tolerant buckwheat absorbs P much more effectively in solutions containing Al than does Al-sensitive barley (Foy and Brown, 1964). Abilities of plant species to tolerate high levels of anionic (aluminate) Al in alkaline fly ash deposits are also associated with abilities to maintain their P status (Rees and Sidrak, 1956; Jones, 1961). Hackett (1967) observed that *Deschampsia flexuosa*, which colonizes acid soils of pH 3.5 to 5.0, was much more tolerant to low P levels than perennial ryegrass, which is less tolerant to acid soils and Al. Clarkson (1967b) found that higher Al tolerance in *Agrostis setacea* compared with *A. canina* and *A. stolonifera* was associated with greater tolerance to low absolute levels of P in the growth medium. Otsuka (1968a, 1968b) reported that the greater acid soil and Al tolerance of datura compared with tomato and eggplant were associated with greater ability to absorb P in the presence of Al, or at low P levels in the absence of Al. Cranberry, which is extremely tolerant to acid soils and Al, can maintain normal P concentrations in its tops when grown with Al concentrations ranging from 2.5 to 150 ppm (pH 3.5) in the nutrient solution (Medappa and Dana, 1970). The detoxification of Al by synthetic chelating agents is closely associated with increases in the P concentrations of plant tops, in addition to growth increases (Rees and Sidrak, 1961; Foy and Brown, 1964). In the latter reference EDDHA [ethylenediamine di(o-hydroxypheny-lacetic acid)] also decreased the Al concentration in plant tops and roots. Ikeda *et al.* (1965) noted a correlation between acid soil tolerance and tolerance to low P levels in wheat varieties.

MacLean and Chiasson (1966) found that Al-induced reductions in the P concentrations of plant tops and increases in those of plant roots were more pronounced in Al-sensitive Herta barley than in Al-tolerant Charlottetown 80. Goodman (1969) stated that P seems to be the most important major nutrient in determining the ecological distribution of plants. Foy *et al.* (1965b, 1967) found that the tops of Al-sensitive Monon wheat contained lower concentrations of P than the tops of the Al-tolerant Atlas 66 when both were grown on Al toxic soil. However, this was not consistently true for plants grown in nutrient cultures. Aluminum-sensitive Kearney barley tops and roots contained higher P concentrations than those of Al-tolerant Dayton when both were grown with Al in nutrient cultures (Foy *et al.*, 1967). There is a possibility, however, that Al affects P metabolism differently in the two varieties. This question merits further study.

In contrast to the evidence presented above, Ouellette and Dessureaux (1958) concluded that differential Al tolerance among alfalfa clones was not due to Al-P interactions. In this same vein, Clarkson (1966a) reported that P deficiency symptoms in shoots of *Agrostis*

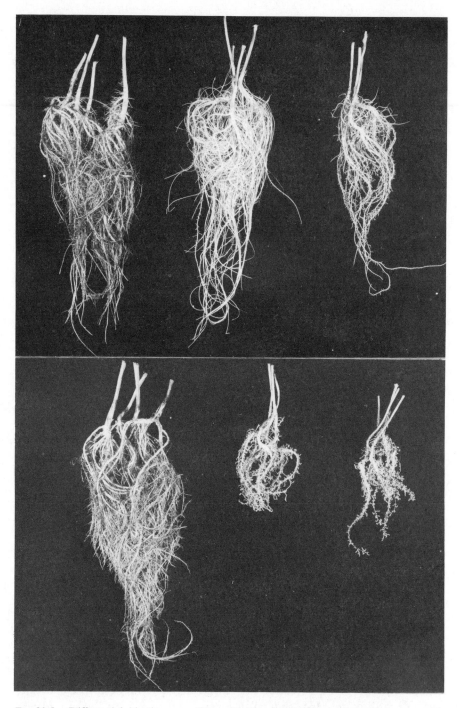

Fig. 20.5. Differential Al tolerance of Perry (*top*) and Chief (*bottom*) soybean roots with 2 ppm Ca in the nutrient solution. *Left to right*: 0, 8, and 12 ppm Al added. (Foy *et al.*, 1969)

Fig. 20.6. Differential Al tolerance of Perry (*top*) and Chief (*bottom*) soybean roots with 50 ppm Ca in the solution. *Left to right*: 0, 8, 10, and 12 ppm Al added. (Foy *et al.*, 1969)

canina, A. stolonifera, and *A. tenuis* were not correlated with root abnormalities produced by Al.

Jones (1961) suggested that organic acids in tolerant species chelate Al and thus prevent the Al-P precipitation that would normally occur in plants at physiological pH values. According to Small (1946), acidophilic plants generally have strong organic acid buffer systems in their cells, in contrast to alkaphiles, in which phosphate buffers dominate the system. However, the alkatolerant plants may have (1) a strong phosphate buffer system dominating a weak organic acid system or (2) a strong organic acid buffer system which can resist neutralization by an alkaline growth medium. Barley, which is sensitive to cationic Al at low pH and anionic Al at high pH, is known to have a phosphate type buffer system. In contrast, *Beta maritime*, which colonizes alkaline, aluminate-toxic fly ash deposits, has an organic acid buffer system. Such differences in plant buffer systems may be important in regulating Al-P interactions and in determining the differential Al tolerance of plant species and varieties (Jones, 1961).

In support of the chelation hypothesis of Al tolerance proposed by Jones (1961), Chamura and Koike (1960) reported that acid soil tolerance in several plant species was positively correlated with the citric acid contents of roots; however, the acid-soil-tolerant species had lower concentrations of citric and several other organic acids in their leaf blades than did the acid-soil-sensitive species. Some Al-tolerant species are known to accumulate high concentrations of Al in their tops, and many Al-sensitive plants accumulate Al in roots but not in tops. Such differences in Al mobility within the plant may partially explain the observed differences in Al toxicity. Clarkson (1966a) showed that Al in the cationic form inhibited cell division in adventitious roots of *Agrostis stolonifera* but chelated Al did not. Grime and Hodgson (1969) suggested that the Al resistance of certain calcifuge species is due to a chelating mechanism which also has an affinity for Fe. Johnson and Jackson (1964) found that Al-EDTA reduced Al uptake and increased Ca uptake by excised barley roots, when compared with $AlCl_3$ at pH 4.0.

A chelation mechanism for the detoxification of Al in tolerant species seems compatible with one hypothesis for explaining the differential P-feeding powers of plants. This hypothesis is that certain plant roots produce exudates that complex Fe and Al, reverse or prevent P fixation, and increase the availability of P to the plant (Drake and Steckel, 1955). Struthers and Sieling (1950) found that the naturally occurring organic acids (citric, oxalic, tartaric, malonic malic, and lactic) were effective in preventing the precipitation of Fe and Al in the pH range at which P fixation occurs in soils. For example, 1 millimole of citric acid completely complexed 1 millimole of Al and prevented P fixation over the pH range of 4 to 9. The efficiency of organic acids in this process increased with the number of carboxyl and functional hydroxyl groups and decreased with the length of the carbon chain. These investigators pointed out that such acids are produced in considerable quantities in soils during the microbial decomposition of organic matter. Harris (1961) suggested that plants absorb Al and return it to the soil, possibly in a chelated form.

Woodhouse (1969) found that the adenosine triphosphatase activity of cell wall preparation from roots of an acid soil ecotype of *Agrostis tenuis* was inhibited less by Al than was that of a preparation from a calcareous soil ecotype of the same species. He suggested the possibility that differential Al tolerances within this species are related to structural changes in these enzymes. In this connection, Bieleski (1970) has suggested that low P levels in root zones may induce the activity of acid phosphatases in roots and enable plants to extract P from organic forms such as phytin.

Clarkson (1969) has pointed out that the observed Al-induced inhibition of oxidative phosphorylation in plant roots should result in decreased respiration. Norton (1966–67) found that 54 ppm Al reduced respiration rates in Al-sensitive sanfoin roots by 5% within 30 minutes and by 35% within 3 hours after exposure, long before inhibited cell division was detected. Respiration rate in Al-tolerant lupine (*Lupinus luteus*) was not affected by the same treatment. In contrast, Clarkson (1969) found that in Al-sensitive onion roots the inhibition of cell division induced by Al occurred much earlier than the small reductions in respiration observed. Furthermore, the inhibition of cell division by Al did not seem to be affected by the presence or absence of P in the medium and could not be overcome by subsequent treatments with P for one week. He concluded that cell division was much more easily disturbed by Al than was general metabolism. Clarkson (1969) found that a 24-hour pretreatment with 27 ppm Al at pH 4.0 decreased respiration rates in the roots of four barley varieties. The decrease was 17% for Al-tolerant Dayton and 6% for Al-sensitive Kearney.

Aluminium is known to interfere with DNA replication in the mitotic cycle, fix P in inorganic form at root surfaces, reduce sugar phosphorylation, decrease respiration, and interfere with the uptake and utilization of Ca, P, and other elements and water. Such information suggests many fertile areas of research for identifying the physiological and biochemical processes responsible for differential Al tolerance in plant species and varieties. This information will, in turn, aid the plant breeder in tailoring plants to fit specific soil situations.

C. Breeding Aluminum-tolerant Plant Varieties

Because liming a soil to pH 5.5 to 6.0 will precipitate Al from solution, and eliminate its toxicity to plants, why not apply lime and forget about Al-tolerant varieties? Listed below are some acid soil situations in which liming may not be an adequate solution to the problem of Al toxicity.

1. Acid Subsoils

Excess soluble Al in acid subsoils reduces root proliferation and drought tolerance. Lime applied to surface soils does not move at a satisfactory rate, and liming the subsoil is difficult and generally not economically feasible. (Soil management practices are needed to prevent the further decrease of subsoil pH to values at which no plant can grow.)

2. Acid Surface Soils and Subsoils in Developing Countries

In many developing countries adequate liming will probably not be practiced for many years because of economic factors, such as lack of lime source, processing and transporting machinery, or suitable roads. In this connection Ikeda *et al.* (1965) has mentioned the necessity for having acid-soil-tolerant wheat and barley varieties for newly reclaimed lands on which adequate liming and P fertilization are not practiced for economic reasons. Ota (1968) proposed the adoption of Al-tolerant rice varieties in Ceylon to prevent bronzing.

3. Acid Surface Soils That Must Be Kept below pH 5.5

Chiasson (1964) has pointed out that most of the barley acreage in eastern Canada follows Irish potatoes. In this potato-barley rotation it is necessary to maintain the soil pH below about 5.4 to control the potato scab disease. However, at these pH levels Al is frequently soluble in toxic concentrations for the standard barley varieties used. The pH factor is so critical that good barley is sometimes produced at pH 5.4 and poor barley at 5.0. Aluminum injury is commonly expressed as a P deficiency symptom.

4. Acid Mine Spoils

Pyrites in acid mine spoils are oxidized to sulfuric acid, producing pH values as low as 2.5 to 3.0. In general, such spoils must be limed to about pH 4.0 (or higher for legumes) to reduce H ion toxicity before plants can be grown. However, between pH 4.0 and about 5.5, toxicities of Al and perhaps other elements, such as Mn, and their interactions with essential elements may limit growth. It is often not economically feasible to apply sufficient lime to neutralize all of the harmful Al. Preliminary studies indicate that a good compromise in such cases is to lime the spoils to pH 4.0 to 4.5, to use plants with greater Al tolerance, and to fertilize with rock phosphate (USDA, 1971a, 1971b). The rock phosphate supplies P and Ca that may be needed, raises the soil pH somewhat, and probably also supplies sufficient soluble P to precipitate Al at a lower pH than can be accomplished with lime. Bradshaw (1970) suggested breeding plants for tolerance to excess lead, zinc, tin, and copper as the best possible approach in reclaiming mine wasteland contaminated with these metals.

In the acid soil situations mentioned above, we need crop species and varieties that can tolerate more Al than the standard varieties, extract P more effectively from Fe and Al compounds, and utilize Ca and P more efficiently in the presence of excess Al. In some cases greater tolerance to excess Mn is also needed. Present crop varieties that show some tolerance to Al are largely the indirect result of selecting for overall yield in the presence of various insect, disease, and climatic factors, with little attention to the soil. However, it is now clear that plant varieties reflect the surface and subsoil conditions under which they were developed. For example, wheat and barley varieties selected in North Carolina, where acid subsoils and Al toxicity are widespread, are generally more tolerant to Al than those selected in Kansas, where acid soils are less common. As another example, Ohio wheat and barley varieties apparently have been selected for higher Al tolerance than Indiana varieties (Foy *et al.*, 1965a). When varieties developed in one soil region are planted on slightly different soils within the same climatic zone, problems of toxicity or nutrient element deficiency may appear. Selecting plants more directly to fit the soil would reduce such problems.

In some plant species considerable progress has been made in breeding for increased Al tolerance. Dessureaux (1969) has studied methods for determining the genetic nature of differential Al tolerance among alfalfa clones. Reid (1971) found that Al tolerance in certain winter barley populations is controlled by a single dominant gene or factor, and he has developed methods for rapid screening of plants in soils and nutrient solutions in the greenhouse (see Table 20.5; Reid *et al.*, 1971) Aluminum tolerance determined in the greenhouse has been well correlated with grain yield on one acid, Al-toxic soil in the field (Reid *et al.*, 1969). The biochemical mechanism of such genetic control remains to be determined. Similar work on Al tolerance in spring barley is being conducted in Canada (MacLean and Chiasson, 1966). Gorsline *et al.* (1968) postulated that three major genes regulate Al-Fe accumulation by certain corn inbred lines.

Kerridge and Kronstad (1968) concluded that at least one degree of Al tolerance in certain wheat populations was regulated by a single dominant gene. Kerridge *et al.* (1971) have developed a nutrient culture technique for separating degrees of Al tolerance within this species. Mesdag *et al.* (1970) have studied the possible genetic linkage between acid soil tolerance and protein content of wheat grain. Atlas 66 wheat, developed in North Carolina, has both high kernel protein content and high tolerance to soil acidity and Al. They concluded that these two traits are probably different genetically, although linked to some extent. Nevertheless, these investigators suggested the possibility of screening lines for soil acidity tolerance and selecting within segregating populations for high protein content in the kernel, provided that positive selection is used and that one parent combines both characteristics. Reid[1] has pointed out that there is no evidence to date that Al tolerance in winter barley is genetically linked with low absolute yield potential in the absence of Al. The Al tolerance trait can therefore be combined with other desirable plant characteristics.

The identification of physiological and biochemical plant properties associated with Al tolerance may provide useful screening tools for the plant breeder and increase our understanding of mineral nutrition in general. Sprague (1969) has emphasized that very little is known about the biochemical causes of differential yield potentials in plants. If we knew why plants adapt to certain soils, we could make more intelligent decisions about whether to change the soil to fit the plant, to change the plant to fit the soil, or to use a combination of the two approaches.

---

[1] D. A. Reid (USDA, ARS, Western Region, University of Arizona, Tucson, Arizona 85721), personal communication.

References

Abruna, Fernando, Robert W. Pearson, and Charles B. Elkins. 1958. Quantitative evaluation of soil reaction and base status changes resulting from field application of residually acid-forming nitrogen fertilizers. *Soil Sci. Soc. Amer. Proc.* 22: 539–42.

Abruna-Rodriguez, Fernando, Jose Vicente-Chandler, Robert W. Pearson, and Servando Silva. 1970. Effect of factors relating to soil acidity on yields and foliar composition of crops growing on typical soils of the humid tropics. I. Tobacco. *Soil Sci. Soc. Amer. Proc.* 34: 629–35.

Adams, Fred. 1968. *Response of Cotton to Lime in Field Experiments*. Auburn Univ. Agr. Exp. Sta. Bull. 376.

———. 1969. *Response of Corn to Lime in Field Experiments on Alabama Soils*. Auburn Univ. Agr. Exp. Sta. Bull. 391.

Adams, F., and Z. F. Lund. 1966. Effect of chemical activity of soil solution aluminum on cotton root penetration of acid subsoils. *Soil Sci.* 101: 193–98.

Adams, F., and R. W. Pearson. 1967. Crop response to lime in the Southern United States and Puerto Rico. In *Soil Acidity and Liming*, ed. R. W. Pearson and Fred Adams, pp. 161–206. Agronomy 12. Madison, Wis.: Amer. Soc. Agron.

———. 1970. Differential response of cotton and peanuts to subsoil acidity. *Agron. J.* 62: 9–12.

Aimi, Reizo, and Taka Murakami. 1964. Cell physiological studies on the effect of aluminum on the growth of crop plants. *Nat. Inst. Agr. Sci. Bull.*, ser. D, 11: 331–96 (Jap.) (*Biol. Abstr.* 46: 77198).

Anderson, I., and H. J. Evans. 1956. Effect of manganese and certain other metal cations on isocitric dehydrogenase and malic enzyme activities in *Phaseolus vulgaris*. *Plant Physiol.* 31: 22–28.

Armiger, W. H., C. D. Foy, A. L. Fleming, and B. E. Caldwell. 1968. Differential tolerance of soybean varieties to an acid soil high in exchangeable aluminum. *Agron. J.* 60: 67–70.

Asen, S., N. W. Stuart, and E. L. Cox. 1963. Sepal color of *Hydrangea macrophylla* as influenced by the source of nitrogen available to plants. *Proc. Amer. Soc. Hort. Sci.* 82: 504–7.

Avdonin, N. S., E. P. Milovidova, and E. D. Maksimova. 1957. The effect of aluminum and manganese on the exchange of substances in plants and on growth. *Vestn. Moskov. Univ. Ser. Biol. Pochvoved. Geol. Georg.* 2: 89–97 (Russ.) (from *Bibliography on Tolerance of Forage Crops, Cereals and Potatoes to Aluminum (1964–1931)*. Harpenden, England: Com. Bur. Soils).

Bertrand, Didier, and Andre deWolf. 1968. Aluminum, trace element necessary for corn? *C. R. HeBD Seances Acad. Sci.*, ser. D, *Natur.* (Paris) 267(26): 2325–27 (*Biol. Abstr.* 50: 133592).

Bhumbla, D. R., and E. O. McLean. 1965. Aluminum in soils: VI. Changes in pH dependent acidity, cation exchange capacity and extractable aluminum with additions of lime to acid surface soils. *Soil Sci. Soc. Amer. Proc.* 29: 370–74.

Bieleski, R. J. 1970. Enzyme change in plants following changes in their mineral nutrition. In *Plant Analyses and Fertilizer Problems*. Abstracts 6th Int. Colloq., Tel Aviv, Israel, March 13–17, 1970. Geneva, New York: W. F. Humphrey Press.

Bohm-Tuchy, E. 1960. Plasmalemma and aluminiumsalz. *Protoplasma* 52: 108–42.

Bradshaw, A. D. 1970. Pollution and plant evolution. *New Scientist*, Dec. 15, pp. 497–500.

Brogan, J. C. 1967. The effect of humic acid on aluminum toxicity. *Trans. 8th Int. Congr. Soil Sci.* (Bucharest) 3: 227–34.

Cate, R. B., Jr., and A. P. Sukhai. 1964. A study of aluminum in rice. Brit. Guiana Soil Survey. *Soil Sci.* 98: 85–93.

Chamura, Shugo. 1962. Studies on the relation between the tolerance of crops to soil acidity and that to low pH. 9. Absorption of basic matters to root crops. *Crop Sci. Soc. Jap. Proc.* 30: 350–54 (Jap.).

Chamura, Shugo, and Akira Hoshi. 1960. Studies on the relation between the tolerance of crops to soil acidity and that to low pH. 10. On the uptake of K ion in passive and active processes. *Crop Sci. Soc. Jap. Proc.* 31: 108–11 (Jap.).

Chamura, Shugo, and E. Koike. 1960. Studies on the relationship between the tolerance of crops to soil acidity and low pH. 6. The composition of organic acids in crops. *Crop Sci. Soc. Jap. Proc.* 28: 345–46 (Jap.).

Chenery, E. M. 1948. Aluminum in plants and its relation to plant pigments. *Ann. Bot.*, n.s. 12: 121–36.

Chiasson, T. C. 1964. Effects of N, P, Ca and Mg treatments on yield of barley varieties grown on acid soils. *Can. J. Plant Sci.* 44: 525–30.

Clarkson, D. T. 1965. The effect of aluminum and some other trivalent metal cations on cell division in the root apices of *Allium cepa. Ann. Bot.* 29: 309–15.

——. 1966a. Aluminum tolerance in species within the genus *Agrostis. J. Ecol.* 54: 167–78.

——. 1966b. Effect of aluminum on the uptake and metabolism of phosphorus by barley seedlings. *Plant Physiol.* 41: 165–72.

——. 1967a. Interactions between aluminum and phosphorus on root surfaces and cell wall materials. *Plant Soil* 27: 347–56.

——. 1967b. Phosphorus supply and growth rate in species of *Agrostis* L. *J. Ecol.* 55: 111–18.

——. 1969. Metabolic aspects of aluminum toxicity and some possible mechanisms for resistance. In *Ecological Aspects of the Mineral Nutrition of Plants*, ed. I. H. Rorison *et al.*, pp. 381–97. Symp. Brit. Ecol. Soc. no. 9, Sheffield, 1968. Oxford and Edinburgh: Blackwell Scientific Publications.

Clarkson, D. T., and John Sanderson. 1969. The uptake of a polyvalent cation and its distribution in the root apices of *Allium cepa*. Tracer and autoradiographic studies. *Planta* (Berl.) 89: 136–54.

——. 1971. Inhibition of the uptake and long distance transport of calcium by aluminum and other polyvalent cations. *J. Exp. Bot.* 23: 837–51.

Clymo, R. S. 1962. An experimental approach to part of the calcicole problem. *J. Ecol.* 50: 707–31.

Coleman, N. T., and G. W. Thomas. 1967. The basic chemistry of soil acidity. In *Soil Acidity and Liming*, ed. R. W. Pearson and F. Adams, pp. 1–14. Agronomy 12. Madison, Wis.: American Society of Agronomy.

Coulter, B. S. 1969. The chemistry of hydrogen and aluminum ions in soils, clay minerals and resins. *Soils Fert.* 32: 215–23.

Dessureaux, L. 1969. Effect of aluminum on alfalfa seedlings. *Plant Soil* 30: 93–97.

DeWard, P. W. F., and C. D. Sutton. 1960. Toxicity of aluminum to black pepper (*Piper nigrium* L.) in Sarawak. *Nature* 188: 1129–30.

Dios, Vidal R., and T. C. Broyer. 1962. Effect of high levels of Mg on the Al uptake and growth of maize in nutrient solutions. *Ann. Edafol. Agribiol.* 21: 13–30 (Span.) (*Soils Fert.* 26: 44).

Drake, M., and J. E. Steckel. 1955. Solubilization of soil and rock phosphate as related to root cation exchange capacity. *Soil Sci. Soc. Amer. Proc.* 19: 449–50.

Eichhorn, G. L. 1962. Metal ions as stabilizers or de-stabilizers of deoxyribonucleic acid structure. *Nature* 197: 474.

Evans, C. E. 1968. Ion exchange relations of aluminum and calcium in soils as influenced by organic matter. Ph.D. Thesis. North Carolina State Univ., Raleigh, N.C. University Microfilms, Ann Arbor, Mich., *Diss. Abstr.*, B, 29 (4): 1233.

Evans, Clyde E., and E. J. Kamprath. 1970. Lime response as related to percent aluminum saturation, solution aluminum and organic matter content. *Soil Sci. Soc. Amer. Proc.* 34: 893–96.

Fleming, A. L., and C. D. Foy. 1968. Root structure reflects differential aluminum tolerance in wheat varieties. *Agron. J.* 60: 172–76.

Foy, C. D., W. H. Armiger, L. W. Briggle, and D. A. Reid. 1965a. Aluminum tolerance of wheat and barley varieties in acid soils. *Agron. J.* 57: 413–17.

Foy, C. D., and J. C. Brown. 1963. Toxic factors in acid soils. I. Characterization of aluminum toxicity in cotton. *Soil Sci. Soc. Amer. Proc.* 27: 403–7.

——. 1964. Toxic factors in acid soils. II. Differential aluminum tolerance of plant species. *Soil Sci. Soc. Amer. Proc.* 28: 27–32.

Foy, C. D., G. R. Burns, J. C. Brown, and A. L. Fleming. 1965b. Differential aluminum tolerance of two wheat varieties associated with plant induced pH changes around their roots. *Soil Sci. Soc. Amer. Proc.* 29: 64–67.

Foy, C. D., A. L. Fleming, and W. H. Armiger. 1969. Aluminum tolerance of soybean varieties in relation to calcium nutrition. *Agron. J.* 61: 505–11.

Foy, C. D., A. L. Fleming, G. R. Burns, and W. H. Armiger. 1967. Characterization of differential aluminum tolerance among varieties of wheat and barley. *Agron. J.* 31: 513–21.

Foy, C. D., A. L. Fleming, and G. C. Gerloff. 1972. Differential aluminum tolerance in two snapbean varieties. *Agron. J.* 64:815–18.

Goletiani, G. E. 1965. The tea plant and available soil aluminum. *Khimiya sel Khoz* 3 (2): 7–12 (Russ.) (English abstr. in *Bibliography on Aluminum in Plant Nutrition (1967–1960)*. Harpenden, England: Com. Bur. Soils).

Goodman, P. J. 1969. Intra-specific variation in mineral nutrition of plants from different habitats. In *Ecological Aspects of the Mineral Nutrition of Plants*, ed. I. H. Rorison *et al.*, pp. 237–53. Symp. Brit. Ecol. Soc. no. 9, Sheffield, 1968. Oxford and Edinburgh: Blackwell Scientific Publication.

Gorsline, G. W., W. I. Thomas, and D. E. Baker. 1968. *Major Gene Inheritance of Sr-Ca, Mg, K, P, Zn, Cu, B, Al-Fe and Mn Concentrations in Corn (Zea mays L.)*. Pa. Agr. Exp. Sta. Bull. 746.

Greene, H. 1963. Prospects in soil science. *J. Soil Sci.* 14: 1–11.

Grime, J. P., and J. G. Hodgson. 1969. An investigation of the ecological significance of lime chlorosis by means of large scale comparative experiments. In *Ecological Aspects of the Mineral Nutrition of Plants*, ed. I. H. Rorison *et al.*, pp. 67–69. Symp. Brit. Ecol. Soc. no. 9, Sheffield, 1968. Oxford and Edinburgh: Blackwell Scientific Publications.

Hackett, C. 1962. Stimulative effects of aluminum on plant growth. *Nature* 195 (4840): 471–72.

——. 1967. Ecological aspects of the nutrition of *Deschampsis flexuosa* (L.) Trin. III. Investigation of phosphorus requirement and response to Al in water culture and a study of growth in soil. *J. Ecol.* 55: 831–40.

Harris, S. A. 1961. Soluble aluminum in plants and soils. *Nature* 189: 513–14.

Harward, M. E., W. A. Jackson, W. L. Lott, and D. D. Mason. 1955. Effects of Al, Fe, and Mn upon the growth and composition of lettuce. *Proc. Amer. Soc. Hort Sci.* 66: 261–66.

Hesse, P. R. 1963. Phosphorus relationships in a mangrove swamp mud with particular reference to aluminum toxicity. *Plant Soil* 19: 205–18.

Hester, J. B. 1935. The amphoteric nature of three coastal plains soils. I. In relation to plant growth. *Soil Sci.* 39: 237–43.

Hewitt, E. J. 1948. The resolution of the factors in soil acidity. IV. The relative effects of aluminum and manganese toxicities on some farm and market garden crops (cont.). *Long Ashton Res. Sta. Ann. Rep.*, pp. 58–65.

Hiatt, A. J., D. F. Amos, and H. F. Massey. 1963. Effect of aluminum on copper sorption by wheat. *Agron. J.* 55: 284–87.

Hofler, K. 1958. Aluminiunisaltz-Wirkung auf Spirogyren und zygumen. *Protoplasma* 49: 248–58.

Hoyt, Paul B., and Marion Nyborg. 1971. Toxic metals in acid soil. I. Estimation of plant-available aluminum. *Soil Sci. Soc. Amer. Proc.* 35: 236–40.

Hu, S. Yu, H. S. Lin, and V. L. Chang. 1957. The chemical composition of some acid indicator plants, calciphiles and halophytes of China. *Pochvovendenie*, no. 12, pp. 52–61 (Russ.).

Huck, M. G. 1972. Impairment of sucrose utilization for cell wall formation in the roots of aluminum damaged cotton seedlings. *Plant and Cell Physiol.* 13: 7–14.

Ikeda, Toshiro, Shuji Higashi, Satoru Kagohashi, and Takao Morya. 1965. Studies on the adaptability of wheat and barley on acid soil especially in regard to its varietal difference and laboratory detection. *Bull. Tokai-Kinki Nat. Agr. Exp. Sta.* 12: 64–79 (Jap.).

Jackson, William A. 1967. Physiological effects of soil acidity. In *Soil Acidity and Liming*, ed. R. W. Pearson and Fred Adams, pp. 43–124. Agronomy 12. Madison, Wis.: American Society of Agronomy.

Johnson, R. E., and W. A. Jackson. 1964. Calcium uptake and transport by wheat seedlings as affected by aluminum. *Soil Sci. Soc. Amer. Proc.* 28: 381–86.

Jones, L. H. 1961. Aluminum uptake and toxicity in plants. *Plant Soil* 13: 297–310.

Kamprath, E. J. 1970. Exchangeable aluminum as a criterion for liming leached mineral soils. *Soil Sci. Soc. Amer. Proc.* 34: 252–54.

Kerridge, Peter C., M. D. Dawson, and David P. Moore. 1971. Separation of degrees of aluminum tolerance in wheat. *Agron. J.* 63: 586–91.

Kerridge, P. C., and W. E. Kronstad. 1968. Evidence of genetic resistance to aluminum toxicity in wheat (*Triticum aestivum* Vill, host). *Agron J.* 60: 710–11.

Kliewer, W. M. 1961. The effects of varying combinations of molybdenum aluminum, manganese, phosphorus, nitrogen, calcium, hydrogen ion concentration, lime and rhizobium strain on growth composition and nodulation of several legumes. Ph.D. Thesis. Cornell University, Ithaca, N.Y.

Lance, J. C., and R. W. Pearson. 1969. Effect of low concentrations of aluminum on growth and water and nutrient uptake by cotton roots. *Soil Sci. Soc. Amer. Proc.* 33: 95–98.

Lee, C. R. 1971a. Influence of aluminum on plant growth and tuber yield of potatoes. *Agron. J.* 63: 363–64.

——. 1971b. Influence of aluminum on plant growth and mineral nutrition of potatoes. *Agron. J.* 63: 604–8.

Levan, A. 1945. Cytological reactions induced by inorganic solutions. *Nature* 156: 751–52.

Lignon, W. S., and W. H. Pierre. 1932. Soluble aluminum studies. II. Minimum concentrations of aluminum found to be toxic to corn, sorghum and barley in culture solutions. *Soil Sci.* 34: 307–21.

Long, F. L., and C. D. Foy. 1970. Plant varieties as indicators of aluminum toxicity in the A$_2$ horizon of a Norfolk soil. *Agron. J.* 62: 679–81.

Lund, Z. F. 1970. The effect of calcium and its relation to some other cations on soybean root growth. *Soil Sci. Soc. Amer. Proc.* 34: 456–59.

Lunt, O. R., and A. M. Kofranek. 1970. Manganese and aluminum tolerance of azalea (C. V. Sweetheart Supreme). In *Plant Analyses and Fertilizer Problems*. Abstracts of 6th Int. Colloquium, Tel Aviv, Israel, March 13–17. Geneva, New York: W. F. Humphrey Press.

McCart, G. D., and E. J. Kamprath. 1965. Supplying Ca and Mg for cotton on sandy, low cation exchange capacity soils. *Agron. J.* 57: 404–6.

MacLean, A. A., and T. C. Chiasson. 1966. Differential performance of two barley varieties to varying aluminum concentrations. *Can. J. Soil Sci.* 46: 147–53.

McLean, F. T., and B. E. Gilbert. 1927. The relative aluminum tolerance of crop plants. *Soil Sci.* 24: 163–74.

——. 1928. Aluminum toxicity. *Plant Physiol.* 3: 293–303.

MacLeod, L. B., and L. P. Jackson. 1965. Effect of concentration of the Al ion on root development and establishment of legume seedlings. *Can. J. Soil Sci.* 45: 221–34.

Medappa, K. C., and M. N. Dana. 1968. Influence of pH, calcium, iron and aluminum on the uptake of radiophosphorus by cranberry plants. *Soil Sci. Soc. Amer. Proc.* 32: 381–83.

——. 1970. Tolerance of cranberry plants to manganese, iron and aluminum. *J. Amer. Soc. Hort. Sci.* 95: 107–10.

Magistad, O. C. 1925. The aluminum content of the soil solution and its relation to soil reaction and plant growth. *Soil Sci.* 20: 181–225.

Mattson, Sante, and J. B. Hester. 1933. The laws of soil colloidal behavior. XII. The amphoteric nature of soils in relation to aluminum toxicity. *Soil Sci.* 36: 229–44.

Mercado, B. T., and J. R. Velasco. 1961. Effect of aluminum on the growth of coconut and other plants. *Philippine Agr.* 45: 268–74.

Mesdag, J., L. A. J. Slootmaker, and J. Post, Jr. 1970. Linkage between tolerance to high soil acidity and genetically high protein content in the kernel of wheat and its possible use in breeding. *Euphytica* 19: 163–74.

Moomaw, J. C., M. T. Nakamura, and G. D. Sherman. 1959. Aluminum in some Hawaiian plants. *Pacific Sci.* 13: 335–41.

Moschler, W. W., G. D. Jones, and G. W. Thomas. 1960. Lime and soil acidity effects on alfalfa growth in a red-yellow podzolic soil. *Soil Sci. Soc. Amer. Proc.* 24: 507–9.

Munns, D. N. 1965a. Soil acidity and the growth of a legume. II. Reactions of aluminum and phosphate in solution and effects of Al, phosphate, Ca and pH on *Medicago sativa* L. and *Trifolium subterraneum* L. in solution culture. *Aust. J. Agr. Res.* 16: 743–55.

——. 1965b. Soil acidity and the growth of a legume. III. Interactions of lime and phosphate in growth of *Medicago sativa* L. in relation to aluminum toxicity and phosphate fixation. *Aust. J. Agr. Res.* 16: 757–66.

Neenan, M. 1960. The effects of soil acidity on the growth of cereals with particular reference to the differential reaction of varieties thereto. *Plant Soil* 12: 324–28.

Norton, G. 1966–67. Some aspects of aluminum toxicity on plant growth. *Univ. Nottingham School of Agr. Rep.*, pp. 99–103.

Ota, Yasuo. 1968. Studies on the occurrence of the physiological disease of rice called "bronzing." *Bull. Nat. Inst. Agr. Sci. Nishigahara, Tokyo, Japan*, ser. D (Plant Physiology, genetics and crops in general), no. 18 (Jap.).

Otsuka, K. 1968a. Aluminum and manganese toxicity in plants. II. Effects of aluminum on growth of barley, wheat, oats, and rye seedlings. *J. Sci. Soil Manure* (Tokyo) 39: 469–74.

——. 1968b. Aluminum and manganese toxicities for plants (Part III). Effect of aluminum-ion concentration on growth and phosphorus uptake of grafted tomatoes. *J. Sci. Soil Manure* 39: 475–78, abstr. in *Soil Sci. and Plant Nutr.* 15 (3): 130.

——. 1968c. Studies on nutritional physiology of grafted plants (Part II). Influence of phosphorus concentration in nutrient medium on growth and phosphorus uptake of grafted plants. *J. Sci. Soil Manure* 39: 479–83, abstr. in *Soil Sci. and Plant Nutr.* 15 (3): 130.

Ouellette, G. J., and L. Dessureaux. 1958. Chemical composition of alfalfa as related to degree of tolerance to manganese and aluminum. *Can. J. Plant Sci.* 38: 206–14.

Paterson, J. W. 1965. The effect of aluminum on the absorption and translocation of calcium and other elements in young corn. Ph.D. Thesis. Pennsylvania State University, 1964. University Microfilms, Ann Arbor, Mich., *Diss. Abstr.* 25: 6142–43.

Pearson, R. W., F. Abruna, and J. Vicente-Chandler. 1962. Effect of lime and nitrogen applications on downward movement of calcium and magsium in two tropical soils of Puerto Rico. *Soil Sci.* 93: 77–82.

Pierre, W. H., J. R. Webb, and W. D. Shrader. 1971. Quantitative effects of nitrogen fertilizer on the development and downward movement of soil acidity in relation to level of fertilization and crop removal in a continuous corn cropping system. *Agron. J.* 63: 291–97.

Pitmann, M. G. 1970. Active H^+ efflux from cells of low salt barley roots during salt accumulation. *Plant Physiol.* 45: 787–90.

Ragland, J. L., and N. T. Coleman. 1962. Influence of aluminum on phosphorus uptake by snapbean roots. *Soil Sci. Soc. Amer. Proc.* 26: 88–90.

Randall, P. J., and P. B. Vose. 1963. Effect of aluminum on uptake and translocation of phosphorus by perennial ryegrass. *Plant Physiol.* 38: 403–9.

Rasmussen, H. P. 1968. Entry and distribution of aluminum in *Zea mays*. The mode of entry and distribution of aluminum in *Zea mays*: Electron microprobe X-ray analysis. *Planta* (Berlin) 81: 28–37.

Rees, W. J., and G. H. Sidrak. 1961. Plant nutrition on fly ash. *Plant Soil* 8: 141–59.

Reeve, N. G., and M. E. Summer. 1970. Effects of aluminum toxicity and phosphorus fixation on crop growth in oxisols in Natal. *Soil Sci. Soc. Amer. Proc.* 34: 263–67.

Reid, D. A. 1971. Genetic control of reaction to aluminum in winter barley. In *Barley Genetics II*, ed. R. A. Nilan, pp. 409–13. Proc. 2nd Int. Barley Genetics Symp., 1969. Pullman, Wash.: Washington State University Press.

Reid, D. A., A. L. Fleming, and C. D. Foy. 1971. A method for determining aluminum response of barley in nutrient solution in comparison to response in Al-toxic soil. *Agron. J.* 63: 600–603.

Reid, D. A., G. D. Jones, W. H. Armiger, C. D. Foy, E. J. Koch, and T. M. Starling. 1969. Differential aluminum tolerance of winter barley varieties and selections in associated greenhouse and field experiments. *Agron. J.* 61: 218–22.

Richburg, John S., and Fred Adams. 1970. Solubility and hydrolysis of aluminum in soil solutions and saturated past extracts. *Soil Sci. Soc. Amer. Proc.* 34: 728–34.

Riley, D., and S. A. Barber. 1969. Bicarbonate accumulation and pH changes at the soybean (*Glycine max* L.), Merr.) root-soil surface. *Soil Sci. Soc. Amer. Proc.* 33: 905–8.

——. 1971. Effect of ammonium and nitrate fertilization on phosphorus uptake as related to root induced pH changes at the root-soil interface. *Soil Sci. Soc. Amer. Proc.* 35: 301–6.

Rios, M. A., and R. W. Pearson. 1964. Some chemical factors in cotton root development. *Soil Sci. Soc. Amer. Proc.* 28: 232–35.

Rorison, I. H. 1958. The effect of aluminum on legume nutrition. In *Nutrition of the Legumes*, ed. E. C. Hallsworth, pp. 43–58. London: Butterworth's Scientific Publications.

——. 1965. The effect of aluminum on the uptake and incorporation of phosphate by excised sanfoin roots. *New Phytol.* 64: 23–27.

Ruschel, A. P., R. Alvahydo, and I. B. M. Sampaio. 1968. Effect of excess aluminum on growth of beans (*Phaseolus vulgaris* L.) in nutrient culture. *Pesq. Agropec. Bras.* 3: 229–33 (Port.); English abstr. in *Soils Fert.* 33 (1): 95.

Sampson, M., D. T. Clarkson, and D. D. Davies. 1965. DNA synthesis in aluminum treated roots of barley. *Science* 148: 1476–77.

Schnitzer, M., and S. I. M. Skinner. 1963. Organo-metallic interactions in soils: I. Reactions between a number of metal ions and the organic matter of a podzol B_h horizon. *Soil Sci.* 96: 86–93.

Small, J. 1946. *pH and Plants*. London: Bailliere, Tindall and Cox.

Small, J., and T. Jackson. 1949. Buffer index values in relation to soil pH tolerances. *Plant Physiol.* 24: 75–83.

Sprague, G. A. 1969. Germ plasm manipulations of the future. In *Physiological Aspects of Crop Yields*, ed. J. D. Eastin, pp. 375–89. Madison Wis.: American Society of Agronomy.

Stoklasa, J. 1911. Catalytic fertilizers for sugar beets. *Be. Zuckerrubenban* 18: 193–97, abstr. in *Exp. Sta. Rec.*, 26: 225 (ref. from Magistad, 1925).

Struthers, P. H., and D. H. Sieling. 1950. Effect of organic anions on phosphate precipitation by iron and aluminum as influenced by pH. *Soil Sci.* 69: 205–13.

Subramoney, N., and S. Saukaranarayanan. 1964. Effect of germination of rice on the pH of the soil. *Inst. Rice Comm. News Lett.* 13: 22–27, abstr. in *Soils Fert.* 27 (a): 3229.

Suchting, H. 1948. Investigations on nutrition conditions in the forest. X. The effect of soluble aluminum on pines in two forest soils. *Z. Pflanzenernahr. Dung. Bodenk.* 42: 193–218.

Szucs, J. 1912. Experimental contributions to a theory of antagonistic activity of ions. *Jähr. Wiss. Bot.* 52: 85.

Tanaka, Akira, and Shuichi Yoshida. 1970. *Nutritional Disorders of the Rice Plant in Asia.* Int. Rice Res. Inst. Tech. Bull. 10 (Los Banos, Manila, P.I.).

Turner, R. G. 1969. Heavy metal tolerance in plants. In *Ecological Aspects of the Mineral Nutrition of Plants*, ed. I. H. Rorison *et al.*, pp. 399–410. Symp. Brit. Ecol. Soc. no. 9 Sheffield, 1968. Oxford and Edinburgh: Blackwell Scientific Publications.

USDA. 1971a. Project recovery. *Agr. Res.*, January, pp. 8–9. USDA, ARS.
——. 1971b. *After the Coal-Lush Grass.* USDA, ARS Picture Story 234. March.

Varvarro, V. 1912. The action of manganese dioxide and other metallic compounds on the germination of seed. *Sta. Sper. Agr. Ital.* 45: 917–29. abstr. in *Exp. Sta. Rec.* 29: 528 (ref. from Magistad, 1925).

Velasco, J. R., A. Holaso, R. S. dela Pena, *et al.* 1959. Aluminum and its possible relationship to the cadang-cadang of coconut. *Philippine. Agr.* 43: 177–99.

Vose, P. B., and P. J. Randall. 1962. Resistance to aluminum and manganese toxicities in plants related to variety and cation exchange capacity. *Nature* 196: 85–86.

Waisel, Y., A. Hoffen, and A. Eshel. 1970. The localization of aluminum in the cortex cells of bean and barley roots by X-ray microanalysis. *Physiol. Plant.* 23: 75–79.

Wallace, Arthur, Sadek M. Sufi, and Evan M. Romney. 1970. Regulation of heavy metal uptake and responses in plants. In *Plant Analyses and Fertilizer Problems.* Abstracts 6th Int. Colloquium, Tel Aviv, Israel, March 13–17, 1970. Geneva, New York: W. F. Humphrey Press.

Wallihan, E. F. 1958. The influence of aluminum on the phosphorus nutrition of plants. *Amer. J. Bot.* 35: 106–12.

Welch, C. D., and W. L. Nelson. 1950. Calcium and magnesium requirements of soybeans as related to the degree of base saturation of the soil. *Agron. J.* 42: 9–13.

Wilkinson, H. F. 1970, Nutrient movement in the vicinity of plant roots. Ph.D. Thesis, Univ. Western Australia, Dept. Soil Science and Plant Nutrition, Institute of Agriculture, Perth, Western Australia.

Williams, D. E., and James Vlamis. 1957. Manganese and boron toxicities in standard culture solutions. *Soil Sci. Soc. Amer. Proc.* 21: 205–9.

Wolcott, A. R., H. D. Foth, J. F. Davis, and J. C. Shickluna. 1965. Nitrogen carriers: Soil effects. *Soil Sci. Soc. Amer. Proc.* 29: 405–10.

Woolhouse, H. W. 1969. Differences in the properties of the acid phosphatases of plant roots and their significance in the evolution of edaphic ecotypes. In *Ecological Aspects of the Mineral Nutrition of Plants*, ed. I. H. Rorison *et al.*, pp. 357–80. Symp. Brit. Ecol. Soc. no. 9, Sheffield, 1968. Oxford and Edinburgh: Blackwell Scientific Publications.

Wright, K. E. 1937. Effects of lime and phosphorus in reducing aluminum toxicity of acid soils. *Plant Physiol.* 12: 173–81.

——. 1948. Internal precipitation of phosphorous in relation to aluminum toxicity. *Plant Physiol.* 18: 708–12.

——. 1952. Aluminum toxicity studies with radioactive phosphorous. *Plant Physiol.* 28: 674–80.

21. Effects of Nitrogen on Phosphorus Absorption by Plants

M. H. Miller

THE increased absorption of phosphorus by plants when nitrogen is added to the soil has been recognized as a significant phenomenon in soil-plant relations since the early 1950's. As early as 1939 it was observed that N in a fertilizer band promoted phosphate uptake (Scarseth *et al.*, 1942). With the advent of radioactive tracers for use in biological research in the late 1940's, experimentation on this phenomenon in both the field and the greenhouse increased rapidly.

Robertson *et al.* (1954) in field experiments in Indiana in 1949 found that absorption of banded fertilizer P by corn (*Zea mays*) was greater when N was included in the band than when it was applied separately. In Nebraska N increased uptake of fertilizer P from bands by oats (*Avena sativa*) and wheat (*Triticum aestivum*) (Olson *et al.*, 1956; Olson and Dreier, 1956). In Saskatchewan, Canada, Rennie and Mitchell (1954) and Rennie and Soper (1958) obtained a similar result for wheat. Grunes *et al.* (1958a) in field experiments in North and South Dakota found that N increased the relative absorption of banded fertilizer P by sugar beets (*Beta vulgaris*) and potatoes (*Solanum tuberosum*).

Greenhouse experiments have shown that N increases the fertilizer P absorption by barley (*Hordeum vulgare*) (Grunes *et al.*, 1958b), corn (Miller and Ohlrogge, 1958), sugar beets (Werkhoven and Miller, 1960), and oats (Bouldin and Sample, 1958).

It is thus clear that the effect of N on P absorption is a general phenomenon in soil-plant relations, occurring with numerous crops under widely varying soil and climatic conditions.

I. CHARACTERIZATION OF THE EFFECT

Many studies have demonstrated the effect of various factors on the extent to which N increases the absorption of fertilizer P.

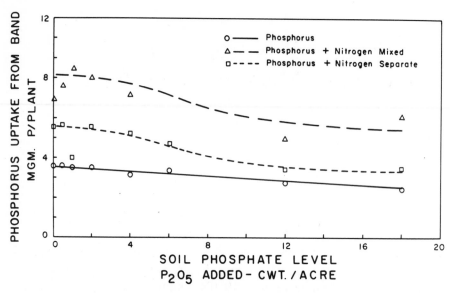

Fig. 21.1. Influence of soil phosphate level and N additions on the P absorption by corn from band-applied P fertilizer. (Miller and Ohlrogge, 1958)

A. Placement of Nitrogen Relative to Phosphorus

Miller and Ohlrogge (1958) observed that N had a much greater effect on P absorption by corn when the N was applied in a band with the P than when it was applied in a separate band (Fig. 21.1). Increased growth could largely account for the increase in fertilizer P absorption when N was applied separately. Olson and Dreier (1956) observed in both field and greenhouse studies that N influenced P absorption by oats much more when it was applied with the banded P than when it was mixed with the soil.

Rennie and Soper (1958), in greenhouse trials, observed a marked increase in absorption of fertilizer P by wheat when NH_4NO_3 was added to a fertilizer P band. No increase was observed when the NH_4NO_3 was either mixed throughout the soil or placed in a band 2.5 cm from the P band. The influence of N on fertilizer P absorption by sugar beets was found to be greater when the N was applied with the P in a band than when it was applied as a separate band (Grunes *et al.*, 1958b) or mixed with the soil (Werkhoven and Miller, 1960).

Leonce and Miller (1966) placed a fertilizer pellet 1.25 cm to the side and 5.0 cm above the tip of a corn root growing down the sloping front of a growth box (Fig. 21.2). Inclusion of $(NH_4)_2SO_4$ with the P increased the absorption of P threefold. Placement of an $(NH_4)_2SO_4$ pellet 5.0 cm below the P pellet and 1.25 cm to the side of the tip of the same root did

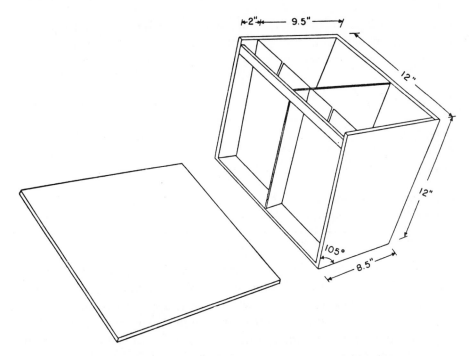

Fig. 21.2. Growth box used in soil experiments by Leonce and Miller (1966)

not influence the absorption of P. They concluded that the NH_4^+ ion must enter the root at the same point as the P to cause increased P absorption.

Thus it appears that the addition of N will increase the absorption of fertilizer P only when the N is in intimate association with the P. Any exceptions to this can probably be explained by increased plant growth due to the N addition resulting in increased absorption of P.

B. Volume of Soil Fertilized and Concentration of N and P

Duncan and Ohlrogge (1958) found that the volume of soil in which the fertilizer was placed influenced the effect of N on P uptake (Table 21.1). When the P was mixed throughout the soil, N had no effect. As the volume of soil in which the fertilizer P was mixed decreased, the effect of N increased, with the greatest effect occurring when the fertilizer was mixed with a 10-cm^3 volume of soil or placed in a compact 0.5-cm cylinder.

The addition of $(NH_4)_2SO_4$ increased the percentage of the plant P that sugar beets derived from the fertilizer when the fertilizer P was banded but not when it was mixed throughout the soil (Grunes *et al.*, 1958b). Werkhoven and Miller (1960) found that N affected sugar beet

Table 21.1. Uptake of labeled phosphate by 16-day-old corn plants as affected by nitrogen and by volume of fertilizer placement

Placement volume	Counts per minute	
	No nitrogen	Plus nitrogen
Total soil	3,761	3,750
90 cc	3,028	7,657
30 cc	3,763	11,090
10 cc	2,549	9,524
Flat 4-cm disk, not mixed with soil	3,206	5,998
Compact 0.5-cm cylinder, not mixed with soil	2,555	8,030

Source: Duncan and Ohlrogge, 1958.

absorption of fertilizer P more when the P was applied in a band than when it was mixed with a 5-cm layer of soil. However, the absorption from the 5-cm layer was greater when N was applied with the P than when it was mixed with the entire soil volume. The main factor that differed in these comparisons was the concentration of N and P. Duncan and Ohlrogge (1959) found that concentration of N and P bore little relation to the effect of N on fertilizer P absorption by corn. The lowest rates they used were, however, in excess of 700 ppm N and 1,500 ppm P. They suggested that there was an optimum ratio of N and P in the fertilizer placement. Mamaril and Miller (1970) also investigated the effect of concentration in a series of experiments with corn. The addition of NH_4^+ (1 meg/100 g soil) did not increase P absorption from a 10-cm-thick layer of soil to which 44 ppm P had been added. However, when 440 ppm P was added, the influence of NH_4^+ was as great in a 10-cm layer as in a 1-cm layer. They concluded that concentration of P, rather than volume of fertilized soil *per se*, was important. Increasing the NH_4^+ concentration in a 1-cm layer of soil fertilized with 440 ppm P increased the effect of N, with the greatest effect occurring at the highest rate used (5 meg NH_4^+/100 g soil).

It is therefore apparent that relatively high levels of both N and P are required for the increased P absorption to occur.

C. Source of Nitrogen

One of the earliest reports of an effect of the source of N on P availability was that of Lorenz and Johnson (1953), who found that the yields of potatoes on nine different soils were considerably greater when $(NH_4)_2SO_4$ was used instead of $Ca(NO_3)_2$. $NaNO_3$ produced yields similar to these produced by $Ca(NO_3)_2$, whereas the yields produced by NH_4NO_3 were intermediate. The authors attributed this effect to an increased availability of soil and fertilizer P due to the acidifying effect of $(NH_4)_2SO_4$.

Olson and Dreier (1956) found that the NH_4^+ and NH_4^+-producing materials created maximum effect of N in promoting fertilizer P uptake by oats. The effect of NO_3^- compounds was comparatively small at early growth stages but approached that of NH_4^+ compounds near crop maturity in the greenhouse. Only NH_4^+ salts had any influence on P absorption by wheat in studies by Rennie and Soper (1958). Neither KNO_3 nor K_2SO_4 had an effect.

Leonce and Miller (1966) reported that addition of $(NH_4)_2SO_4$ or NH_4Cl to a fertilizer P pellet placed 1.25 cm to the side of a corn root tip increased the absorption of fertilizer P during absorption periods of 15 days, whereas addition of KNO_3 decreased it. They concluded that the NH_4^+ ion had a specific effect on the absorption of P by plants.

Thus it appears that NH_4^+ is the form of N that has the greatest effect on fertilizer P absorption. Nitrate has shown some effect, particularly at later growth stages.

II. MECHANISMS RESPONSIBLE FOR EFFECT

The effect under consideration in this section is the increased fertilizer P content of plant shoots when N is added with the P. A schematic presentation of this effect and the factors that may be responsible for it is shown in Figure 21.3.

The increased fertilizer P content of shoots may be due to either (*a*) increased fertilizer P absorption by the root or (*b*) an increase in the proportion of the absorbed P that is transferred to the shoot. The transfer to the shoot involves two major steps: (*a*) transfer across the endodermis and (*b*) transfer in the xylem vessels. An influence of N on either of these steps would result in an increased P content of shoots for a given quantity of P absorbed by the root.

The fertilizer P absorbed by the root is the product of the absorption per unit root surface area and the root surface area in contact with the fertilizer P. Each of these is dependent on additional soil and/or plant factors as indicated in Figure 21.3. In the following sections, the evidence relating to the effect of N on each of these factors is presented.

A. Root Surface in Contact with Fertilizer P

Phosphorus is regarded as a relatively immobile nutrient in soil. Therefore the absorption of P is highly dependent on root-soil contact.

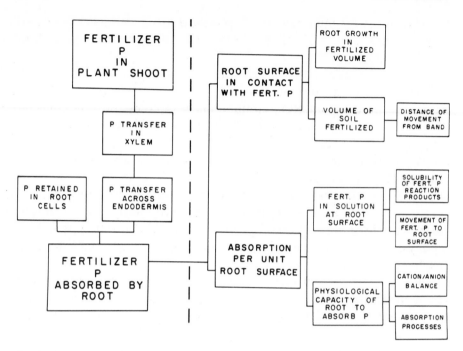

Fig. 21.3. Processes and factors involved in the effect of N on absorption of P

Increased contact between root surface and fertilized soil could occur either through an increased growth of roots in a given volume of fertilized soil or an increase in the volume of soil fertilized. If the addition of N to a fertilizer band increased either of these factors, an increased absorption of fertilizer P would be expected.

1. Root Growth

Miller and Ohlrogge (1958) observed that the addition of N to a fertilizer P band caused a proliferation of corn roots in the band volume. The ratio of the observed weight of roots in the band volume divided by the expected weight if equal distribution occurred was 1.5 in the presence of N. Olson and Dreier (1956) found no evidence of proportionately greater oat root concentration in greenhouse pots in the zone of N-P placement than with singular nutrient applications. They did, however, observe a very definite indication of a lateral growth of roots to a side band containing N and P in field plots. This effect was not apparent with a P-only band.

Duncan and Ohlrogge (1958, 1959) have investigated the effect of N and P on root growth quite extensively. They grew corn in split root systems that permitted treatment of single roots independent of the remainder of the root system. Table 21.2 presents a portion of their

Table 21.2. Effects of fertilizer treatments on the development of a single corn root

Treatment	Single root weight as % of total
Both root systems treated alike	9.3
Single root in phosphate	14.4
Single root in nitrogen plus phosphate	57.5
Single root in phosphate, balance in nitrogen	8.0

Source: Duncan and Ohlrogge, 1958.

results when using vermiculite as a growth medium. It is apparent that a combination of N and P greatly increased the growth of the single root in relation to the remainder of the root system. The authors state that root weights alone do not give the full picture. The roots that developed in the presence of nitrogen and phosphate in combination were much finer and silkier in appearance, and the number of roots was obviously much greater. Clearly the difference in root surface area between treated and untreated roots was very much greater than the difference in root weights would indicate.

Further studies (Duncan and Ohlrogge, 1959) indicated that when $(NH_4)_2SO_4$ was included with P in placements of relatively small volume, a dense mass of roots grew in and around the volume of fertilized soil. When the entire soil was fertilized, there were no noticeable effects on root growth. They emphasized that the stimulation of root growth did not result in a larger ratio of roots to tops. The result was more of a changed distribution than a true stimulation. This fact would explain the observation that N increased fertilizer P absorption only when a portion of the soil was fertilized. Duncan and Ohlrogge (1959) concluded that "when the volume of soil fertilized is small in relation to the total soil volume, nitrogen increases the uptake of phosphorus in part by increasing root growth in the fertilized soil."

Miller and Vij (1962) found that inclusion of $(NH_4)_2SO_4$ in a fertilizer P band more than doubled the sugar beet root weight in the band volume 8 and 10 weeks after planting. There was a similar increase in fertilizer P content in the shoots. Root weight in the band volume accounted for 87% of the variability in fertilizer P content of the shoots at the 8-week stage. These authors concluded that the major factor responsible for the increased absorption of P was the increased growth of roots in the fertilized volume.

Miller (1965) used a split root system in which a single corn root was grown in a soil medium to which the fertilizer treatments were applied. The remainder of the root system was grown in a sand medium to which nutrient solution varying in N content was supplied (Fig. 21.4). In the first experiment the addition of $(NH_4)_2SO_4$ to the fertilizer P band increased both the root weight in the soil volume and the fertilizer P content of shoots at the low (0.022 g/l) and intermediate (0.112 g/l) level of N in the sand culture. The root weight was not increased at the high level of N (0.224 g/l), but there was an increase in fertilizer P content of shoots (Fig. 21.5). The increase in root weight was due to increased branching of the single root growing in the fertilized soil rather than increased length of the root itself. In this experiment the N content of the sand culture did not influence growth, indicating that the lowest level was sufficient. In a second experiment the levels of N in the sand culture were reduced so that the lowest level (.004 g/l) was deficient. The intermediate and high levels corresponded to the low and intermediate levels of the first experiment. The effect of $(NH_4)_2SO_4$ on absorption of fertilizer P was essentially the same as in the first experiment. However, no increase in root weight in the fertilized soil resulted from the addition of $(NH_4)_2SO_4$ at any level of N in the sand culture.

A third experiment established that neither light intensity in the greenhouse nor time of harvest was responsible for the difference in the effect of N on root growth in the first two experiments. The addition of $(NH_4)_2SO_4$ increased the absorption of fertilizer P at each harvest date and at each of three light intensities created by shading. In no case was there an increase in root weight in the fertilized soil. Miller (1965) concluded that although increased root growth may result from the addition of N, it is not a prerequisite for increased P absorption. He stated that the most probable explanation for the increased P absorption was that $(NH_4)_2SO_4$ exerted a specific influence on the physiological activity that controls P absorption.

It is apparent that the combination of N and P in a fertilizer band has a stimulating effect on root growth in the band. This increased root growth will result in an increase in P absorption from the band and is therefore one mechanism responsible for the effect. It is also apparent, however, that increased root growth is not sufficient to explain all the observed effects of N on P absorption.

2. Volume of Soil Fertilized

When monocalcium phosphate (MCP) is added to a soil in a concentrated band, water enters the band due to a vapor pressure gradient, and a concentrated P solution moves into the soil in a blotting type of movement (Lawton and Vomocil, 1954; Lindsay and Stephenson, 1959a). This solution is initially very acidic (Lindsay and Stephenson

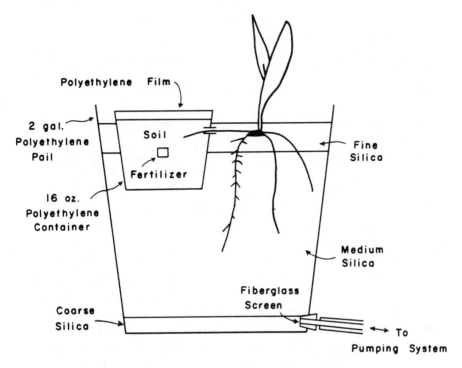

Fig. 21.4. Diagram of plant growth system used by Miller (1965)

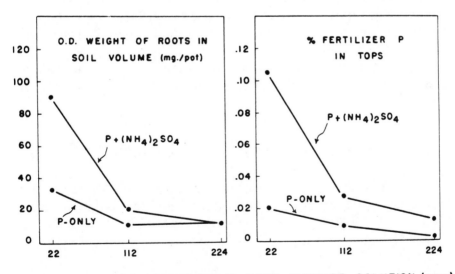

NITROGEN CONCENTRATION IN SAND CULTURE SOLUTION (p.p.m.)

Fig. 21.5. Weight of roots in soil medium and percentage of fertilizer P in corn tops, in experiment No. 1 of Miller (1965)

1959a) and thus dissolves large quantities of iron, aluminum, and manganese. As the P moves out, it is precipitated in various reaction products, principally iron, aluminum, and calcium phosphates (Lindsay and Stephenson, 1959b, 1959c; Lindsay *et al.*, 1959). Nitrogen salts could increase the volume of soil fertilized by increasing the distance to which the P moves.

Bouldin and Sample (1958) measured the distance of movement of P from pellets of MCP, MCP + NH_4NO_3, MCP + KNO_3, and MCP + KCl, respectively. The maximum distance of P movement from the pellets after 3 weeks differed among soils, but pellet composition had little effect on the movement.

Miller and Vij (1962), in studies similar to those of Bouldin and Sample (1958), found that $(NH_4)_2SO_4$ increased the distance of movement from the pellet by 0.25 cm at 3 weeks and 0.18 cm at 6 weeks but had no effect after 8 weeks. Because $(NH_4)_2SO_4$ had no effect on fertilizer P absorption by sugar beets after 6 weeks but did after 8 weeks, the authors concluded that the effect of $(NH_4)_2SO_4$ on distance of movement of fertilizer P was not a significant factor.

Blanchar and Caldwell (1966) placed relatively large pellets of MCP, MCP + NH_4Cl, and MCP + KCl, respectively, in soil, allowed them to react for 14 days, then leached the soil columns with 4 cm of water. Water-soluble fertilizer P distribution around each pellet was nearly the same for all three treatments. They concluded that the increased fertilizer P absorption observed in the presence of NH_4Cl was not due to differences in the distribution pattern of the fertilizer P around the pellets.

From these data we can conclude that the effect of N on fertilizer P absorption is not due to an increased movement of fertilizer P from the band.

B. Absorption per Unit of Root Surface Area

The absorption of P per unit of root surface area is a function of the physiological capacity of the root to absorb and the fertilizer P concentration in solution at the root surface. Either of these could be affected by N.

1. Fertilizer P in Solution at the Root Surface

The fertilizer P in solution at the root surface is a function of the solubility of the fertilizer P reaction products and the movement to the root surface.

a. Solubility of Fertilizer P Reaction Products. As indicated in the previous section, when fertilizer P is added in a band, a concentrated P solution moves into the soil, and the P is precipitated in various reaction products. Any factor that affected this precipitation could be expected to influence the solubility of the fertilizer P in the soil and hence the amount absorbed by the plant.

Starostka and Hill (1954) studied the effect of soluble salts on the solubility of dicalcium phosphates in an effort to explain the greater availability of this material when it was a component of mixtures than when used alone. They found that, in general, salts such as $(NH_4)_2SO_4$, which in laboratory tests markedly increased the solubility of dicalcium phosphate with increased ionic strength (Fig. 21.6), gave the greatest response of alfalfa (*Medicago sativa*) in the greenhouse. Those salts such as NH_4NO_3, which increased the solubility only moderately (Fig. 21.6), only occasionally gave increased crop response; while salts such as $Ca(NO_3)_2$, which decreased the solubility, usually gave a negative response.

Rennie and Mitchell (1954) attributed the increased availability of P in the presence of NH_4^+ salts to a lowering of pH in the fertilizer band volumes by the acid NH_4NO_3 or by nitrification of NH_4^+. Rennie and Soper (1958), however, found that K_2SO_4 had as great an acidifying effect as $(NH_4)_2SO_4$ but did not increase the fertilizer P absorption. They concluded that the increased fertilizer P absorption in the presence of $(NH_4)_2SO_4$ was due to an influence of NH_4^+ on the plant's ability to take up P rather than to any alteration of the availability of the applied P.

Bouldin and Sample (1958) measured the water-soluble fertilizer P content of the soil at varying distances from a fertilizer pellet containing MCP and various additional salts. They found that the pellet composition had a marked influence on the solubility of the soil-fertilizer P reaction products in Hartsells fine sandy loam but a lesser effect in Mountview silt loam. These authors observed that the water-soluble fertilizer P concentration was approximately a linear function of distance from the pellet. The composition of the pellet had only a slight effect on distance of movement. The water-soluble P concentration in each segment of soil was summed over all volumes to determine the total water-soluble fertilizer P in the volume of soil influenced by the fertilizer. They found that KNO_3 and NH_4NO_3 increased the water-soluble fertilizer P and that this fertilizer P was well correlated with the uptake of P by oats in a greenhouse experiment.

In a subsequent series of experiments, Bouldin and Sample (1959) compared monocalcium phosphate, monoammonium phosphate, and diammonium phosphate with regard to the solubility of reaction products and availability of the fertilizer P to oat plants. They found

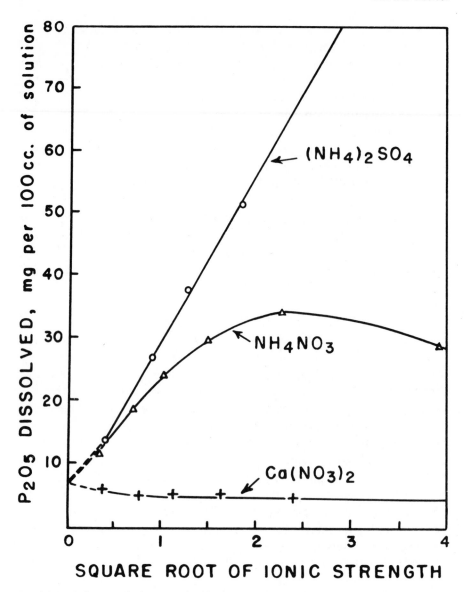

Fig. 21.6.   Influence of salt concentration on the solubility of dicalcium phosphate in solutions of type salts. (Starostka and Hill, 1954)

that the integrated solubility, i.e., the summation of the water-soluble fertilizer P over the total soil volume influenced by the fertilizer, was highly correlated with the average P content of the oat plants ($r^2 = 0.85$). They concluded that although there were notable exceptions, the major influence of intimate association between fertilizer N and P could be explained on the basis of integrated solubility. In these studies (Bouldin and Sample, 1958, 1959), three different soils were used, Hartsells fine sandy loam (pH 5.2), Webster silty clay loam (pH 8.3), and Mountview silt loam (pH 6.6). Although the associated salt had an effect on the integrated solubility in all three soils, only in the Mountview soil was there an appreciable increase in fertilizer P absorption upon addition of N. The correlation ($r^2 = 0.85$) was obtained by grouping the data for the Hartsells and Webster soils, neither of which showed an appreciable influence of N on P absorption. The correlation for the Mountview soil, which showed the greatest effect of N on P absorption, was much poorer than that for the Hartsells and Webster soils. This indicates that the integrated solubility was not directly responsible for the effect of N on P absorption.

A similar technique was used to measure concentration of water-soluble fertilizer P at three distances from a pellet as a function of time (Table 21.3). The water-soluble fertilizer P was very high next to the pellet after 2 days but decreased rapidly during the first week and more slowly thereafter. The concentration farther from the pellet increased with time as the fertilizer moved out. The addition of $(NH_4)_2SO_4$ to the pellet did not significantly affect the amount of water-soluble fertilizer P at any distance. There was, however, a tendency for the amount to be lower close to the pellet.

Miller and Vij (1962) found the integrated solubility of fertilizer P, measured in the same manner as Bouldin and Sample (1958) used, was decreased by the addition of $(NH_4)_2SO_4$ to the pellet after 3-, 6-, and 8-week reaction periods. They concluded that if water-soluble

Table 21.3. Water-soluble fertilizer phosphorus at three distances from pellet of monocalcium phosphate

Reaction time (days)	Treatment	Amount of P (μg/g soil)		
		1.6 mm from pellet	6.4 mm from pellet	12.7 mm from pellet
2	P only	3,872	1,003	34
	$(NH_4)_2SO_4$ + P	4,789	866	29
7	P only	1,519	869	44
	$(NH_4)_2SO_4$ + P	1,527	755	24
14	P only	1,019	536	76
	$(NH_4)_2SO_4$ + P	641	479	104
28	P only	653	340	107
	$(NH_4)_2SO_4$ + P	465	320	187
56	P only	623	320	77
	$(NH_4)_2SO_4$ + P	338	260	90

Source: M. H. Miller, unpublished data, Dept. of Soil Science, University of Guelph.

fertilizer P is the best measure of availability, the addition of $(NH_4)_2SO_4$ would tend to decrease rather than increase the absorption of fertilizer P. In their experiments, however, the addition of $(NH_4)_2SO_4$ more than doubled the fertilizer P content of sugar beet tops at the 8-week stage. As discussed previously, they related this effect to increased root growth in the fertilizer band. Fertilizer P solubility had no influence on fertilizer P absorption.

Blanchar and Caldwell (1966) found that, after leaching, the water-soluble fertilizer P in the fertilized zone was the same for MCP, MCP + NH_4Cl, and MCP + KCl pellets. They concluded that NH_4^+ increased the fertilizer P absorption by increasing the capacity of the plant to absorb P rather than by affecting the availability of P in the soil.

From these studies, it can be concluded that although there are chemical effects of salts in a fertilizer band, the influence of N salts on the solubility of the fertilizer P has not been shown to be a major factor responsible for the increased absorption of fertilizer P.

b. *Movement of Fertilizer P to the Root Surface.* Phosphorus is regarded as being relatively immobile in the soil and is considered to move to the plant root primarily by diffusion (Barber, 1962). However, in a concentrated fertilizer band, it is feasible that the concentration in solution would be sufficient to allow a significant amount of fertilizer P to move to the root surface by mass-flow.

Minshall (1964) observed that exudation from detopped tomato (*Lycopersicon esculentum*) plants was increased by the addition of NH_4^+ or NH_4^+-producing compounds. This increased water movement in the xylem could result in an increased water movement and hence increased mass-flow of fertilizer P to the plant root. However, Leonce and Miller (1966) found the same effect of NH_4^+ on fertilizer P absorption by corn grown at 35% and at 95% relative humidity. The water transpired, and hence the mass-flow in the soil, was much greater at 35% relative humidity. This indicates that increased mass-flow in the soil is not a major causal factor in the increased absorption of fertilizer P.

There have been numerous studies on the diffusion of nutrients to plant roots (Olsen and Kemper, 1968). Codiffusing and counterdiffusing ionic species have a mutual diffusion coefficient which depends on the individual coefficients and on the respective concentrations. In general the species present in lowest concentration controls the rate of diffusion (Olsen and Kemper, 1968). Since $H_2PO_4^-$ ions are normally very low in concentration in the soil solution, one would not expect the presence of other ionic species to affect the rate of diffusion of P.

Grunes *et al.* (1958a, 1958b) and Grunes (1959) hypothesized that N increased the proportion of plant P that was derived from the fertilizer because of an increased root efficiency which resulted in an increased rate of movement of fertilizer P to the root surface relative to soil P. This suggested mechanism was dependent on an increased shoot growth due to an increased P stress caused by the N fertilizer and hence an increased efficiency of the root system to absorb P. However, no direct evidence was given in support of this hypothesis.

Although the evidence available relating to this aspect of the problem is sketchy, there is very little to suggest that the increased fertilizer P absorption in the presence of N is due to a greater rate of movement of fertilizer P to the root surface.

2. Physiological Capacity of the Root to Absorb P

The absorption of ions by plant roots is a complex phenomenon which is not yet completely understood. The ions must be transported across one or more membranes that act as barriers to absorption. Metabolic energy is required for this transport. Numerous examples of competitive and synergistic effects of associated ions on the transport have been reported. Some of these are related to cation-anion balance effects, while others indicate an effect on absorption processes.

a. Cation-Anion Balance. Electrical neutrality must be maintained across the membrane during absorption. Therefore, either (1) the same number of anions and cations must be absorbed, (2) there must be an exchange of similarly charged ions in the root for those ions absorbed, or (3) there must be a shift of the organic acid balance within the cell to maintain neutrality.

Arnon (1939) found that barley plants growing in nutrient solution with an NH_4^+ source of N had a higher content of P than those grown with a nitrate source. He believed that a rapidly absorbable cation such as the ammonium ion would favor the absorption of anions such as $H_2PO_4^-$ ions.

Blair *et al.* (1970) observed a significantly higher P content in tops and roots of corn grown in a nutrient culture with NH_4^+ rather than NO_3^- as the source of N. They believed this was the result of a stimulated anion uptake in response to cation N (NH_4^+) uptake. This increase in P absorption was, however, small in comparison to that reported in many soil studies. They also observed an increased absorption of sulfate ions and a decrease in solution culture pH when NH_4^+ was the source of N. On the other hand, when NO_3 was the source

of N, higher plant levels of Ca^{2+} and Mg^{2+} and an increase in nutrient
solution pH were measured, probably due to exchange of OH^- and/or
HCO_3^- ions from the plant.

From this evidence it is clear that the source of N, NH_4^+ or NO_3^-,
affects the ability of the plant root to absorb P. This in part explains
the increased absorption when NH_4^+ salts are applied with a fertilizer
P band.

b. Absorption Processes. Cole *et al.* (1963) found that inclusion of
NO_3^- or NH_4^+ did not influence the absorption of P by corn from
solution during a 2-hour absorption period. However, pretreatment
of the plants with N caused a very pronounced increase in P absorption
per unit weight of roots. Nitrogen pretreatments also stimulated translocation of P to the plant tops. These effects suggested to the authors
that there was a connection between P uptake and N metabolism.
They hypothesized that the increases in P uptake rates may reflect
higher levels of N intermediates whose syntheses have processes in
common with those of P uptake. This effect, they felt, might explain
the higher proportion of fertilizer P to soil P taken up by plants when
N is added.

Thien and McFee (1970) confirmed the results of Cole *et al.* (1963).
They pretreated 13-day-old corn plants for 24 hours in NH_4^+, NO_3^-,
P, and H_2O solutions. Plants were then transferred to treatment
solutions containing labeled P alone or with a N source. Uptake and
translocation of labeled P were continuously monitored for 6 hours.
Nitrogen pretreatment significantly increased P absorption and translocation rates during the 6-hour absorption period. These effects were
absent when N was included only in the absorption solution. The
authors concluded, as did Cole *et al.*, that this indicated there was no
companion ion effect of N on P absorption as suggested by Arnon
(1939). The influence of pretreatment N on labeled P content of the
tops was greater than on that of the roots, indicating a different effect
of N on translocation than on absorption. This suggested to the authors
the existence of an N intermediate acting in two transport systems,
across the initial cellular barrier and in the system responsible for
moving P to the xylem.

Leonce and Miller (1966) also hypothesized an effect of the NH_4^+
ion on the rate at which the phosphate-carrier complex releases the
phosphate ion into the xylem. They placed fertilizer pellets containing
MCP both alone and in combination with $(NH_4)_2SO_4$, KNO_3, or
NH_4Cl to the side of the tip of a corn root growing down the sloping
front of a growth box (Fig. 21.2). Autoradiographs of the fertilized zone
showed an accumulation of labeled P in the roots of the roots treated
with P only and with P + KNO_3. No such accumulation was observed

in the roots treated with $P + (NH_4)_2SO_4$ or $P + NH_4Cl$ (Fig. 21.7). The fertilizer P content of the tops was more than doubled by the addition of $(NH_4)_2SO_4$. NH_4Cl caused a significant but somewhat lower increase, whereas KNO_3 decreased the fertilizer P content of the tops. Leonce and Miller (1966) assumed that the concentration of P observed in the autoradiographs was in the roots and hypothesized that the effect of NH_4^+ was to increase the rate of translocation across the root symplast into the xylem.

However, Miller *et al.* (1970), using the soft β^- emitter, ^{33}P, clearly showed that the fertilizer P was accumulated on, rather than in, the root (Fig. 21.8). Electron microprobe scans of root cross sections indicated that the fertilizer P was precipitated on the root surface in association with Ca^{++} in the absence of NH_4^+. No such precipitation was observed when NH_4^+ was present in the fertilizer pellet (Fig. 21.9). This indicated that the NH_4^+ ion caused a reaction at the soil-root interface which prevented the precipitation of fertilizer P and increased its absorption.

3. Reactions at Soil-Root Interface

Miller *et al.* (1970) observed that the pH of the roots and the adhering soil removed from the vicinity of the fertilizer pellets was 6.7 when

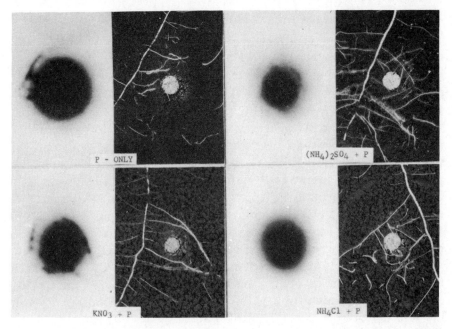

Fig. 21.7. Photographs and autoradiographs of area surrounding pellets of ^{32}P-labeled concentrated superphosphate and three N salts. Photographs and autoradiographs at same scale; diameter of pellet = 0.25 in. (Leonce and Miller, 1966)

MCP   MCP + K₂SO₄   MCP + (NH₄)₂SO₄

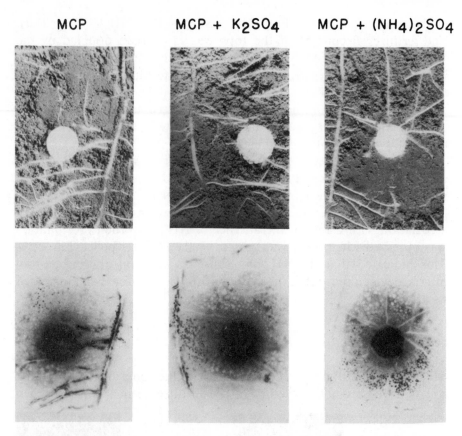

Fig. 21.8. Photographs and autoradiographs of area surrounding pellets of ^{33}P-labeled concentrated superphosphate. (Miller *et al.*, 1970)

$(NH_4)_2SO_4$ was included in the pellet, 6.9 with K_2SO_4, and 7.3 with P only. Since the $H_2PO_4^- : HPO_4^{2-}$ ratio at pH 7.3 is 0.8, compared to a ratio of 3.2 at pH 6.7, they concluded that the lowering of the pH in the presence of $(NH_4)_2SO_4$ would reduce the tendency for precipitation of $CaHPO_4 \cdot 2H_2O$ at a given P concentration. The higher proportion of P in the $H_2PO_4^-$ form at the lower pH would also increase the absorption by the root because of the greater rate of absorption of $H_2PO_4^-$ ions compared to HPO_4^{2-} ions (Hagen and Hopkins, 1955). This was substantiated by the fact that the absorption of fertilizer P was greatest in the presence of $(NH_4)_2SO_4$, intermediate with K_2SO_4, and lowest with P only. Miller *et al.* (1970) hypothesized that the reduction in pH at the soil-root interface was due to the exchange of H^+ ions from within the root for NH_4^+ or K^+ ions in the soil. In both the P + $(NH_4)_2SO_4$ and P + K_2SO_4 treatments, the absorption of NH_4^+ and K^+ would greatly exceed the absorption of anions, thus releasing an excess of H^+ ions to the soil.

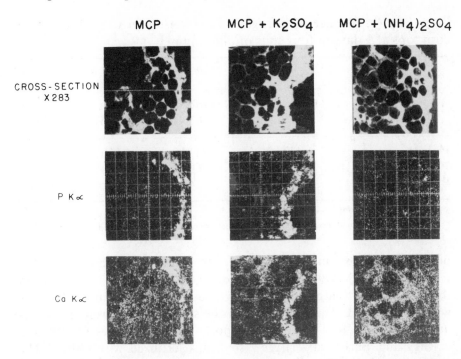

Fig. 21.9. Electron microprobe scans of root cross sections. *Top row*: Electron back-scatter photograph. *Middle and bottom rows*: P distribution and Ca distribution by K $\propto$ X-ray emission. (Miller *et al.*, 1970)

Riley and Barber (1969) found an accumulation of HCO_3^- and an increase in pH at the soil-root interface of soybean (*Glycine max*) roots compared to the original soil. The magnitude of the HCO_3^- accumulation and pH increases was related to the NO_3^- level of the soil solution. They attributed this effect to a greater uptake of anions than cations with a concomitant release of HCO_3^-. They hypothesized that if the plant roots were supplied with NH_4^+-nitrogen, the pH of the soil-root interface would decrease because of the greater cation than anion uptake.

Blair *et al.* (1971) hypothesized that the application of NO_3^- with the P would increase P accumulation on the root surface and decrease P uptake by the plant, whereas NH_4^+ would have the reverse effect. They further hypothesized that because at a pH of 5.0 essentially all of the P in the soil solution is in the $H_2PO_4^-$ form, any changes in pH of the soil-root interface in acid soils would have markedly less influence on P absorption than a similar change in soil-root interface pH in more alkaline soils, where the HPO_4^{2-} ion predominates. They tested these hypotheses in a greenhouse experiment in which fertilizer bands of MCP, MCP + $(NH_4)_2SO_4$, and MCP + KNO_3 were placed in each of four soils varying in pH from 4.2 to 8.2. In situ ^{32}P activity

Table 21.4. Influence of nitrogen source on the *in situ* ^{32}P count of corn tops at 9 days in four soils of differing pH

Treatment	^{32}P (cpm per pot)			
	Donnybrook (pH 8.2)	Burford (pH 7.4)	Oneida (pH 5.5)	Welland (pH 4.2)
P-alone	1,479 b†	391 b	947 a	109 a
P + KNO$_3$	434 a	60 a	218 a	39 a
P + (NH$_4$)$_2$SO$_4$	4,392 c	1,114 c	3,212 b	115 a

Source: Blair *et al.*, 1971.

† Values in each column followed by a different letter are significantly different at 95% probability.

of the corn tops sampled 9 days after planting was highest in the P + (NH$_4$)$_2$SO$_4$ treatment, intermediate in the P only, and lowest in the P + KNO$_3$ treatment. This effect was most significant in the soils with the high pH. The differences in the soil at pH 4.2 were not significant (Table 21.4). These data support both hypotheses formulated.

Riley and Barber (1971) grew soybeans fertilized with either NH$_4^+$ or NO$_3^-$ on soil with four different initial pH levels obtained by liming a soil with an initial pH of 5.5. Fertilization of the soil with NH$_4^+$ decreased the pH of the soil-root interface, whereas fertilization with NO$_3^-$ increased it. The difference in soil-root interface pH between the NH$_4^+$- and NO$_3^-$- fertilized soils was as much as 1.9 pH units with an initial pH of 5.2 and as small as 0.2 units when soil pH prior to N application was 7.8. Ammonium fertilized soybeans absorbed more P and had a higher P concentration than NO$_3^-$-fertilized soybeans regardless of the pH of the bulk soil. The P content was closely correlated with the pH of the soil-root interface but not with that of the bulk soil. They concluded that the increased availability of P from the soil where NH$_4^+$ was used was mainly due to the effect of the NH$_4^+$ on the pH of the soil-root interface. In their experiments P uptake was approximately equal at a given soil-root interface pH for NH$_4^+$- and NO$_3^-$- fertilized soybeans, suggesting that NH$_4^+$ and NO$_3^-$ do not have specific effects on absorption. In these experiments the NH$_4^+$, NO$_3^-$, and P were mixed with the entire soil volume. As discussed previously, earlier results on this phenomenon indicated that N increased P absorption only when they were applied together in a localized volume of soil (Duncan and Ohlrogge, 1958). Two factors may have been responsible for the difference in results. As indicated previously (Mamaril and Miller, 1970), concentration of NH$_4^+$ is a significant factor. Riley and Barber (1971) used concentrations approximating the 1 meq/100 g of soil of Mamaril and Miller (1970), whereas

Duncan and Ohlrogge (1958) used much lower concentrations. In addition, when NH_4^+ is mixed throughout the soil, it is normally quickly converted to nitrate. Riley and Barber (1971) prevented this conversion by the addition of a nitrification inhibitor, 2-chloro-6-trichloromethyl pyridine. In a band application, the NH_4^+ may remain as such for an appreciable time, resulting in a decrease in soil-root interface pH and hence an increase in P absorption.

The data (Table 21.5) obtained by Blair *et al.* (1971) from plant analysis after 17 days were not as easily interpreted as those after 9 days (Table 21.4). The authors felt that at 17 days the data reflect the net result of the many phenomena that contribute to the increased P uptake in the presence of N.

The concentration of fertilizer P with the $P + (NH_4)_2SO_4$ treatment was significantly greater than with the P-alone or $P + KNO_3$ treatments in all but the Welland soil with a pH of 4.2. However, concentration with the $P + KNO_3$ treatment was greater than with the P-alone treatment in the Donnybrook soil and was not significantly different in the other three soils. The greater absorption from the $P + KNO_3$ treatment in the Donnybrook soil may have been due to the increased

Table 21.5. Influence of nitrogen source on growth and fertilizer P content of corn after 17 days of growth on four soils of differing pH

Soil		Treatment	Tops			Roots† yield (mg)
			Yield (g/pot)	Fert. P content (μg/mg)	N content (μg/mg)	
Donnybrook	pH 8.2	P alone	3.13 a‡	1.59 a	21.0 a	93.1 a
		$P + KNO_3$	4.61 b	2.72 b	34.0 b	139.2 b
		$P + (NH_4)_2SO_4$	5.28 c	6.16 c	42.0 c	151.0 b
Burford	pH 7.4	P alone	4.98 b	1.85 a	31.3 a	126.6 a
		$P + KNO_3$	3.18 a	1.91 a	34.0 a	90.9 a
		$P + (NH_4)_2SO_4$	7.93 c	5.76 b	35.3 a	192.4 b
Oneida	pH 5.5	P alone	3.35 a	2.02 a	17.0 a	124.3 b
		$P + KNO_3$	5.41 a	2.48 a	31.0 b	39.0 a
		$P + (NH_4)_2SO_4$	8.92 b	5.18 b	33.0 b	197.3 c
Welland	pH 4.2	P alone	1.86 a	1.13 a	23.7 a	79.6 a
		$P + KNO_3$	2.20 a	1.27 a	39.0 b	52.3 a
		$P + (NH_4)_2SO_4$	2.51 a	2.13 a	39.0 b	61.7 a

Source: Blair *et al.*, 1971.

† Root weight in volume of soil influenced by fertilizer.

‡ Values in each column within soils followed by a different letter are significantly different at 95% probability.

root growth, the increased top growth, or the increased N content of the tops, all of which have been shown to influence fertilizer P content. The $(NH_4)_2SO_4$ treatment significantly increased the fertilizer P content in comparison to the KNO_3 treatment in all soils except the Welland. The increase in the two calcareous soils could be due to a reduction in pH at the soil-root interface and the change in the ratio of $H_2PO_4^-$: HPO_4^{2-} ions. However, in the Oneida soil with an initial pH of 5.5, the P would be essentially all in the $H_2PO_4^-$ form. Therefore, unless the NO_3^- ion absorption caused a large increase in pH, there would not be a major change in the ratio. Riley and Barber (1971) reported an increase in pH of the rhizocylinder of from 5.2 to 6.6 with NO_3^- addition. An increase of this magnitude in the experiments of Blair *et al.* (1971) could account for the difference in fertilizer P absorption in the Oneida soil with the $(NH_4)_2SO_4$ and KNO_3 treatments. However, it would not explain the difference in absorption with the $(NH_4)_2SO_4$ and P-only treatments. Also, in the Oneida soil there was significantly greater root growth in the P + $(NH_4)_2SO_4$ than in the P-alone or P + KNO_3 treatment. This greater root growth, possibly in association with a finer root system as suggested by Duncan and Ohlrogge (1958), may explain the effect on the Oneida soil. Electron-microprobe analysis of root cross sections showed an accumulation of Ca^{++} and P at the soil-root interface in the Donnybrook and Burford soils in the P-only treatment and, to a lesser extent, in the Oneida soil. The fact that some accumulation was observed in the Oneida soil indicated that precipitation of P at the root surface was occurring even in a soil with this low pH. Some accumulation of Fe was also observed on the Oneida soil. No accumulations were observed on the Waterloo soil with a pH of 4.2. Electron microprobe scans were not obtained for the P + KNO_3 treatment.

Blair *et al.* (1971) concluded that a decrease in the pH of the soil-root interface brought about by the unequal uptake of cations and anions and the resulting increase of $H_2PO_4^-$ ions was responsible for the increased P uptake during the very early growth period when NH_4^+ accompanied the P. Other factors such as increased root growth and increased plant N content may also have contributed to the observed increase in P uptake at later growth stages.

VI. SUMMARY

Increased fertilizer P absorption on addition of N is a general phenomenon in the soil-plant system. This effect is most pronounced when the N is in the NH_4^+ form and is applied in intimate association with the fertilizer P in a concentrated band.

Several mechanisms have been shown to be responsible, in part, for this effect. The principal ones are (*a*) increased root growth in the fertilized volume; (*b*) increased physiological capacity of the root to absorb P, probably due to an involvement of N-intermediates in the absorption process; and (*c*) alteration of the pH at the soil-root interface due to excess cation over anion absorption and a consequent release of H^+ ions. The last mechanism is probably responsible for the increased absorption of P upon addition of NH_4^+ at very early growth stages. The increased root growth, and increased physiological capacity of the root to absorb P, would be a factor after the first week or so of growth. Although it has not been conclusively shown, there is little evidence to show that NH_4^+ and NO_3^- differ in their effects on root growth or physiological capacity of the root to absorb P.

The increased absorption due to a reduction in pH of the soil-root interface would occur only with an NH_4^+ source of N; NO_3^- would tend to increase the pH and hence reduce the absorption. This phenomenon, however, is not limited to NH_4^+; whenever cation absorption exceeds anion absorption a reduction in pH would occur. Conversely when anion absorption exceeds cation absorption an increase in pH would occur. The importance of these reactions at the soil-root interface is almost certainly not limited to P. They would have a very significant effect on absorption of several ions.

References

Arnon, D. I. 1939. The effect of ammonium and nitrate nitrogen on the mineral composition and sap characteristics of barley. *Soil Sci.* 48: 295–307.

Barber, S. A. 1962. A diffusion and mass-flow concept of soil nutrient availability. *Soil Sci.* 93: 39–49.

Blair, G. J., C. P. Mamaril, and M. H. Miller. 1971. Influence of nitrogen source on phosphorus uptake by corn from soils differing in pH. *Agron. J.* 63: 235–38.

Blair, G. J., M. H. Miller, and W. A. Mitchell. 1970. Nitrate and ammonium as sources of nitrogen for corn and their influence on the uptake of other ions. *Agron. J.* 62: 530–32.

Blanchar, R. W., and A. C. Caldwell. 1966. Phosphate-ammonium-moisture relationships in soils. II. Ion concentrations in leached fertilizer zones and effects on plants. *Soil Sci. Soc. Amer. Proc.* 30: 43–48.

Bouldin, D. R., and E. C. Sample. 1958. The effect of associated salts on the availability of concentrated superphosphate. *Soil Sci. Soc. Amer. Proc.* 22: 124–29.

——. 1959. Laboratory and greenhouse studies with monocalcium monoammonium and diammonium phosphates. *Soil Sci. Soc. Amer. Proc.* 23: 338–42.

Cole, C. V., D. L. Grunes, L. K. Porter, and S. R. Olsen. 1963. The effects of nitrogen on short-term phosphorus absorption and translocation in corn (*Zea mays*). *Soil Sci. Soc. Amer. Proc.* 27: 671–74.

Duncan, W. G., and A. J. Ohlrogge. 1958. Principles of nutrient uptake from fertilizer bands. II. Root development in the band. *Agron. J.* 50: 605–8.

——. 1959. Principles of nutrient uptake from fertilizer bands. III. Band volume, concentration and nutrient composition. *Agron. J.* 51: 103–6.

Grunes, D. L. 1959. Effect of nitrogen on the availability of soil and fertilizer phosphorus to plants. In *Advances in Agronomy*, ed. A. G. Norman, 11: 369–96. New York: Academic Press.

Grunes, D. L., H. R. Haise, and L. O. Fine. 1958a. Proportionate uptake of soil and fertilizer phosphorus by plants as affected by nitrogen fertilization. II. Field experiments with sugarbeets and potatoes. *Soil Sci. Soc. Amer. Proc.* 22: 49–52.

Grunes, D. L., F. G. Viets, Jr., and S. H. Shih. 1958b. Proportionate uptake of soil and fertilizer phosphorus by plants as affected by nitrogen fertilization. I. Growth chamber experiment. *Soil Sci. Soc. Amer. Proc.* 22: 43–48.

Hagen, C. E., and H. T. Hopkins. 1955. Ionic species in orthophosphate absorption by barley roots. *Plant Physiol.* 30: 193–99.

Lawton, K., and J. A. Vomocil. 1954. Dissolution and migration of phosphorus from granular superphosphate in some Michigan soils. *Soil Sci. Soc. Amer. Proc.* 18: 26–32.

Leonce, F. S., and M. H. Miller. 1966. A physiological effect of nitrogen on phosphorus absorption by corn. *Agron. J.* 58: 245–49.

Lindsay, W. L., J. R. Lehr, and H. F. Stephenson. 1959. Nature of the reactions of monocalcium phosphate monohydrate in soils. III. Studies with metastable triple-point solution. *Soil Sci. Soc. Amer. Proc.* 23: 342–45.

Lindsay, W. L., and H. F. Stephenson. 1959a. Nature of the reactions of monocalcium phosphate monohydrate in soils. I. The solution that reacts with the soil. *Soil Sci. Soc. Amer. Proc.* 23: 12–17.

——. 1959b. Nature of the reactions of monocalcium phosphate monohydrate in soils. II. Dissolution and precipitation reactions involving iron, aluminum, manganese and calcium. *Soil Sci. Soc. Amer. Proc.* 23: 18–22.

——. 1959c. Nature of the reactions of monocalcium phosphate monohydrate in soils. IV. Repeated reactions with metastable triple point solution. *Soil Sci. Soc. Amer. Proc.* 23: 440–45.

Lorenz, O. A., and C. M. Johnson. 1953. Nitrogen fertilization as related to the availability of phosphorus in certain California soils. *Soil Sci.* 75: 119–29.

Mamaril, C. P., and M. H. Miller. 1970. Effects of ammonium on the uptake of phosphorus, sulfur and rubidium by corn. *Agron. J.* 62: 753–58.

Miller, M. H. 1965. Influence of $(NH_4)_2SO_4$ on root growth and P absorption by corn from a fertilizer band. *Agron. J.* 57: 393–96.

Miller, M. H., C. P. Mamaril, and G. J. Blair. 1970. Ammonium effects on phosphorus absorption through pH changes and phosphorus precipitation at the soil-root interface. *Agron. J.* 62: 524–27.

Miller, M. H., and A. J. Ohlrogge. 1958. Principles of nutrient uptake from fertilizer bands. I. Effect of placement of nitrogen fertilizer on the uptake of band-placed phosphorus at different soil phosphorus levels. *Agron. J.* 50: 95–97.

Miller, M. H., and V. N. Vij. 1962. Some chemical and morphological effects of ammonium sulphate in a fertilizer phosphorus band for sugarbeets. *Can. J. Soil Sci.* 42: 87–95.

Minshall, W. H. 1964. Effect of nitrogen-containing nutrients on the exudation from detopped tomato plants. *Nature* 202: 925–26.

Olsen, S. R., and W. D. Kemper. 1968. Movement of nutrients to plant roots. In *Advances in Agronomy*, ed. A. G. Norman, 20: 91–151. New York: Academic Press.

Olson, R. A., and A. F. Dreier. 1956. Nitrogen, a key factor in fertilizer phosphorus efficiency. *Soil Sci. Soc. Amer. Proc.* 20: 509–14.

Olson, R. A., A. F. Dreier, G. W. Lowrey, and A. D. Flowerday. 1956. Availability of phosphate carriers to small grains and subsequent clover in relation to: II. Concurrent soil amendments. *Agron. J.* 48: 111–16.

Rennie, D. A., and J. Mitchell. 1954. The effect of nitrogen additions on fertilizer phosphate availability. *Can. J. Agr. Sci.* 34: 353–63.

Rennie, D. A., and R. J. Soper. 1958. The effect of nitrogen additions on fertilizer-phosphorus availability II. *J. Soil Sci.* 9: 155–67.

Riley, D., and S. A. Barber. 1969. Bicarbonate accumulation and pH changes at the soybean (*Glycine max* (L.) merr) root-soil interface. *Soil Sci. Soc. Amer. Proc.* 33: 905–8.

——. 1971. Effect of ammonium and nitrate fertilization on phosphorus uptake as related to root-induced pH changes at the root-soil interface. *Soil Sci. Soc. Amer. Proc.* 35: 301–6.

Robertson, W. K., P. M. Smith, A. J. Ohlrogge, and D. M. Kinch. 1954. Phosphorus utilization by corn as affected by placement and nitrogen and potassium fertilization. *Soil Sci.* 77: 219–26.

Scarseth, G. D., H. L. Cook, B. A. Krantz, and A. J. Ohlrogge. 1942. *How to Fertilize Corn Effectively in Indiana.* Ind. Agr. Exp. Sta. Bull. 482.

Starostka, R. W., and W. L. Hill. 1954. Influence of soluble salts on the solubility of and plant response to dicalcium phosphate. *Soil Sci. Soc. Amer. Proc.* 18: 193–98.

Thien, S. J., and W. W. McFee. 1970. Influence of nitrogen on phosphorus absorption and translocation in *Zea mays. Soil Sci. Soc. Amer. Proc.* 34: 87–90.

Werkhoven, C. H. E., and M. H. Miller. 1960. Absorption of fertilizer phosphorus by sugarbeets as influenced by placement of phosphorus and nitrogen. *Can. J. Soil Sci.* 40: 49–58.

22. Toxic Effects of Aqueous Ammonia, Copper, Zinc, Lead, Boron, and Manganese on Root Growth

A. C. Bennett

GENERALLY, an ion or a compound which is toxic to roots adversely affects physiological functions and growth at very low concentrations. The assessment of morphological and physiological changes in roots brought about by toxic substances is the most definitive and accurate means of defining toxicity of a substance. However, the visible morphological symptoms of a number of toxicities to roots are likely to be similar. For example, browning of the root tip accompanied by a cessation of growth has been associated with aqueous ammonia $[NH_3(aq)]$ and copper toxicity as well as with calcium deficiency. Because the root tip is the most active physiological part of the root, it exhibits morphological toxicity symptoms first. Since initial symptoms often occur in such a small segment of the root, identification of a specific toxicity is difficult without extensive specimen examination and prior knowledge of the rooting medium.

Identification of toxic effects by growth measurement is hazardous since the possibility of measuring indirect effects on root growth exists. Toxic effects that are initially manifested in reduced shoot growth can rapidly influence root growth. Such indirect effects are minimized by the use of short-term and split root type experiments. Awareness of the possibility of secondary growth influences when designing experiments can help prevent drawing of erroneous conclusions concerning toxic effects to roots. Though growth rate is not the most definitive indicator of toxicities, it usually is the most obvious and easiest to assess quantitatively. Accurate identification of a particular toxicity is possible by relating growth and symptoms to the composition of the soil or nutrient solution in which the plant grew.

This section deals with the toxicities of ammoniacal-nitrogen, copper, lead, zinc, boron, and manganese to plant roots.

I. AMMONIACAL-NITROGEN

A. Background Information

The toxic effect of ammoniacal-nitrogen to roots is of considerable practical importance because most nitrogen fertilizer applied to soil is this
form. Thus plant roots frequently encounter a soil solution containing
appreciable quantities of ammoniacal-N. Ammoniacal-N has been
shown to restrict root growth in both acid and alkaline media (Barker
et al., 1966b; Bennett and Adams, 1970). Also, I have reviewed several
scientific communications wherein apparently incorrect conclusions
were drawn because the researchers did not carefully consider the
ammoniacal-N situation of their growth media.

Ammoniacal-N in aqueous solutions is composed of the cation NH_4^+
plus the uncharged form, written $NH_3(aq)$. Elsewhere in the literature
$NH_3(aq)$ is synonymous with NH_4OH. The same material is sold in
$29\%-30\%$ aqueous solutions as ammonium hydroxide. The existence
of a real NH_4OH molecule is doubtful, though an aquated form of
ammonia, $NH_3 \cdot H_2O$, has been identified under conditions of temperature and pressure not encountered by living plants.

The phytotoxicity of ammoniacal-N has been known for many years
(Darwin, 1882; Naftel, 1931; Willis and Piland, 1931). It is likely that
ammoniacal-N toxicity was first suspected when the smell emanating
from rotting manure was associated with poor plant growth in soil
treated with fresh barnyard material. Charles Darwin (1882) published
lengthy papers on the toxicity of carbonate of ammonia to plant roots
and leaves. The subject has been investigated extensively since the introduction of anhydrous ammonia as a major source of fertilizer N. Reduced
germination and seedling growth when urea was used as an N source
has been attributed to toxic ammonia (Brage *et al.*, 1960; Cooke, 1962;
Court *et al.*, 1964). The toxicity of diammonium phosphate to germination and root growth is at least partially attributable to toxicity of ammonia and ammoniacal-N (Bennett and Adams, 1970; Willis and Piland,
1931). The toxicity of ammoniacal-N is usually attributed to the $NH_3(aq)$
molecule (Blanchar, 1967; Allred and Ohlrogge, 1964; Megie *et al.*, 1967),
which occurs in significant quantities only in alkaline media. However,
there are reports of disruption of plant growth in acid media traceable
to NH_4^+ as the predominant N form in the nutrient medium (Barker *et
al.*, 1966b; Naftel, 1931).

B. Relationship between Ammoniacal-Nitrogen Forms
and Solution Acidity

The form in which ammoniacal-N exists in aqueous media such as soil
and nutrient solutions is controlled by solution acidity or H^+ activity.

The distribution of NH_4^+ and $NH_3(aq)$ in aqueous solutions is described by the equilibrium relationships

$$NH_3(aq) \rightleftharpoons NH_4^+ + OH^- \qquad [1]$$

and

$$[NH_4^+][OH^-]/[NH_3(aq)] = K_b, \qquad [2]$$

where brackets denote ionic activities and K_b is the dissociation constant of $NH_3(aq)$. The relationship of H^+ to NH_4^+ and $NH_3(aq)$ is derived by substituting $K_w/[H^+]$ for $[OH^-]$ in equation [2]:

$$[NH_4^+]/[NH_3(aq)] = (K_b/K_w)[H^+], \qquad [3]$$

where K_w is the ionization constant of water.

C. Evaluation of NH_3(aq) in Solutions

Direct quantitative evaluation of $NH_3(aq)$ by analytical means is not practical. There are two primary approaches to indirect $NH_3(aq)$ determination.

1. Solution Analysis

Equation [3] provides a basis for $NH_3(aq)$ evaluation through analytical determination of H^+ and total ammoniacal-N, i.e., $NH_4^+ + NH_3(aq)$. The H^+ activity is directly assessable by electrometric pH measurement. The total ammoniacal-N of a solution can be determined by various methods such as Nesslerization or steam distillation from alkaline solutions into boric acid and subsequent titration.

In most instances the concentration of $NH_3(aq)$ is much less than that of NH_4^+; thus its contribution to total ammoniacal-N can be ignored. In this case the analytical value for ammoniacal-N can be assumed to be the NH_4^+ concentration. To estimate $NH_3(aq)$ accurately, the NH_4^+ activity must be calculated by multiplying its concentration by the activity coefficient, $f_{NH_4^+}$, a function of the ionic strength of the solution. The activity coefficient is obtained from the Debye-Hückel equation as outlined in Chapter 15. Known H^+ and NH_4^+ activities can be substituted into equation [3], after which the $NH_3(aq)$ concentration can be calculated.

If the calculated $NH_3(aq)$ is 3% of the total ammoniacal-N or greater, it is necessary to reestimate it after subtracting the $NH_3(aq)$ found from the analytically determined ammoniacal-N to obtain a better estimate of NH_4^+ concentration. Time-consuming iterations are probably not necessary since the contribution to ionic strength of overestimated NH_4^+

is usually negligible. Simply subtract the first estimate of $NH_3(aq)$ from total ammoniacal-N, reapply the same activity coefficient for NH_4^+, and proceed to a more accurate estimate of $NH_3(aq)$. However, if significant concentrations of divalent anions such as SO_4 and HPO_4 exist in the solution, it is necessary to allocate some of the total ammoniacal-N to ion pairs. This procedure is described in Chapter 15; it is applicable to both nutrient and soil solutions. If the $NH_3(aq)$ content of a soil system is to be determined accurately, it is mandatory that the displaced soil solution or an accurate approximation thereof be analyzed. Use of soil solution approximations usually requires extensive knowledge about the soil in question.

2. Gas Analysis

The relationship between gaseous NH_3 and $NH_3(aq)$ can be used to determine $NH_3(aq)$. This relationship is

$$P_{NH_3(g)} = 12.9 \, M \, [NH_3(aq)], \qquad [4]$$

where $P_{NH_3(g)}$ is in millimeters of Hg and $NH_3(aq)$ is in moles per liter. The pressure of $NH_3(g)$ in the equilibrium soil air or equilibrium atmosphere above a nutrient solution is measured by trapping the $NH_3(g)$ from a known volume of air in an acid and then determining the ammoniacal-N by a standard method. The moles of ammoniacal-N are then multiplied by the molar volume of $NH_3(g)$ at the atmospheric pressure of equilibration—about 22.4 l at one atm if $NH_3(g)$ is assumed ideal—to obtain the volume of $NH_3(g)$ in the air analyzed. Then $NH_3(aq)$ concentration can be calculated using the following equation:

$$M_{NH_3(aq)} = \left(\frac{\text{volume } NH_3(g)}{\text{volume air analyzed}} \times \frac{\text{barometric pressure}}{\text{in mm of Hg}} \right) \Big/ 12.9.$$
$$[5]$$

The constant 12.9 of equation [4] is only applicable at low concentrations of $NH_3(g)$; it is acceptable in the range of $NH_3(g)$ pressure that produces toxicity of NH_3 to roots.

D. Symptoms of Ammoniacal-Nitrogen Toxicity

1. $NH_3(aq)$ Toxicity

Darwin (1882) reported that immersion of the roots of *Euphorbia peplus* into a solution of carbonate of ammonia resulted in precipitation of brown granular matter concentrated in the area just back of the root tip. This matter also formed stripes that extended to near the stem. No root hairs arose from cells containing the brown matter. Bennett and Adams (1970) reported that cotton (*Gossypium hirsutum*) roots turned brown in nutrient solutions that contained toxic quantities of $NH_3(aq)$; the tips of roots affected by $NH_3(aq)$ secreted a muscilagenous material. An abrupt irreversible cessation of root growth occurs if $NH_3(aq)$ reaches a critical concentration; i.e., the roots are killed outright (Darwin, 1882). Allred and Ohlrogge (1964) noted that the susceptibility of seedling components to ammonium environments was: primary seminal root > lateral seminal root > plumule. Though the damage of $NH_3(aq)$ is initiated in the root system, the first obvious symptom of $NH_3(aq)$ toxicity is often wilting of the leaves, which become flaccid and abnormally dark-green (Stuart and Haddock, 1968; Bennett and Adams, 1970).

2. NH_4^+ Toxicity

Barker *et al.* (1966a, 1966b) and Puritch and Barker (1967) have investigated the effects of NH_4^+ in acid media on root and shoot growth of beans (*Phaseolus vulgaris* L.). They described NH_4^+ toxicity symptoms as curling and drying at the leaf margins and yellowing and necrosis of the laminae. Ammonium toxicity to cucumber (*Cucurbita sativa* L.) caused curling of the leaf edges and marginal chlorosis (Matsumoto *et al.*, 1968). None of these three communications mention toxicity symptoms to roots. Root growth was depressed, but it is probable that the depression resulted from decreased photosynthetic activity associated with damaged leaves.

E. Physiological Aspects of Ammoniacal-Nitrogen Toxicity

There is some evidence that NH_4^+ and $NH_3(aq)$ toxicities are the same physiological phenomenon, though the symptoms of the respective effects certainly do not substantiate such a conclusion. The $NH_3(aq)$ form appears to have a unique and dramatic effect on roots. Stuart and Haddock (1968) reported that water uptake by excised sugar beet (*Beta vulgaris*) roots was reduced 95% when they were placed in 1 mM solutions of $(NH_4)_2CO_3$ at pH 8.4. It can be shown that the $NH_3(aq)$ concentration of this solution was about 0.11 mM.

. It is likely that $NH_3(aq)$ is physiologically active in very low concentrations in the nutrient media because it is able to surmount the barrier constituted by the cell membrane (Macmillan, 1956; Warren and Nathan,

1958; Warren, 1962). Its lack of charge and the reaction of alkali-like ions with fat would facilitate its entry through cell membranes. It is logical that $NH_3(aq)$ might be swept into the root in the transpiration stream since it is uncharged. After entry into the root, the physiological pH partitions the absorbed $NH_3(aq)$ into NH_4^+ and $NH_3(aq)$. When a critical quantity of $NH_3(aq)$ is present in the ambient medium, the amination mechanisms of the plant are unable to modify all reduced N that is absorbed. In the plant, unmetabolized ammoniacal-N has been shown to uncouple photosynthetic phosphorylation (Krogman et al., 1959), inhibit ferricyanide reduction (Krogman et al., 1959), reduce respiration (Vines and Wedding, 1960), decrease photosynthetic oxygen evolution (Puritch and Barker, 1967), reduce the rate of the reaction NADH → NAD (Vines and Wedding, 1960), and inhibit isocitrate dehydrogenase activity (Katsunuma et al., 1965). Barker et al. (1966a, 1966b) found that bean plants grown in acid media dominated by ammoniacal-N were high in soluble amino and ammonium nitrogen. They suspected that acidity of the ambient medium reduced amination of absorbed ammoniacal-N and/or promoted protein degradation. Matsumoto et al. (1968) concluded that ammonium toxicity resulted in inhibited translocation of glucose in cucumber leaves.

There are few reports of generally applicable parameters of $NH_3(aq)$ toxicity. Bennett and Adams (1970) reported that a reduced rate of cotton primary root elongation resulted when $NH_3(aq)$ concentrations of the ambient media were greater than about 0.17 to 0.24 mM in soil solutions and 0.16 to 0.22 mM in nutrient solutions. Stuart and Haddock (1968) reported reduced water uptake by excised sugar beet roots when $NH_3(aq)$ concentration was about 0.11 mM (estimated by the author). Krogman et al. (1959) reported that ferricyanide reduction by isolated spinach (Spinacia oleracea) chloroplasts was reversibly stimulated by ammoniacal-N concentrations of about 0.6 mM but irreversibly inhibited by solutions containing from 6×10^{-3} to 2×10^{-2} M of ammoniacal-N at pH 7.8. Estimated $NH_3(aq)$ concentrations of these latter solutions are 0.17 to 0.56 mM, the same magnitude reported by Bennett and Adams (1970) and Stuart and Haddock (1968) in widely different experiments. However, Blanchar (1967) noted that corn (Zea mays) seed germination was less in soil that equilibrated with air in which the partial pressure of $NH_3(g)$ ranged from 0.077 to 0.156 mm Hg during the experiment. These pressures are equivalent to $NH_3(aq)$ concentrations of 5.97 to 12.1 mM. Allred and Ohlrogge (1964) found that $PNH_3(g)$ of 0.063 [$NH_3(aq)$ = 4.88 mM] had a significant injurious effect on early physiological development of corn. The primary seminal root appeared to reach a length of 2–3 mm during a 7-day germination period in the atmosphere described; seminal roots of seed germinated over NH_3-free H_2O reached 4 cm during the same period. Ensminger et al. (1965) found that soaking

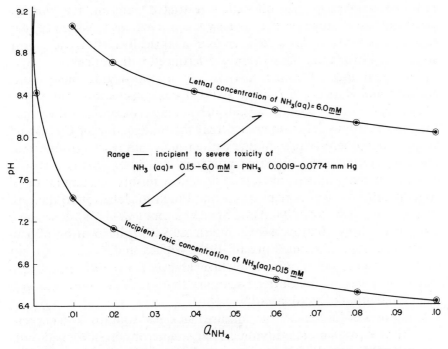

Fig. 22.1. Equimolarity diagram for toxic and lethal quantities of $NH_3(aq)$

cottonseed in 0.25 M $(NH_4)_2SO_4$ at pH 7.8, 1.0 M $(NH_4)_2HPO_4$ at pH 7.2, and 0.1 M $(NH_4)_2HPO_4$ at pH 8.2 significantly reduced the germination percentage. Estimated $NH_3(aq)$ concentrations for these solutions are 10.4 mM, 3.5 mM, and 13.4 mM, respectively. The data of Krogman *et al.* (1959), Stuart and Haddock (1968), and Bennett and Adams (1970) seem to define an incipiently toxic level of $NH_3(aq)$. The data of Allred and Ohlrogge (1964), Ensminger *et al.* (1965), and Blanchar (1967) probably more nearly define a lethal toxic concentration of $NH_3(aq)$.

Figure 22.1 is a graphic representation of the combinations of pH and NH_4^+ activity that produce incipient toxic levels of $NH_3(aq)$, i.e., 0.15 mM, and lethal levels of $NH_3(aq)$, i.e., about 6.0 mM.

II. COPPER

Copper is universally present in soils; however, deficiencies can occur in organic soils and soils with pH in the 7–8 range. Virtually all of Florida's citrus lands were originally deficient in Cu. Copper is highly toxic to roots, and toxic concentrations are translocated only negligibly via the stems. Thus Cu is a widely used toxicant for destroying roots that have encroached into sewer lines and septic tank fields. Roots are eradicated with little damage to trees other than restricted water uptake from the destroyed roots. However, Cu toxicity has proved costly where the element and its compounds have been used historically as fungicides or in areas of acid soils that have been overfertilized with copper.

The physiology of copper toxicity is not thoroughly understood. McBrien and Hassal (1967) noted that Cu was considerably more toxic to *Chlorella vulgaris* when it was absorbed under anaerobic rather than aerobic conditions. They suggested that the toxic action of Cu was related to its propensity to combine with protein sulphydryl groups, thus disrupting tertiary and quaternary protein structure. This is a typical mode of toxicity of heavy metals to living organisms. Accumulation of large quantities of Cu in plant roots and little translocation to the tops is reported by Ishizuka (1942) and Struckmeyer *et al.* (1969). Elsewhere it has been shown that excess Cu interfered with translocation of iron in several crops (Smith and Sprecht, 1953; Wallace and Kock, 1966), but this phenomenon was not noted by Struckmeyer *et al.* (1969) in tobacco (*Nicotiana tabacum*). The latter also report that excess Cu interfered with magnesium uptake by tobacco.

Symptoms of Cu toxicity to roots are blackened root tips and darkening of the root system; laterals appear very numerous on shortened roots (Broyer and Furnstal, 1945; Reuther and Smith, 1953; Sowell *et al.*, 1957; Struckmeyer *et al.*, 1969). The symptoms are usually not sufficiently unique to identify Cu toxicity without the knowledge that possible excessive Cu is present in the growth medium. Struckmeyer *et al.* (1969) pointed out that anatomical examination of roots assists in the identification of the disorder. They reported that tobacco roots grown in solutions containing toxic concentrations of Cu had flattened and elongated cortical cells; some phloem cells were disarranged; walls of the xylem were thin and misshapen; lateral roots were short with little parental tissue remaining between root primordia and differentiating roots; there was less differentiation of vascular tissue than in normal roots. They further reported that cells of the affected roots were misshapen, abnormally oriented, and frequently surrounded by a fine particulate precipitate along the external root surface.

Symptoms on the aerial portions of plants are wilt (Sowell *et al.*, 1957), Fe deficiency chlorosis (Reuther and Smith, 1953; Smith and Sprecht, 1953), and interveinal chlorosis of younger leaves (Struckmeyer *et al.*, 1969). Iron deficiency was verified when addition of chelated Fe (KFe-EDTA) overcame the chlorosis in citrus but did not prevent Cu toxicity to roots (Smith and Sprecht, 1953).

The level of Cu that is toxic to plant roots varies with the nutrient medium as well as with species or even varieties of plants. Lund[1] noted

---

[1] Z. F. Lund (USDA, ARS, SWCRD, Auburn Univ., Auburn, Ala.), personal communications, 1971.

that early taproot elongation of cotton was significantly depressed by as little as 0.04 ppm Cu in 1/4-strength Hoaglands nutrient solution; 0.08 ppm Cu killed the roots within 5 days after Cu addition. Smith and Sprecht (1953) reported damage to mandarin and sweet-orange (*Citrus* spp.) seedlings at concentrations as low as 0.1 ppm in nutrient solutions similar to 1/4-strength Hoaglands. Sowell *et al.* (1957) reported that cotton roots which grew in nutrient solutions were damaged by a concentration of 0.8 ppm Cu but not by 0.4 ppm. Struckmeyer *et al.* (1969) reported Cu toxicity symptoms on tobacco roots grown in Hoaglands nutrient solutions containing 0.32 ppm Cu; no symptoms appeared on roots that grew in solutions containing 0.16 ppm Cu. Derose *et al.* (1937) reported that 1.0 ppm Cu in nutrient solution was toxic to the tomato (*Lycopersicon esculentum*) plant. The differences in toxic concentrations cited here are partially attributable to age and type of root system. In addition, activity of Cu is probably a more generally applicable measure of the toxic parameter than concentration. Thus the values reported by Lund[2] and by Smith and Sprecht (1953) are comparable because the experiments were all conducted in 1/4-strength Hoaglands solution. The work of Struckmeyer *et al.* (1969) was conducted with full Hoaglands solution (estimated ionic strength of 0.029), whereas the data of Sowell *et al.* (1957) was from experiments conducted in a solution of ionic strength of 0.031. However, the nutrient solutions of Sowell *et al.* (1957) contained 2.25 times as much SO_4^{2-} as those of Struckmeyer *et al.* (1969). More extensive $CuSO_4^0$ pair formation in the former solutions accounts for a part of the difference in toxic concentrations observed. Thus I have estimated the activity of Cu that caused observable symptoms to be $4.5\mu M$ for the experiments of Sowell *et al.* (1957) and $2.6\mu M$ for the experiments of Struckmeyer *et al.* (1969). Considering the range between toxic and nontoxic concentrations reported by these workers, it seems likely that the same activities were toxic in both cases. The lower concentrations shown to be toxic by Lund[3] and by Smith and Sprecht (1953) might have been due to age of roots and duration of experiments, respectively. Less than 1.0 ppm of Cu is highly toxic to growing roots of most crops. This figure refers to concentration in nutrient or displaced soil solutions.

III. LEAD

To warm-blooded animals, lead is an accumulative poison. Thus any factor that might result in abnormal Pb intake by man must be under-

[2] *Ibid.*
[3] *Ibid.*

stood. Concern with environmental pollution has intensified interest in the dangers of Pb poisoning. The abnormal content of Pb in soils and plants near heavily traveled highways is the consequence of the burning of tetraethyl lead in high-performance gasolines by automobiles. Thus Singer and Hanson (1969) found that Pb content of soil was highest 1.5 meters from the roadway, Pb content of soil was influenced up to 15 meters from the roadway, and Pb content was directly related to the density of traffic and age of the road. Cannon and Bowles (1962) found as much as 700 ppm Pb in grass growing within 5 feet of a main traffic artery near Denver, Colorado, and in homegrown vegetables from gardens within 50 feet of a main artery in Conandaigua, N.Y.; samples of vegetables collected within 25 feet of any road surveyed contained an average of 80 ppm Pb, whereas only 20 ppm were found in vegetables that grew more than 500 feet from roads. Thus an understanding of the Pb-plant interrelationship would be helpful in evaluating the impact of the burning of tetraethyl lead on the environment. However, very little contemporary research has been published that defines parameters of Pb toxicity to plant roots. There are a few early communications by perceptive authors. From these it can be concluded that Pb is not translocated appreciably from roots to shoots when toxic quantities are present (Hevesy, 1923). Hammett (1928a) reported that Pb absorbed from solutions of $Pb(NO_3)_2$, $PbCl_2$, and $Pb(C_2H_3O_2)_2$ by onions (*Allium cepa*), beans, and corn was accumulated largely in the area of maximum mitotic activity of the root, i.e., just behind the rootcap. He also noted that the area of greatest Pb accumulation was blackened but that the black deposit was soluble in very dilute nitric acid. Hammett (1928b) in a subsequent report found that 10 ppm Pb severely restricted primary root elongation of beans; 25 ppm Pb appeared to have completely halted growth. He concluded that Pb concentrations of from $10\mu M$ to $50\mu M$ retarded root growth of seedlings.

IV. ZINC

Reports of incidence of zinc toxicity to plants are generally restricted to three situations:

1. The effect of Zn on growth of plants in acid soil in galvanized containers (Conner, 1920; Crosier *et al.*, 1937)
2. The effect of Zn in or from mine debris on plant growth (Davies, 1941)
3. The effect of Zn in acid orchard soils on plant growth (Lee and Page, 1967)

Only the latter two cases are of significant economic and/or ecological importance at the current time.

Reclamation of deposits of mine tailings to retard runoff and subsequent pollution of waterways by establishing suitable vegetative cover
on the site frequently encounters such problems as Zn toxicity. Zinc
toxicity may be a factor in such reclamation efforts since Zn is associated
with the mining of several important metals. Zinc toxicity to cotton has
been identified in acid peach (*Prunus persica*) orchard soils in South
Carolina where disease control had been accomplished over a period
of years with zinc and sulfur sprays. Liming the soil to about pH 6.5
is usually sufficient to cause precipitation of toxic quantities of Zn.

There is limited information concerning the physiology of Zn toxicity
to roots. Lingle *et al.* (1963) noted that as little as 5×10^{-7} M Zn in
nutrient solutions strongly interfered with Fe uptake and transport by
both decapitated and intact soybean plants. Smith and Sprecht (1953)
report that Zn interfered with upward translocation of Fe, as evidenced
by the onset of chlorosis in citrus seedlings. Zinc accumulated in large
quantities in citrus roots but was less restricted to the roots than Cu.

The symptoms of Zn toxicity are blackening of the root system. The
toxicity also usually results in chlorosis of leaves caused by Fe deficiency.

Lund[4] noted normal elongation of cotton taproots in 1/4-strength
Hoaglands nutrient solution that contained 0.5 and 1.0 ppm Zn; 5 ppm
Zn resulted in immediate cessation of elongation. Roots of soybean
(*Glycine max*) varieties Hudson, Manchu, and Peking were damaged by
0.8 ppm and 0.2 ppm Zn, respectively, in nutrient culture; roots eventually died in solutions that contained 1.6 ppm and 0.4 ppm, respectively
(Earley, 1943). Polson and Adams (1970) also report varietal differences
among soybeans in response to excessive Zn levels. Tukuoka and Morooka (1938) noted that 0.5 ppm Zn in water soil culture was toxic to
paddy rice (*Oryza sativa*). Three ppm Zn were toxic to sweet orange and
mandarin in nutrient solutions (Smith and Sprecht, 1953). Rose (*Rosa*
sp.) roots showed Zn toxicity symptoms after 4 months in nutrient solution fortified with 1 ppm Zn (Marmon, 1937). Thus it seems that concentrations of Zn in the 1 ppm range are toxic to roots of several species.

V. BORON AND MANGANESE

There is little substantial evidence that either boron or manganese is
toxic to roots *per se*. There is ample work that shows that both are toxic
to aerial parts of most plants. There are reports of reduced root growth
when the concentration of Mn^{2+} or B was high; the effect probably
resulted from reduced shoot growth.

Rios and Pearson (1964) concluded that Mn is not toxic to young
primary roots of cotton. The data of Hiatt and Ragland (1963) showed

---

[4] *Ibid.*

that the weight ratio of shoot : root decreased as Mn concentration of a nutrient solution increased. Thus Mn was highly toxic to shoots but not directly to roots. Concomitantly, a lower percentage of the total plant Mn accumulated in the roots when solution Mn was higher.

Chatterton *et al.* (1969) report that desert saltbush (*Atriplex polycarpa*) germinated without inhibition in nutrient solutions containing up to 50 ppm B; one seed group collected in a low B area had shorter roots when grown in 80 ppm B solutions. However, root : top ratios were not significantly altered by B. Mortvedt and Osborn (1965) reported that root weight of 7- and 14-day-old oat (*Avena sativa*) seedlings was less in soils that contained more than 1.4 ppm B. They did not report top weights, but B was as much as 9 times as high in tops as in roots. Thus it would seem that the roots were not damaged sufficiently to preclude uptake and translocation of B.

VI. SUMMARY

Aqueous ammonia [$NH_3(aq)$], Cu, Zn, and Pb are highly toxic to plant roots *per se*. The toxicities are usually expressed as abruptly curtailed root growth and cessation of root function, indicated by depressed water uptake. Thus immersion of the whole root system into a nutrient environment that contains a toxic concentration of one of these factors is likely to result in wilt of a rapidly transpiring plant. Subsequent toxicity symptoms to roots usually consist of darkening or browning beginning just above the root tip. Since the visible symptoms of toxicity of all these factors are similar, positive identification of the specific toxicity is facilitated by solution and plant analysis.

A true toxicity to roots such as that produced by concentrations of $NH_3(aq)$, Cu, Zn, and Pb is characterized by accumulation of the toxic factor in the roots; very little translocation to the aerial parts of the plant occurs. Thus early symptoms characteristic of the toxic factor are not usually expressed in the shoots.

In contrast, it appears that B and Mn are not toxic to roots *per se*. Both these toxicants are accumulated in aerial parts of plants when toxic concentrations are present in the root medium. Translocation of the toxic factor continues after symptoms appear on aerial plant parts, and greater concentrations of the toxic factor are found in the shoots than in roots of affected plants.

References

Allred, S. E., and A. J. Ohlrogge. 1964. Principles of nutrient uptake from fertilizer bands. III. Germination and emergence of corn as affected by ammonia and ammonium phosphate. *Agron. J.* 56: 309–13.

Barker, A. V., R. J. Volk, and W. A. Jackson. 1966a. Growth and Nitrogen distribution patterns in bean plants (*Phaseolus vulgaris* L.) subjected to ammonium nutrition: I. Effects of carbonates and acidity control. *Soil Sci. Soc. Amer. Proc.* 30: 228–32.

———. 1966b. Root environment acidity as a regulator factory in ammonium assimilation by the bean plant. *Plant Physiol.* 41: 1193–99.

Bennett, A. C., and Fred Adams. 1970. Concentration of NH_3(aq) required for incipient NH_3 toxicity to seedlings. *Soil Sci. Soc. Amer. Proc.* 34: 259–63.

Blanchar, R. W. 1967. Determination of the partial pressure of NH_3 in soil air. *Soil Sci. Soc. Amer. Proc.* 31: 791–95.

Brage, B. L., W. R. Zich, and L. O. Fine. 1960. The germination of small grains and corn as influenced by urea and other nitrogenous fertilizers. *Soil Sci. Soc. Amer. Proc.* 24: 294–96.

Broyer, T. C., and A. H. Furnstal. 1945. Note on the effects of copper impurities in distilled water on growth of plants. *Plant Physiol.* 20: 690–91.

Cannon, H. L., and J. M. Bowles. 1962. Contamination of vegetation by tetraethyl lead. *Science* 137: 765–66.

Chatterton, N. J., C. M. McKell, J. R. Goodwin, and F. T. Bingham. 1969. *Atriplex polycarpa*: II. Germination and growth in water cultures containing high levels of B. *Agron. J.* 61: 451–53.

Conner, S. D. 1920. The effect of zinc in soil tests with zinc and galvanized iron pots. *J. Amer. Soc. Agron.* 12: 61–64.

Cooke, J. J. 1962. Damage to plant roots caused by urea and anhydrous ammonia. *Nature* 194: 1262–63.

Court, M. N., R. C. Stephen, and J. S. Waid. 1964. Toxicity as a cause of the inefficiency of urea as a fertilizer I. *J. Soil Sci.* 15: 42–47.

Crosier, W. F., S. R. Patrick, and M. T. Munn. 1937. Abnormal germination resulting from improperly galvanized trays. *Phytopathology* 27: 867–68.

Darwin, Charles. 1882. The action of carbonate of ammonia on the roots of certain plants. *J. Linn. Soc.* (London) 19: 239–61.

Davies, G. N. 1941. An investigation of the effect of zinc sulphate on plants. *J. Appl. Biol.* 28: 81–84.

Derose, H. R., W. S. Eisenmenger, and W. S. Ritchie. 1937. The comparative nutritive effects of copper, zinc, chromium, and molybdenum. *Mass. Agr. Exp. Sta. Bull.* 339: 11–12.

Earley, E. B. 1943. Minor element studies with soybeans: I. Variental reaction to concentrations of zinc in excess of the nutritional requirements. *J. Amer. Soc. Agron.* 35: 1012–23.

Ensminger, L. E., J. T. Hood, and G. H. Willis. 1965. The mechanism of ammonium phosphate injury to seeds. *Soil Sci. Soc. Amer. Proc.* 29: 320–22.

Hammett, F. S. 1928a. Studies in the biology of metals: I. The localization of lead by growing roots. *Protoplasma* 4: 183–86.

——. 1928b. Studies in the biology of metals: II. The retardative influence of lead on root growth. *Protoplasma* 4: 187–91.

Hevesy, George. 1923. The absorption and translocation of lead by plants. *Biochem. J.* 17: 439–45.

Hiatt, A. J., and J. L. Ragland. 1963. Manganese toxicity of burley tobacco. *Agron. J.* 55: 47–49.

Ishizuka, Y. 1942. Causal nature of toxic action of copper ions on the growth of rice plants. II. The abnormal accumulation of copper ions near the growth point of the root. *J. Sci. Soil Manure* 16: 43–45 (*Chem. Abstr.* 46: 4064).

Katsunuma, N., M. Okada, and Y. Nishi. 1965. Regulation of the urea cycle and TCA cycle by ammonia. In *Advances in Enzyme Regulation*, ed. G. Weber, 4: 317–35. Elmsford, N. Y.: Pergamon Press.

Krogman, David W., Andre T. Jagendorf, and Mordhay Avron. 1959. Uncouplers of spinach chloroplast photosynthetic phosphorylation. *Plant Physiol.* 34: 272–77.

Lee, C. R., and N. R. Page. 1967. Soil factors influencing the growth of cotton following peach orchards. *Agron. J.* 59: 237–40.

Lingle, J. C., L. O. Tiffin, and J. C. Brown. 1963. Iron uptaketransport of soybeans as influenced by other cations. *Plant Physiol.* 38: 71–76.

McBrien, D. C. H., and K. A. Hassal. 1967. The effect of toxic doses of copper upon respiration, photosynthesis, and growth of *Chlorella vulgaris. Physiol. Plant.* 20: 113–17.

Macmillan, A. 1956. The entry of ammonia into fungal cells. *J. Exp. Bot.* 7: 113.

Marmon, A. H. 1937. Effect of zinc on rose growth. *Florists Rev.* 80 (2069): 19–20.

Matsumoto, H., N. Wakiuchi, and E. Takahashi. 1968. Changes of sugar levels in cucumber leaves during ammonium toxicity. *Physiol. Plant.* 21: 1210–16.

Megie, C. A., R. W. Pearson, and A. E. Hiltbold. 1967. Toxicity of decomposing crop residues to cotton germination and seedling growth. *Agron. J.* 59: 197–99.

Mortvedt, J. J., and G. Osborn. 1965. Boron concentration adjacent to fertilizer granules in soil and its effect on root growth. *Soil Sci. Soc. Amer. Proc.* 29: 187–91.

Naftel, James A. 1931. The absorption of ammonium and nitrate nitrogen by various plants at different stages of growth. *J. Amer. Soc. Agron.* 23: 142–58.

Polson, D. E., and M. W. Adams. 1970. Differential response of navy beans (*Phaseolus vulgaris* L.) to zinc. I. Differential growth and elemental composition at excessive zinc levels. *Agron. J.* 62: 557–60.

Puritch, George S., and Allen V. Barker. 1967. Structure and function of tomato leaf chloroplasts during ammonium toxicity. *Plant Physiol.* 42:1229–38.

Reuther, W., and P. F. Smith. 1953. Effects of high copper content of sandy soil on growth of citrus seedlings. *Soil Sci.* 75: 219–24.

Rios, M. A., and R. W. Pearson. 1964. The effect of some chemical environmental factors on cotton root behavior. *Soil Sci. Soc. Amer. Proc.* 28:232–35.

Singer, M. J., and L. Hanson. 1969. Lead accumulation in soils near highways in the twin cities metropolitan area. *Soil Sci. Soc. Amer. Proc.* 33: 152–53.

Smith, P. F., and A. W. Sprecht. 1953. Heavy metal nutrition and iron chlorosis of citrus seedlings. *Plant Physiol.* 28: 371–82.

Sowell, W. F., R. D. Rouse, and J. I. Wear. 1957. Copper toxicity of the cotton plant in solution cultures. *Agron. J.* 49: 206–7.

Struckmeyer, Ester B., L. A. Peterson, and Florence Hse-Mei Tai. 1969. Effects of copper on the composition and anatomy of tobacco. *Agron. J.* 61: 932–36.

Stuart, Darrel M., and Jay L. Haddock. 1968. Inhibition of water uptake in sugarbeet roots by ammonia. *Plant Physiol.* 43: 345–50.

Tukuoka, M., and H. Morooka. 1938. Effect of zinc on growth of paddy rice. *J. Soc. Trop. Agr.* (Taihoku Imp. Univ.) 10: 24–37 (*Bibliography on Minor Elements*, vol. 1, 1940, New York: Chilean Nitrate Educational Bureau).

Vines, H. M., and R. T. Wedding. 1960. Some effects of ammonia on plant metabolism and a possible mechanism for ammonia toxicity. *Plant Physiol.* 35: 820–25.

Wallace, A., and P. C. de Kock. 1966. Translocation of iron in tobacco, sunflower, soybean, and bushbean plants. In *Current Topics in Plant Nutrition*, ed. Arthur Wallace, pp. 3–9. Ann Arbor, Mich.: Edwards Brothers.

Warren, K. S. 1962. Ammonia toxicity and pH. *Nature* 195: 47–49.

Warren, K. S., and D. G. Nathan. 1958. The passage of ammonia across the blood brain barrier and its relation to blood pH. *J. Clin. Invest.* 37: 1724–28.

Willis, L. G., and J. R. Piland. 1931. Ammonium calcium balance: a concentrated fertilizer problem. *Soil Sci.* 31: 5–23.

Index

Index